Handbook of Organophosphorus Chemistry

Handbook of Organophosphorus Chemistry

Editor: Deacon Palmer

MURPHY & MOORE
www.murphy-moorepublishing.com

www.murphy-moorepublishing.com

MURPHY & MOORE

Cataloging-in-Publication Data

Handbook of organophosphorus chemistry / edited by Deacon Palmer.
p. cm.
Includes bibliographical references and index.
ISBN 978-1-63987-738-6
1. Organophosphorus compounds. 2. Chemistry, Organic. I. Palmer, Deacon.
QD305.P46 H36 2023
547.07--dc23

Murphy & Moore Publishing
1 Rockefeller Plaza,
New York City,
NY 10020, USA

ISBN 978-1-63987-738-6

Contents

Preface

Organic compounds that contain phosphorus are called organophosphorus compounds. Organophosphorus compounds have various applications. They can be used as gasoline additives, lubricants, flame-retardants, plasticizers, hydraulic fluids and pharmaceuticals. They have widespread use not only as ligands in coordination chemistry and catalysis, but also in materials science, medicinal chemistry and agrochemicals. Organophosphorus chemistry is a field of chemistry under which the reactivity and characteristics of organophosphorus compounds are studied. Some of the major categories of organophosphorus compounds studied within this field are phosphate esters and amides, phosphonium salts and phosphoranes, and phosphines. This book explores all the important aspects of organophosphorus chemistry in the present day scenario. The various studies that are constantly contributing towards advancing technologies and evolution of this area of chemistry are examined in detail. The book will serve as a valuable source of reference for graduate and postgraduate students.

This book unites the global concepts and researches in an organized manner for a comprehensive understanding of the subject. It is a ripe text for all researchers, students, scientists or anyone else who is interested in acquiring a better knowledge of this dynamic field.

I extend my sincere thanks to the contributors for such eloquent research chapters. Finally, I thank my family for being a source of support and help.

Editor

1

Application of the Microwave Technique in Continuous Flow Processing of Organophosphorus Chemical Reactions

Erika Bálint, Ádám Tajti and György Keglevich *

Department of Organic Chemistry and Technology, Budapest University of Technology and Economics, 1521 Budapest, Hungary; ebalint@mail.bme.hu (E.B.); tajti.adam@mail.bme.hu (A.T.)

* Correspondence: gkeglevich@mail.bme.hu

Abstract: The microwave (MW) technique is an efficient tool in the realization of organic reactions, as well as in the analytical field and in the food industry. The continuous flow approach is of special interest as a promising way to scale-up MW-assisted syntheses. Besides summarizing the batch precedents, this review focuses on the utilization of the MW technique in the continuous-flow realization of organophosphorus transformations. The advantages of the continuous flow technique against the batch accomplishment are also shown. A few materials chemistry-related applications are also mentioned.

Keywords: microwave irradiation; microwave reactors; continuous flow technique; organophosphorus syntheses

---

1. Introduction

1.1. Batch vs. Continuous Flow Systems

Continuous procedures have been playing a significant role in important fields like the oil, plastic, and fine chemical industries, or metal processing for a long time [1]. The paradigm shift from batch reactions to continuous accomplishments is the consequence of progress in organic chemistry, and came to the front in the last decade [2].

As compared to a step-by-step batch reaction (Figure 1/I), where the most important parameters are the concentration of the starting materials, the temperature, and the reaction time, in continuous flow operations (Figure 1/II) the outcome can be controlled by the flow rate of the reagents and the average residence time set in the reactor. A continuous flow reaction can reach a steady state when the intake and the output are stationary. This means that the product is obtainable continuously in a permanent quality [3]

Flow reactors offer a basically different processing environment [4]. In a typical flow reactor, relatively small quantities are present in the reacting volume, allowing for a better mixing and a faster response on parameter adjustments. The precise parameter control is the key factor to obtain the target molecules with better purity and selectivity, as well as in higher yields. Continuous processes also mean an improvement regarding safety, as the handling of toxic or unstable intermediates becomes easier when smaller amounts are used [5]. Due to the low volume ratio and small quantities, the temperature regulation is simple, which makes the overall reaction safer. Moreover, flow reactors do not contain any gas volume, which enables reaction temperatures exceeding the boiling point of the solvent more easily.

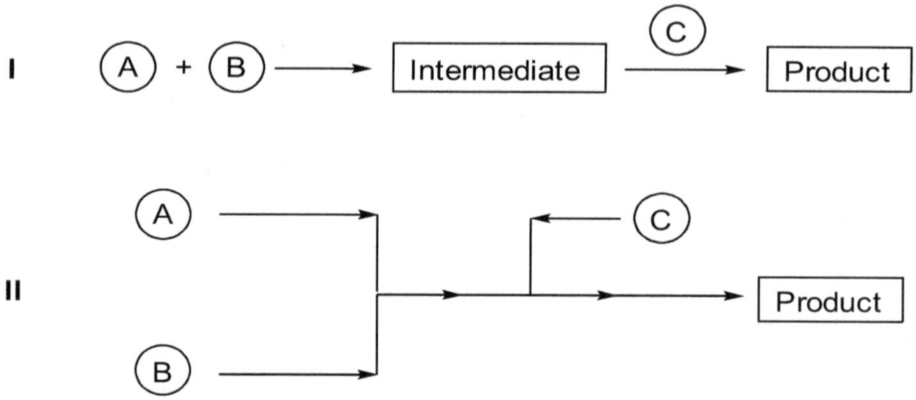

Figure 1. Schematic drawings of a batch and a continuous flow system.

An up-to-date and safe synthesis requires a fully controlled process [6]. Flow reactors can be readily coupled with tools of the process analytical technology (PAT), allowing real-time monitoring and the possibility for automation. Due to the easy monitoring and constant product quality, a continuous flow reactor can be an ideal choice for syntheses in the pharmaceutical industry [7].

The scale-up of flow reactions should be mentioned as a possible drawback [8]. This problem may be overcome by operating a series of parallel flow reactors of the same size. The flow reactions under discussion require a homogeneous medium with low viscosity, which may also mean a challenge in certain applications. The price of the necessary equipment for a continuous process is usually higher compared to that of a batch reactor [9].

1.2. Batch Microwave Chemistry

Microwave (MW) irradiation is widely used as a heat source in different fields, such as organic, inorganic, and analytical chemistry, finding applications in the medicinal, polymer, and food industries, as well as in the areas of material processing [10,11].

The evolution of MW-assisted processes started in the 1950s [12]. For three decades, the MW technique was mainly used in the food industry or as a heat source for drying. The first chemistry-related applications, e.g., sample preparation and digestion in analytical chemistry, also appeared during this period. The first MW-assisted chemical syntheses were introduced in 1986 [13,14]. Since then, thousands of publications have been published in this area.

Early synthetic attempts using MW irradiation were carried out in kitchen MW ovens [12]. In these devices, the MW irradiation is controlled by turning on and off the magnetron, which does not allow continuous irradiation and makes the process uncontrollable. The lack of precise measurement and control of the most important parameters (temperature and pressure) makes these experiments unreproducible. The problem was recognized, and this initiated the development of professional MW reactors designed for laboratory use. The first commercially available MW reactors appeared on the market in the early 2000s. The improvement from household MW ovens was significant. These MW reactors were equipped with a specially designed MW cavity, a built-in magnetic stirrer, and a precise temperature sensor, and also offered a connection to a PC for the monitoring and control of the experiments. The use of professional MW chemistry became broadly available, and this multiplied the amount of publications in the area under discussion.

In contrast to conventional heat sources, MW can directly heat the molecules in the bulk of a mixture [15]. Besides the energy efficiency, MW heating has several other benefits, such as accelerating reactions and offering milder reaction conditions, higher yields, and better selectivities, as well as allowing solvent- and/or catalyst-free methods.

As the mechanism of MW heating requires the interaction of MWs and the material, the dielectric constants of the reagents have to be checked before designing a MW-assisted synthesis [10]. At least one of the starting materials should be polar enough to be able to absorb the MW energy. In certain cases, polar additives (e.g., polar catalysts, salts, or ionic liquids) may be added to increase the polarity of the mixture [16–18].

Over the years, the MW technique has proven to be efficient in a wide variety of syntheses. Among these, a few reaction types like couplings, multicomponent reactions, condensations, and cycloadditions turned out to be especially advantageous under MW conditions [19].

Although the MW technique has several benefits, the scale-up suffers from a serious limitation [12]. The MW irradiation is generated by a magnetron coil, which has a defined geometry. Structural materials, like borosilicate glass or Teflon, are needed for MW equipment requiring a special design, not allowing for the establishment of larger-scale reactors. Manufacturers of MW devices offer multimode reactors, where 6–12 samples can be heated parallelly. However, this is far away from a real scale-up, which is inevitable for industrial-like applications. A more suitable approach allowing MW-assisted syntheses in larger quantities is discussed in the next chapter.

1.3. Continuous Flow Microwave Attempts

According to a review in the field of MW-assisted material chemistry [20], continuous processes have been developed for the preparation of carbides, sintering of ceramics, or the continuous treatment of oil-contaminated drill cuttings. These instances differ from the usual flow reactions discussed in this subchapter, as the feed of the reactor is not a homogenous solution.

In MW-assisted organic chemistry, the continuous flow technique is of special interest. Combining MW heating with the continuous flow technique creates a promising way to scale-up MW-assisted syntheses [12,15,21–23]. The first systems coupling MWs and continuous flow were applied to polymer heating and solid drying in the 1980s [24,25]. The continuous flow MW reactors for chemical reactions appeared in the 1990s [26]. Since then, the number of publications in this field has been constantly increasing, and until the present more than 300 papers have been published. The advantages of continuous flow MW reactors were reported for a series of transformations. However, in many cases, fabricated domestic MW ovens were used. Due to the uncontrolled temperature and pressure in these non-professional MW "reactors", it is practically impossible to reproduce these transformations [23,27]. Gradually, professional MW-flow systems with accurate temperature measuring devices were also developed [28,29].

The continuous flow MW systems usually consist of three main parts: the dispensing units for the starting reagents, the MW cavity, and the product collector (Figure 2). The reagents are fed into the reactor using one or more pumps that may be HPLC or syringe pumps. The pressure is usually controlled by a back-pressure regulator while the temperature is monitored using a built-in IR sensor or a fiber optic sensor. In most cases, the flow reactor is made of MW transparent Pyrex or Teflon. Further scale-up of flow MW systems can be realized by applying parallel reactors.

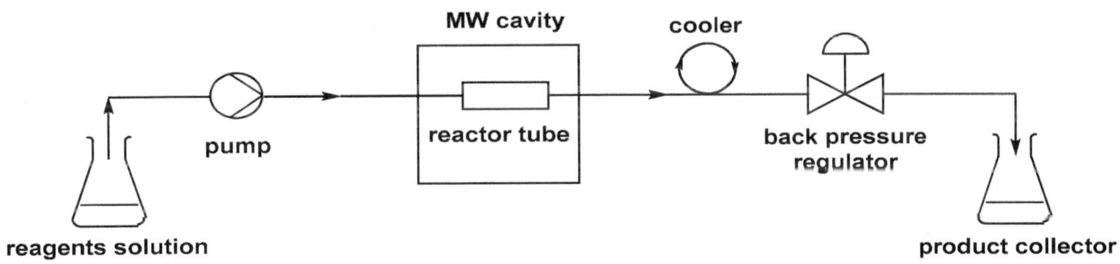

Figure 2. Schematic drawing of continuous flow microwave (MW) systems.

There are many types of continuous flow MW reactors [21], which contain a normal flask or tube [30], an Ω- or U-shaped tube [31,32], a spiral glass tube [30,33–35], or a mixed tube [36] to accommodate the reactive volume. Furthermore, reactor designs of fixed bed tubular coils [37,38], filled columns [37–39], and capillary reactors [40–42] are also available. Coupling MW heating and microreactors is also a challenging endeavour [23].

There are several papers, which reported on continuous flow MW accomplishments in the g or kg scales [27,43–57], or even larger scales (500 kg product per day) [58].

The spread of MW and continuous flow techniques have resulted in an enormous development in synthetic chemistry. Continuous flow MW reactors make possible efficient, fast, and selective transformations, as well as precise monitoring and control of the reaction parameters.

1.4. Microwaves in Batch Organophosphorus Syntheses

Guenin summarized the early results obtained in MW-promoted P-chemistry [55].

It was found by us that P-hydroxy phospholene oxides, -phospholane oxides, and -hexahydrophosphinine oxides (**1**) could be esterified with alcohols under MW conditions to provide the respective phosphinates (**2**) in yields of 31–82% [56–60]. Although this protocol means a green way to prepare the phosphinates, the temperatures of 200–235 °C and the times of up to 5 h are disadvantageous. Later on, we observed that, in the presence of 10% of ionic liquids (ILs), the esterifications were more efficient. This was shown via the esterification of different ring phosphinic acids (**1**) with pentanol (Scheme 1). Among the imidazoliaun salts tested as catalysts, [bmim][PF_6] was found to be the most appropriate [61]. The use of this additive made almost complete conversions possible, as well as higher yields (72–94%) at lower temperatures of 180–220 °C after shorter reaction times of 0.5–2 h. The positive effect of the IL additives is the consequence of their better MW absorbing ability.

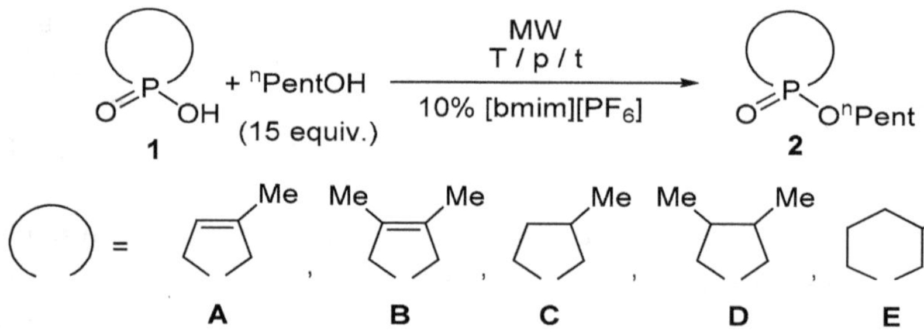

Scheme 1. Direct esterification of ring phosphinic acids (**1**) by pentanol using [bmim][PF_6] as the catalyst.

The solid–liquid heterogeneous phase alkylation of ring phosphinic acids (**1**) was studied by alkyl halides under MW irradiation, using K_2CO_3 to remove the liberated hydrogen halogenide (Scheme 2) [57,62]. The effect of triethylbenzylammonium chloride (TEBAC) as a phase transfer catalyst (PTC) was also investigated. In the cases of the alkylation of 1-hydroxy-3-phospholene oxides (**1A** or **1B**), it was found that by using benzyl bromide, which is of increased reactivity, the respective phosphinates (**3A** and **3B**) were obtained in yields of 92%/84% (also in the absence of a catalyst). At the same time, when alkyl halides of normal (^{n}Pr- and ^{n}Bu-bromide) or decreased reactivities (^{i}Pr-bromide) were applied, the presence of TEBAC enhanced the reactions. Yields of 90–96% were obtained for derivatives with normal alkyl chains. In these cases, the effect of the catalyst was synergetic with the MWs. The alkylations were then extended to the esterification of 1-hydroxyphospholane oxides (**1C** and **1D**) and a 1-hydroxy-hexahydrophosphinine oxide (**1E**). The respective esters were obtained in rather good yields (81–98%).

Aryl phosphonates (**5**) may be intermediates for arylphosphonic acids. A MW-promoted, $NiCl_2$-catalyzed Arbuzov reaction of aryl bromides (**4**) with triethyl phosphite was developed by our team (Scheme 3). In this case, a catalytic transformation was assisted by MWs [63,64].

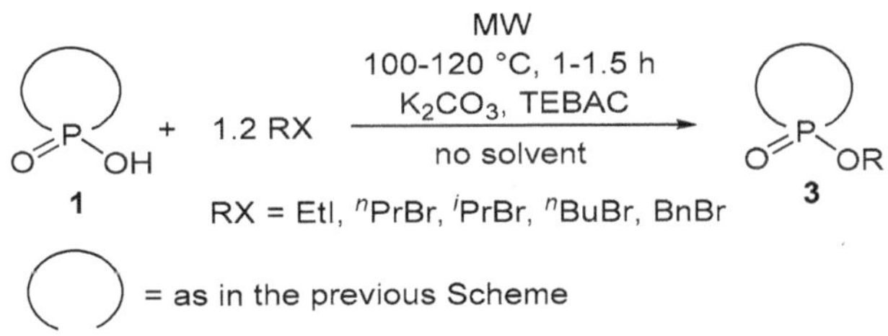

Scheme 2. The reaction of ring phosphinic acids (**1**) with alkyl halides.

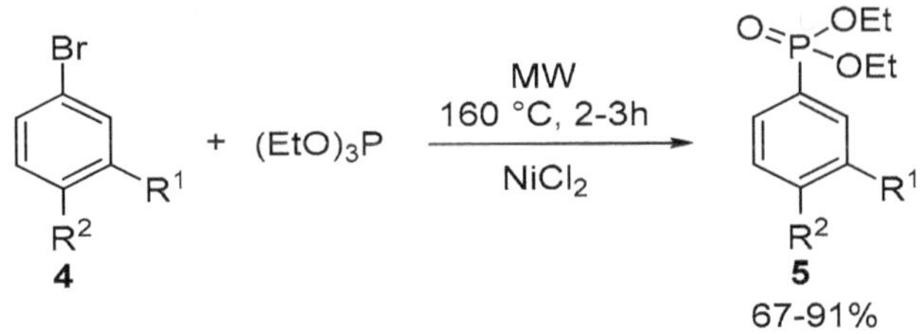

R^1	H	H	H	H	H	H	H	H	H	H	H	H	H	OMe	F	Ph	Cl	C(O)Me	COOH
R^2	H	OMe	^{t}Bu	^{c}Hex	^{n}Pr	Me	F	Ph	OCF_3	Cl	C(O)Me	COOH	CF_3	H	H	H	H	H	H

Scheme 3. Arbuzov reaction of aryl bromides and triethyl phosphite.

The solid–liquid phase alkylation of CH acidic P-derivatives, such as diethyl cyanomethylphosphonate (**6**), diethyl ethoxycarbonylmethylphosphonate (**7**), and tetraethyl methylenebisphosphonate (**8**), providing products **9–11**, was developed under MW irradiation (Scheme 4) [65–67]. There was no need to apply a PTC under MW irradiation [68,69]. A protocol for the di-substitution of diethyl ethoxycarbonylmethylphosphonate (**7**) was also worked out [70].

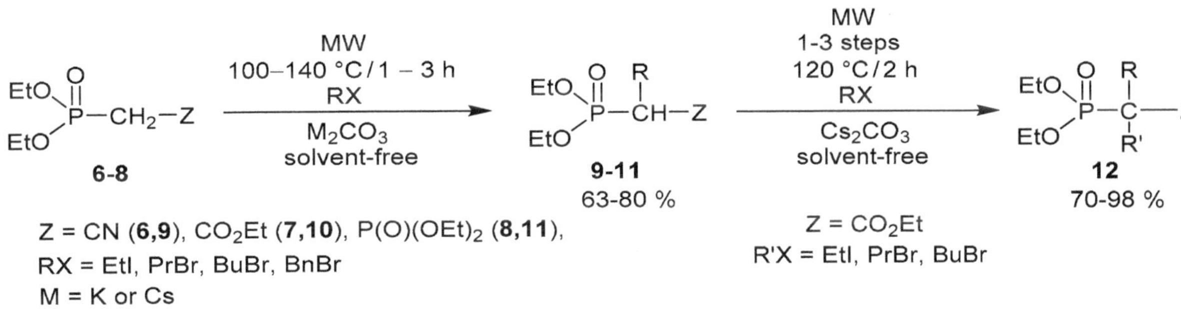

Scheme 4. Alkylation of active methylene-containing phosphonates.

It was found that there was no need for a catalyst in the Kabachnik–Fields reactions carried out under MW conditions [71,72]. A series of α-aminophosphonate derivatives (**13**) was synthesized by the three-component reactions of primary or secondary amines, oxo-compounds, and different >P(O)H reagents, like dialkyl phosphites, alkyl phenyl-*H*-phosphinates, and secondary phosphine oxides (Scheme 5) [71–76].

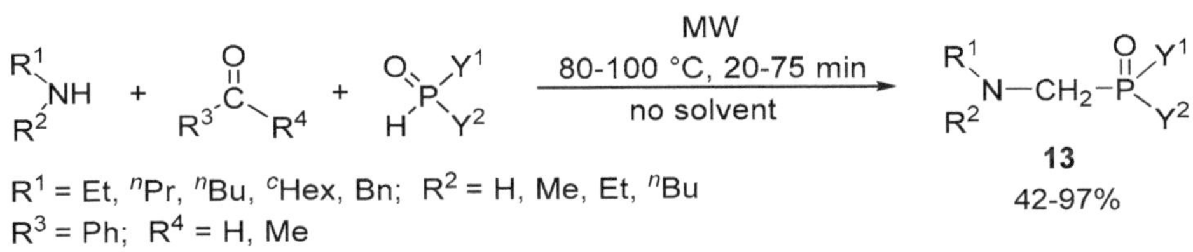

Y^1	MeO	EtO	BuO	BnO	EtO	EtO	PrO	BuO	Bn	Ph	$4\text{-}MeC_6H_4$
Y^2	MeO	EtO	BuO	BnO	OctO	Ph	Ph	Ph	Bn	Ph	$4\text{-}MeC_6H_4$

Scheme 5. Simple Kabachnik–Fields reaction.

Double Kabachnik–Fields reactions were also developed, where the primary amines were reacted with two equivalents of paraformaldehyde and the same amounts of the >P(O)H species under MW irradiation in a catalyst-free manner (Scheme 6) [73–79]. The respective products (**14**) could be prepared in variable yields. After double deoxygenation of the bis(aminophosphine oxides) (**14**, $Y^1 = Y^2$ = Bn, Ph or 4-MeC_6H_4), the bisphosphines obtained were used as bidentate P-ligands in the preparation of ring Pt complexes [76–79].

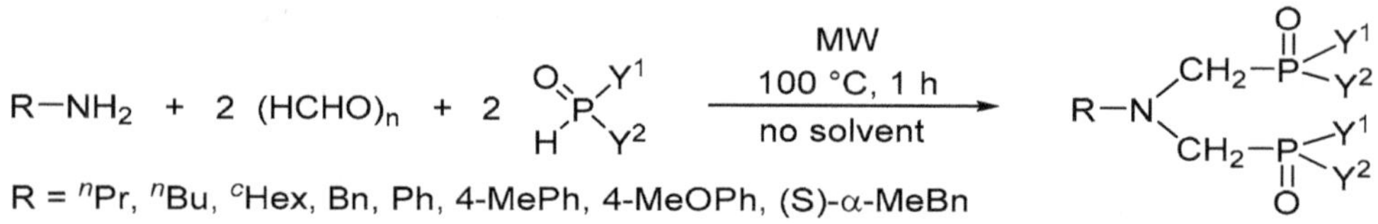

Y^1	MeO	EtO	BuO	BnO	EtO	EtO	PrO	BuO	Bn	Ph	4-MeC_6H_4
Y^2	MeO	EtO	BuO	BnO	OctO	Ph	Ph	Ph	Bn	Ph	4-MeC_6H_4

Scheme 6. Double Kabachnik–Fields reaction starting from primary amines.

The deoxygenation of phosphine oxide by-products is challenging. Cl_3SiH or $PhSiH_3$ are suitable reducing agents, however, the previous reagent is corrosive, while the latter one is expensive. In replacement of these reagents, the cheap $(Me_2SiH)_2O$ (TMDS) and polymethylsiloxane ($-(OSiMeH)_n-$; PMHS) were used under MW conditions. It was proved that under MW irradiation, there was no need for any catalyst proposed by Beller et al. [80]. This is exemplified by the reduction of 1-phenyl-3-methyl-3-phospholene 1-oxide (**15**) using $PhSiH_3$, TMDS and PMHS, as summarized in Scheme 7 [81–84].

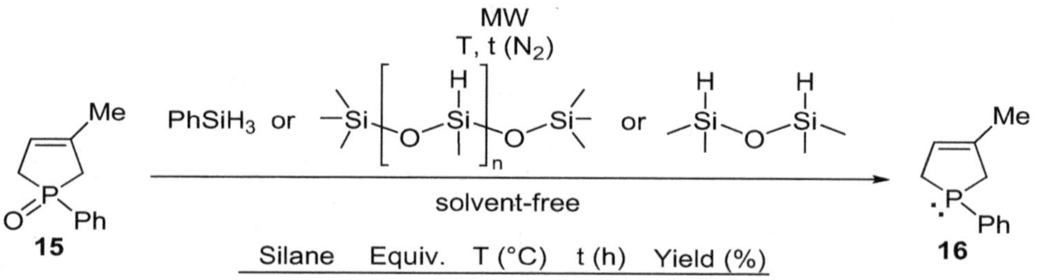

Silane	Equiv.	T (°C)	t (h)	Yield (%)
$PhSiH_3$	3	80	1	91
TMDS	4	110	3	92
PMHS	2	110	2	92

Scheme 7. Reduction of a 3-phospholene oxide by cheap silanes.

The first P–C coupling reaction between aryl- or vinyl-halides and dialkyl phosphites furnishing phosphonates was described by Hirao using $Pd(PPh_3)_4$ as the catalyst [85,86]. Later on, other methods were also reported to replace the Pd(0) catalyst with Pd(II) or another metal (e.g., Ni(II) salts applied together with mono- or bidentate P-ligands). The P–C coupling was then extended by utilizing other >P(O)H reagents, bases, and solvents. Our research group found that the P–C coupling of aryl-bromides and different >P(O)H species, like dialkyl phosphites, ethyl phenyl-*H*-phosphinate, and diphenylphosphine oxide may take place in the presence of $Pd(OAc)_2$ as the catalyst precursor under MW conditions without the addition of the usual P-ligands, but using the P-reagent in some excess (Scheme 8) [87,88]. The ArP(O)< products (**17**) were obtained in variable yields. Our idea was that the excess of the >P(O)H compounds existing under a tautomeric equilibrium (with the >P-OH form) may act as P-ligands [89].

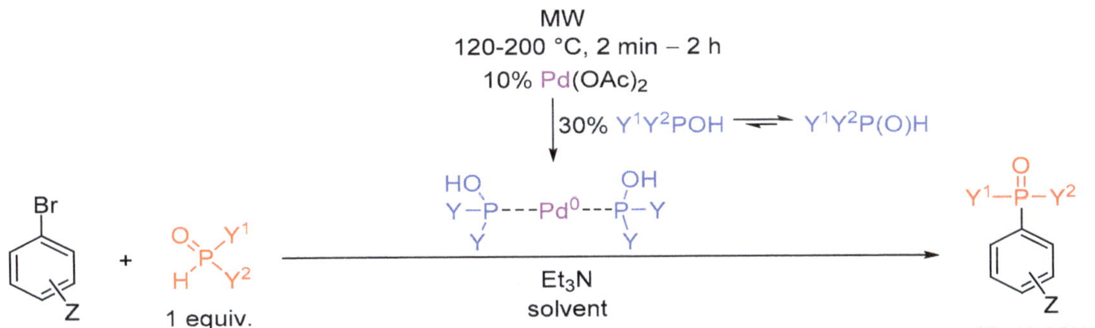

17, 41-95%

Z = H, 4-MeO, 3-MeO, 4-Pr, 4-Et, 4-Me, 4-Cl, 3-Cl, 4-F, 3-F, 4-CO_2Et, 3-CO_2Et, 4-C(O)Me, 3-C(O)Me

Y^1	OEt	OBu	OEt	O^{n}Pr	O^{i}Pr	O^{n}Bu	O^{i}Bu	O^{i}Pent	Ph	4-MeOC_6H_4	4-^{t}BuC_6H_4	4-MeC_6H_4	3-MeC_6H_4	4-ClC_6H_4	4-FC_6H_4
Y^2	OEt	OBu	Ph	Ph	Ph	Ph	Ph	Ph	Ph	4-MeOC_6H_4	4-^{t}BuC_6H_4	4-MeC_6H_4	3-MeC_6H_4	4-ClC_6H_4	4-FC_6H_4

solvent: –, MeCN, EtOH, BuOH, BnOH

Scheme 8. The P–C coupling reaction of bromobenzenes and >P(O)H reagents.

2. Microwave-Assisted Continuous Flow Applications

2.1. Development of the Continuous Flow Microwave Device

Among continuous flow MW-assisted syntheses, there are examples for esterifications [27,44,46], acylations [27,44,46], multicomponent [27] and rearrangement reactions [27,28,44], couplings [27,28,43,44], and polycondensations [90]. The simplest model is the esterification of carboxylic acids with alcohols, where different types of flow cells (e.g., teflon coils, glass tubes, or ceramic tubes) were applied in kitchen MW ovens [91,92] or in professional MW reactors [93,94].

A continuous flow MW system (Figure 2, Section 1.3) was developed by us, using a commercially avialable flow cell (Figure 3) immersed into a CEM® MW reactor [95]. The mixture of the reagents is fed into the reactor by a HPLC pump. The temperature of the mixture is monitored and controlled by an IR sensor built into the reactor. After irradiation at the desired temperature, the leaving reaction mixture is cooled down in a spiral-like cooler, and is passed through a back pressure regulator operating at 250 PSI.

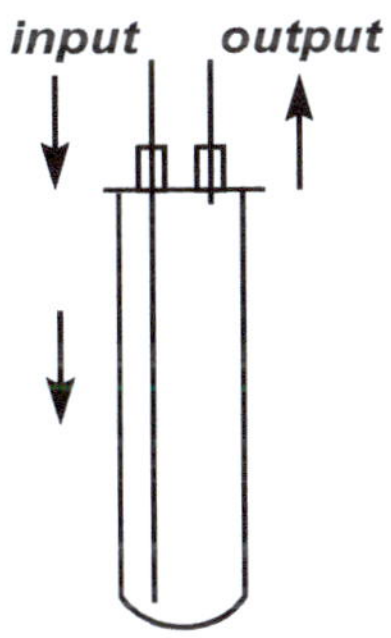

Figure 3. Sketch of the continuous flow cell.

The first model reaction to be investigated by us in the flow system developed was the direct esterification of benzoic acid with various aliphatic alcohols [95]. At first, batch reactions were carried out to investigate the reaction. The reactivity of the alcohols towards benzoic acid was mapped and the parameters were optimized. Based on the batch results, the esterification rate of benzoic acid depends on the boiling point (from EtOH to ^{n}PentOH) and the steric properties of the alcohols (^{n}Pr vs ^{i}Pr). Then, the esterifications were optimized further in the continuous flow MW reactor. It was found that when using 15 equivalents of alcohol at 140 °C at a flow rate of 0.35–0.25 mL/min (with a residence time of 20–30 min), the direct esterifications were complete and the corresponding alkyl benzoates were obtained in yields of 95–98%.

2.2. Elaboration of the Continuous Flow Transesterification of Dialkyl Phosphites

The MW-assisted transesterification of dialkyl phosphites (dialkyl *H*-phosphonates) was studied with simple alcohols (Scheme 9) under catalyst-free conditions [96]. These compounds may serve as starting materials, for example in the Kabachnik–Fields condensations or in the aza-Pudovik reactions, to provide potentially bioactive α-aminophosphonates [97]. As the first step, the transesterifications were performed under batch conditions. It was found that, depending on the conditions (molar ratio, temperature, and reaction time), the alcoholysis of dialkyl phosphites (**18**) resulted in the formation of dialkyl phosphites with two different (**19**) or with two identical (**20**) alkyl groups in different ratios. Dialkyl phosphites with two different alkyl groups (**19**) are valuable building-blocks in the synthesis of chiral organophosphorus derivatives [73].

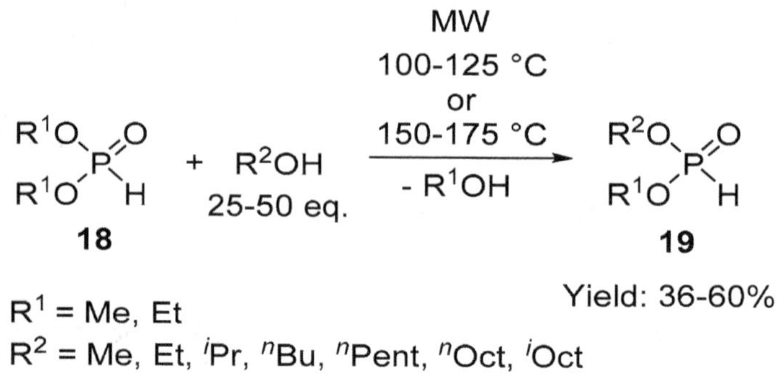

Scheme 9. Batch alcoholysis of dialkyl phosphites by simple alcohols.

Hydroxyalkyl- or aminoalkyl-functionalized *H*-phosphonates could be synthesized by the alcoholysis of dialkyl phosphites with glycols or amino alcohols (Scheme 10) [98,99]. During the investigation of this reaction, it was found that mixed phosphites (**21**) and/or fully transesterified products (**22**) were formed as major components, depending on the ratio of the starting materials, the temperature, and the reaction time. These derivatives may be starting materials in the synthesis of P-containing polymers, which form a special group within plastics [100,101]. In certain cases, a small amount of by-products (**23**–**25**) was also present in the reaction.

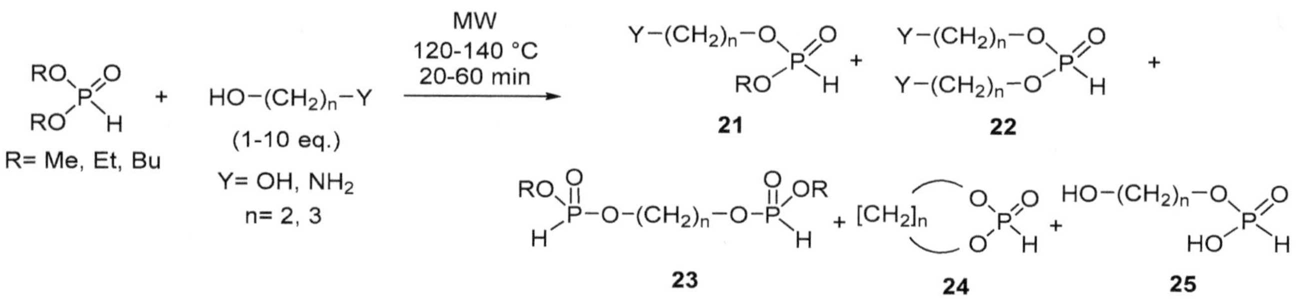

Scheme 10. Batch alcoholysis of dialkyl phosphites by diols and aminoalcohols.

The continuous flow alcoholysis of dialkyl phosphites by aliphatic alcohols in the absence of catalyst was also elaborated using the same continuous flow MW system applied in the esterification of benzoic acid (Scheme 11) [102]. By the precise control of the reaction parameters, the alcoholysis could be fine-tuned towards dialkyl phosphites with two different (**19**) or with two identical alkyl groups (**20**). Although the selectivity of the reaction was rather similar under flow and batch conditions, in the continuous flow reactor, the dialkyl phosphites with two different alkyl groups were obtained at a shorter reaction time in a higher proportion with a somewhat higher productivity.

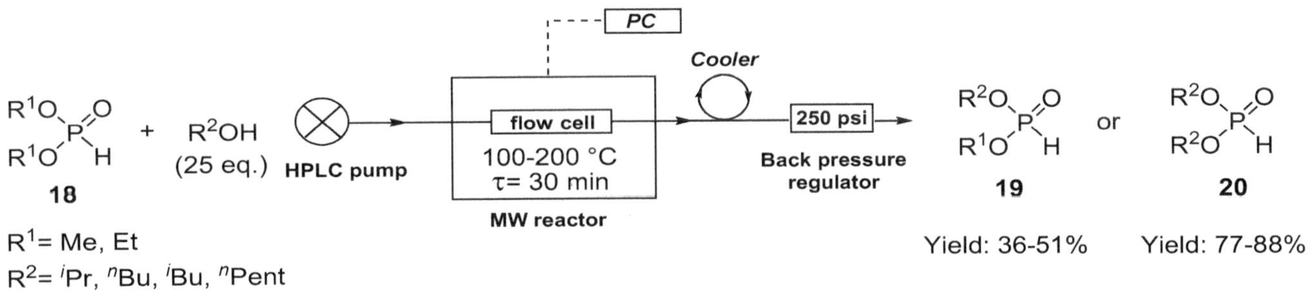

Scheme 11. Continuous flow MW system for the alcoholysis of dialkyl phosphites.

2.3. Continuous Flow Synthesis of α-Aminophosphonates by the aza-Pudovik Reaction

We elaborated the MW-assisted continuous flow synthesis of potentially biologically active α-aminophosphonates by the aza-Pudovik reaction, involving the addition of dialkyl phosphites to the C=N double bond of imines (Scheme 12) [103]. Before optimization, we carried out the flow-compatible addition based on our previous experiences [104]. In this simple two-component model reaction, the reagents were fed together into the flow cell. The MW-assisted catalyst-free continuous flow reaction of *N*-benzylidene-butylamine or *N*-benzylidene-cyclohexylamine and diethyl phosphite in ethanol afforded the corresponding α-aminophosphonates (**26**) in yields of 90–92%.

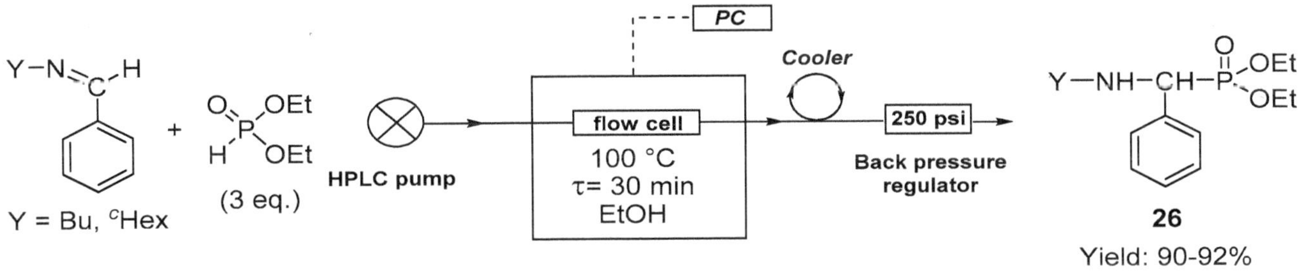

Scheme 12. Continuous flow synthesis of α-aminophosphonates by the aza-Pudovik reaction.

2.4. The Synthesis of α-Aminophosphonates by Continuous Flow Kabachnik–Fields Reaction

An efficient synthetic route towards α-aryl-α-aminophosphonates is the three-component Kabachnik–Fields reaction of primary amines, benzaldehyde derivatives, and dialkyl phosphites. The continuous flow variation was elaborated by us (Scheme 13) [103]. To avoid the possible side reactions of the starting materials at room temperature resulting in the formation of an imine or an α-hydroxyphosphonate, a dual-pump continuous flow system had to be developed. The mixtures of the primary amine and diethyl phosphite in ethanol (pump A), and benzaldehyde in ethanol (pump B) were fed separately into the mixer of the HPLC pump. After optimization of the conditions, the continuous flow method was proved to be more efficient than the batch approach, as the corresponding α-aminophosphonates (**27**) were obtained in a shorter reaction time, and a lower excess of phosphite was needed.

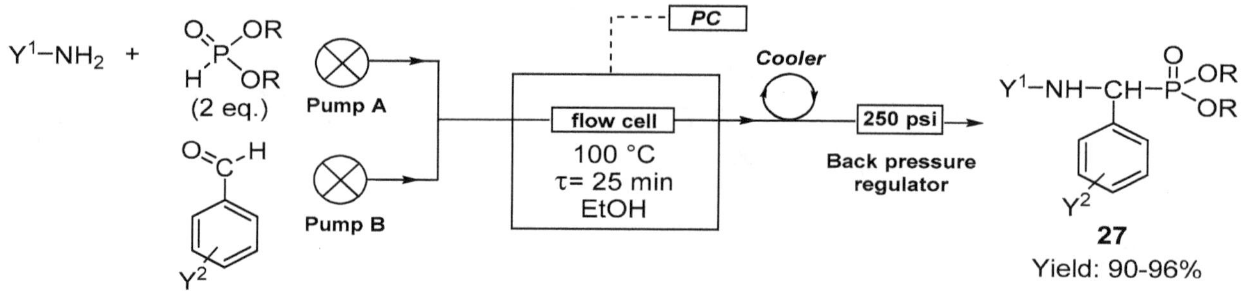

Y^1= Ph, 4-ClC_6H_4, 4-$MeOC_6H_4$, $PhCH_2$, 4-$MeOPhCH_2$
Y^2= H, 2-Cl, 3-Cl, 4-Cl, 4-OH, 4-Me, 4-MeO

Scheme 13. Continuous flow synthesis of α-aminophosphonates by the Kabachnik–Fields reaction.

3. Conclusions

The recent paradigm shift from batch processes to continuous flow operations has brought about a significant change in organic chemistry, and meant a challenge in organophosphorus chemistry. The continuous flow technique means improvements in respect to safety, selectivity, and efficiency, and makes possible the development of automated syntheses more easily. MW chemistry has come to the front in the 1980s. Since then, the MW technique has spread in various fields, including organic, analytical, and material chemistry. Many types of reactions were proved to be more efficient under MW conditions. In the field of organophosphorus chemistry, several transformations, such as the esterification of phosphinic acids, O- and C-alkylations, Kabachnik–Fields condensation, the reduction of phosphine oxides, and P–C couplings all take advantage of MW irradiation. By combining flow chemistry and the MW technique, besides the benefits mentioned above, the easy scale-up of MW-assisted reactions has also become possible. The elaboration of the alcoholysis of dialkyl phosphites and the synthesis of α-aminophosphonates by aza-Pudovik reactions, as well as Kabachnik–Fields reactions in a continuous flow MW reactor, allowed the preparation of the target organophosphorus products more efficiently, meaning an improved productivity as compared to the batch approaches.

Author Contributions: G.K., E.B. and Á.T. wrote the paper.

Funding: The project was supported by the Hungarian Research Development and Innovation Fund (K119202 and FK123961), by the National Research, Development and Innovation Fund of Hungary in the frame of the FIEK_16-1-2016-0007 (Higher Education and Industrial Cooperation Center) project. E.B. was supported by the János Bolyai Research Scholarship of the Hungarian Academy of Sciences (BO/00278/17/7), and by the ÚNKP-18-4-BME-131 New National Excellence Program of the Ministry of Human Capacities. Á.T. was supported by the ÚNKP-18-3-III-BME-251 New National Excellence Program of the Ministry of Human Capacities.

References

1. *Ullmann's Encyclopedia of Industrial Chemistry*, 7th ed.; Wiley-VCH: Weinheim, Germany, 2000; ISBN 978-3-527-30385-4.
2. Plutschack, M.B.; Pieber, B.; Gilmore, K.; Seeberger, P.H. The Hitchhiker's guide to flow chemistry. *Chem. Rev.* **2017**, *117*, 11796–11893. [CrossRef] [PubMed]
3. Glasnov, T. *Continuous-Flow Chemistry in the Research Laboratory*; Springer International Publishing: Basel, Switzerland, 2016; ISBN 978-3-319-32194-3.
4. Vaccaro, L. *Sustainable Flow Chemistry: Methods and Applications*; Wiley: Weinheim, Germany, 2017; ISBN 978-3-527-338528.
5. Movsisyan, M.; Delbeke, E.I.P.; Berton, J.K.E.T.; Battiloccihio, C.; Ley, S.V.; Stevens, C.V. Taming hazardous chemistry by continuous flow technology. *Chem. Soc. Rev.* **2016**, *45*, 4892–4928. [CrossRef] [PubMed]
6. Bakeev, K.A. *Process Analytical Technology: Spectroscopic Tools and Implemented Strategies for the Chemical and Pharmaceutical Industries*; Wiley-VCH: Chichester, UK, 2010; ISBN 978-0-470-72207-7.

7. May, S.A. Flow Chemistry, Continuous Processing, and Continuous Manufacturing: A Pharmaceutical Perspective. *J. Flow Chem.* **2017**, *7*, 137–145. [CrossRef]
8. Strying, P.; Parracho, A.I.R. From discovery to production: Scale-out of continuous flow meso reactors. *Beilstein J. Org. Chem.* **2009**, *5*, 29. [CrossRef] [PubMed]
9. Pashkova, A.; Greiner, L. Towards Small-Scale Continuous Chemical Production: Technology Gaps and Challenges. *Chem. Ing. Tech.* **2011**, *83*, 1337–1342. [CrossRef]
10. Cravotto, G.; Carnaroglio, D. *Microwave Chemistry*; De Gruyter: Berlin, Germany, 2017; ISBN 978-3-11-047993-5.
11. Horikoshi, S.; Schiffmann, R.F.; Fukushima, J.; Serpone, N. *Microwave Chemical and Materials Processing*; Springer: Singapore, 2018; ISBN 978-981-10-6465-4.
12. Bálint, E.; Keglevich, G. The Spread of the Application of the Microwave Technique in Organic Synthesis. In *Milestones in Microwave Chemistry*; Keglevich, G., Ed.; Springer: Cham, Switzerland, 2016; pp. 1–10.
13. Gedye, R.; Smith, F.; Westaway, K.; Ali, H.; Baldisera, L.; Laberge, L.; Rousell, J. The use of microwave ovens for rapid organic synthesis. *Tetrahedron Lett.* **1986**, *27*, 279–282. [CrossRef]
14. Giguere, R.J.; Bray, T.L.; Duncan, S.M.; Majetich, G. Application of commercial microwave ovens to organic synthesis. *Tetrahedron Lett.* **1986**, *27*, 4945–4948. [CrossRef]
15. Kappe, C.O.; Stadler, A.; Dallinger, D. *Microwaves in Organic and Medicinal Chemistry*, 2nd ed.; Wiley: Weinheim, Germany, 2012; Volume 52, ISBN 978-3-527-33185-7.
16. Zhao, X.; Zhang, Z.; Wang, L.; Xi, K.; Cao, Q.; Wang, D.; Yang, Y.; Du, Y. Excellent microwave absorption property of Graphene-coated Fe nanocomposites. *Sci. Rep.* **2013**, *3*, 3421. [CrossRef] [PubMed]
17. Guo, H.; Yang, J.; Pu, B.; Chen, H.; Li, Y.; Wang, Z.; Niu, X. Excellent microwave absorption of lead halide perovskites with high stability. *J. Mater. Chem. C* **2018**, *6*, 4201–4207. [CrossRef]
18. Gong, J.; Yang, F.; Shao, Q.; He, X.; Zhang, X.; Liu, S.; Tang, L.; Deng, Y. Microwave absorption performance of methylimidazolium ionic liquids: Towards novel ultra-wideband metamaterial absorbers. *RSC Adv.* **2017**, *7*, 41980–41988. [CrossRef]
19. Kiss, N.Z.; Bálint, E.; Keglevich, G. Microwave-Assisted Syntheses in Organic Chemistry. In *Milestones in Microwave Chemistry*; Keglevich, G., Ed.; Springer: Cham, Switzerland, 2016; pp. 11–45.
20. Kitchen, H.J.; Vallance, S.R.; Kennedy, J.L.; Tapia-Ruiz, N.; Carassiti, L.; Harrison, A.; Whittaker, A.G.; Drysdale, T.D.; Kingman, S.W.; Gregory, D.H. Modern Microwave Methods in Solid-State Inorganic Materials Chemistry: From Fundamentals to Manufacturing. *Chem. Rev.* **2014**, *114*, 1170–1206. [CrossRef] [PubMed]
21. Keglevich, G.; Sallay, P.; Greiner, I. Continuous flow microwave reactors. *Hung. Chem. J.* **2008**, *63*, 278–283.
22. de la Hoz, A.; Díaz-Ortiz, A. Nonconventional Techniques in Sustainable Flow Chemistry. In *Sustainable Flow Chemistry: Methods and Applications*; Vaccaro, L., Ed.; Wiley: Weinheim, Germany, 2017; pp. 219–248.
23. Estel, L.; Poux, M.; Benamara, N.; Polaert, I. Continuous flow-microwave reactor: Where are we? *Chem. Eng. Process.* **2016**, *113*, 56–64. [CrossRef]
24. Wang, C.S. Processing parameters of continuous microwave heating of ethylene-propylene terpolymer. *Rubber Chem. Technol.* **1984**, *57*, 134–144. [CrossRef]
25. Gunasekaran, S. Grain drying using continuous and pulsed microwave energy. *Dry. Technol.* **1990**, *8*, 1039–1047. [CrossRef]
26. Strauss, C.R. A Strategic, 'Green' Approach to Organic Chemistry with Microwave Assistance and Predictive Yield Optimization as Core, Enabling Technologies. *Aust. J. Chem.* **1999**, *52*, 83–96. [CrossRef]
27. Baxendale, I.; Hayward, J.; Ley, S. Microwave reactions under continuous flow conditions. *Comb. Chem. High Throughput Screen.* **2007**, *10*, 802–836. [CrossRef] [PubMed]
28. Öhrngren, P.; Fardost, A., Russo, F.; Schanche, J. S.; Fagrell, M.; Larhed, M. Evaluation of a nonresonant microwave applicator for continuous-flow chemistry applications. *Org. Proc. Res. Dev.* **2012**, *16*, 1053–1063. [CrossRef]
29. Rydfjord, J.; Svensson, F.; Fagrell, M.; Savmarker, J.; Thulin, M.; Larhed, M. Temperature measurements with two different IR sensors in a continuous-flow microwave heated system. *Beilstein J. Org. Chem.* **2013**, *9*, 2079–2087. [CrossRef] [PubMed]
30. Bonaccorsi, L.; Proverbio, E. Influence of process parameters in microwave continuous synthesis of zeolite LTA. *Microporous Mesoporous Mater.* **2008**, *112*, 481–493. [CrossRef]
31. Khadilkar, B.M.; Madyar, V.R. Scaling up of dihydropyridine ester synthesis by using aqueous hydrotrope solutions in a continuous microwave reactor. *Org. Process Res. Dev.* **2001**, *5*, 452–455. [CrossRef]

32. Pillai, U.R.; Sahle-Demessie, E.; Varma, R.S. Hydrodechlorination of chlorinated benzenes in a continuous microwave reactor. *Green Chem.* **2004**, *6*, 295–298. [CrossRef]
33. Correa, R.; Gonzalez, G.; Dougar, V. Emulsion polymerization in a microwave reactor. *Polymer* **1998**, *39*, 1471–1474. [CrossRef]
34. Cáceres, A.; Jaimes, M.; Chávez, G.; Bravo, B.; Ysambertt, F.; Márquez, N. Continuous system with microwave irradiation to obtain alkyl benzoates. *Talanta* **2005**, *68*, 359–364. [CrossRef] [PubMed]
35. Vámosi, P.; Matsuo, K.; Masuda, T.; Sato, K.; Narumi, T.; Takeda, K.; Mase, N. (2018) Rapid Optimization of Reaction Conditions Based on Comprehensive Reaction Analysis Using a Continuous Flow Microwave Reactor. *Chem. Rec.* **2018**, *18*, 1–9. [CrossRef]
36. Bonnet, C.; Estel, L.; Ledoux, A.; Mazari, B.; Louis, A. Study of the thermal repartition in a microwave reactor: Application to the nitrobenzene hydrogenation. *Chem. Eng. Proc.* **2004**, *43*, 1435–1440. [CrossRef]
37. Bo, L.; Quan, X.; Chen, S.; Zhao, H.; Zhao, Y. Degradation of p-nitrophenol in aqueous solution by microwave assisted oxidation process through a granular activated carbon fixed bed. *Water Res.* **2006**, *40*, 3061–3068. [CrossRef] [PubMed]
38. Bagley, M.C.; Jenkins, R.L.; Lubinu, M.C.; Mason, C.; Wood, R. A simple continuous flow microwave reactor. *J. Org. Chem.* **2005**, *70*, 7003–7006. [CrossRef] [PubMed]
39. Kabza, K.G.; Chapados, B.R.; Getswicki, J.; McGrath, J.L. Microwave-induced esterification using heterogeneous acid catalyst in a low dielectric constant medium. *J. Org. Chem.* **2000**, *65*, 1210–1214. [CrossRef] [PubMed]
40. He, P.; Haswell, S.J.; Fletcher, P.D.I. Efficiency, monitoring and control of microwave heating within a continuous flow capillary reactor. *Sens. Actuator B Chem.* **2005**, *105*, 516–520. [CrossRef]
41. Shore, G.; Morin, S.; Organ, M.G. Catalysis in capillaries by Pd thin films using microwave-assisted continuous-flow organic synthesis (MACOS). *Angew. Chem. Int. Ed.* **2006**, *45*, 2761–2766. [CrossRef] [PubMed]
42. Comer, E.; Organ, M.G. A microreactor for microwave-assisted capillary (continuous flow) organic synthesis (MACOS). *J. Am. Chem. Soc.* **2005**, *127*, 8160–8167. [CrossRef] [PubMed]
43. Singh, B.K.; Kaval, N.; Tomar, S.; Van der Eycken, E.; Parmar, V.S. Transition metal-catalyzed carbon–carbon bond formation Suzuki, Heck, and Sonogashira reactions using microwave and microtechnology. *Org. Process Res. Dev.* **2008**, *12*, 468–474. [CrossRef]
44. Glasnov, T.N.; Kappe, C.O. Microwave-assisted synthesis under continuous-flow conditions. *Macromol. Rapid Commun.* **2007**, *28*, 395–410. [CrossRef]
45. Ullah, F.; Samarakoon, T.; Rolfe, A.; Kurtz, R.D.; Hanson, P.R.; Organ, M.G. Scaling out by microwave-assisted, continuous flow organic synthesis (MACOS): Multi-gram synthesis of bromo- and fluoro-benzofused sultams benzthiaoxazepine-1,1-dioxides. *Chem. Eur. J.* **2010**, *16*, 10959–10962. [CrossRef] [PubMed]
46. Dressen, M.H.C.L.; van de Kruijs, B.H.P.; Meuldijk, J.; Vekemans, J.A.J.M.; Hulshof, L.A. Flow processing of microwave-assisted (heterogeneous) organic reactions. *Org. Process Res. Dev.* **2010**, *14*, 351–361. [CrossRef]
47. Bergamelli, F.; Iannelli, M.; Marafie, J.A.; Moseley, J.D. A commercial continuous flow microwave reactor evaluated for scale-up. *Org. Process Res. Dev.* **2010**, *14*, 926–930. [CrossRef]
48. Bagley, M.C.; Fusillo, V.; Jenkins, R.L.; Lubinu, M.C.; Mason, C. Continuous flow processing from microreactors to mesoscale: The Bohlmann-Rahtz cyclodehydration reaction. *Org. Biomol. Chem.* **2010**, *8*, 2245–2251. [CrossRef] [PubMed]
49. Moseley, J.D.; Lawton, S.J. Initial results from a commercial continuous flow microwave reactor for scale-up. *Chem. Today* **2007**, *25*, 16–19.
50. Benaskar, F.; Hessel, V.; Krtschil, U.; Löb, P.; Stark, A. Intensification of the capillary-based Kolbe-Schmitt synthesis from resorcinol by reactive ionic liquids, microwave heating or a combination thereof. *Org. Process Res. Dev.* **2009**, *13*, 970–982. [CrossRef]
51. Leadbeater, N.E.; Barnard, T.M.; Stencel, L.M. Batch and continuous-flow preparation of biodiesel derived from butanol and facilitated by microwave heating. *Energy Fuels* **2008**, *22*, 2005–2008. [CrossRef]
52. Smith, C.J.; Iglesias-Siguenza, F.J.; Baxendale, I.R.; Ley, S.V. Flow and batch mode focused microwave synthesis of 5-amino-4-cyanopyrazoles and their further conversion to 4-aminopyrazolopyrimidines. *Org. Biomol. Chem.* **2007**, *5*, 2758–2761. [CrossRef] [PubMed]

53. Organ, M.G.; Hanson, P.R.; Rolfe, A.; Samarakoon, T.B.; Ullah, F. Accessing stereochemically rich sultams via microwave-assisted, continuous flow organic synthesis (MACOS) scale-out. *J. Flow Chem.* **2011**, *1*, 32–39. [CrossRef] [PubMed]
54. Morschhäuser, R.; Krull, M.; Kayser, C.; Boberski, C.; Bierbaum, R.; Püschner, P.A.; Glasnov, T.N.; Kappe, C.O. Microwave-assisted continuous flow synthesis on industrial scale. *Green Process Synth.* **2012**, *1*, 281–290. [CrossRef]
55. Guenin, E.; Meziane, D. Microwave Assisted Phosphorus Organic Chemistry: A Review. *Curr. Org. Chem.* **2011**, *15*, 3465–3485. [CrossRef]
56. Kiss, N.Z.; Ludányi, K.; Drahos, L.; Keglevich, G. Novel Synthesis of Phosphinates by the Microwave-Assisted Esterification of Phosphinic Acids. *Synth. Commun.* **2009**, *39*, 2392–2404. [CrossRef]
57. Keglevich, G.; Bálint, E.; Kiss, N.Z.; Jablonkai, E.; Hegedűs, L.; Grün, A.; Greiner, I. Microwave-Assisted Esterification of Phosphinic Acids. *Curr. Org. Chem.* **2011**, *15*, 1802–1810. [CrossRef]
58. Keglevich, G.; Kiss, N.Z.; Mucsi, Z.; Körtvélyesi, T. Insights into a surprising reaction: The microwave-assisted direct esterification of phosphinic acids. *Org. Biomol. Chem.* **2012**, *10*, 2011–2018. [CrossRef] [PubMed]
59. Kiss, N.Z.; Böttger, É.; Drahos, L.; Keglevich, G. Microwave-assisted direct esterification of cyclic phosphinic acids. *Heteroat. Chem.* **2013**, *24*, 283–288. [CrossRef]
60. Mucsi, Z.; Kiss, N.Z.; Keglevich, G. A quantum chemical study on the mechanism and energetics of the direct esterification, thioesterification and amidation of 1-hydroxy-3-methyl-3-phospholene 1-oxide. *RSC Adv.* **2014**, *4*, 11948–11954. [CrossRef]
61. Kiss, N.Z.; Keglevich, G. Microwave-assisted direct esterification of cyclic phosphinic acids in the presence of ionic liquids. *Tetrahedron Lett.* **2016**, *57*, 971–974. [CrossRef]
62. Bálint, E.; Jablonkai, E.; Bálint, M.; Keglevich, G. Alkylating esterification of 1-hydroxy-3-phospholene oxides under solventless MW conditions. *Heteroat. Chem.* **2010**, *21*, 211–214. [CrossRef]
63. Keglevich, G.; Grün, A.; Bölcskei, A.; Drahos, L.; Kraszni, M.; Balogh, G.T. Synthesis and proton dissociation properties of arylphosphonates: A microwave-assisted catalytic arbuzov reaction with aryl bromides. *Heteroat. Chem.* **2012**, *23*, 574–582. [CrossRef]
64. Dargó, G.; Bölcskei, A.; Grün, A.; Béni, S.; Szántó, Z.; Lopata, A.; Keglevich, G.; Balogh, G.T. Proton dissociation properties of arylphosphonates: Determination of accurate Hammett equation parameters. *J. Pharm. Biomed. Anal.* **2017**, *143*, 101–109. [CrossRef] [PubMed]
65. Greiner, I.; Grün, A.; Ludányi, K.; Keglevich, G. Solid–liquid two-phase alkylation of tetraethyl methylenebisphosphonate under microwave irradiation. *Heteroat. Chem.* **2011**, *22*, 11–14. [CrossRef]
66. Keglevich, G.; Grün, A.; Blastik, Z.; Greiner, I. Solid-liquid phase alkylation of P=O-functionalized CH acidic compounds utilizing phase transfer catalysis and microwave irradiation. *Heteroat. Chem.* **2011**, *22*, 174–179. [CrossRef]
67. Grün, A.; Blastik, Z.; Drahos, L.; Keglevich, G. Microwave-assisted alkylation of diethyl ethoxycarbonylmethylphosphonate under solventless conditions. *Heteroat. Chem.* **2012**, *23*, 241–246. [CrossRef]
68. Keglevich, G.; Grün, A. Microwave Irradiation as a Substitute for Phase Transfer Catalyst in CAlkylation Reactions. *Curr. Green Chem.* **2015**, *2*, 254–263. [CrossRef]
69. Grün, A.; Bálint, E.; Keglevich, G. Solid-Liquid Phase C-Alkylation of Active Methylene Containing Compounds under Microwave Conditions. *Catalysts* **2015**, *5*, 634–652. [CrossRef]
70. Grün, A.; Blastik, Z.; Drahos, L.; Keglevich, G. Dialkylation of diethyl ethoxycarbonylmethylphosphonate under microwave and solventless conditions. *Heteroat. Chem.* **2014**, *25*, 107–113. [CrossRef]
71. Keglevich, G.; Szekrényi, A. Eco-Friendly Accomplishment of the Extended Kabachnik–Fields Reaction; a Solvent- and Catalyst-Free Microwave-Assisted Synthesis of α-Aminophosphonates and α-Aminophosphine Oxides. *Lett. Org. Chem.* **2008**, *5*, 616–622. [CrossRef]
72. Keglevich, G.; Bálint, E. The Kabachnik–Fields Reaction: Mechanism and Synthetic Use. *Molecules* **2012**, *17*, 12821–12835. [CrossRef] [PubMed]
73. Tajti, Á.; Bálint, E.; Keglevich, G. Synthesis of ethyl octyl α-aminophosphonate derivatives. *Curr. Org. Synth.* **2016**, *13*, 638–645. [CrossRef]

74. Bálint, E.; Tóth, R.E.; Keglevich, G. Synthesis of alkyl α-aminomethyl-phenylphosphinates and *N,N*-bis(alkoxyphenylphosphinylmethyl)amines by the microwave-assisted Kabachnik–Fields reaction. *Heteroat. Chem.* **2016**, *27*, 323–335. [CrossRef]
75. Bálint, E.; Tajti, Á.; Kalocsai, D.; Mátravölgyi, B.; Karaghiosoff, K.; Czugler, M.; Keglevich, G. Synthesis and utilization of optically active α-aminophosphonate derivatives by Kabachnik-Fields reaction. *Tetrahedron* **2017**, *73*, 5659–5667. [CrossRef]
76. Bálint, E.; Tripolszky, A.; Jablonkai, E.; Karaghiosoff, K.; Czugler, M.; Mucsi, Z.; Kollár, L.; Pongrácz, P.; Keglevich, G. Synthesis and use of α-aminophosphine oxides and *N,N*-bis(phosphinoylmethyl)amines—A study on the related ring platinum complexes. *J. Organomet. Chem.* **2016**, *801*, 111–121. [CrossRef]
77. Keglevich, G.; Szekrényi, A.; Szöllősy, Á.; Drahos, L. Synthesis of bis(phosphonatomethyl)-, bis(phosphinatomethyl)-, and bis(phosphinoxidomethyl)amines, as well as related ring bis(phosphine) platinum complexes. *Synth. Commun.* **2011**, *41*, 2265–2272. [CrossRef]
78. Bálint, E.; Fazekas, E.; Pintér, G.; Szöllősy, Á.; Holczbauer, T.; Czugler, M.; Drahos, L.; Körtvélyesi, T.; Keglevich, G. Synthesis and utilization of the bis(>P(O)CH2)amine derivatives obtained by the double Kabachnik–Fields reaction with cyclohexylamine; Quantum chemical and X-ray study of the related bidentate chelate platinum complexes. *Curr. Org. Chem.* **2012**, *16*, 547–554. [CrossRef]
79. Bálint, E.; Fazekas, E.; Pongrácz, P.; Kollár, L.; Drahos, L.; Holczbauer, T.; Czugler, M.; Keglevich, G. *N*-Benzyl and *N*-aryl bis(phospha-Mannich adducts): Synthesis and catalytic activity of the related bidentate chelate platinum complexes in hydroformylation. *J. Organomet. Chem.* **2012**, *717*, 75–82. [CrossRef]
80. Li, Y.H.; Das, S.; Zhou, S.L.; Junge, K.; Beller, M. General and Selective Copper-Catalyzed Reduction of Tertiary and Secondary Phosphine Oxides: Convenient Synthesis of Phosphines. *J. Am. Chem. Soc.* **2012**, *134*, 9727–9732. [CrossRef] [PubMed]
81. Kovács, T.; Keglevich, G. The reduction of tertiary phosphine oxides by silanes. *Curr. Org. Chem.* **2017**, *21*, 569–585. [CrossRef]
82. Keglevich, G.; Kovács, T.; Csatlós, F. The deoxygenation of phosphine oxides under green chemical conditions. *Heteroat. Chem.* **2015**, *26*, 199–205. [CrossRef]
83. Kovács, T.; Urbanics, A.; Csatlós, F.; Binder, J.; Falk, A.; Uhlig, F.; Keglevich, G. A Study on the Deoxygenation of Phosphine Oxides by Different Silane Derivatives. *Curr. Org. Synth.* **2016**, *13*, 148–153. [CrossRef]
84. Kovács, T.; Urbanics, A.; Csatlós, F.; Keglevich, G. A study on the deoxygenation of trialkyl-, dialkyl-phenyl- and alkyl-diphenyl phosphine oxides by hydrosilanes. *Heteroat. Chem.* **2017**, *28*, e21376. [CrossRef]
85. Hirao, T.; Masunaga, T.; Yamada, N.; Ohshiro, Y.; Agawa, T. Palladium-catalyzed New Carbon-Phosphorus Bond Formation. *Bull. Chem. Soc. Jpn.* **1982**, *55*, 909–913. [CrossRef]
86. Hirao, T.; Masunaga, T.; Ohshiro, Y.; Agawa, T. A Novel Synthesis of Dialkyl Arenephosphonates. *Synthesis* **1981**, *1981*, 56–57. [CrossRef]
87. Jablonkai, E.; Keglevich, G. P-ligand-free, microwave-assisted variation of the Hirao reaction under solvent-free conditions; The P-C coupling reaction of >p(O)H species and bromoarenes. *Tetrahedron Lett.* **2013**, *54*, 4185–4188. [CrossRef]
88. Keglevich, G.; Jablonkai, E.; Balázs, L.B. A "green" variation of the Hirao reaction: The P–C coupling of diethyl phosphite, alkyl phenyl-*H*-phosphinates and secondary phosphine oxides with bromoarenes using a P-ligand-free $Pd(OAc)_2$catalyst under microwave and solvent-free conditions. *RSC Adv.* **2014**, *4*, 22808–22816. [CrossRef]
89. Keglevich, G.; Henyecz, R.; Mucsi, Z.; Kiss, N.Z. The Palladium Acetate-Catalyzed Microwave-Assisted Hirao Reaction without an Added Phosphorus Ligand as a "Green" Protocol: A Quantum Chemical Study on the Mechanism. *Adv. Synth. Catal.* **2017**, *359*, 4322–4331. [CrossRef] [PubMed]
90. Ergan, B.T.; Bayaromoglu, M. Poly (L-lactic acid) synthesis using continuous microwave irradiation–simultaneous cooling method. *Chem. Eng. Commun.* **2018**, *205*, 1665–1677. [CrossRef]
91. Chen, S.-T.; Chiou, S.-H.; Wang, K.-T. Preparative scale organic synthesis using a kitchen microwave oven. *J. Chem. Soc. Chem. Commun.* **1990**, 807–809. [CrossRef]
92. Pipus, G.; Plazl, I.; Koloini, T. Esterification of benzoic acid in microwave tubular flow reactor. *Chem. Eng. J.* **2000**, *76*, 239–245. [CrossRef]
93. Cablewski, T.; Faux, A.F.; Strauss, C.R. Development and Application of a Continuous Microwave Reactor for Organic Synthesis. *J. Org. Chem.* **1994**, *59*, 3408–3412. [CrossRef]

94. Krull, M.; Moschhaeuser, R. Continuous Method for Producing Esters of Aromatic Carboxylic Acids. U.S. Patent 0088918, 12 April 2012.
95. Tajti, Á.; Tóth, N.; Bálint, E.; Keglevich, G. Esterification of benzoic acid in a continuous flow microwave reactor. *J. Flow Chem.* **2018**, *8*, 11–19. [CrossRef]
96. Bálint, E.; Tajti, Á.; Drahos, L.; Ilia, G.; Keglevich, G. Alcoholysis of dialkyl phosphites under Microwave conditions. *Curr. Org. Chem.* **2013**, *17*, 555–562. [CrossRef]
97. Bálint, E.; Tripolszky, A.; Tajti, Á. Synthesis of α-aminophosphonates by the Kabachnik–Fields reaction. In *Organophosphorus Chemistry*; Keglevich, G., Ed.; De Gruyter: Berlin, Germany, 2018; pp. 108–147.
98. Keglevich, G.; Bálint, E.; Tajti, Á.; Mátravölgyi, B.; Balogh, G.T.; Bálint, M.; Ilia, G. Microwave-assisted alcoholysis of dialkyl phosphites by ethylene glycol and ethanolamine. *Pure Appl. Chem.* **2014**, *86*, 1723–1728. [CrossRef]
99. Tajti, Á.; Keglevich, G.; Bálint, E. Microwave-assisted alcoholysis of dialkyl *H*-phosphonates by diols and amino alcohols. *Phosphorus Sulfur Silicon Relat. Elem.* **2017**, *192*, 769–775. [CrossRef]
100. Wang, Y.-C.; Yuan, Y.-Y.; Du, J.-Z.; Yang, X.-Z.; Wang, J. Recent Progress in Polyphosphoesters: From Controlled Synthesis to Biomedical Applications. *Macromol. Biosci.* **2009**, *9*, 1154–1164. [CrossRef] [PubMed]
101. Troev, K.D. *Polyphosphoesters: Chemistry and Application*; Elsevier: Oxford, UK, 2012.
102. Bálint, E.; Tajti, Á.; Tóth, N.; Keglevich, G. Continuous Flow Alcoholysis of Dialkyl *H*-Phosphonates with Aliphatic Alcohols. *Molecules* **2018**, *23*, 1618. [CrossRef] [PubMed]
103. Bálint, E.; Tajti, Á.; Ladányi-Pára, K.; Tóth, N.; Mátravölgyi, B.; Keglevich, G. Continuous flow synthesis of α-aryl-α-aminophosphonates. *Pure Appl. Chem.* **2019**, *91*, 67–76. [CrossRef]
104. Bálint, E.; Tajti, Á.; Ádám, A.; Csontos, I.; Karaghiosoff, K.; Czugler, M.; Ábrányi-Balogh, P.; Keglevich, G. The synthesis of α-aryl-α-aminophosphonates and α-aryl-α-aminophosphine oxides by the microwave-assisted Pudovik reaction. *Beilstein J. Org. Chem.* **2017**, *13*, 76–86. [CrossRef] [PubMed]

2

Advances and Challenges in the Creation of Porous Metal Phosphonates

Bharadwaj Mysore Ramesha * and Vera Meynen

Laboratory of Adsorption and Catalysis (LADCA), Department of Chemistry, University of Antwerp, Universiteitsplein 1, 2610 Wilrijk, Belgium; vera.meynen@uantwerpen.be

* Correspondence: bharadwaj.mysoreramesha@uantwerpen.be

Abstract: In the expansive world of porous hybrid materials, a category of materials that has been rather less explored than others and is gaining attention in development is the porous metal phosphonates. They offer promising features towards applications which demand control over the inorganic–organic network and interface, which is critical for adsorption, catalysis and functional devices and technology. The need to establish a rationale for new synthesis approaches to make these materials in a controlled manner is by itself an important motivation for material chemists. In this review, we highlight the various synthetic strategies exploited, discussing various metal phosphonate systems and how they influence the properties of porous metal phosphonates. We discuss porous metal phosphonate systems based on transition metals with an emphasis on addressing challenges with tetravalent metals. Finally, this review provides a brief description of some key areas of application that are ideally suited for porous metal phosphonates.

Keywords: hybrid porous materials; layered metal phosphonates; phosphonate MOFs; supramolecular-templated porous metal phosphonates; separation and extraction; heterogeneous catalysis; proton conduction

1. Introduction

There has been a surge in the pursuit to create inorganic–organic hybrid materials in the past two decades, with the success of metal–organic frameworks (MOFs) and periodic mesoporous organosilicas (PMOs). Hybrid materials are interesting for material chemists and in application as the integration of both organic and inorganic functionalities allows to modify the matrix and interface properties of materials. The unique combination of inorganic and organic features creates particular properties and control with respect to the structure, physical properties and/or specific chemical reactivity and/or interactions. This has indeed resulted in the extensive and intensive creation of hybrid porous materials such as MOFs [1–5] and PMOs [6–8]. An interesting, more specific and less studied class of materials are porous metal phosphonates, built up from the interaction between metals and organophosphonic units. These hybrid porous metal phosphonate materials are particularly interesting where stable performance under sometimes harsh conditions and specific surface interactions are an important requirement for application, like the sorption of metal ions, heterogeneous catalysis, thin-film-based devices, chromatography or membranes technology.

We describe in this review the evolution of metal phosphonates over the past couple of decades from layered metal phosphonates to phosphonate-based MOFs and finally to templated (supramolecular) mesoporous metal phosphonates. All of these materials fall under the category of coordination polymeric networks and can be subdivided further based on their long-range order and/or pore morphology into: phosphonate MOFs, layered metal phosphonates and templated (supramolecular) metal phosphonates, as illustrated by Figure 1. This classification is thus only based on the final structure/texture of the porous coordination networks, which is primarily determined by the interaction of the metal precursors and phosphonic moieties in solution during synthesis, resulting in structures featuring a variety of coordination modes. We would like to point out that this is not a detailed review concerning each of these classes of materials, but rather an attempt to put together some of the similarities and differences when it comes to the choice of metal, metal valency and phosphonic linkers with respect to specific interesting morphologies of porous coordination networks that can be obtained. The critical aspects required to obtain particular metal phosphonate architectures are recognized by analyzing the reaction parameters and their consequences on the porous hybrid frameworks obtained. For a more detailed reading on the synthesis and/or applications for each class of metal phosphonates, we refer the readers to some of the interesting reviews [9–17].

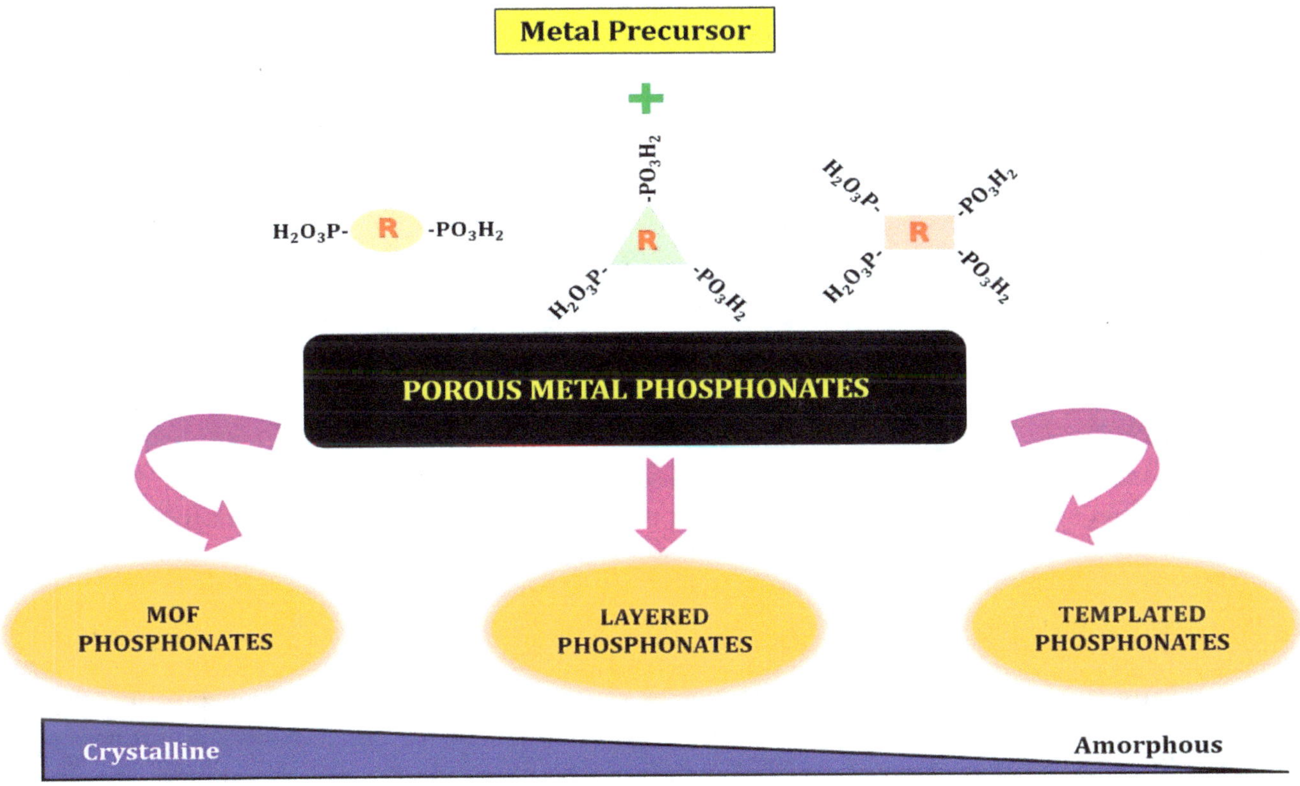

Figure 1. Scheme for classifying porous metal phosphonates based on structural/morphological properties.

Many of the interesting advances and developments in the field of porous metal phosphonates hybrids have taken place in the last couple of decades. A substantial part of the progress in porous metal phosphonates has been built on the knowledge stemming from the pioneering works from Clearfield et al. and Alberti et al. on layered metal phosphates/phosphonates [9,10,18–20]. We describe an overview of the different synthesis methods undertaken for various transition metals and then funnel down towards the metal (IV) phosphonate system. The focus of this review is addressing the challenges posed for creating porous metal (IV) phosphonate networks. We restrict ourselves to materials synthesized purely from transition metal precursors and organic precursors containing only phosphonate moieties. Therefore, mixed ligand systems such as the metal carboxyphosphonates are not a part of this review [21]. This review consists of two parts, where we first summarize on the synthesis methods employed for obtaining porous metal phosphonates and subsequently provide an overview of the potential benefit of porous metal phosphonates in some key areas of application.

2. Synthesis of Porous Metal Phosphonates

In this section, we will discuss some of the metal phosphonate networks that have been created in the past couple of decades leading up to the metal (IV) phosphonates and the challenges faced in obtaining these metal phosphonate networks. The synthetic conditions of the porous metal phosphonates together with the choice of precursor(s) and type of organophosphonic acid define the type of material that is ultimately desired, which we discuss in this review as a part of the three different structural/morphology classes (Figure 1):

(1) Layered metal phosphonates;
(2) Phosphonate–metal organic frameworks;
(3) Supramolecular templated porous metal phosphonates.

The synthetic details for each class of materials, namely layered metal phosphonates, phosphonate-based MOFs and templated porous metal phosphonates, have been summarized in the Supplementary Information (Tables S1, S2 and S3, respectively).

2.1. Layered Metal Phosphonates

Layered metal phosphonates were synthesized by the direct precipitation between metal precursors and organophosphonic acids in a similar way as known for metal phosphates [9,10]. The resulting materials were found to be analogous in structure to layered metal phosphates, but with the interlayer distance depending on the organic linker group [22,23]. As in the literature "metal organophosphonates" are often described as "metal phosphonates", we will from here on leave out "organo" even though these materials are hybrid materials synthesized from metal and organophosphonic precursors.

In this section, we will discuss the syntheses of metal phosphonates of divalent/trivalent metals and end by discussing tetravalent metal phosphonates. The key difference between those two relates to the solubility of metal ions in solutions. Transition metals strongly interact with phosphonic acid precursors and form an insoluble coordination polymer. However, a prerequisite to obtain a crystalline structure is a controlled, ordered buildup of molecular precursors into a hybrid network, requiring the metal moieties to be solubilized in sufficient quantity, allowing homogeneous interactions with the phosphonic precursors. In the case of metals that are divalent/trivalent, this is achieved by lowering the pH of the reaction medium, and thus synthesizing in an acidic condition. However, with tetravalent metal precursors, the formation of a metal phosphonate coordination network is kinetically too fast, resulting in the precipitation of the mixtures in an uncontrolled way, even in high acidic conditions and therefore hampering controlled structural properties [24]. One of the ways to overcome this has been by the introduction of HF as a solubilizing/mineralizing agent, whereby F^- ions form a complex with metal ions, causing a slow release of metal ions into the solution available for interaction with the phosphonic precursor upon heating under pressure [20].

With respect to divalent metals, zinc (II) diphosphonates formed by phenylene and bisphenylene phosphonic linkers resulted in a structure that was found to be dependent on the length of the aryl linker. In the case of the monophenylene linker, the phenylene phosphonate groups functioned as pillars between the octahedral zinc oxide layers. However, in the presence of biphenylene as a linker, the inorganic unit formed chains rather than a layered structure with the zinc being coordinated tetrahedrally. The same phenomena were observed with copper (II) diphosphonates with phenylene and bisphenylene linkers [25,26]. Moreover, when using non-aromatic linkers, it was reported that the structure of the materials was sensitive to the length of the linker/spacer in copper (II) and zinc (II) diphosphonates made from ethylene diphosphonic acid or propylene diphosphonic acid. The presence of an extra methylene unit leads to a change in the coordination of copper and zinc from an octahedral layered into a tetrahedral chain structure. Hence, the length of linker moiety (alkyl or aryl) has pronounced consequences on the coordination of the linkers to the metal-oxo centers and thus dictates the final structure of the material. These key structural features in layered materials were determined by single-crystal X-ray diffraction and was therefore important to understand the details of the coordination geometry with different linkers [27].

Different types of precursors in combination with variations in synthesis conditions were found to influence the structural properties of layered metal phosphonates. The use of diphosphonic acids containing an amine group allows to tune the coordination with the metal, based on the reaction conditions and the choice of metal precursor, which influences the formation of the network. Zinc (II) and manganese (II) diphosphonates were synthesized with N-ethylimino bis(methylene phosphonic acid), yielding a 2D-layered structure with Mn (II) and a 3D network when Zn (II) was the metal source. The degree of deprotonation of the phosphonic linker and the affinity of the metal for the amine group was found to dictate the way the hybrid structure evolved during the synthesis [28]. Another example illustrating the importance of reaction conditions was demonstrated by the reaction of cobalt (II) and octylene diphosphonic acid. The room temperature synthesis of the precursors in HF resulted in a clathrate structure made of 1D chains featuring a cobalt atom and undissociated diphosphonic acid in a polymeric network held together by hydrogen bonds. However, when the same system was subjected to a hydrothermal treatment (in the absence of HF), layered cobalt (II) diphosphonate with octyl chains in the inter-layer were formed. Therefore, depending on the synthesis conditions and presence or absence of HF, different structures were formed. The degree of protonation plays an important role in the hydrogen bonded networks in such systems [29].

Due to the closed packing of the phosphonic linkers as pillars, the metal phosphonates did not possess any porosity. Nevertheless, porous systems are known to have higher surface areas and provide better interfacial properties that are crucial for applications such as catalysis, gas adsorption, separation/extraction and others. Therefore, porosity is often a highly desired property and we therefore describe how porosity can be introduced in these layered metal phosphonate systems. These methods involved the use of additives or mixing solvent systems as will be discussed in the following paragraphs.

A strategy to impart porosity in the layered materials was by spacing out the closely packed inter-layer spacers through the use of organic linkers with over-crowded diphosphonic acids. This created inorganic layers (e.g., Cd (II) and Ni (II) oxides) with anchored organic groups pointing inwards between the layers but without close packing. Nevertheless, these materials were still not possessing any (accessible) porosity [30]. In another approach to achieve porosity in layered metal phosphonates, zinc (II) precursor and phenylene and biphenylene diphosphonic acid were reacted in the presence of varying amounts of phosphoric acid in the synthesis mixture. It was shown that indeed, a three-component system with a suitable phosphonate/phosphate (1:4 and 1:7) ratio resulted in a material that had inter-particle slit-shaped porosity in the mesopore region (~6.6 nm). It was noted that an equal phosphonate/phosphate ratio resulted in a material that was purely zinc(II) phosphonate, while higher phosphate incorporation led to the formation of zinc phosphate. However, with the right amount of phosphoric acid, phosphate moieties were incorporated between the phenylene linkers at random places without altering the crystal structure, resulting in a mesoporous texture [31].

The use of metals like Ca, Zn, or Mn (valency < 3) with phosphonic acids usually leads to materials that exhibit a layered morphology [32]. Therefore, the same strategy as in divalent metals was employed to impart porosity in zirconium (IV)-based metal phosphonates. The Zr (IV) precursor was reacted with 4,4′-biphenyldiphosphonic acid in the presence of phosphoric acid that would act as a spacer between the aryl linkers [20]. A mechanism of a house-of-cards arrangement was proposed, creating porosity between the layered particles of zirconium bisphosphonate–phosphite materials. The inter-particle mesoporosity was achieved by balancing the amount of phosphoric acid/diphosphonic acid ratio [33]. However, as tetravalent metals have the problem of insolubility (e.g., zirconium and titanium metals) caused by the rapid hydrolysis and condensation of metal (IV) precursors resulting in rapid precipitation when they have reacted with phosphonic acid species, the crystallinity of these layered metal phosphonates seems to be improved only when the synthesis was performed in the presence of HF as the solubilizer/mineralizer [20]. The porosity in these materials was mainly dictated by the packing density of the organic linker between the inorganic metal oxide layers, acting as spacers, and on the presence of ingredients such as solvent/additives in the synthesis recipe [9,10]. Attempts to make narrow pore size lamellar metal phosphonates were achieved by Alberti et al. by replacing around 30% of the butane and benzene diphosphonic acid moieties by a phosphite moiety in between the zirconium oxide layer. This way, mesoporosity was observed with pore sizes around 2–4 nm, depending on the amount of phosphite incorporated [18]. Similarly, complex pillaring moieties were incorporated, such as the 3,3′,5,5′—tetramethylbiphenyldiphosphonic acid that has bulky ends with narrow biphenyl units [19]. This material possessed a high surface area of 375 m^2/g with inter-layer microporosity (with a pore size of ~0.5 nm) [34].

Tin (IV) phosphonates were prepared that consist of small nanoparticles with microporosity due to inter-particle voids that created tunnels in dimensions of 1.1–2.4 nm, depending on the size of the organic group of the phosphonate [35]. A similar attempt to synthesize porous layered material with tin (II) was unsuccessful even with the inclusion of mixed phosphonic ligands [36]. In contrast, a mixed phosphonic acid precursor of phenyl phosphonic acid and methyl phosphonic acid with tin (IV), resulted in mixed micro/mesoporosity. The stacking of tiny nanoparticles of tin (IV) phosphonates that was arranged in a house-of-cards fashion contributed to the mixed porosity. The porosity was found to be dependent on the ratio of phenyl phosphonic/methyl phosphonic acid amount and on the presence or absence of HF. In the absence of HF, the materials were mesoporous and in the presence of HF, the materials showed microporosity [37]. In another study, layered tin (IV) phosphonates that were synthesized from diphosphonic acids had completely different structures in comparison with tin (IV) phosphonates from monophosphonic acids. Tin (IV) phosphonates with diphosphonic acids presented larger particles with sizes ~20 μm, while tin (IV) phosphonates with monophosphonic acids were ~2 μm in size. Thus, the nature of the phosphonic acids is crucial to particle growth in forming tin (IV) phosphonates. Along with the mono- or di-functionality of the phosphonic acid, the metal source plays an important role. With zirconium (IV), diphosphonic acids were inserted between the inorganic layers as pillars, whereas in the case of tin (IV) they were found to be incorporated in between inorganic layer as pillars as well as inserted as a part of the inorganic–organic chains. Because of this dual role, when tin (IV) was the metal, not all diphosphonic moieties were present as pillars and therefore voids were present in the layers which resulted in porosity in the final material [38]. In another study, the growth of porous layered tin (IV) phosphonates was explained by random cross-linking of layered species providing inter-layer porosity. A possible hypothesis for the underlying mechanism that caused the random stacking instead of layer growth was the presence of stacking defects by the combination of diphosphonate and monophosphonate ligands or the presence of protonated phosphonic acids [39]. Based on these results, a generalized growth model for metal (IV)diphosphonate coordination networks was proposed, resulting from rapid precipitation of unevenly grown capped plates, leading to formation of structure as illustrated by Figure 2 [39,40].

Apart from other factors such as the choice of the phosphonate ligand and the presence or absence of HF, also the type of solvent used plays a role in obtaining porous metal phosphonates. The obtained porosity was found to be dependent on the solvent system employed. Zr (IV) precursors are soluble in DMSO but poorly soluble in an alcohol/water medium, and the vice versa for Sn (IV) precursors. Based on the choice of alcohol (EtOH, PrOH, or BuOH) the porosity could be tuned towards microporous Sn (IV) bisphosphonates. However, for zirconium (IV) bisphosphonates, it was only possible when DMSO was used as the solvent [41].

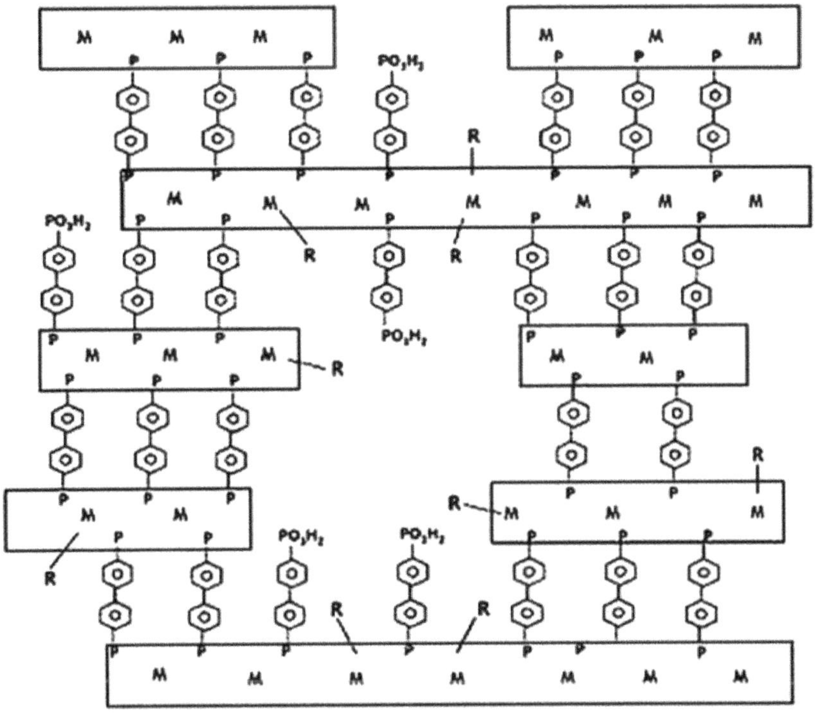

Figure 2. Schematic drawing of how micropores are formed in the metal biphenylene bis(phosphonates). The horizontal rectangles represent the inorganic layers (M = Zr or Sn) which are cross-linked by the organic moieties. R stands for F^-, if HF is used in the preparation, and the surfaces are capped by M because an excess of metal is used in the syntheses. Reprinted with permission from [39].

2.2. MOFs–Phosphonates

Metal-organic frameworks are one of the prominent classes of coordination polymeric networks that are built-up from metal nodes and organo-functional units resulting in porous structures with very high surface areas. Here, we focus on a specific type of MOFs that consists of metallic units with phosphonate linkers. Frameworks based on metal phosphonates are rather an emerging class of MOFs when compared to the well established carboxylates, whereas the surge in carboxylate MOFs has been on a continuous rise ever since their inception in the late 1990s. The successful synthesis of frameworks based on metal phosphonates (albeit small in number) has only been possible in this decade. This can largely be attributed to the stronger coordination of the phosphonate moiety towards metals than the carboxylate counterparts [11]. Therefore, hindering the orderly growth of a porous metal phosphonate framework.

We begin this section by discussing some of the phosphonate frameworks, similar to layered metal phosphonates: divalent/trivalent metal phosphonate frameworks that have led to the development of tetravalent metal phosphonate frameworks. One can expect that similar to layered structures, the problem of solubility is very apparent and impedes creating porous architectures. One finds in the literature that MOFs that have been deemed as the 'ultra stable' systems are mostly composed of Ti (IV) and Zr (IV) nodes [24]. Thus, it is crucial in studying how these frameworks could be obtained with phosphonic linkers. In this section, we build on the knowledge of layered metal phosphonates and describe how their lamellar motif can be altered, going towards open porous architectures.

A 3D framework synthesized from zinc and 1,4-dihydroxy-2,5-benzenediphosphonic acid in DMF resulted in the formation of 1D chains of zinc tetrahedra and RPO_3^{2-} with a surface area of 209 m^2/g (microporous with 1 nm × 1 nm pores), once DMF was removed from the framework. The key component of the phosphonic linker, the dihydroxy group, prevented the formation of a dense layered motif and thus resulted in an open framework [42]. The reaction of ethylenediaminetetrakis (methylenephosphonic acid) with lead (II) and zinc (II) also resulted in two different types of 3D open frameworks that were microporous. This phosphonic ligand containing eight phosphonic acid sites provided myriad coordination modes that were dependent on the choice of metal and the pH of the reaction medium. The framework based on lead was made up of helical chains of square-pyramidal lead and RPO_3 phosphonate units. However, in the case of zinc, the amino group was not involved in binding with the metal and resulted in a square-wave type arrangement of zinc and the phosphonate layer. Both these frameworks possessed open channels in the micropore region (0.68 nm × 1 nm) [43].

Based on the reaction conditions and the choice of metal and phosphonic precursors, a wide variation in coordination modes between the metal units to the phosphonic acid units can be obtained (see for example Figures 3 and 4 versus Figure 5). When additives are added, novel coordination modes can be formed. For example, a framework with 1D pores was synthesized from benzene diphosphonic acid and copper (II) in the presence of amino triazole as an additive. This framework features a pillared structure with apparent pores of ~0.45 nm. It could only be formed due to the amino group of the triazole coordinating the copper atoms and producing 1D columns of 6-coordinated copper. The absence of the amino group of the triazole lead to the classical pillared copper benzene diphosphonate structure with no interlayer porosity [44]. In another study, the reaction of lead (II) with organo-piperazine as a diphosphonic ligand resulted in an initial 2D layered structure. This 2D structure was then transformed into a 3D framework via supramolecular assembly during the synthesis, due to the weak interaction of Pb–O(N) contacts. The formation of the supramolecular assembly leading to the obtained framework was attributed to hydrogen bonds, van der Waals and some electrostatic interactions [45].

Modifying the phosphonate ligands into $R[PO(OH)(OR)]_2$, by derivatizing one of the acidic P–OH groups into a P–OR group, mimics a carboxylate ligand. Thus, this type of precursors is expected to allow the formation of secondary building units (SBU) that are well known from carboxylate MOFs. This concept was first employed with zinc(1,4-benzenediphosphonate-bis(monoethyl ester)), to create a pillared structure with the ester moiety pointing inwards between the layers and with permanent porosity (aperture ~0.6 nm) due to the Van der Waal forces between the alkyl chains in the interlayer [46]. The same phosphonic linker was later used to synthesize frameworks based on copper, and in this case, the linker possessed both methyl and ethyl ester groups. The resulting structure of the copper phosphonate framework with the ethyl ester was similar to the previous example with zinc. The use of these phosphonate linkers, mimicking a carboxylate linker, has led to the creation of porous frameworks built up via SBUs. These phosphonate MOFs also have advantages in application as they, for example, showed good affinity for CO_2 with the high isosteric heat of adsorption of 45 kJ/mol similar to carboxylate MOFs such as HKUST-1 (35 kJ/mol), MIL-101 (44 kJ/mol), MIL-100 (63 kJ/mol) and Mg-MOF-74 (47 kJ/mol) [47].

Larger pore metal phosphonate framework synthesized from nickel (II) and copper (II) with N,N′-piperazinebis(methylenephosphonic)acid as the linker have been prepared, consisting of a hexagonal array of channels with a pore size of 1 nm that was filled with physisorbed water. However, upon dehydration, the framework distorted. When using a 2-methylated derivative of the same phosphonic linker, this structural distortion could be prevented and the framework retained its structure [48]. Reticular syntheses are described as the process of assembling rigid molecular building blocks into ordered networks, held by strong bonds [1]. Thus, to attain larger pores, isoreticular methods have been employed to synthesize large pore metal phosphonate MOFs in materials such as STA-12 with an isoreticular pore size of 0.9 nm. They were produced by the reaction of N,N′-piperazine bis(methylenephosphonic acid) with divalent metals (Mn, Fe, Co and Ni) [49,50]. To obtain larger pores, pushing towards the mesopore region, linkers such as the N,N′-4,4′-bipiperazine bis(methylenephosphonic acid) were reacted with cobalt to obtain 3D frameworks that have pores of 1.8 nm (STA-16). This 3D framework was composed of $CoON_5$ octahedra as a spiral, which resulted in hexagonal pores with a 1.8 nm pore opening as shown in Figure 3 [51].

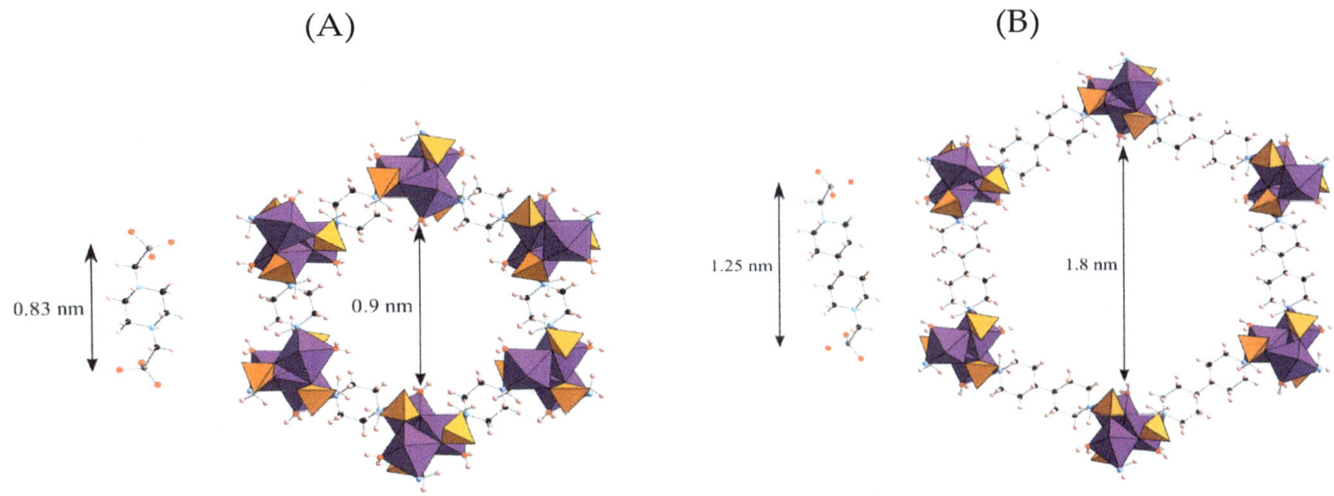

Figure 3. Frameworks of STA-12 (**A**) and STA-16 (**B**) viewed along the channel axis. Color code: cobalt: purple, phosphorus: orange, oxygen: red, nitrogen: blue, carbon: black, hydrogen: pink. Reprinted with permission from [51].

Phosphonate MOFs stable in water were synthesized from barium and octaethyl pyrene-1,3,6,8-tetraphosphonate resulting in a permanently porous framework denoted CALF-25. It is a three-dimensional framework made up of 1D inorganic–organic chains of barium phosphonate that are cross-linked by pyrene ligands. The pores of this framework consist of 1D rectangular openings of 0.46 nm × 0.39 nm containing hydrophobic ethyl chains [52]. Alternatively, a framework built up from tin and a triphosphonate ligand resulted in an amorphous, highly robust, and porous system having a surface area of 500 m^2/g and pore width of 0.85 nm (CALF-28) [53,54]. In another study, the use of flexible tritopic phosphonate linkers provided the platform to synthesize materials of 2D and 3D frameworks or even 0D dimers with various metals (Cu, Ni, and Mn). Benzene-1,3,5-triyl-tris(methylene)triphosphonic-acid (BttP) or 2,4,6-trimethyl-BttP were used as the flexible linkers with a rigid aromatic core. It was shown that flexible tritopic phosphonate linkers allow a tunable structure of the framework, based on the choice of metal, pH, and the synthesis conditions, to end up with either a 2D or 3D architecture. The flexibility in the triphosphonate moiety allowed the linkers to adopt various coordination environments in bonding to different metals, influencing the conformation of the triphosphonate moiety and the resulting structure of the metal phosphonate framework [55].

Finally, phosphonate MOFs made from metal (IV) precursors are summarized in the following paragraphs. Phosphonate-based MOFs based on metal (IV) have been extremely difficult to prepare since they usually form a dense, amorphous layered structure. MIL-22 is a titanium (IV)-based 3D layered framework, wherein hydrous titanium oxide was reacted hydrothermally with ethylene diphosphonic acid. The inorganic–organic chain built up by Ti–O–Ti and Ti–O–P bonds, containing narrow channels that are filled with water in the interlayer region, provide minimal to no porosity. However, an important property was the excellent thermal stability of MIL-22 compared to other open-frameworks [56]. These MIL-22 structures were derived from the metal phosphonate frameworks synthesized with vanadium and alkyl diphosphonic acids [57–59]. However, by employing N,N′-piperazine bismethyl phosphonic acid, the frameworks MIL-91 (Ti (IV) and Al (III)) were synthesized. The MIL-91-Ti framework consists of MO_6 octahedra and bisphosphonate groups having a surface area of ~500 m^2/g, pore volume of 0.19 cm^3/g and pores of size 0.35 nm × 0.40 nm. This was the first successful open phosphonate framework made from titanium(IV) as the metal node, shown in Figure 4 [60].

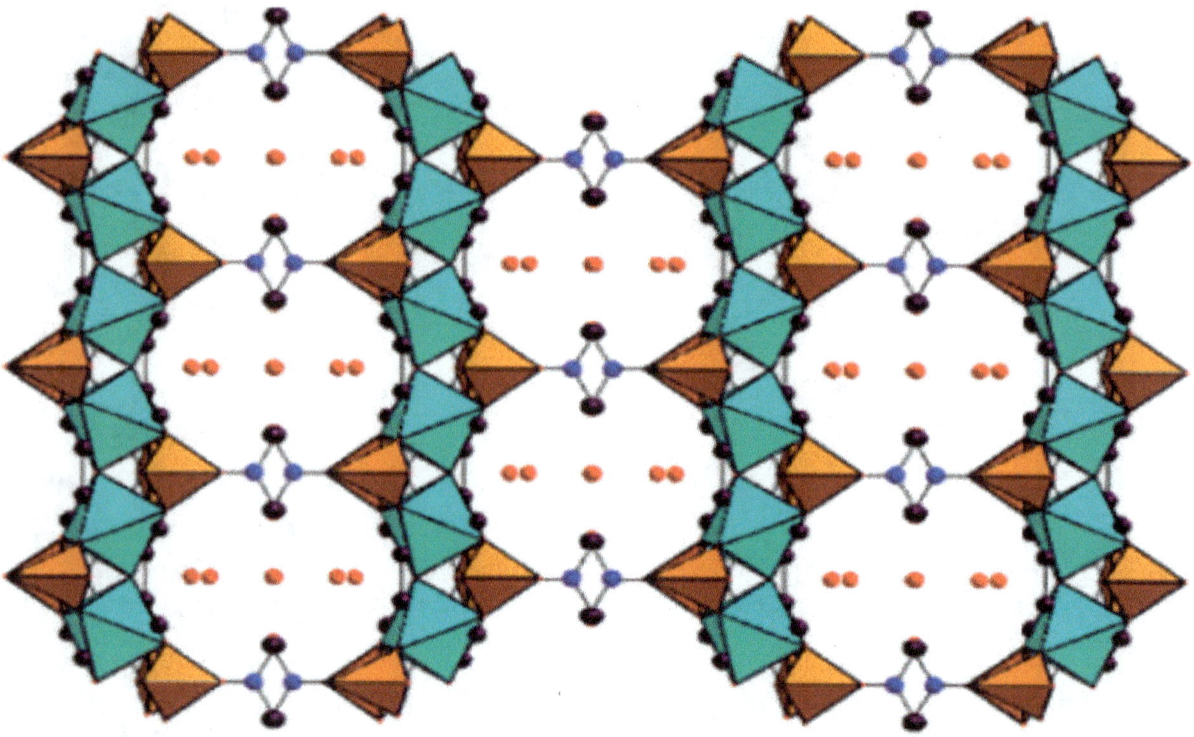

Figure 4. Structure of MIL-91(Ti) viewed along the c axis. Color code: titanium octahedra: cyan, phosphorus tetrahedra: orange, oxygen: red, nitrogen: blue, and carbon atoms: purple. Reprinted with permission from [60].

In another study using the same linker, piperazine-N,N′-bis(methylenephosphonic acid) and zirconium(IV), two 3D frameworks, were synthesized. The structure of these frameworks was found to be dependent on the pH of the reaction medium. At higher pH, a 3D framework was created by zirconium octahedra bonded to RPO_3, forming composite building units (CBU), linking the inorganic–organic chains through the piperazine units resulting in the formation of a zirconium phosphonate framework. At a lower pH, the formation of the classic pillared-layered structure was observed [61]. Alternatively, a robust honeycomb-layered framework of zirconium triphosphonate was synthesized with 1,3,5-tris(4-phosphonophenyl)benzene as the linker by direct precipitation in the presence of HF. Each layer consisted of zirconia octahedra linked to RPO_3 tetrahedra that acted as the SBU to form the honeycomb-layered framework. Even though the framework had good hydrolytic stability, it lacked porosity due to the capping of the layers [62]. Therefore, in order to induce porosity, the linker was replaced by 2,4,6-tris(4-phosphonomethyl)-phenyl 1,3,5-triazine (ttbmp), which is a flexible phosphonate ligand compared to the previous rigid linker. The resulting open framework UPG-1 was composed of 1D polymeric chains of zirconia octahedra linked by ttbmp, creating two kinds of channels: a 1 nm channel lined with free P–O groups and a smaller channel of 0.5 nm lined by uncoordinated PO_3C groups as shown in Figure 5 [63]. In a similar manner, an aluminum phosphonate framework with a pore size of 1.24 nm was obtained [64].

The importance of hydrogen bonding in metal phosphonate structures has recently been clearly illustrated by Taddei et al. [65]. In this work, the conformational effect of phosphonic linkers on the evolution of the structure of the framework was shown. Two distinct structures were obtained as a result of the position of the ($-CH_2-PO_3H_2$) moiety in either para- or meta- location on the novel triazine linker, which influenced the hydrogen bonding and therefore, resulted in a new zirconium phosphonate framework, UPG-2 with unique proton conducting behavior.

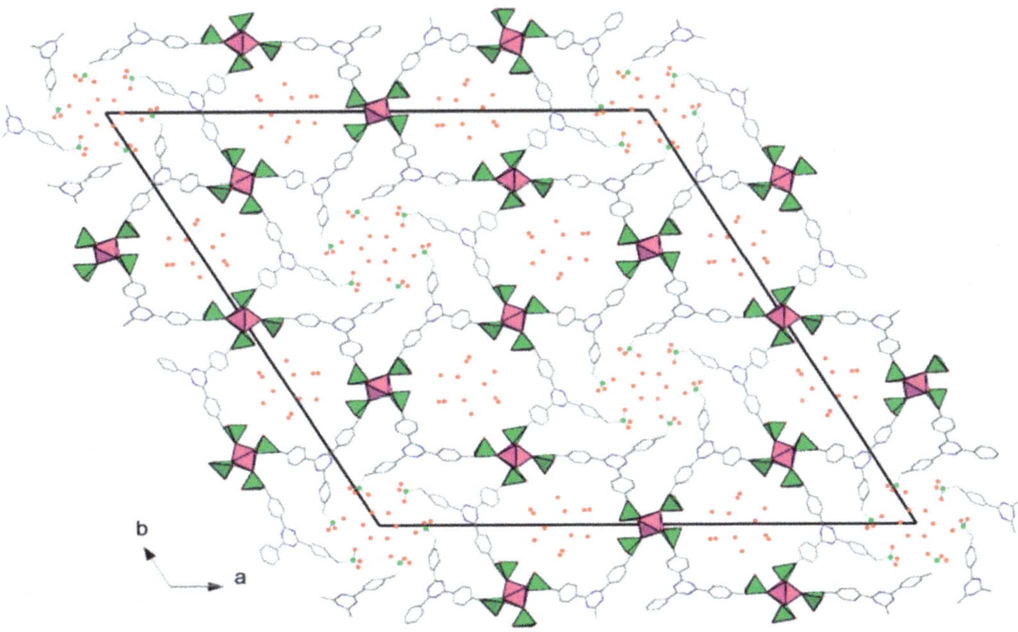

Figure 5. The crystal structure of UPG-1 viewed along the c axis. Color code: zirconium octahedra: purple, coordinating PO_3C tetrahedra: green. Reprinted with permission from [63].

2.3. Templated Porous Metal Phosphonates

The final part of the synthesis section is committed to porous metal phosphonates wherein the porous texture is created by the incorporation of templates (supramolecular) as structure-directing agents (SDAs). These materials take inspiration from the PMO hybrid system but follow a different approach as the metal oxide and organophosphonate precursors are incorporated into a hybrid network by a heterocondensation step. The protocol to make these mesoporous metal phosphonates is mostly based on the methodologies of porous metal oxide systems [66]. In general, porous metal oxides are obtained by the formation of metal–oxo oligomers around SDAs in solution or alternatively an evaporation-induced oligomer assembly with SDA triggered by the evaporation of the solvent, denoted as the evaporation-induced self-assembly method (EISA). Finally, a thermal step is applied to crystallize the metal oxide and to remove SDA molecules [67]. Alternatively, the SDA can be removed by extraction with/without a thermal step to enhance the condensation/crystallization [66,67]. Other synthesis methods for this class of materials have been adapted from the sol–gel approaches used to synthesize metal oxides. The sol–gel based materials form in most cases an amorphous metal oxide network that requires a final calcination step for crystallization [66,67]. However, in the case of hybrid materials possessing organic moieties embedded in the inorganic–organic network, calcination often cannot be applied for crystallization when the temperature required exceeds the thermal stability of the organic functional group. In these cases, extraction with or without a mild thermal treatment step are required.

Templated metal phosphonate networks are often characterized with a texture possessing no long-range order (amorphous). One of the earliest mesoporous metal oxides in the presence of an inorganic phosphorus precursor was created by reaction of titanium isopropoxide with phosphoric acid in the presence of octadecyltrimethyl ammonium chloride/bromide as the SDA. This material consisted of an amorphous network of Ti–O–P with a surface area of 700 m^2/g and pore sizes of 2.7–3.3 nm [68]. There have been many mesoporous metal phosphates made by the use of phosphoric acid without the presence of any organic functional groups [69–71]. Although valuable materials, these types of materials are not the subject of this review, and focus will be devoted to those materials made from a combination of metal and organophosphonic precursors.

Templated mesoporous aluminum phosphonate can be synthesized via the reaction of aluminum isopropoxide and methylene diphosphonic acid with octadecyltrimethylammonium chloride (ODTMACl) as the SDA in the presence of tetramethylammonium hydroxide. The hybrid structure had an amorphous network with surface area of 738 m^2/g and a pore size of 1.8 nm. The pores were periodically arranged in a 2D hexagonal mesophase as analyzed by low-angle XRD and TEM. However, these materials were soluble in acidic ethanol solutions and had to be calcined in order to remove the surfactant and enhance condensation, which caused the loss of periodicity of the porous network in the final material [72]. Another method to produce mesoporous aluminum diphosphonates was by the "atrane route" where aluminum butoxide is heated with nitrilotriethanol to obtain aluminum atrane complexes and subsequently reacted with ethylene diphosphonic acid in the presence of CTAB as the SDA. This hybrid mesoporous material had a surface area of 674 m^2/g with a pore size of 3.3 nm. These materials were formed on the basis of $S^+ I^-$ ionic self-assembly, leading to the creation of a mesoporous aluminum diphosphonate [73,74]. Some examples of other aluminum diphosphonates based on the atrane route protocol in the presence of different templates such as F127, P123, Brij-56 and F68 are shown in Figure 6A [75,76].

Interestingly, mesoporous metal phosphonates have also been synthesized with the use of monophosphonates. The synthesis of tin phenylphosphonate in the presence of sodium dodecyl sulfate (SDS) as the template led to the formation of a material that displayed microporosity and mesoporosity, exhibiting a combined type-I and type-IV isotherm. This material was built-up from a layered structure consisting of tin (IV) phosphonate nanoparticles with microporosity. During calcination, thermal shrinkage resulted in the reduction in the inter-particle voids that were between 5 and 8 nm [77]. When the choice of metal was altered to zirconium and reacted with phenyl phosphonic acid in the absence of any supramolecular templates, a mesoporous material was obtained. The origin of the mesoporosity was attributed to the formation of a hydrophobic core by the phenyl groups, which led to the formation of pores with sizes of 5–10 nm [78]. Unique phosphonic linkers such as 1-phosphonomethylproline (H_3PMP) have been utilized with the purpose of synthesizing mesoporous aluminum phosphonates whose pores were lined by chiral amino acids. The atrane route was utilized to prepare this hybrid material, which possessed a surface area of 422 m^2/g and a pore size of 3.7 nm [79]. Another methodology that was followed to obtain mesoporous metal phosphonates without the use of templates was based on the use of a complex tetraphosphonic linker that prevented the formation of layered motifs and made it possible to achieve open structured hybrid metal phosphonates. Tetrakis-1,3,5,7-(4-phosphonatophenyl) adamantane (TPPhA) was one such complex phosphonic linker, which upon non-hydrolytic condensation with titanium and vanadium precursors, resulted in the formation of a hybrid mesoporous material with a surface area of 550 and 118 m^2/g, respectively [80,81].

Mesoporous metal phosphonates have also been synthesized by using polyphosphonic acids that feature a hetero-atom bridging the two or more phosphonate units. The hybrid materials synthesized with this strategy allowed to prevent the formation of a lamellar morphology and obtain interesting porous architectures. The 1-hydroxy ethylidene-1,1-diphosphonic acid (HEDP), a diphosphonate containing a hydroxy group in the bridging unit of the diphosphonic acid, was reacted with titanium butoxide in the presence/absence of β-cyclodextrin and polystyrene (PS) spheres by a hydrothermal method. This material showed no porosity apart from being macroporous due to PS spheres as shown in Figure 6C. This resulted in a low surface area around 23–25 m^2/g [82]. A higher surface area was achieved by reacting the same inorganic and phosphonic acid reagents under milder synthesis conditions (room temperature) and in the absence of the PS spheres and β-cyclodextrin. The result was a hybrid porous titanium phosphonate that showed a combination of a type-II and type-IV isotherm, with a surface area of 158 m^2/g. The texture of the material showed inter-particle voids with a layered structure as observed by electron microscopy [83]. However, the incorporation of soft templates like pluronic templates (F127/P123) in the condensation reaction of HEDP and titanium butoxide at room temperature, led to a material with a larger surface area of ~377–511 m^2/g, respectively. The results from the nitrogen sorption isotherm and SEM/TEM implied that the texture of the material was arising from inter-particle voids in the mesopore and macropore region [84]. To enhance the surface area further, claw type polyphosphonic acids containing hetero-atoms with titanium precursors can be used, resulting in materials that form hybrid networks with pores containing unreacted P–OH groups, as shown by Figure 6B [85]. When reacting titanium tetrachloride and ethylenediamine tetramethylene phosphonic acid (EDTMP) in the presence of Brij-56 as the template, a hybrid titanium phosphonate having hexagonally ordered mesopores (2–3 nm in size) with type-IV isotherm and a surface area ~1000 m^2/g was obtained. By changing the SDA from Brij-56 to CTAB, a tunable mesopore assembly ranging from a hexagonal phase to a cubic phase was achieved [86]. Moreover, template-free methods have been applied in which EDTMP was reacted with titanium butoxide, resulting in the formation of a porous material with hierarchical inter-particle porosity [87,88].

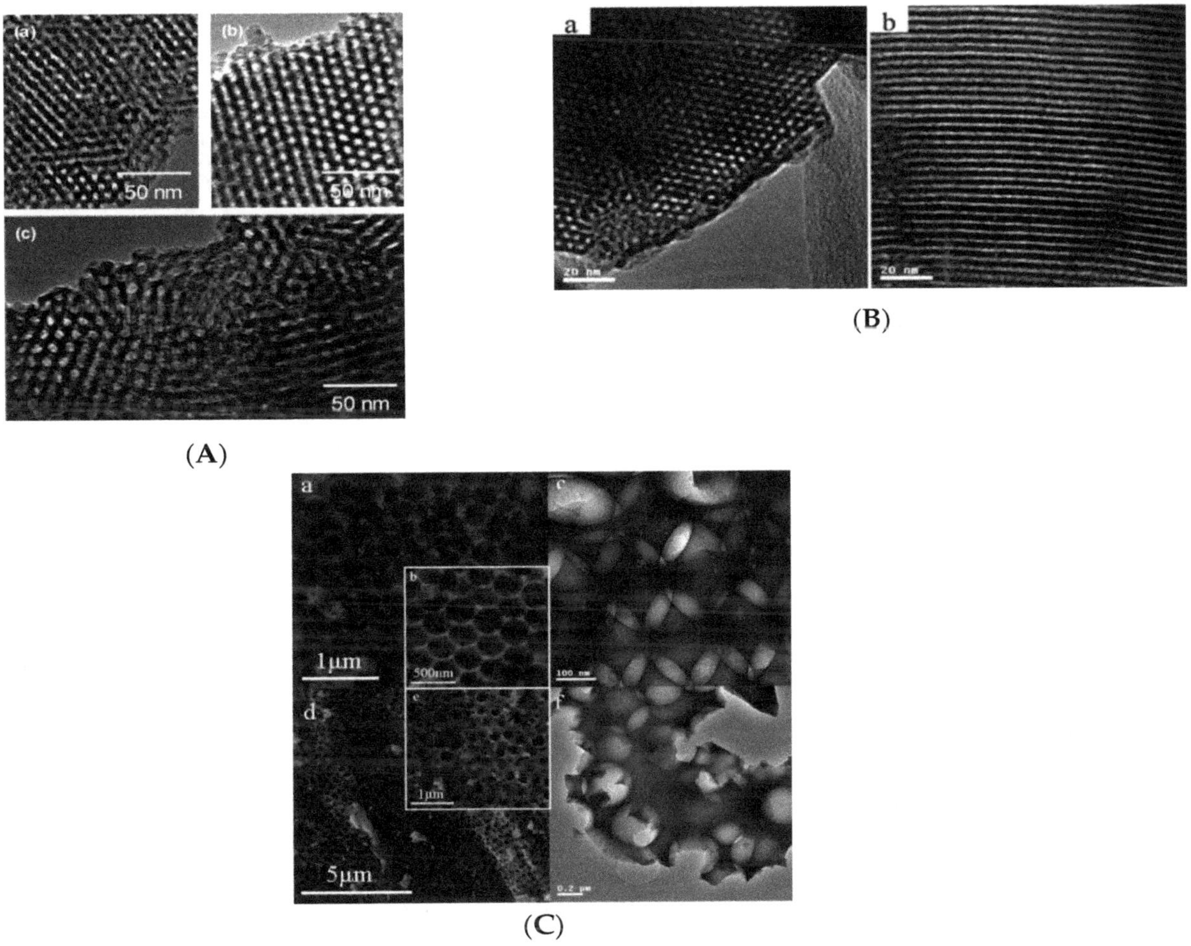

Figure 6. (**A**) TEM images of calcined aluminum methylene diphosphonates prepared in the presence of: (a) $EO_{80}PO_{30}EO_{80}$ (Pluronic F68), (b) $EO_{106}PO_{70}EO_{106}$ (Pluronic F127), and (c) $EO_{20}PO_{70}EO_{20}$ (Pluronic P123) tri-block co-polymers, respectively. Reprinted with permission from [75]. Copyright (2005), American Chemical Society. (**B**) (a,b) TEM images of periodic mesoporous titanium phosphonate-1 hybrid material. Reprinted with permission from [85]. Copyright (2010), The Royal Society of Chemistry and (**C**) (a,b) SEM and (c) TEM images of titanium phosphonate hybrid, (d,e) SEM and (f) TEM images of titanium phosphonate hybrid with β-cyclodextrin. Reprinted with permission from [82].

New concerted efforts in constructing hybrid porous metal phosphonates have been directed towards achieving control over the hetero-condensation of metal and phosphonic precursors. Recently, porous aluminum phosphonate synthesized with partially hydrolyzed acidic intermediates, formed from pre-hydrolyzed bisphosphonate ester precursors were reported. The partial hydrolysis allows to tune the reactivity towards aluminum chloride [89]. This synthesis methodology allowed the incorporation of a variety of organic linkers such as amides, thiophene, and ether moieties into porous metal phosphonate networks [90]. Another novel approach has been utilizing the non-hydrolytic sol–gel methodology to synthesize hybrid titanium phosphonate by the use of biphenyl bisphosphonate ester and titanium isopropoxide in acetic anhydride. These materials showed titanium oxide nanoparticles that were crosslinked with the bisphosphonate esters creating a high surface area material of 830 m^2/g [91].

3. Applications

Porous hybrid metal phosphonates show great potential in applications covering a wide diversity of domains. In this section, however, we will discuss those key application areas that we believe benefit most from the synergic combination of the organophosphonate functionality and the metal units. We will focus on those materials with high specific surface areas induced by a porous network, enabling the possibility of creating a tunable interface between the substrate (hybrid material) and the reactive/interactive species. Tailorable materials with surface-specific functionalities are vital in applications such as: (i) separation/extraction processes, (ii) heterogeneous catalysis and (iii) proton conduction. These areas are the target domains wherein porous metal phosphonates would provide a good choice and hence we focus on these fields here.

3.1. Separation and Extraction

A summary of the application of porous metal phosphonates in separation and extraction are displayed in Table 1. An example is the mesoporous hybrid system based on aluminum with BHMT (N-based pentaphosphonic) and ATMP (N-based triphosphonic) phosphonate moieties in the presence of F127 as a surfactant, providing a hierarchical meso/macroporous texture (type-II isotherm) with inter-particle mesoporosity and surface areas of 128 and 154 m^2/g, respectively. They were evaluated for the extraction of copper ions from an aqueous solution (10–50 mg/L). The adsorption of Cu^{2+} ions (10 mg/L) resulted in a K_d (distribution coefficient) of 277,485 and 17,281 mL/g for BHMT and ATMP aluminum phosphonate hybrids, respectively, at a pH ~6–8. However, with higher Cu^{2+} concentrations (50 mg/L) the K_d value decreased significantly to 6370 mL/g and 2744 mL/g for BHMT and ATMP aluminum frameworks, respectively. The adsorption was attributed to the copper chelation to nitrogen and RPO_3^{2-} groups present in the hybrid framework and their K_d (at lower Cu^{2+} concentration) was comparable to mesoporous silica [92,93]. Next to metal adsorption, the same materials were applied for the adsorption of large size molecules such as proteins, in this example lyzozyme. The hybrid frameworks followed Langmuir type adsorption with a monolayer capacity of 11.8 μmol/g (pH = 11), 10.42 μmol/g (pH = 12), 9.23 μmol/g (pH = 9.6), and 8.24 μmol/g (pH = 6.5). The maximum adsorption capacity was observed at the isoelectric point of the protein (pI = 11). Therefore, the hydrophobic interactions between the protein molecules and the alkyl chains lining the pores of the hybrid BHMT–aluminum phosphonate are promising for development of adsorbents for large molecule-like proteins [94]. However, these hybrid materials display lower uptake in comparison to mesoporous silica (>100 μmol/g) such as SBA-15 and PMOs [95–98].

A hybrid framework based on zirconium phosphonate created from zirconium(IV) precursors and ATMP has been evaluated as cation-exchange adsorbent. This hybrid material was chosen for the separation of three groups of metals; transition metals (Cu^{2+}, Zn^{2+}, Ni^{2+} and Co^{2+}), heavy metals (Pb^{2+}, Hg^{2+}, and Cd^{2+}) and f-block metals (Ce^{3+} and Th^{4+}). The preferential order of adsorption amongst the transition metals were $Cu^{2+} > Ni^{2+} > Co^{2+} > Zn^{2+}$ ($K_{d,Cu}$ = 9879.7, $K_{d,Ni}$ = 1550.0, $K_{d,Co}$ = 600.0, and $K_{d,Zn}$ = 584.6 mL/g), for heavy metals $Pb^{2+} > Cd^{2+} > Hg^{2+}$ ($K_{d,Pb}$ = 1261.5, $K_{d,Cd}$ = 1037.5, and $K_{d,Hg}$ = 56.4 mL/g), and for f-block metals $Ce^{3+} > Th^{4+}$ ($K_{d,Ce}$ = 2000.0, and $K_{d,Th}$ = 114.7 mL/g) in aqueous solution. However, in acetic acid, HNO_3, and ammonium nitrate solutions, the distribution coefficients were reduced due to ion competitions. Amongst the metal ions studied, only Cu^{2+} (in aqueous, 0.02 and 0.2 M acetic acid) and Ce^{3+} (in 0.02 M acetic acid) showed complete sorption with a high separation factor (SF) between Cu^{2+}/Zn^{2+} (16.9) and Ce^{3+}/Th^{4+} (17.54) [99]. Similarly, hybrid manganese phosphonate synthesized from $MnCl_2$ and EDTMP was evaluated for metal ion uptake and protein adsorption. The optimal pH for the Cu^{2+} uptake was ~6 and the adsorption followed Langmuir behavior with a maximum monolayer capacity of 79.5 mg/g at 20 °C. This value is in the same order of magnitude as that of mesoporous silica adsorbents (~24.8 mg/g) and hybrid titanium phosphonates (also made with EDTMP showed ~28.8 mg/g uptake) [100].

Zirconium hybrid phosphonate frameworks made from 1-phosphomethylproline (H_3PMP) and 1,4-bis(phosphomethyl) piperazine (BPMP) phosphonic ligands resulted in a pH-sensitive porous framework. This framework was employed for the adsorption and delivery of nucleic acids without further modification/post-treatment. The adsorption of salmon sperm DNA onto the framework followed Langmuir type adsorption. Among the various hybrids created by this method, the ZrBF-2 showed the largest equilibrium monolayer capacity of 238.6 mg/g. The increased uptake was attributed to the presence of uncondensed P–OH groups and pyridinic N-atoms present along the porous framework of the hybrid material [101]. This uptake by the hybrid framework is comparable to salmon sperm DNA uptake of 375 and 110.7 mg/g by magnetic mesoporous silica adsorbents [102,103]. The mechanism of adsorption followed pseudo-second order kinetics. Moreover, the framework was pH responsive, allowing adsorption–desorption cycles by changes in the pH. The structure exhibited a cationic surface under acidic and neutral pH where the surface could adsorb negatively charged DNA molecules by electrostatic interactions. Subsequently, the release of the nucleic acid was triggered by increasing the pH, changing the surface charge and resulting in desorption. This allowed for the pH controlled loading and unloading of nucleic acids as illustrated by Figure 7 [101].

In another application, zirconium phosphonate frameworks synthesized by zirconium(IV) and phenyl phosphonic acid in the presence of an SDS template were employed as immobilized metal affinity chromatography (IMAC) for the selective enrichment of phosphopeptides. The specificity for phosphopeptides was evaluated with use of a tryptic digest consisting of a mixture of bovine serum albumin (BSA) and bovine-β-casein (B-β-C) as the sample peptides (mass ratio B-β-C:BSA = 1:5). The specificity of the bovine-β-casein adsorption was attributed to the chelation of phosphopeptide to the Zr(IV) metal centers present in the framework [104].

Finally, we shall go through the use of porous metal phosphonates for the solid-phase extraction of lanthanides and actinides. Tin phosphonate synthesized from various polyphosphonic acids (HEDP, ATMP, EDTMP and DTPMP) was employed for radionuclide solid-phase separation. The study consisted of: a batch type sorption of metal ions Cs^+, Sr^{2+}, Al^{3+}, Fe^{3+}, Th^{4+}, $UO_2{}^{2+}$, and rare earth elements (REE) (all lanthanides) in 0.1, 1 and 3 M HNO_3 solutions and a competitive sorption kinetics study for the metal ions Eu^{3+}, Fe^{3+}, La^{3+}, Lu^{3+}, Sr^{2+}, Th^{4+}, and $UO_2{}^{2+}$ in 1M HNO_3 solution. In the 0.1 M HNO_3 solution, tin phosphonate hybrids containing HEDP displayed a high uptake of middle lanthanides ($K_d > 6000$ mL/g), whereas the EDTMP displayed a high uptake of late lanthanides ($K_d > 4000$ mL/g), showing the importance of the type of phosphonic linker. However, a high uptake of Fe^{3+} (K_d ~1000–10,000 mL/g) was observed among all the hybrids in 0.1 M HNO_3 solution. An interesting observation was that all the hybrid materials showed extreme affinity for the Th^{4+} radionuclide in all nitric acid solutions (K_d ~127,171 mL/g and 164,399 mL/g for ATMP and EDTMP hybrids). Nevertheless, although these hybrids showed promising capabilities in use as a solid-phase adsorbent, they displayed slow kinetics for the uptake of radionuclides [105].

Mixed hybrids of zirconium phosphonate–phosphate (phosphonate/phosphate ratio from 1.5:1 to 1:8) were evaluated for solid-phase lanthanide extraction from an aqueous mixture of metal ions. In this study, the choice of metal ions were Nd^{3+}, Sr^{2+}, and Cs^+. The Nd^{3+} ions were always adsorbed onto the hybrid materials >98% at all phosphate ratios in the presence of Sr^{2+} and Cs^+ ions. The uptake capacity increased with higher phosphate amounts for Nd^{3+} but at very high phosphate amounts, the uptake of Sr^{2+} and Cs^+ ions increased, and led to a lower separation factor (SF) with respect to Nd^{3+} ions [106]. In the same study, hybrid tin phosphonate–phosphate were compared with a Zr(IV) phosphonate–phosphate system. The Zr-hybrid showed clear affinity only towards M^{3+} ions, whereas the Sn-hybrid showed affinity to M^{3+} and slightly for M^{2+} ions. Thus, showing the importance of the choice of metal and phosphonate/phosphate ratio while creating hybrid material used for the solid-phase extraction of lanthanides [106].

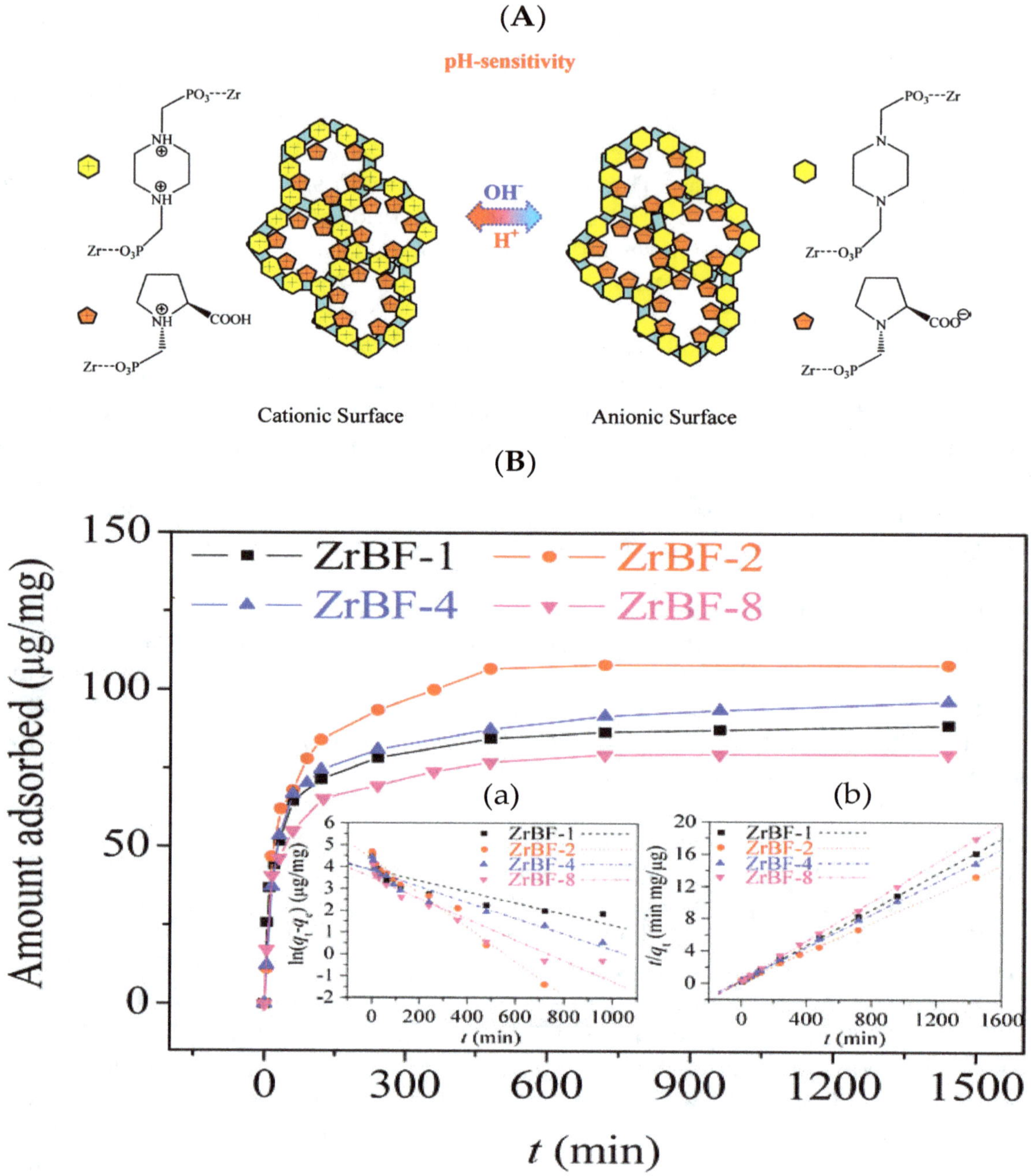

Figure 7. (**A**)—illustrative scheme for ZrBF's biofunctionality, pH sensitivity, and functionalizability; (**B**)—adsorption kinetics of ZrBF-n (n = 1, 2, 4 and 8) for DNA. Insets—linear fit according to the pseudo-first order rate equation (a) and pseudo-second order rate equation (b), respectively. Reprinted with permission from [101].

In another study, Zr/Sn–phosphonates containing no inorganic phosphate were also evaluated for selective separation. These were subjected to the lanthanide extraction of Nd^{3+}, Eu^{3+}, and Ce^{3+} in the presence of Cs^{+} and Sr^{2+} at pH = 3. Additionally, here, the Zr-based hybrid phosphonates showed the preferential uptake of only +3 metal ions, whereas the Sn-based hybrids had a slight affinity towards Sr^{2+} along with +3 metal ions. These hybrids were also evaluated for the removal of Am^{3+} from spent nuclear fuel. Both Zr and Sn hybrids showed >99% removal of Am^{3+} from the solution (pH = 3) with

K_d of 78,000 and 30,000 mL/g, respectively. To evaluate the radiolytic stability of these hybrid materials, the sorbents were subjected to an irradiation with a Co-60 gamma source 3.18×10^6 Gray at a dose of 1.617×10^5 rad/hr. The hybrid materials retained their structure as evidenced by XRD and showed a similar uptake of Am^{3+} ions after being irradiated [107].

Zirconium phosphonate were synthesized hydrothermally, using the benzene-1,3,5-triphosphonic acid (BTP) ligand to obtain both amorphous and crystalline (in the presence of HF) frameworks. These materials were evaluated for metal separation from a mixed solution containing the following ions: Cs^+, Sr^{2+}, La^{3+}, Ce^{3+}, Pr^{3+}, Nd^{3+}, Gd^{3+}, Dy^{3+}, Ho^{3+}, Co^{2+} and Th^{4+} with an ionic strength of 1 mmol/L and the nominal concentration of individual ions ~20 mg/L in 0.1 M HNO_3 solution. Figure 8, shows the extraction of the different ions as a function of ionic radii of the metal ions by the hybrid Zr-phosphonate in function of the phosphate/zirconium ratio in the material varying from 0.5 (low) to 0.8 (high). The order of the selectivity of ions followed: $Th^{4+} > Ln^{3+} > Co^{2+}, Sr^{2+} > Cs^+$, and the selectivity was dependent on the cation valence and polarizability. With higher molar amounts of phosphonic ligands, the K_d increased, suggesting that the binding of the cations is achieved via the phosphonic acid groups distributed in the framework. A competitive kinetics study showed that at equilibrium (concentration of metal ions > 1 mg/L), Dy^{3+} and Ho^{3+} were strongly extracted in mixed cation solution containing the rest of the (REE) ions along with Cs^+, Sr^{2+} and Co^{2+}. The authors highlighted in the same work that the materials that were amorphous and free of fluoride performed with enhanced selectivity and better kinetics than the crystalline materials in highly acidic solutions. This study showed that the distribution of unreacted phosphonic acid moieties was one of the important criteria for improved cation binding to the hybrid framework [108].

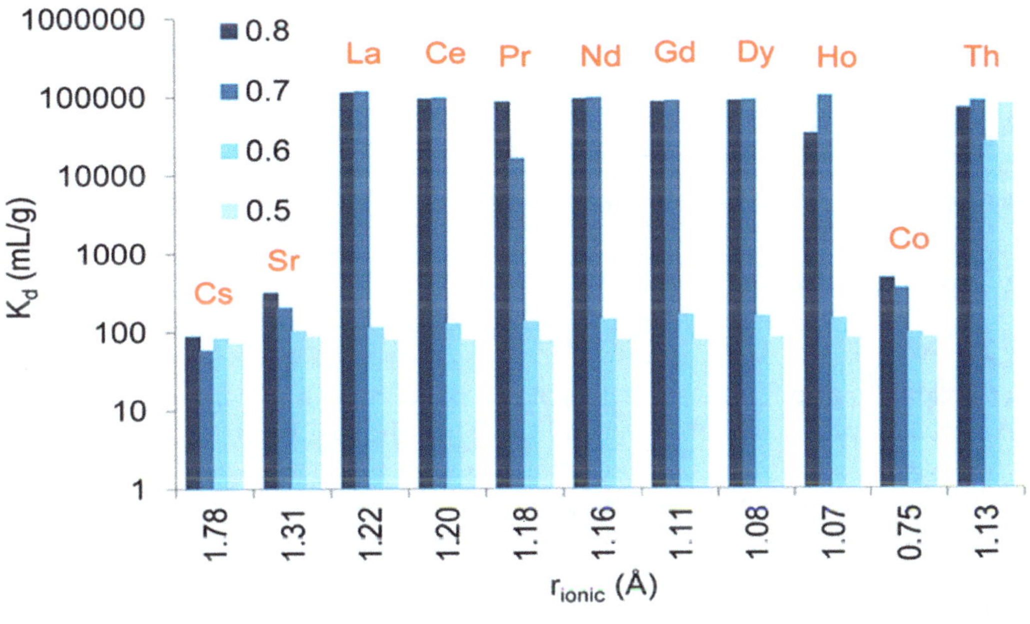

Figure 8. Extraction of cations of different ionic radius by poorly crystalline ZrBTP-0.8, ZrBTP-0.7, ZrBTP-0.6, ZrBTP-0.5, prepared in the absence of HF from mixed cation solutions with each cation at nominal 20 mg/L in 0.10 M HNO_3. Reprinted with permission from [108].

In another study, porous adsorbents based on zirconium phosphonate have been prepared hydrothermally with $ZrCl_4/ZrOR_4$ and ATMP. These hybrid sorbents were evaluated for competitive lanthanide separation from a mixture of cations. All zirconium phosphonate hybrids showed a selective uptake of lanthanides over Cs^+, Sr^{2+} and Co^{2+}. Among the lanthanides, Eu^{3+} was the most strongly sorbed lanthanide across all the hybrids. Interestingly, the amorphous hybrids in this study also displayed a higher uptake compared to the crystalline hybrid materials. The presence of fluoride ions had a detrimental effect on the uptake of lanthanide ions resulting in the lower capacity presented by the crystalline hybrids. Finally, the choice of metal precursor was evaluated, the zirconium chloride-based hybrids led to materials that displayed better lanthanides (>99% extraction of La^{3+}, Nd^{3+}, Eu^{3+}, Ho^{3+} and Yb^{3+}) uptake than the materials prepared from zirconium isopropoxide (~40–70% extraction of La^{3+}, Nd^{3+}, Eu^{3+}, Ho^{3+} and Yb^{3+}). However, the zirconium isopropoxide-based hybrids showed more selectivity among the lanthanides over Cs^{+} and Sr^{2+}. The changes in sorption properties were attributed to the differences in the incorporation of the phosphonate moiety (ATMP) into the hybrid network [109].

Table 1. Separation and extraction application list for porous metal phosphonates.

Metal Precursor	Phosphonic Linker	Synthesis	Application—Separation/Extraction	Reference
$Al(O^{s}Bu)_3$	ATMP BHMT	Hydrothermal	Cu^{2+} adsorption and lyzozyme adsorption	[94]
$ZrOCl_2$	ATMP	Precipitation	Cation exchange—divalent metal ions	[99]
$ZrCl_4$	H_3PMP BPMP	Hydrothermal	Adsorption and delivery of DNA	[101]
$ZrCl_4$	PPA	Hydrothermal	Immobilized metal affinity chromatographic adsorbent for phosphopeptide enrichment	[104]
$MnCl_2$	EDTMP	Hydrothermal	Cu^{2+} sorption and selective protein adsorption of Cyt–C over BSA	[100]
$SnCl_4$	HEDP ATMP EDTMP	Hydrothermal	Radionuclide separation of Th^{4+}	[105]
$ZrOCl_2$	BTP BDP BMP	Hydrothermal	Lanthanide and actinide separation	[108]
$ZrOCl_2$	BDPA H_3PO_4	Hydrothermal	Ion-exchange metal ions	[106]
$ZrOCl_2$ $SnCl_4$	BDPA H_3PO_4	Hydrothermal	Actinides separation	[107]
$ZrCl_4/Zr(O^{i}Pr)_4$	ATMP	Hydrothermal	Lanthanides extraction	[109]

Abbreviation: ATMP = aminotris(methylene phosphonic acid); BHMT = bis(hexamethylene-triamine)-penta (methylene phosphonic acid); H_3PMP = 1-phosphomethylproline; BPMP = 1,4-bis(phosphomethyl) piperazine; PPA = phenyl phosphonic acid; EDTMP = ethylenediamine tetra(methylene phosphonic acid); HEDP = 1-hydroxyethane 1,1-diphosphonic acid; BTP = benzene 1,3,5-triphosphonic acid; BDP = 3,5-diphosphonobenzoic acid; BMP = 5-phosphonoisophthalic acid; BDPA = 1,4-phenylenediphosphonic acid.

3.2. Catalysis

A list of the catalytic applications for which metal phosphonates have been used for is summarized in Table 2. As an example, mesoporous vanadium phosphonate has been employed as catalyst for the oxidation of benzyl alcohols. The oxidation reaction resulted in high conversion (>99%) and selectivity (96%) in the formation of benzaldehyde. There was no formation of side reaction products such as benzoic acid or ethers due to the over-oxidation or acid condensation, respectively. Benzyl alcohols containing electron-donating groups ($4\text{-MeOPhCH}_2\text{OH}$ and $4\text{-MePhCH}_2\text{OH}$) and electron withdrawing groups ($4\text{-NO}_2\text{PhCH}_2\text{OH}$ and $4\text{-CF}_3\text{PhCH}_2\text{OH}$) were also successfully converted to the respective aldehydes with similar conversion and selectivity as in the case of benzaldehyde. Evidence of some shape selectivity was observed, as 2,4,6-trimethyl benzyl alcohol oxidation resulted in poor conversion (56%) and low selectivity (79%) compared to other benzyl alcohol substrates [81].

One of the benefits of hybrid inorganic–organic materials is that they can introduce chirality in the material's porous network through the organic moiety on the phosphonate linker. The hydrolysis of styrene oxide under confined homochiral environment available in the hydrophobic pores of hybrid titanium phosphonate containing leucine moieties was evaluated. The yield of the reaction in the presence of the hybrid catalyst was 30% S-styrene oxide with an enantiomer efficiency (EE) of >99%. Moreover, other epoxides such as 4-methyl styrene oxide and α-4-methyl styrene oxide reacted in an identical manner, with a yield of ~20% having ~98% and ~95% EE, respectively. The authors postulated that the epoxide was activated by moieties having a partial positive charge in the titanium phosphonate framework such as the phosphonate, or Ti(IV) centers, followed by the attack of water directed by the homochiral leucine peptide to yield S-styrene glycol [110]. Similarly, the enantioselective addition reaction to obtain optically active phenylpropanol has also been catalyzed by hybrid titanium phosphonates [111].

Zirconium phosphonates were employed as a solid-acid catalyst in the synthesis of methyl-2,3-o-isopropylidene-β-d-ribofuranoside. The hybrid material was compared alongside commercial acidic resin, NKC-9 and concentrated hydrogen chloride. The reaction was carried out at 70°C and yielded 28% after 2 h when using the hybrid zirconium phosphonate, which levelled off at 35.6% after 3 h. With the commercial resin NKC-9, the yield of 21.1% and 33% was reached after 2 h and 3 h of reaction, respectively. Thus, the hybrid zirconium phosphonate performed similar to the commercial resin even though the resin had substantially higher proton concentrations [112]. In another example, porous zirconium phosphonates were evaluated as a solid-acid catalyst for the esterification of acetic acid with cyclohexanol as well as for the hydrolysis of ethyl acetate in aqueous media. The results were compared with commercial resin NKC-9 and $ZrPO_4$. After 10 h of reaction between acetic acid and cyclohexanol at 100 °C, the NKC-9 showed the highest yield of 87.9% followed by the zirconium hybrid phosphonates synthesized with HEDP, ATMP, and EDTMP phosphonic acids providing yields of 76.8%, 65.4% and 75.3%, respectively. The inorganic $ZrPO_4$ gave a yield of 72.9% under similar reaction conditions. However, the TOF (measured at 100 °C) for the hybrid zirconium phosphonates based on EDTMP, HEDP and ATMP were 310.2, 234.0 and 306.1, respectively. These TOF were higher than those of the commercial resin, NKC-9 (118.7) and the inorganic $ZrPO_4$ (210.1). As a side note: these reported TOF values were calculated based on the products formed, hence their activity calculation also includes the occurring diffusion/adsorption effects that might be rate determining. Moreover, the authors seem to have calculated turnover number (TON) according to the definition of Kuzoch et al., 2012 rather than TOF values [113]. The acid content of the hybrid zirconium phosphonates was 1.92, 1.25, and 1.42 mmol/g for the HEDP, ATMP and EDTMP-based hybrids (NKC-9 = 4.33 and $ZrPO_4$ = 2.03 mmol/g), respectively. The higher TOF values for the hydrolysis reaction in the presence of hybrid zirconium phosphonates was attributed to higher specific surface areas and the surface hydrophobicity of the catalyst. When the same hydrolysis reaction of ethyl acetate was performed in aqueous media, the catalytic activity per acidic proton was 51.1, 21.5, 30.6, 35.2, and 15.1 mmol/mol_{acid}/min for Zr-EDTMP, Zr-HEDP, Zr-ATMP, NKC-9 and inorganic $ZrPO_4$, respectively [114].

In another example, hybrid iron (III) phosphonate, created by the reaction of iron (III) chloride with BTP precursors, was employed as a solid-acid catalyst for transesterification reactions. Amongst many reactants, ethyl cyanoacetate showed the highest conversion of 88.3% with a selectivity of 100% (at 60 °C, 6 h). It was observed that as the electron withdrawing effect (CN > C(=O)CH_3 > Cl) of the ester reactant decreased, the conversion to the product decreased. The electrophilicity of the carbonyl carbon of the ester reactant was attributed as the main driving force for the reaction. Negatively charged free phosphonic acids in the framework of the hybrid material could prevent molecules enriched with p-electron clouds entering the porous channels. Thus, providing the catalyst with selectivity for the transesterification of esters that contain fewer mobile electrons [115].

Next to the phosphonate moiety discussed in the previous paragraph, the organic group and metal center can also play an important functional role in catalysis. Porous tin phosphonate made from tin(IV) chloride and pentaethylenehexamine–octakis-(methylphosphonic acid) hexadecenoic sodium was employed as Baeyer–Villiger catalyst for the oxidation of cyclohexanone in the absence of peroxides as an oxidant. The product of this reaction was the formation of adipic acid with a yield of ~74%. When the hybrid material was replaced with an inorganic tin(IV) phosphate, the yield was only of the order ~22%, suggesting the key role played by the organic group providing free –NH_2 groups in the porous framework, aiding in stabilizing the keto-enol tautomerization. In addition, the enol form is also stabilized by the Sn-centers. Furthermore, the Sn-center was found to activate the molecular oxygen and helped to form the cyclic six-membered transition state compound that later rearranged to form a cyclic ester as a second intermediate [116]. In another example, hybrid tin(IV) phosphonates synthesized from tin (IV) chloride and BTP were used as catalysts for the synthesis of 1,4-dihydropyridines from 1,3-dicarbonyls and ammonium acetate under solvent-free conditions (Hantzsch ester reaction). It was observed that the 1,4-addition was selectively facilitated over the 1,2-addition to the unsaturated C=O compound in the presence of the hybrid tin(IV) phosphonate catalyst. The authors attributed this to the adsorption of unsaturated C=O to the hydrophilic surface of the porous framework containing Sn-centers. Since there were no free C=O groups, the enamine exclusively attacked the C=C bond resulting in the formation of only 1,4-dihydropyridine as illustrated in Figure 9A [117].

Alternatively, porous molybdenum(V) phosphonate made with $MoCl_5$ and BTP was employed as a catalyst for the one-pot synthesis of 2-aryl-benzimidazoles at room temperature by the condensation of o-phenylenediamine and substituted aromatic aldehydes (yield for 2-phenyl benzimidazole = 93%). When the same reaction was catalyzed by pure H_3PO_4 and $MoCl_5$, lower yields of 53% and 40%, respectively, were obtained. Therefore, the presence of vacant d-orbitals of the Mo metal centers along with the free P–OH groups were correlated to the activation of the C=O group of the aldehydes [118].

Fenton-like heterogeneous catalysts from hybrid cobalt phosphonates, made from $CoCl_2$ and DTPMP, were employed for the catalytic oxidation of methylene blue (MB). The optimal pH for the degradation reaction was around ~7, and the complete degradation of MB occurred within 25 min, with a degradation rate of 0.136 min^{-1}. Decreasing the pH to 5 and 2 resulted in a notable deterioration of the MB degradation rate of 0.0331 and 0.0203 min^{-1}, respectively. Similarly, at higher pH ~9, the degradation reduced to a rate of 0.0931 min^{-1}. Under acidic conditions, electrostatic repulsion between the positively charged hybrid framework and MB molecules prohibited effective adsorption. Under alkaline conditions, the lower degradation was attributed to the fact the sulfate radicals generated in the reaction system were quenched by the OH^- ions and possibly also by the formation of $Co(OH)_2$ that might form on the surface of the hybrid cobalt phosphonates [119].

In another example, hybrid iron phosphonate was employed as a Fenton-like catalyst for the catalytic oxidation of cyclohexanone to adipic acid in an aqueous medium at 75 °C for 10 h under oxygen atmosphere and in the absence of peroxide. The conversion to adipic acid was determined to be 72% with a selectivity of 96%. The hybrid iron phosphonate catalyst had comparable results to other Fe-based catalysts with the important difference that other classical heterogeneous Fenton catalysts required hydrogen peroxide as the oxidant (Figure 9B). The authors explained that the Fe(II) sites are responsible for the activation of molecular oxygen, in contrast to the strong oxidizing capability of Fe(III) [120].

In a later study, the above hybrid materials were evaluated as catalysts for the cycloaddition of CO_2 with aziridines to obtain oxazolidinones at 120 °C under 3 MPa pressure of CO_2 for 6 h under solvent-free conditions. The three zirconium phosphonate hybrids showed a high conversion of >98% and regioselectivity with yields of 85%, 79% and 90% for Zr-HEDP, Zr-ATMP, and Zr-EDTMP, respectively. Among these three hybrid catalyst systems, the Zr-HEDP displayed the highest regioselectivity of 96:4 in the formation of 3-ethyl-5-phenyloxazolidin-2-one with respect to 3-ethyl-4-phenyloxazolidin-2-one, in comparison to the Zr-ATMP (95:5) and Zr-EDTMP (90:10) hybrids. The authors proposed a mechanism wherein the proton from the -P-OH group coordinated with the N-containing aziridine and the adsorbed CO_2 interacted with the -OH group of the HEDP moiety. This way, the aziridine was attacked by the activated CO_2 (nucleophilic), leading to the formation of carbamate, followed by intramolecular ring closure resulting in oxazolidinone formation [121].

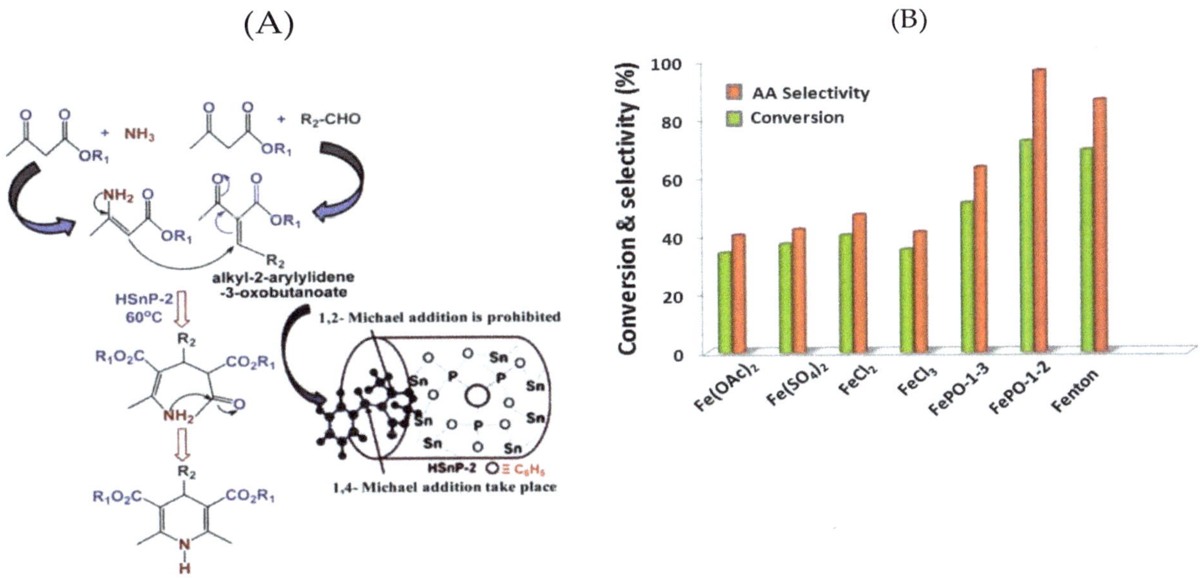

Figure 9. (**A**)—probable reaction mechanism and the role of the HSnP-2 catalyst in selective synthesis of 1,4-dihydropyridines. Reprinted with permission from [117]. Copyright (2013), The Royal Society of Chemistry and (**B**)—dependence of the product yield on various catalysts for the aerobic oxidation of cyclohexanone at 75 °C. Reprinted with permission from [120].

Table 2. Summary of the porous metal phosphonates for catalytic application.

Metal Precursor	Phosphonic Linker	Synthesis	Application—Catalysis	Reference
$V(O)(i\text{-}PrO)_3$	TPPhA	Non-hydrolytic condensation	Oxidation of benzylic alcohols	[81]
$Ti(SO_4)_2$	(1R, 2S)-(-)-2-P (1R, 2S)-(+)-2-P	Precipitation (Hydrothermal)	Enantioselective addition of benzaldehyde	[110]
$Ti(O^iPr)_4$	Polypeptide capped with phosphonic acid moiety	Non-hydrolytic condensation	Enantioselective hydration of styrene oxide	[111]
$ZrCl_4$	HEDP	Precipitation/Hydrothermal	Solid-acid catalyst for synthesis of methyl-2,3-*O*-isopropylidene-β-D-ribofuranoside	[112]
$SnCl_4$	PEHMP	Precipitation/Hydrothermal	Oxidation of cyclohexanone to adipic acid in absence of peroxides	[116]
$FeCl_3$	BTP	Precipitation/Hydrothermal	Transesterification for synthesis of biofuels	[115]
$SnCl_4$	BTP	Precipitation/Hydrothermal	Synthesis of 1,4-dihydropyridines	[117]
$MoCl_5$	BTP	Precipitation/Hydrothermal	Synthesis of benzimidazoles	[118]
$CoCl_2$	DPTMP	Hydrothermal	Catalytic oxidation of methylene blue	[119]
$ZrOCl_2$	HEDP ATMP EDTMP	Precipitation/Hydrothermal	Hydrolysis of ethyl acetate; esterification of acetic acid with ethanol and cyclohexanol	[114]
$ZrOCl_2$	HEDP ATMP EDTMP	Precipitation/Hydrothermal	Cycloaddition of aziridnes and CO_2	[121]
$FeCl_3$	HEDP	Precipitation/Hydrothermal	Oxidation of cyclohexanone to adipic acid	[120]

Abbreviation: TPPhA = tetrakis-1,3,5,7-(4-diethylphosphonatophenyl)adamantane; (1R,2S)-(-)-P = (1R,2S)-(-)-2-[N-(diethylethylphosphosphonate)]imino-1,2-diphenylethanol; (1R, 2S)-(+)-P = (1R,2S)-(+)-2-[N-(diethylethylphosphosphonate)]imino-1,2-diphenylethanol; HEDP = 1-hydroxyethane 1,1-diphosphonic acid; PEHMP = pentaethylenehexamine-octakis-(methylphosphonic acid); BTP = benzene 1,3,5-triphosphonic acid; DTPMP = diethylenetriamine penta(methylenephosphonic acid); ATMP = aminotris(methylene phosphonic acid); EDTMP = ethylenediamine tetra(methylene phosphonic acid).

3.3. Proton Conduction

As a final application in which porous metal phosphonates can have a particular benefit, proton conduction is discussed. A summary of the proton conducting frameworks has been listed in Table 3. The framework built-up by zinc precursor and BTP as the phosphonate ligand (denoted as PCMOF-3) was evaluated for proton conduction. The aquo ligands on zinc were ascribed as the main source of protons to be transported. The proton conductivity of the framework was strongly dependent on the humidity, with 4.5×10^{-8} S/cm at 44% RH (25 °C) and 1×10^{-5} S/cm at 98% RH (25 °C). The activation energy (E_a) for the proton transfer ~0.17 eV was derived from the bulk conductivity of the PCMOF-3 framework. In comparison, the activation energy for 1 mol/L HCl has been reported as 0.11 eV and that of Nafion as 0.22 eV. Thus, PCMOF-3 represents a proton-conducting metal phosphonate framework in which the conductivity does not involve secondary SO_3H sites. The polar interlayer of the framework serves as a scaffold for highly ordered water molecules that form a Grotthuss proton transfer pathway, as evidenced by the low E_a. The study concluded with the evidence that both the phosphonate 'O' atoms and Zn aquo ligands contributed to the proton transfer process [122].

In another study, the β-PCMOF-2 framework, prepared by sodium and phloroglucinol in the presence of chlorosulfonic acid was applied. This sulfonate framework displayed a proton conductivity of 1.3×10^{-3} S/cm at 85 °C and 90% RH. In order to improve the conductivity, an isomorphic ligand exchange was performed with a BTP ligand to replace part of the aryl sulfonic moieties with the BTP ligands, forming a new mixed PCMOF-$2^{1}/_{2}$ framework. This resulted in an order of magnitude with improved conductivity to values of 2.1×10^{-2} at 85 °C and 90% RH. This conductivity was the highest equilibrated MOF proton conduction reported, exceeding the sulfuric acid modified MIL-101 framework. The activation energy from the Arrhenius plot was 0.21 eV, indicative for the Grotthuss mechanism of proton conduction. The conductivity in this framework was extremely dependent on the RH and dropped significantly to 2.4×10^{-5} S/cm at 20 °C and 50% RH [123].

In another study, a magnesium phosphonate framework was made from $Mg(NO_3)_2$ and octamethylene diamine-N,N,N′,N′-tetrakis-methylenephosphonic acid (H_8ODTMP) and evaluated for proton conductivity. This magnesium phosphonate framework was possessing 1D channels filled with water molecules running along the c and b axes. The P–OH groups in the framework were found to be pointing inwards to the channel, and thus allowing proton transport to take place, which results in conductivity. The highest proton conductivity for this framework was 1.6×10^{-3} S/cm at 19 °C, 100% RH. The proton conductivity in the framework was attributed to the proton transfer in the framework via the Grotthuss hoping mechanism [124]. The influence of metal choice was studied with hybrid metal phosphonates, MFM-Ni and MFM-Co were studied for their proton conductivity properties. At 25 °C and 98% RH, the proton conductivities of Ni and Co hybrids were 4.5×10^{-4} and 4.4×10^{-5} S/cm, respectively. The differences in conductivities between the isostructural materials arose from differences in the bond strength between the coordination water molecules and the metal cations. The activation energy for proton conduction in the Ni-hybrid was 0.43 eV, this value is at the boundary of the two mechanisms responsible for proton conduction (Grotthuss < 0.4 eV and vehicular > 0.4 eV) [125].

We conclude this section with the illustration of the capabilities of the metal phosphonate frameworks in thin film applications, such as in sensors or devices Hybrid Co–Ca phosphonates consisting of both Co and Ca and 1,4,7-triazacyclononane-1,4,7-triyl-tris-(methylenephosphonic acid) have been synthesized in the form of powders and as single crystals for their use as proton conducting materials. For the powder samples, the proton conductivity was 1.55×10^{-5} S/cm at 25 °C and 95% RH. However, the conductivity was reduced by five orders of magnitude, 0.79×10^{-10} S/cm at 25 °C when the RH was 40%. Single-crystals have been grown for these hybrid framework along the (010), (202) and $(20\bar{1})$ directions to determine the conductivity along different planes. The single-crystal grown along the (010) direction displayed the highest conductivity of 1.00×10^{-3} S/cm at 25 °C and 95% RH. The lowest conductivity was observed along the $(20\bar{1})$ direction with 4.35×10^{-8} S/cm at 25 °C and 95% RH. The higher conductivity along the (010) direction was attributed to a continuous lattice of water molecules and the 'O' from the phosphonate present in the channel. The activation energy for the single-crystal hybrids along the (010), (202) and (20$\bar{1}$) directions were 0.90, 0.86 and 0.58 eV, respectively (Figure 10). These values were too large to assign the Grotthuss mechanism for conduction, instead a vehicle mechanism could be in operation. It was highlighted that the presence of ClO_4^- anions, originating from an additive used during synthesis, stayed incorporated in the structure. These anions together with the phosphonate 'O' atoms played a key role in connecting the water molecules via a hydrogen bonded network during the conduction process [126].

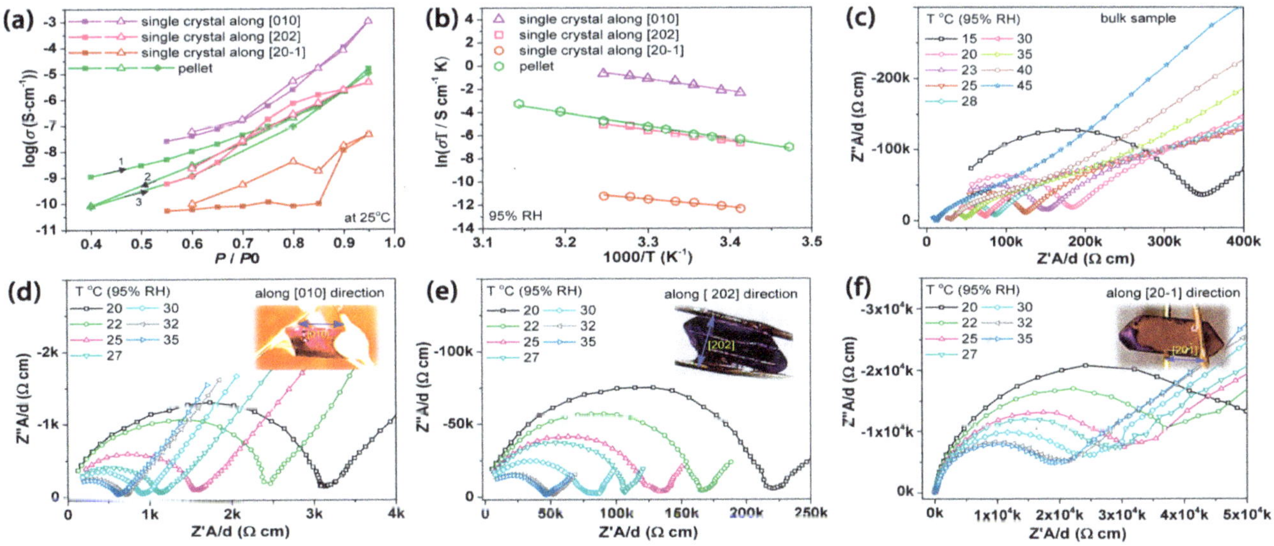

Figure 10. (**a**) Proton conductivity of the single crystals of CoCa·nH_2O measured along different directions at 25 °C upon increasing (filled) or decreasing (open) humidity. The conductivity of the pellet sample is also shown for comparison; (**b**) plots of log(σT) versus 1000/T for bulk and single crystal samples of CoCa·4H_2O at 95% relative humidity. The solid lines represent the best fit of the data. Nyquist plots for CoCa·nH_2O at 95% RH and various temperatures; (**c**) bulk sample; and (**d**–**f**) single crystal sample along the [010], [202], and [20$\bar{1}$] directions, respectively. Reprinted with permission from [126].

Table 3. Summary of proton conduction in the phosphonate frameworks.

Metal Phosphonate	Conductivity (S cm^{-1})	Activation Energy (eV)	References
$ZnCO_3 \cdot 2Zn(OH)_2$ + BTP	1.0×10^{-5} (at 98% RH, 25 °C)	0.17	[122]
$Mg(NO_3)_2$ + H_8ODTMP	1.6×10^{-3} (at 100% RH, 19 °C)	0.31	[124]
β-PCMOF-2$^{1/2}$	2.1×10^{-2} (at 90% RH, 85 °C)	0.21	[123]
CoCa. n H_2O (pellet)	1.55×10^{-5} (pellet) (at 95% RH, 25°C)	0.98 (pellet)	[126]
CoCa. n H_2O (SC)	1.00×10^{-3} (SC) (at 95% RH, 25°C)	0.90 (SC)	
MFM-500-Ni	4.5×10^{-4} (at 98% RH, 25 °C)	0.43	[125]
MFM-500-Co	4.4×10^{-5} (at 98% RH, 25 °C)		
UPG-2	5.7×10^{-4} (at 95% RH, 100 °C)	-	[65]

4. Summary and Future Outlook

The field of porous metal phosphonates is a growing field of research and is not yet as well established as the carboxylate MOFs, organic–inorganic hybrid silicas, or layered hybrid materials. In this exploratory review, we have highlighted the different classes of porous metal phosphonates coordination networks. Based on layered metal phosphonate systems, we saw the evolution of porous metal phosphonates towards the creation of microporous phosphonate MOFs and porous hybrid inorganic–organic systems (with or without supramolecular templating). The kinetic control, in particular when aiming for mesoporous metal phosphonates, still remains a key challenge in attaining interesting porous architectures. Rapid precipitation due to the heterocondensation/phase separation between metal precursors and phosphonic moieties is one of the remaining concerns. Specifically, when one considers the group IV metals, challenges to harmonize the interactions in a controlled manner are still present when designing porous systems without the use of hazardous HF. Key insights on how the different types of (chelated) metal(IV) precursors react with phosphonate moieties is still a largely unexplored domain. It can thus be envisioned that the challenge to create novel coordination networks of metal phosphonates lies in the reactivity control of the interaction between the metal and phosphonic precursors, in such a way that rapid precipitation is avoided and a controlled build-up of a hybrid porous network is achieved. This problem is further complicated when porosity needs to be induced by the use of SDAs, especially in the case of ionic templates due to their interaction with the phosphonic anion or with the metal ion, influencing reactivity and kinetics. Therefore, the role of templating in these metal phosphonate systems is often not straightforward and more challenging compared to porous metal oxide systems. To date, little studies focused on how ionic and/or polymeric SDAs interact with the phosphonic moieties during the formation of porous metal phosphonate networks. Progress in the successful synthesis of metal phosphonate networks that have been described in this review, proof of the benefit of tailoring the interaction of the phosphonate ligand with the metal precursors (and SDAs) and the need to progress further on this path towards new, more controlled and diverse porous metal phosphonates. This serves to shift the strong kinetically favored condensation reaction, that precipitates as a poorly ordered coordination network, towards the thermodynamically or kinetically controlled build-up of the network.

The domains for the application of porous metal phosphonates span a wide array of fields. However, we highlighted the main contenders where porous metal phosphonates lend themselves to be the ideal choice. This is often in those applications where multifunctionality (e.g., the combination of structural features with the particular chemistry of the metal and phosphonate unit) provides benefits. Heterogeneous catalysis by hybrid metal phosphonates could benefit from the synergy of combining metal centers with tunable organo-functionalities to optimize the interface for catalyzing reactions. The presence of hydrophilic/acidic porous architecture makes metal phosphonates a good candidate for applications like proton conduction, solid-acid catalysts and metal sorption. In the case of separation/extraction applications, the porous amorphous coordination networks could have potential as solid-phase adsorbents. Mainly in the fields of separation of lanthanides and actinides from highly acidic solutions, wherein the stability of the material under high acidic conditions and radiolytic stability are key issues, they serve as prime candidates for solid-phase extraction (SPE). Efforts in employing crystalline metal phosphonate MOFs as a thin film would be fruitful in order to realize fuel-cell applications, or as sensors, or other applications where ordered channels with specific functional groups are an important feature, such as in fabricated devices. Porous metal phosphonates are thus an important part of the organic–inorganic coordination networks, that promise tremendous potential for material chemists in creating new porous architectures with multifunctional features that can be of benefit to diverse application domains.

Author Contributions: Conceptualization, B.M.R. and V.M.; writing—original draft preparation, B.M.R.; writing—review and editing, B.M.R. and V.M.; supervision, V.M.; funding acquisition, V.M. All authors have read and agreed to the published version of this manuscript.

References

1. Yaghi, O.M.; O'Keeffe, M.; Ockwig, N.W.; Chae, H.K.; Eddaoudi, M.; Kim, J. Reticular synthesis and the design of new materials. *Nat. Cell Biol.* **2003**, *423*, 705–714. [CrossRef] [PubMed]
2. Eddaoudi, M.; Kim, J.; Rosi, N.; Vodak, D.; Wachter, J.; O'Keeffe, M.; Yaghi, O.M. Systematic Design of Pore Size and Functionality in Isoreticular MOFs and Their Application in Methane Storage. *Science* **2002**, *295*, 469–472. [CrossRef] [PubMed]

3. Rowsell, J.L.C.; Millward, A.R.; Park, K.S.; Yaghi, O.M. Hydrogen Sorption in Functionalized Metal–Organic Frameworks. *J. Am. Chem. Soc.* **2004**, *126*, 5666–5667. [CrossRef] [PubMed]
4. Barthelet, K.; Marrot, J.; Riou, D. A breathing hybrid organic-inorganic solid with very large pores and high magnetic characteristics. *Angew. Chem.* **2002**, *41*, 281–284. [CrossRef]
5. Ferey, G.; Mellot-Draznieks, C.; Serre, C.; Millange, F.; Dutour, J.; Surblé, S.; Margiolaki, I. A Chromium Terephthalate-Based Solid with Unusually Large Pore Volumes and Surface Area. *Science* **2005**, *309*, 2040–2042. [CrossRef] [PubMed]
6. Inagaki, S.; Guan, S.; Fukushima, Y.; Ohsuna, A.T.; Terasaki, O. Novel Mesoporous Materials with a Uniform Distribution of Organic Groups and Inorganic Oxide in Their Frameworks. *J. Am. Chem. Soc.* **1999**, *121*, 9611–9614. [CrossRef]
7. Asefa, T.; MacLachlan, M.J.; Coombs, N.; Ozin, G.A. Periodic mesoporous organosilicas with organic groups inside the channel walls. *Nat. Cell Biol.* **1999**, *402*, 867–871. [CrossRef]
8. Melde, B.J.; Holland, B.T.; Blanford, A.C.F.; Stein, A. Mesoporous Sieves with Unified Hybrid Inorganic/Organic Frameworks. *Chem. Mater.* **1999**, *11*, 3302–3308. [CrossRef]
9. Clearfield, A. Recent advances in metal phosphonate chemistry. *Curr. Opin. Solid State Mater. Sci.* **1996**, *1*, 267–278. [CrossRef]
10. Clearfield, A. Recent advances in metal phosphonate chemistry II. *Curr. Opin. Solid State Mater. Sci.* **2002**, *6*, 495–506. [CrossRef]
11. Shimizu, G.K.H.; Vaidhyanathan, R.; Taylor, J.M. Phosphonate and sulfonate metal organic frameworks. *Chem. Soc. Rev.* **2009**, *38*, 1430–1449. [CrossRef] [PubMed]
12. Gagnon, K.J.; Perry, H.P.; Clearfield, A. Conventional and Unconventional Metal–Organic Frameworks Based on Phosphonate Ligands: MOFs and UMOFs. *Chem. Rev.* **2012**, *112*, 1034–1054. [CrossRef]
13. Ma, T.-Y.; Yuan, Z.-Y. Metal Phosphonate Hybrid Mesostructures: Environmentally Friendly Multifunctional Materials for Clean Energy and Other Applications. *ChemSusChem* **2011**, *4*, 1407–1419. [CrossRef] [PubMed]
14. Zhu, Y.-P.; Ma, T.-Y.; Liu, Y.-L.; Ren, T.-Z.; Yuan, Z.-Y. Metal phosphonate hybrid materials: From densely layered to hierarchically nanoporous structures. *Inorg. Chem. Front.* **2014**, *1*, 360–383. [CrossRef]
15. Bao, S.-S.; Shimizu, G.K.; Zheng, L.-M. Proton conductive metal phosphonate frameworks. *Co-ord. Chem. Rev.* **2019**, *378*, 577–594. [CrossRef]
16. Bhanja, P.; Na, J.; Tang, J.; Lin, J.; Wakihara, T.; Bhaumik, A.; Yamauchi, Y. Nanoarchitectured Metal Phosphates and Phosphonates: A New Material Horizon toward Emerging Applications. *Chem. Mater.* **2019**, *31*, 5343–5362. [CrossRef]
17. Shearan, S.J.; Stock, N.; Emmerling, F.; Demel, J.; Wright, P.A.; Demadis, K.D.; Vassaki, M.; Costantino, F.; Vivani, R.; Sallard, S.; et al. New Directions in Metal Phosphonate and Phosphinate Chemistry. *Crystals* **2019**, *9*, 270. [CrossRef]
18. Alberti, G.; Costantino, U.; Vivani, R.; Zappelli, P. Preparation of Zirconium Diphosphonate-Phosphites With A Narrow Distribution Of Mesopores. *MRS Proc.* **1991**, *233*, 101–106. [CrossRef]
19. Alberti, G.; Casciola, M.; Costantino, U.; Vivani, R. Layered and pillared metal(IV) phosphates and phosphonates. *Adv. Mater.* **1996**, *8*, 291–303. [CrossRef]
20. Clearfield, A. The early history and growth of metal phosphonate chemistry. In *Metal Phosphonate Chemistry: From Synthesis to Applications*; Clearfield, A., Demadis, K., Eds.; Royal Society of Chemistry Publishing: Cambridge, UK, 2012; pp. 1–44.
21. Cabeza, A.; Aranda, M.A.G. Metal Carboxyphosphonates. In *Metal Phosphonate Chemistry: From Synthesis to Applications*; Clearfield, A., Demadis, K., Eds.; Royal Society of Chemistry Publishing: Cambridge, UK, 2012; pp. 107–132.
22. Dines, M.B.; Digiacomo, P.M. Derivatized lamellar phosphates and phosphonates of M(IV) ions. *Inorg. Chem.* **1981**, *20*, 92–97. [CrossRef]
23. Dines, M.B.; Griffith, P.C. Synthesis and characterization of layered tetravalent metal terphenyl mono- and bis-phosphonates. *Polyhedron* **1983**, *2*, 607–611. [CrossRef]
24. Serre, C.; Mouchaham, G.; Steunou, N.; Devic, T.; Serre, C. Titanium coordination compounds: From discrete metal complexes to metal–organic frameworks. *Chem. Soc. Rev.* **2017**, *46*, 3431–3452. [CrossRef]
25. Poojary, D.M.; Zhang, B.; Bellinghausen, P.; Clearfield, A. Synthesis and X-ray structures of covalently pillared zinc bis(phosphonates). *Inorg. Chem.* **1996**, *35*, 5254–5263. [CrossRef]

26. Poojary, D.M.; Zhang, B.; Bellinghausen, P.; Clearfield, A. Synthesis and X-ray Powder Structures of Two Lamellar Copper Arylenebis(phosphonates). *Inorg. Chem.* **1996**, *35*, 4942–4949. [CrossRef]
27. Poojary, D.M.; Zhang, B.; Clearfield, A. Pillared layered metal phosphonates. Synthesis and X-ray powder structures of copper and zinc alkylenebis(phosphonates). *J. Am. Chem. Soc.* **1997**, *119*, 12550–12559. [CrossRef]
28. Mao, J.G.; Wang, Z.; Clearfield, A. Synthesis, characterization, and crystal structures of to divalent metal diphosphonates with a layered and a 3D network structure. *Inorg. Chem.* **2002**, *41*, 2334–2340. [CrossRef]
29. Bakhmutova, E.V.; Ouyang, X.; Medvedev, D.G.; Clearfield, A. Cobalt Phosphonates: An Unusual Polymeric Cobalt Phosphonate Containing a Clathrated Phosphonate Anion and a Layered Bisphosphonate. *Inorg. Chem.* **2003**, *42*, 7046–7051. [CrossRef] [PubMed]
30. Penicaud, V.; Massiot, D.; Gelbard, G.; Odobel, F.; Bujoli, B. Preparation of structural analogues of divalent metal monophosphonates, using bis(phosphonic) acids: A new strategy to reduce overcrowding of organic groups in the interlayer space. *J. Mol. Struct.* **1998**, *470*, 31–38. [CrossRef]
31. Zhang, B.; Poojary, D.M.; Clearfield, A. Synthesis and Characterization of Layered Zinc Biphenylylenebis(phosphonate) and Three Mixed-Component Arylenebis(phosphonate)/Phosphates. *Inorg. Chem.* **1998**, *37*, 1844–1852. [CrossRef]
32. Cao, G.; Hong, H.G.; Mallouk, T.E. Layered metal phosphates and phosphonates: From crystals to monolayers. *Acc. Chem. Res.* **1992**, *25*, 420–427. [CrossRef]
33. Alberti, G.; Marmottini, F.; Vivani, R.; Zappelli, P. Preparation and Characterization of Pillared Zirconium Phosphite-Diphosphonates with Tuneable Inter-Crystal Mesoporosity. *J. Porous Mater.* **1998**, *5*, 221–226. [CrossRef]
34. Alberti, G.; Marmottini, F.; Murcia-Mascarós, S.; Vivani, R. Preparation and Preliminary Characterization of a Covalently Pillared Zirconium Phosphate- Diphosphonate with Interlayer Microporosity. *Angew. Chem. Int. Ed.* **1994**, *33*, 1594–1597. [CrossRef]
35. Subbiah, A.; Pyle, D.; Rowland, A.; Huang, J.; Narayanan, R.A.; Thiyagarajan, P.; Zoń, J.; Clearfield, A. A Family of Microporous Materials Formed by Sn(IV) Phosphonate Nanoparticles. *J. Am. Chem. Soc.* **2005**, *127*, 10826–10827. [CrossRef] [PubMed]
36. Subbaiah, A.; Bhuvanesh, N.; Clearfield, A. A novel inorganic-organic compound: Synthesis and structural characterization of tin(II) phenylbis(phosphonate), $Sn_2(PO_3C_6H_4PO_3)$. *J. Solid State Chem.* **2005**, *178*, 1321–1325. [CrossRef]
37. Huang, J.; Subbiah, A.; Pyle, D.; Rowland, A.; Smith, B.; Clearfield, A. Globular Porous Nanoparticle Tin(IV) Phenylphosphonates and Mixed Methyl Phenylphosphonates. *Chem. Mater.* **2006**, *18*, 5213–5222. [CrossRef]
38. Gómez-Alcántara, M.D.M.; Cabeza, A.; Olivera-Pastor, P.; Fernández-Moreno, F.; Sobrados, I.; Sanz, J.; Morris, R.E.; Clearfield, A.; Aranda, M.A.G.; Cabeza, A.; et al. Layered microporous tin(iv) bisphosphonates. *Dalton Trans.* **2007**, *23*, 2394–2404. [CrossRef]
39. Kirumakki, S.; Huang, J.; Subbiah, A.; Yao, J.; Rowland, A.; Smith, B.; Mukherjee, A.; Samarajeewa, S.; Clearfield, A. Tin(iv) phosphonates: Porous nanoparticles and pillared materials. *J. Mater. Chem.* **2009**, *19*, 2593–2603. [CrossRef]
40. Perry, H.P.; Law, J.; Zon, J.; Clearfield, A. Porous zirconium and tin phosphonates incorporating 2,2′-bipyridine as supports for palladium nanoparticles. *Microporous Mesoporous Mater.* **2012**, *149*, 172–180. [CrossRef]
41. Cabeza, A.; Gómez-Alcántara, M.D.M.; Olivera-Pastor, P.; Sobrados, I.; Sanz, J.; Xiao, B.; Morris, R.E.; Clearfield, A.; Aranda, M.A.G.; Cabeza, A.; et al. From non-porous crystalline to amorphous microporous metal(IV) bisphosphonates. *Microporous Mesoporous Mater.* **2008**, *114*, 322–336. [CrossRef]
42. Liang, J.; Shimizu, G.K.H. Crystalline Zinc Diphosphonate Metal−Organic Framework with Three-Dimensional Microporosity. *Inorg. Chem.* **2007**, *46*, 10449–10451. [CrossRef]
43. Wu, J.; Hou, H. Highly selective ferric ion sorption and exchange by crystalline metal phosphonates constructured from tetraphosphonic acids. *Inorg. Chem.* **2007**, *46*, 7960–7970. [CrossRef] [PubMed]
44. Vaidhyanathan, R.; Liang, J. A route to functionalized pores in coordination polymers via mixed phosphonate and amino triazole linkers. *Supramol. Chem.* **2011**, *23*, 278–282. [CrossRef]
45. Ma, K.-R.; Zhang, D.-J.; Zhu, Y.-L. Structure and Characterization of a Novel 3D Lead Phosphonate Metal—Organic Framework with Cationic Layer Based on Weak Pb—O(N) Contact. *Aust. J. Chem.* **2010**, *63*, 452–457. [CrossRef]

46. Iremonger, S.S.; Liang, J.; Vaidhyanathan, R.; Shimizu, G.K.H. A permanently porous van der Waals solid by using phosphonate monoester linkers in a metal organic framework. *Chem. Commun.* **2011**, *47*, 4430–4432. [CrossRef] [PubMed]
47. Iremonger, S.S.; Liang, J.; Vaidhyanathan, R.; Martens, I.; Shimizu, G.K.H.; Thomas, D.D.; Aghaji, M.Z.; Yeganegi, S.; Woo, T.K. Phosphonate Monoesters as Carboxylate-like Linkers for Metal Organic Frameworks. *J. Am. Chem. Soc.* **2011**, *133*, 20048–20051. [CrossRef] [PubMed]
48. Groves, J.A.; Miller, S.R.; Warrender, S.J.; Mellot-Draznieks, C.; Lightfoot, P.; Wright, P.A. The first route to large pore metal phosphonates. *Chem. Commun.* **2006**, 3305–3307. [CrossRef] [PubMed]
49. Miller, S.R.; Pearce, G.M.; Wright, P.A.; Bonino, F.; Chavan, S.M.; Bordiga, S.; Margiolaki, I.; Guillou, N.; Férey, G.; Bourrelly, S.; et al. Structural Transformations and Adsorption of Fuel-Related Gases of a Structurally Responsive Nickel Phosphonate Metal–Organic Framework, Ni-STA-12. *J. Am. Chem. Soc.* **2008**, *130*, 15967–15981. [CrossRef]
50. Pearce, G.M. Synthesis, Adsorption and Catalysis of Large Pore metal Phosphonates. Ph.D. Thesis, University of St Andrews, St Andrews, UK, 2009.
51. Wharmby, M.T.; Mowat, J.P.S.; Thompson, S.P.; Wright, P.A. Extending the Pore Size of Crystalline Metal Phosphonates toward the Mesoporous Regime by Isoreticular Synthesis. *J. Am. Chem. Soc.* **2011**, *133*, 1266–1269. [CrossRef]
52. Taylor, J.M.; Vaidhyanathan, R. Enhancing the water stability of metal-organic frameworks via phosphonate monoester linkers. *J. Am. Chem. Soc.* **2012**, *134*, 14338–14340. [CrossRef]
53. Mah, R.K.; Lui, M.W. Enhancing order and porosity in a highly robust tin(IV) triphosphonate network. *Inorg. Chem.* **2013**, *52*, 7311–7313. [CrossRef]
54. Mah, R.K.; Gelfand, B.S.; Taylor, J.M.; Shimizu, G.K. Reconciling order, stability, and porosity in phosphonate metal organic frameworks via HF-mediated synthesis. *Inorg. Chem. Front.* **2015**, *2*, 273–277. [CrossRef]
55. Tang, S.F.; Pan, X.B. Fabrication of new metal phosphonates from tritopic triphosphonic acid containing methyl groups and auxillary ligands: Syntheses, structures, and gas adsorption properties. *Cryst. Eng. Chem.* **2013**, *15*, 1860–1873.
56. Serre, C.; Ferey, G. Hybrid open frameworks. 8. Hydrothermal synthesis, crystal structure and thermal behavior of the first three-dimensional titanium(IV) diphosphonate with an open framework structure: $Ti_3O_2(H_2O)_2(O_3P\text{-}(CH_2)\text{-}PO_3)_2{\cdot}(H_2O)_2$, or MIL-22. *Inorg. Chem.* **1999**, *38*, 5370–5373. [CrossRef]
57. Riou, D.; Roubeau, O. Composite microporous compounds. Part 1: Synthesis and structure determination of two new vanadium alkyldiphosphonates (MIL-2 and MIL-3) with three-dimensional open frameworks. *Microporous Mesoporous Mater.* **1998**, *23*, 23–31. [CrossRef]
58. Riou, D.; Serre, C. Composite microporous compounds (MIL-n).II: Hydrothermal synthesis and abinitioResolution by X-ray powder diffraction of MIL-5: A vanado diphosphonates with three-dimensional neutral frameworks. *J. Solid State Chem.* **1998**, *141*, 89–93.
59. Riou, D.; Ferey, G. Hybrid open frameworks (MIL-n). Part 3 crystal structures of HT and LT forms of MIL-7: A new vanadium propylenediphosphonate with an open framework. Influence of the synthesis temperature on the oxidation state of vanadium with the same structural type. *J. Mater. Chem.* **1998**, *8*, 2733–2735. [CrossRef]
60. Serre, C.; Groves, J.A. Synthesis, structure and properties of related microporous N,N′-piperazinebismethylenephosphonates of aluminium and titanium. *Chem. Mater.* **2006**, *18*, 1451–1457. [CrossRef]
61. Taddei, M.; Costantino, F.; Vivani, R. Synthesis and Crystal Structure from X-ray Powder Diffraction Data of Two Zirconium Diphosphonates Containing Piperazine Groups. *Inorg. Chem.* **2010**, *49*, 9664–9670. [CrossRef]
62. Taddei, M.; Costantino, F.; Vivani, R.; Sabatini, S.; Lim, S.-H.; Cohen, S.M. The use of a rigid tritopic phosphonic ligand for the synthesis of a robust honeycomb-like layered zirconium phosphonate framework. *Chem. Commun.* **2014**, *50*, 5737–5740. [CrossRef]
63. Taddei, M.; Costantino, F.; Marmottini, F.; Comotti, A.; Sozzani, P.; Vivani, R. The first route to highly stable crystalline microporous zirconium phosphonate metal-organic frameworks. *Chem. Commun.* **2014**, *50*, 14831–14834. [CrossRef]
64. Tang, S.F.; Cai, J.J. A highly porous three-dimensional aluminum phosphonate with hexagonal channels: Synthesis, structure and adsorption properties. *Dalton Trans.* **2014**, *43*, 5970–5973. [CrossRef] [PubMed]

65. Taddei, M.; Shearan, S.J.; Donnadio, A.; Casciola, M.; Vivani, R.; Costantino, F. Investigating the effect of positional isomerism on the assembly of zirconium phosphonates based on tritopic linkers. *Dalton Trans.* **2020**, *49*, 3662–3666. [CrossRef]
66. Gu, D.; Schuth, F. Synthesis of non-siliceous mesoporous oxides. *Chem. Soc. Rev.* **2014**, *43*, 313–344. [CrossRef] [PubMed]
67. Soler-Illia, G.J.; Azzaroni, O. Multifunctional hybrids by combining ordered mesoporous materials and macromolecular building blocks. *Chem. Soc. Rev.* **2011**, *40*, 1107–1150. [CrossRef] [PubMed]
68. Bhaumik, A.; Inagaki, S. Mesoporous Titanium Phosphate Molecular Sieves with Ion-Exchange Capacity. *J. Am. Chem. Soc.* **2001**, *123*, 691–696. [CrossRef]
69. Guo, X.; Ding, W.; Wang, X.; Yan, Q. Synthesis of a novel mesoporous iron phosphate. *Chem. Commun.* **2001**, 709–710. [CrossRef]
70. Serre, C.; Auroux, A.; Gervasini, A.; Hervieu, M.; Férey, G. Hexagonal and Cubic Thermally Stable Mesoporous Tin(IV) Phosphates with Acidic and Catalytic Properties. *Angew. Chem. Int. Ed.* **2002**, *41*, 1594–1597. [CrossRef]
71. Ren, T.-Z.; Yuan, Z.-Y.; Azioune, A.; Pireaux, A.J.-J.; Su, B.-L. Tailoring the Porous Hierarchy of Titanium Phosphates. *Langmuir* **2006**, *22*, 3886–3894. [CrossRef]
72. Kimura, T. Synthesis of novel mesoporous aluminum organophosphonate by using a bridged diphosphonic acid. *Chem. Mater.* **2003**, *15*, 3742–3744. [CrossRef]
73. Haskouri, J.E.; Guillem, C.; Latorre, J.; Beltrán, A.; Beltrán, D.; Amorós, P. The first pure mesoporous aluminum phosphonates and diphosphonates—New porous hybrid materials. *Eur. J. Inorg. Chem.* **2004**, 1804–1807. [CrossRef]
74. El Haskouri, J.; Guillem, C.; Latorre, J.; Beltrán, A.; Beltrán, D.; Amoros, P. S^+I^- Ionic Formation Mechanism to New Mesoporous Aluminum Phosphonates and Diphosphonates. *Chem. Mater.* **2004**, *16*, 4359–4372. [CrossRef]
75. Kimura, T. Oligomeric Surfactant and Triblock Copolymer Syntheses of Aluminum Organophosphonates with Ordered Mesoporous Structures. *Chem. Mater.* **2005**, *17*, 5521–5528. [CrossRef]
76. Kimura, T. Synthesis of Mesostructured and Mesoporous Aluminum Organophosphonates Prepared by Using Diphosphonic Acids with Alkylene Groups. *Chem. Mater.* **2005**, *17*, 337–344. [CrossRef]
77. Mal, N.K.; Fujiwara, M.; Yamada, Y.; Matsukata, M. Synthesis of Surfactant-assisted Microporous Layered Tin Phenylphosphonate. *Chem. Lett.* **2003**, *32*, 292–293. [CrossRef]
78. Sarkar, K.; Yokoi, T.; Tatsumi, T.; Bhaumik, A. Mesoporous hybrid zirconium oxophenylphosphate synthesized in absence of any structure directing agent. *Microporous Mesoporous Mater.* **2008**, *110*, 405–412. [CrossRef]
79. Shi, X.; Yang, J.; Yang, Q. Mesoporous aluminum organophosphonates functionalized with chiral L-proline groups in the pore. *Eur. J. Inorg. Chem.* **2006**, 1936–1939. [CrossRef]
80. Vasylyev, M.V.; Wachtel, E.J.; Popovitz-Biro, R.; Neumann, R. Titanium Phosphonate Porous Materials Constructed from Dendritic Tetraphosphonates. *Chem. A Eur. J.* **2006**, *12*, 3507–3514. [CrossRef]
81. Vasylyev, M.V.; Neumann, R. Preparation, characterization, and catalytic aerobic oxidation by a vanadium phosphonate mesoporous materials constructed from a dendritic phosphonate. *Chem. Mater.* **2006**, *18*, 2781–2783. [CrossRef]
82. Ma, T.Y.; Zhang, X.J.; Shao, G.S.; Cao, J.L.; Yuan, Z.Y. Ordered microporous titanium phosphonate materials: Synthesis, photocatalytic activity, and heavy metal ion adsorption. *J. Phys. Chem. C* **2008**, *112*, 3090–3096. [CrossRef]
83. Ma, T.; Zhang, X.-J.; Yuan, Z.-Y. High selectivity for metal ion adsorption: From mesoporous phosphonated titanias to meso-/macroporous titanium phosphonates. *J. Mater. Sci.* **2009**, *44*, 6775–6785. [CrossRef]
84. Ma, T.-Y.; Lin, X.-Z.; Zhang, X.-J.; Yuan, Z.-Y. High surface area titanium phosphonate materials with hierarchical porosity for multi-phase adsorption. *New J. Chem.* **2010**, *34*, 1209–1216. [CrossRef]
85. Ma, T.-Y.; Lin, X.-Z.; Yuan, Z.-Y. Periodic mesoporous titanium phosphonate hybrid materials. *J. Mater. Chem.* **2010**, *20*, 7406–7415. [CrossRef]
86. Ma, T.-Y.; Lin, X.-Z.; Yuan, Z.-Y. Cubic mesoporous titanium phosphonates with functionality. *Chem. Eur. J.* **2010**, *16*, 8487–8494. [CrossRef] [PubMed]
87. Ma, T.-Y.; Lin, X.-Z.; Yuan, Z.-Y. Hierarchical meso-/macroporous phosphated and phosphonated titania nanocomposite materials with high photocatalytic activity. *Stud. Surf. Sci. Catal.* **2010**, *175*, 571–574.

88. Zhang, X.J.; Ma, T.Y.; Yuan, Z.-Y. Titania-phosphonate hybrid porous materials: Preparation, photocatalytic activity, and heavy metal ion adsorption. *J. Mat. Chem.* **2008**, *18*, 2003–2010. [CrossRef]
89. Kimura, T. Moleular design of bisphosphonates to adjust their reactivity towards metal sources for the surfactant-assisted synthesis of mesoporous films. *Angew. Chem. Int. Ed.* **2017**, *56*, 13459–13463. [CrossRef]
90. Wakabayashi, R.; Kimura, T. Further Understanding of the Reactivity Control of Bisphosphonates to a Metal Source for Fabricating Highly Ordered Mesoporous Films. *Chem. Eur. J.* **2019**, *25*, 5971–5977. [CrossRef]
91. Wang, Y.; Alauzun, J.G.; Mutin, P.H. Water-Stable, Nonsiliceous Hybrid Materials with Tunable Porosity and Functionality: Bridged Titania-Bisphosphonates. *Chem. Mater.* **2020**, *32*, 2910–2918. [CrossRef]
92. Hossain, K.; Mercier, L. Intraframework Metal Ion Adsorption in Ligand-Functionalized Mesoporous Silica. *Adv. Mater.* **2002**, *14*, 1053–1056. [CrossRef]
93. Zhu, H.; Jones, D.J.; Zajac, J.; Dutartre, R.; Rhomari, A.M.; Rozière, J. Synthesis of Periodic Large Mesoporous Organosilicas and Functionalization by Incorporation of Ligands into the Framework Wall. *Chem. Mater.* **2002**, *14*, 4886–4894. [CrossRef]
94. Ma, T.; Zhang, X.-J.; Yuan, Z.-Y. Hierarchical Meso-/Macroporous Aluminum Phosphonate Hybrid Materials as Multifunctional Adsorbents. *J. Phys. Chem. C* **2009**, *113*, 12854–12862. [CrossRef]
95. Salis, A.; Bhattacharyya, M.S.; Monduzzi, M. Specific Ion Effects on Adsorption of Lysozyme on Functionalized SBA-15 Mesoporous Silica. *J. Phys. Chem. B* **2010**, *114*, 7996–8001. [CrossRef] [PubMed]
96. Steri, D.; Monduzzi, M.; Salis, A. Ionic strength affects lysozyme adsorption and release from SBA-15 mesoporous silica. *Microporous Mesoporous Mater.* **2013**, *170*, 164–172. [CrossRef]
97. Ding, Y.; Yin, G.; Liao, X.; Huang, Z.; Chen, X.; Yao, Y.; Li, J. A convenient route to synthesize SBA-15 rods with tunable pore length for lysozyme adsorption. *Microporous Mesoporous Mater.* **2013**, *170*, 45–51. [CrossRef]
98. Bhattacharyya, M.S.; Hiwale, P.; Piras, M.; Medda, L.; Steri, D.; Piludu, M.; Salis, A.; Monduzzi, M. Lysozyme Adsorption and Release from Ordered Mesoporous Materials. *J. Phys. Chem. C* **2010**, *114*, 19928–19934. [CrossRef]
99. Shah, B.; Chudasam, U. Application of zirconium phosphonate-a novel hybrid material as an ion exchanger, Desalin. *Water Treat.* **2012**, *38*, 227–235. [CrossRef]
100. Zhu, Y.-P.; Liu, Y.-L.; Ren, T.-Z.; Yuan, Z.-Y. Hollow manganese phosphonate microspheres with hierarchical porosity for efficient adsorption and separation. *Nanoscale* **2014**, *6*, 6627–6636. [CrossRef]
101. Tang, Y.; Ren, Y.; Shi, X. Bifunctional mesoporous zirconium phosphonates for delivery of nuclei acids. *Inorg. Chem.* **2013**, *52*, 1388–1397. [CrossRef]
102. Zhang, J.; Li, X.; Rosenholm, J.M.; Gu, H. Synthesis and characterization of pore size-tunable magnetic mesoporous silica nanoparticles. *J. Colloid Interface Sci.* **2011**, *361*, 16–24. [CrossRef]
103. Zhang, J.; Sun, W.; Bergman, L.; Rosenholm, J.M.; Lindén, M.; Wu, G.; Xu, H.; Gu, H.-C. Magnetic mesoporous silica nanospheres as DNA/drug carrier. *Mater. Lett.* **2012**, *67*, 379–382. [CrossRef]
104. Wei, X.; Shi, X. Mesoporous Zirconium Phenylphosphonates for Selective Enrichment of Phosphopeptides. *J. Phys. Chem. C* **2014**, *118*, 4213–4221. [CrossRef]
105. Lv, K.; Han, J.; Yang, C.-T.; Cheng, C.-M.; Luo, Y.-M.; Wang, X.-L. A category of hierarchically porous tin (IV) phosphonate backbone with the implication for radioanalytical separation. *Chem. Eng. J.* **2016**, *302*, 368–376. [CrossRef]
106. Sibernagel, R.; Martin, C.H.; Clearfield, A. Zirconium(IV) phosphonate-phosphate as efficient ion-exchange materials. *Inorg. Chem.* **2016**, *55*, 1651–1656. [CrossRef] [PubMed]
107. Silbernagel, R.; Shehee, T.C.; Martin, C.H.; Hobbs, D.T.; Clearfield, A. Zr/Sn(IV) Phosphonates as Radiolytically Stable Ion-Exchange Materials. *Chem. Mater.* **2016**, *28*, 2254–2259. [CrossRef]
108. Luca, V.; Tejada, J.J.; Vega, D.; Arrachart, G.; Rey, C. Zirconium(IV)–Benzene Phosphonate Coordination Polymers: Lanthanide and Actinide Extraction and Thermal Properties. *Inorg. Chem.* **2016**, *55*, 7928–7943. [CrossRef]
109. Veliscek-Carolan, J.; Rawal, A.; Luca, V.; Hanley, T. Zirconium phosphonate sorbents with tunable structure and function. *Microporous Mesoporous Mater.* **2017**, *252*, 90–104. [CrossRef]
110. Milo, A.; Neumann, R. A Tripodal Peptidic Titanium Phosphonate as a Homochiral Porous Solid Medium for the Heterogeneous Enantioselective Hydration of Epoxides. *Adv. Synth. Catal.* **2010**, *352*, 2159–2165. [CrossRef]

111. Ji, Y.; Ma, X.; Wu, X.; Wang, N.; Wang, Q.; Zhou, X. A novel type of chiral titanium phosphonates and hybrid titanium phosphonates for heterogeneous asymmetric catalysis. *Catal. Lett.* **2007**, *118*, 187–194. [CrossRef]
112. Lin, X.-Z.; Yuan, Z.-Y. Synthesis of Mesoporous Zirconium Organophosphonate Solid-Acid Catalysts. *Eur. J. Inorg. Chem.* **2012**, *2012*, 2661–2664. [CrossRef]
113. Kozuch, S.; Martin, J.M.L. "Turning over" Definitions in Catalytic cycles. *ACS Catal.* **2012**, *2*, 2787–2794. [CrossRef]
114. Lin, X.-Z.; Ren, T.-Z.; Yuan, Z. Mesoporous zirconium phosphonate materials as efficient water-tolerable solid acid catalysts. *Catal. Sci. Technol.* **2015**, *5*, 1485–1494. [CrossRef]
115. Pramanik, M.; Bhaumik, A. Organic-Inorganic Hybrid Supermicroporous Iron(III) Phosphonate Nanoparticles as an Efficient Catalyst for the Synthesis of Biofuels. *Chem. Eur. J.* **2013**, *19*, 8507–8514. [CrossRef] [PubMed]
116. Dutta, A.; Pramanik, M.; Patra, A.K.; Nandi, M.; Uyama, H.; Bhaumik, A. Hybrid porous tin(iv) phosphonate: An efficient catalyst for adipic acid synthesis and a very good adsorbent for CO_2 uptake. *Chem. Commun.* **2012**, *48*, 6738. [CrossRef] [PubMed]
117. Pramanik, M.; Bhaumik, A. Self-assembled hybrid tin phosphonate nanoparticles with bimodal porosity: An insight towards the efficient and selective catalytic process for the synthesis of bioactive 1,4-dihydropyridines under solvent-free conditions. *J. Mater. Chem. A* **2013**, *1*, 11210–11220. [CrossRef]
118. Pramanik, M.; Bhaumik, A. Self-Assembled Hybrid Molybdenum Phosphonate Porous Nanomaterials and Their Catalytic Activity for the Synthesis of Benzimidazoles. *ChemCatChem* **2014**, *6*, 2577–2586. [CrossRef]
119. Zhu, Y.-P.; Ren, T.-Z.; Yuan, Z.-Y. Hollow cobalt phosphonate spherical hybrid as high-efficiency Fenton catalyst. *Nanoscale* **2014**, *6*, 11395–11402. [CrossRef]
120. Bhanja, P.; Ghosh, K.; Islam, S.S.; Patra, A.K.; Islam, S.M.; Bhaumik, A. A new hybrid iron phosphonate materials as an efficient catalyst for the synthesis of adipic acid in air and water. *ACS Sustain. Chem. Eng.* **2016**, *4*, 7147–7157. [CrossRef]
121. Lin, X.-Z.; Yang, Z.-Z.; He, L.-N.; Yuan, Z. Mesoporous zirconium phosphonates as efficient catalysts for chemical CO_2 fixation. *Green Chem.* **2015**, *17*, 795–798. [CrossRef]
122. Taylor, J.M.; Mah, R.K.; Moudrakovski, I.L.; Ratcliffe, C.I.; Vaidhyanathan, R.; Shimizu, G.K.H. Facile Proton Conduction via Ordered Water Molecules in a Phosphonate Metal–Organic Framework. *J. Am. Chem. Soc.* **2010**, *132*, 14055–14057. [CrossRef]
123. Kim, S.; Dawson, K.W.; Gelfand, B.S.; Taylor, J.M.; Shimizu, G.K.H. Enhancing Proton Conduction in a Metal-Organic Framework by Isomorphous Ligand Replacement. *J. Am. Chem. Soc.* **2013**, *135*, 963–966. [CrossRef]
124. Colodrero, R.M.P.; Olivera-Pastor, P.; Losilla, E.R.; Hernández-Alonso, D.; Aranda, M.A.G.; Leon-Reina, L.; Rius, J.; Demadis, K.D.; Moreau, B.; Villemin, D.; et al. High Proton Conductivity in a Flexible, Cross-Linked, Ultramicroporous Magnesium Tetraphosphonate Hybrid Framework. *Inorg. Chem.* **2012**, *51*, 7689–7698. [CrossRef] [PubMed]
125. Pili, S.; Argent, S.P.; Morris, C.G.; Rought, P.; Garcia-Sakai, V.; Silverwood, I.P.; Easun, T.L.; Li, M.; Warren, M.R.; Murray, C.A.; et al. Proton conduction in a phosphonate-based metal-organic framework mediated by intrinsic "Free diffusion inside a sphere". *J. Am. Chem. Soc.* **2016**, *138*, 6352–6355. [CrossRef] [PubMed]
126. Bao, S.-S.; Li, N.-Z.; Taylor, J.M.; Shen, Y.; Kitagawa, H.; Zheng, L.-M. Co-Ca Phosphonate Showing Humidity-Sensitive Single Crystal to Single Crystal Structural Transformation and Tunable Proton Conduction Properties. *Chem. Mater.* **2015**, *27*, 8116–8125. [CrossRef]

3

The Effect of New Thiophene-Derived Aminophosphonic Derivatives on Growth of Terrestrial Plants: A Seedling Emergence and Growth Test

Jarosław Lewkowski [1,*], Zbigniew Malinowski [1], Agnieszka Matusiak [1], Marta Morawska [1], Diana Rogacz [2] and Piotr Rychter [2,*]

1 Department of Organic Chemistry, Faculty of Chemistry, University of Łódź, Tamka 12, Łódź 91-403, Poland; zbigmal@uni.lodz.pl (Z.M.); agusk@poczta.fm (A.M.); martaem88@gmail.com (M.M.)

2 Faculty of Mathematics and Natural Science, Jan Długosz University in Częstochowa, 13/15 Armii Krajowej Av., Częstochowa 42-200, Poland; diana.rogacz@gmail.com

* Correspondence: jlewkow@uni.lodz.pl (J.L.); p.rychter@ajd.czest.pl (P.R.)

Academic Editor: György Keglevich

Abstract: The aim of this work was to synthesize selected thiophene-derived aminophosphonic systems and evaluate the phytotoxicity of newly obtained products according to the OECD 208 Guideline. Seven new thiophene-derived *N*-substituted dimethyl aminomethylphosphonic acid esters **2a–h** were synthesized by the addition of an appropriate phosphite to azomethine bond of starting Schiff bases **1a–h**, and NMR spectroscopic properties of aminophosphonates were investigated. These eight compounds were analyzed in regard to their phytotoxicity towards two plants, radish (*Raphanus sativus*) and oat (*Avena sativa*). On the basis of the obtained results, it was found that tested aminophosphonates **2a–h** showed an ecotoxicological impact against selected plants, albeit to various degrees.

Keywords: thiophene-derived aminophosphonates; phytotoxicity; growth inhibition; environmental protection; ecotoxicology; OECD standard

1. Introduction

Nowadays, since production of food in agricultural areas is in the phase of intensive growth, the use of herbicides is absolutely necessary to manipulate or control undesirable vegetation of plants. Herbicides are used in landscapes throughout all over the world and are generally accepted as a moderately safe compounds. However, widespread use or overdosage of herbicides over long periods of time can result in residues in crops, soil, and land waters and, as a consequence, may lead to health and environmental risks [1–4]. Permanent applications of a given herbicide can result in resistance of a weed to the agent. Consequently, the weeds, while becoming resistant will not respond to the herbicide's active properties [5,6]. With this respect, there is a necessity of designing new compounds with potential herbicidal properties.

Sulfur-containing compounds, which have been developed as agrochemicals for plant protection, are currently being reviewed from the standpoint of their use and biological properties in each field of fungicides, insecticides, or herbicides [7–9].

Sulfur-containing herbicides are successfully and widely used as a foliar (applied to weed foliage) and a soil-applied herbicide absorbed by the root or shoot of emerging seedlings.

Biodegradability, photodegradation, persistence, soil mobility, and the accumulation in living organisms of sulfur-containing herbicides are currently in extensive research phases [10–14].

Designing new sulfur-containing compounds with potential herbicidal activity in soil, their persistence in soil must be taken into consideration. On the one side, herbicides, which are applied to the soil, typically affect seed emergence or the growth of weed seedlings and must remain in the soil to be effective. Unfortunately, on the other side, some herbicides like thifensulfuron-methyl can persist in the environment for a long time, becoming harmful agents for the surrounding environment. Therefore, planning the synthesis, it is difficult to strike a balance between the activity of a compound and its biodegradability in soil [15].

Biological properties of thiophene-deriving aminophosphonates have been largely studied [16–20]. Scientists found several of them to be promising plant protection agents [16] or to have rather strong antimicrobial [17], cytotoxic [18], antifungal [19,20], or even antiviral [20] properties. Therefore, working on new variations of methods for synthesis of aminophosphonic systems, chemists include derivatives bearing thiophene moiety as examples [17,21–26]. It is to stress that all those methodologies are mostly based on the Kabachnik–Fields reaction [27–29].

Working on our large project aimed at the search of a new class of herbicidal agents, we feel obliged to perform tests of phytotoxicity of newly synthesized compounds, especially when these compounds are aminophosphonic derivatives, which are known generally to have moderate-to-strong phytotoxic action on higher plants [30–33].

Our previous investigation provided results demonstrating that aminophosphonic acids bearing furfuryl moiety (derivatives of *C*-furfurylphosphonoglycine) are compounds of moderate phytotoxicity being able to kill both monocotyledonous oat (*Avena sativa*) and dicotyledonous radish (*Raphanus sativus*) with an amount of 100–200 mg in 1 kg of soil [34].

The presented research is concerned with a screening test of a synthesized series of variously *N*-substituted dimethyl amino(2-thienyl)methylphosphonates (**2a–g**) and dibenzyl *N*-furfurylamino(2-thienyl)methylphosphonate (**2h**) with respect to their phytotoxic properties. Our assumption was that the potential hazard of thiophene-derived aminophosphonates is important enough in the light of their possible applications and therefore should be investigated.

Apart from that, we performed the preliminary evaluation of aminophosphonates **2a–h** as potential soil-applied herbicides for agricultural/horticultural purposes. The potential effects of herbicides strongly results from their mechanism of action and the way they are applied. Since some types of herbicides are non-selective, which means the chemicals kill all classes of plants, not only unwanted weeds, for the proposed experiment, but also two types of plants (mono- and dicotyledonous) have been chosen as experimental objects.

2. Results and Discussion

2.1. Chemistry

Schiff bases **1a–h** were synthesized following the published and commonly known procedure of simple mixing thiophene-2-carboxaldehyde with appropriate amine in methanol and stirring them at room temperature for 24 h [35]. This procedure produced imines **1a–h**, which were isolated and used for further conversions without any purification. ^{1}H-NMR spectra were made only to verify the identity, based on the ^{1}H-NMR diagnostic singlet of a proton of the azomethine group (–CH=N–) above 8 ppm [31,32].

Aminophosphonates **2a–h** were synthesized basing on the aza-Pudovik reaction—a slightly modified procedure of a dimethyl or dibenzyl phosphite addition to the azomethine bond of corresponding Schiff bases **1a–h** in boiling toluene for 5 h. After the workup described in Section 3, the resulting aminophosphonates **2a–h** were obtained in 60%–70% yields (Scheme 1).

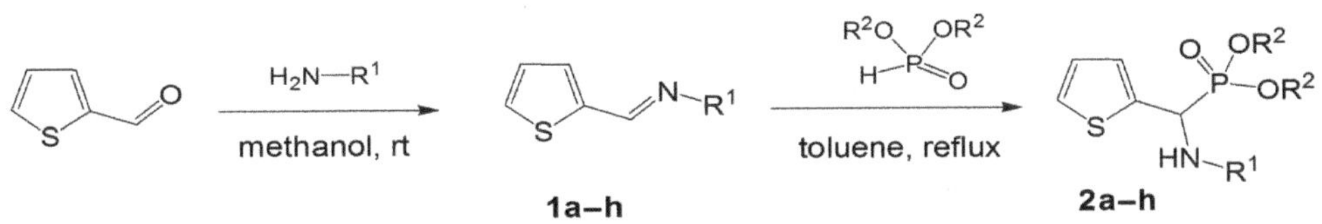

1a: $R^1 = 2\text{-}CH_3C_6H_4$; **1b**: $R^1 = 3\text{-}CH_3C_6H_4$; **1c**: $R^1 = 4\text{-}CH_3C_6H_4$; **1d**: $R^1 = 3\text{-}OCH_3C_6H_4$;
1e: $R^1 = 4\text{-}OCH_3C_6H_4$; **1f**: $R^1 = CH_2C_6H_5$; **1g**: $R^1 = CH(C_6H_5)_2$; **1h**: $R^1 = CH_2Fur$.

2a: $R^1 = 2\text{-}CH_3C_6H_4$; $R^2 = CH_3$; **2b**: $R^1 = 3\text{-}CH_3C_6H_4$; $R^2 = CH_3$; **2c**: $R^1 = 4\text{-}CH_3C_6H_4$; $R^2 = CH_3$;
2d: $R^1 = 3\text{-}OCH_3C_6H_4$; $R^2 = CH_3$; **2e**: $R^1 = 4\text{-}OCH_3C_6H_4$; $R^2 = CH_3$; **2f**: $R^1 = CH_2C_6H_5$; $R^2 = CH_3$;
2g: $R^1 = CH(C_6H_5)_2$; $R^2 = CH_3$; **2h**: $R^1 = CH_2Fur$; $R^2 = CH_2Ph$.

Scheme 1. Synthesis of thiophene-derived aminophosphonates **2a–h**.

2.2. *Evaluation of Phytotoxicity of Aminophosphonates* **2a–h**

The rate of growth and development of plants as well as their quality, including increased number and size of leaves and stems, is strongly dependent (as commonly known) on the ground composition. Since plants have a high amount of water, its level depends on the contents of water in its environment, so taking dry mass as a measure of plant growth seems to be more credible.

Water content is given as a percentage of the dry or fresh weight. To assess the phytotoxicity of the analyzed compounds, the emergence and weight (dry and green) of control plant seedlings, with the emergence and mass (dry and green) of plant seedlings growing on the soil with a given admixture of the examined substances **2a–h**, were determined and compared. Dry weight changes of tested plants are shown in Figure 1. The visual assessment of any damage of tested plants concerns mainly the degree of plant growth inhibition, which is evaluated by visible or other signs of necrosis of chlorosis. Necrosis is the symptom of the death of plant tissues or organs caused by infection. Leaf and/or stem deformation may also be observed, while chlorosis is an inhibition of chlorophyll biosynthesis, resulting in yellowing of tissue.

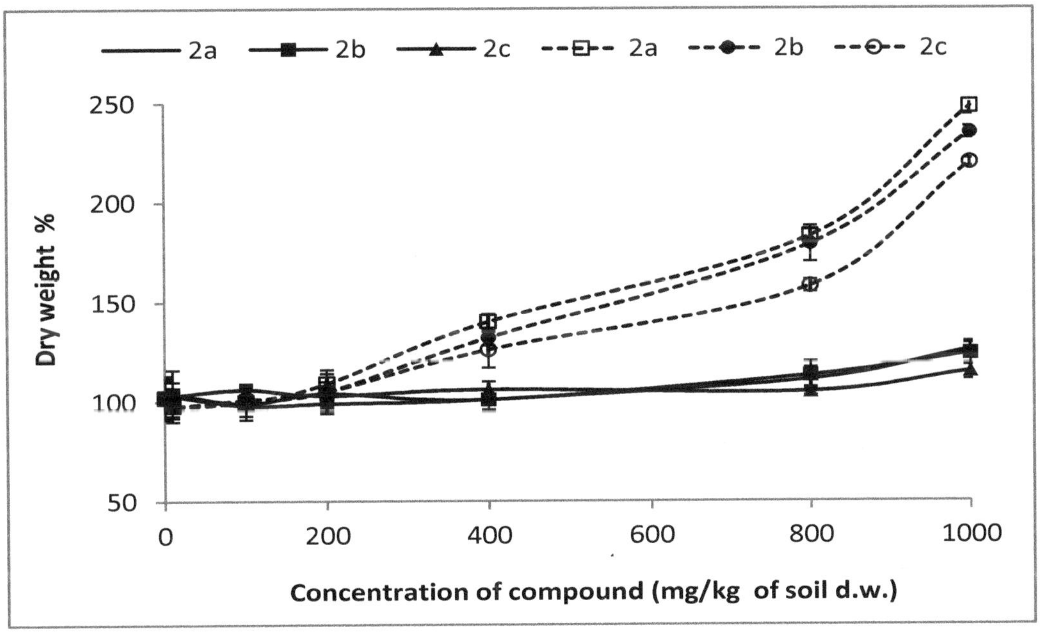

Figure 1. Changes of dry weight of treated plants expressed as percent to the value in untreated plants (control plants = 100% of dry weight). Solid lines represent changes of oat dry weight. Dotted lines represent changes of radish dry weight.

Preliminary tests for aminophosphonates **2a–c** for both oat and radish have shown inhibitive effects for plants. The dicotyledonous radish was noticed as the more sensitive plant when compared to sprouts of oat. Comparing results of toxicity of aminophosphonates **2a–c** for both types of tested plants, it was noticed that Compounds **2a** and **2b** are more toxic as compared with **2c** (Tables 1 and 2 Figures 1–4).

Table 1. Average changes (mean of three replicates) in basic parameters of the plant growth test for oat (Avena sativa) treated with **2a–c**. Least significant difference for samples (LSDS) and concentration (LSDC) is given for each tested parameter. %F.M. refers to plant biomass (fresh weight) expressed as percent of untreated control.

Sample Concentration in Soil (mg/kg of Soil Dry Matter)	Emerged Seedlings Number	% of Germination	Fresh Matter (g/pot)	% F.M.
0	20	100	2.780	100
1	19	97	2.682	96
10	19	95	2.654	95
100	19	97	2.919	105
200	19	95	2.644	95
400	19	95	2.715	98
800	18	88	2.100	76
1000	18	90	1.847	66
0	20	100	2.780	100
1	19	95	2.740	99
10	19	97	2.663	96
100	19	97	2.729	98
200	19	97	2.651	95
400	20	98	2.661	96
800	18	92	2.021	73
1000	18	92	1.659	60
0	20	100	2.780	100
1	19	97	2.759	99
10	20	98	2.704	97
100	20	98	2.801	101
200	19	97	2.862	103
400	20	98	2.762	99
800	19	95	2.643	95
1000	19	95	2.253	81
	LSDS = 1 LSDC = 1		LSDS = 0.401 LSDC = 0.245	

All the other results, to avoid confusion concerned with a large amount of data, have been placed in Supplementary Materials. To describe results, moreover, the demonstrated values of no observed effect concentration (NOEC) and lowest observed effect concentration (LOEC), changes in the fresh mass of plants were expressed as % inhibition compared to control and presented as EC_{50} (Figure 2). Effective concentration EC_{50} was calculated using GraphPad Prism software (Version 7, GraphPad Software, Inc., La Jolla, CA 92037, USA).

Table 2. Average changes (mean of three replicates) in basic parameters of the plant growth test for radish (*Raphanus sativus*) treated with **2a–c**.

Sample Concentration in Soil (mg/kg of Soil Dry Matter)	Emerged Seedlings Number	% of Germination	Fresh Matter (g/pot)	% F.M.
0	20	100	4.951	100
1	20	100	4.908	99
10	19	97	4.831	98
100	19	98	4.728	95
200	18	93	3.928	79
400	16	83	1.752	35
800	18	90	1.136	23
1000	15	75	0.772	16

Table 2. *Cont.*

Sample Concentration in Soil (mg/kg of Soil Dry Matter)	Emerged Seedlings Number	% of Germination	Fresh Matter (g/pot)	% F.M.
0	20	100	4.951	100
1	20	100	4.803	97
10	19	95	4.843	98
100	19	97	4.728	95
200	19	97	3.928	79
400	17	86	2.142	43
800	17	88	1.546	31
1000	15	78	1.106	22
0	20	100	4.951	100
1	19	98	4.914	99
10	19	98	4.798	97
100	19	98	4.873	98
200	19	97	4.787	97
400	18	93	3.934	79
800	18	93	3.328	67
1000	17	86	2.733	55
	$LSD_S = 1$ $LSD_C = 1$		$LSD_S = 0.435$ $LSD_C = 0.267$	

Least significant difference for samples (LSD_S) and concentration (LSD_C) is given for each tested parameter. %F.M. refer to plant biomass (fresh weight) expressed as percent of untreated control.

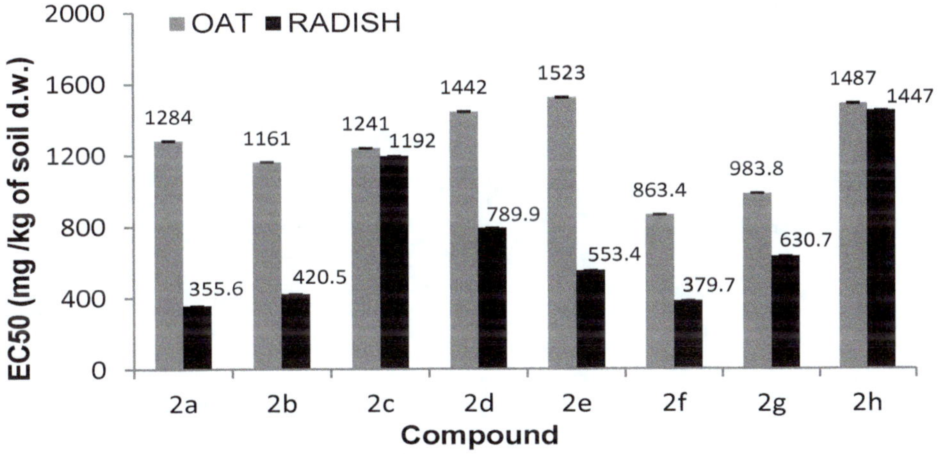

Figure 2. The EC_{50} values of fresh weight inhibition following exposure to tested compounds.

Figure 3. *Cont.*

Figure 3. Digital photographs of oat treated with **2a–c** (concentration in mg/kg of soil dry weight) on the 14th day of growth.

Figure 4. Digital photographs of radish treated with **2a–c** (concentration in mg/kg of soil dry weight) on the 14th day of growth.

N-methylphenyl aminophosphonates with *ortho* and *meta* substituted phenyl ring (**2a**, **2b**) have a significantly stronger influence on the inhibition of plant growth as compared with *N*-*para*-methylphenyl aminophosphonate (**2c**). In a case of oat, the values of NOEC and LOEC for Compounds **2a** and **2b** were 400 and 800 mg/kg of soil dry weight, respectively. These values for Compound **2c** were higher (NOEC and LOEC 800 and 1000 mg/kg d.w. of soil), indicating lower toxicity and subsequently higher tolerance of oat against this substance (Figure 3).

At the highest concentration (1000 mg), a *ca.* 40% decrease in fresh matter of oat sprouts for Compounds **2a** and **2b** when compared to control plants is observed (Table 1).

To make the description of the obtained results more readable and user-friendly, we only present selected tables and figures to the main text here (Tables 1 and 2, Figures 1–4).

Tested substances were much more toxic for the seedlings of radish, resulting in a decrease in NOEC and LOEC values. The first observed negative symptoms of tested **2a–c** compounds against radish were observed for the concentrations of 200 mg/kg of dry weight of soil (Samples **2a**, **2b**) and of 400 mg/kg of dry weight of soil for Compound **2c** (Figure 4). A very high drop of fresh weight of radish seedling was noticed for all compounds (**2a–c**), but aminophosphonates **2a** and **2b** were more toxic. At the concentration of 1000 mg/kg of dry weight of soil, a decrease in radish fresh weight, resulting from the action of Compounds **2a–c**, was 16%, 22%, and 55%, respectively. High values of dry matter at higher concentrations of the tested substances in pots correspond with a stronger inhibition of water uptake through the root system of plants (Figure 1). The increase in values of oat dry matter when compared to control plants is not as high as in the case of radish, which indicates better tolerance of oat roots against tested substances and moderately observed disturbances in the water uptake process.

The number of emerged seedlings at higher concentrations of tested compounds also decreased and was more visible for radish.

A plant growth test conducted for aminophosphonates bearing methoxy groups substituted in position *meta* (**2d**) and *para* (**2e**) in a phenyl ring have proved their toxic impact against both types of plants. Contrary to Compounds **2b–c** substituted with *meta*-methyl and *para*-methylphenyl groups, respectively, the examined *N*-*m*-methoxyphenyl derivative **2d** was less toxic for both radish and oat as compared with *N*-*p*-methoxyphenyl derivative **2e** (Tables S1 and S2, Figures S1–S3). The NOEC and LOEC values in the oat growth test for Compounds **2d** were much higher (800 and 1000 mg respectively) when compared to **2e** (100 and 200 mg respectively). The same values of NOEC and LOEC of Sample **2e** were revealed for radish. Again, as in the case of Compounds **2a–c**, the much more sensitive plant was radish. In comparison, the yield of the fresh matter of oat at the highest concentration (1000 mg) of **2d** and **2e** was 19% and 31% (respectively) lower compared to non-treated plants. The yield of the fresh weight of radish at the highest concentration did not exceed 29% (for **2d**) and 13% (for **2e**) of fresh matter of control plants. The high increase in dry matter values point out the dysfunction of roots (Figure S1).

Comparing the phytotoxicity results of aminophosphonates **2f**, **2g**, and **2h**, Compound **2h** was the most toxic agent for both plants. NOEC and LOEC values of Compound **2h** were the same for radish and oat—40 and 80 mg/kg of dry matter, respectively. Among Compounds **2f** and **2g**, the former had a higher negative impact on treated radish and oat. It was shown again that the more sensitive specimen of plant was radish. A very low yield of fresh matter for radish at the highest concentration (1000 mg) when compared to non-treated plants with simultaneously very high dry matter of plants indicates a strong inhibition of plant growth (both sprouts and roots) (Tables S3–S5, Figures S4–S8)

It is worth comparing the plant growth of Samples **2a–c** with **2f**, the non-aromatic amine derivative bearing the non-substituted phenyl ring. Sample **2f** is much more toxic against oat when compared to **2a–c**. In the case of radish, results are comparable to those obtained for Samples **2a,b** (the same values of NOEC and LOEC). Samples **2a** and **2b**, as well as *N*-benzyl derivative **2f**, in the highest applied concentration (1000 mg/kg), harmfully influenced radish seeds especially, while for oat, at the same concentration, Sample **2f** (without substituted phenyl ring with methyl group) caused almost total inhibition of sprout germination. As a result, contrary to dimethyl *N*-benzylamino(2-thienyl)methylphosphonate (**2f**), which kills both plants in a non-selective way, dimethyl *N*-(2-methylphenyl)amino(2-thienyl)methylphosphonate (**2a**) and dimethyl *N*-(3-methylphenyl)amino(2-thienyl)methylphosphonate (**2b**) were found to be selectively toxic towards radish, slightly harming oat.

3. Materials and Methods

3.1. Chemistry

All solvents (POCh, Gliwice, Poland) were routinely distilled and dried prior to use. Amines, dimethyl and dibenzyl phosphites, as well as 2-thiophenecarboxaldehyde (Aldrich, Poznań, Poland) were used as received. Melting points were measured on a MelTemp II apparatus (Bibby Scientific Limited, Staffordshire, UK) and were not corrected. NMR spectra were recorded on a Bruker Avance III 600 MHz (Billerica, MA, USA) operating at 600 MHz (^{1}H-NMR), 150 MHz (^{13}C-NMR), and 243 MHz (^{31}P-NMR). TMS was used as the internal standard for ^{1}H- and ^{13}C-NMR, and phosphoric acid was used as the external standard for ^{31}P-NMR. The following abbreviations were used for listing NMR signals: s—singlet, d—doublet, dd—doublet of doublets, ddd—doublet of doublet of doublets, and m—multiplet. Elemental analyses were carried out at the Centre for Molecular and Macromolecular Science of the Polish Academy of Science in Łódź, Poland.

General Procedures of Preparaing Amino(2-thienyl)methylphosphonates **2a–h**

2-Thiophenecarboxaldehyde (1.12 g, 10 mmol) was dissolved in methanol (15 mL), and a solution of an appropriate amine (10 mmol) in methanol (15 mL) was added. The obtained solution was stirred at room temperature for 24 h. Then, a small portion of anhydrous potassium carbonate was added, the mixture stirred for additional 15 min. Then, the inorganic salt was filtered off, and the filtrate was evaporated to achieve imines in quantitative yields, which were used for the further reaction without any purification.

Thus, the obtained imine (10 mmol) was dissolved in acetonitrile (15 mL), and an appropriate phosphite (10 mmol) in acetonitrile (15 mL) was added. The mixture was refluxed during the day and stirred at room temperature during the night. Total time of the reaction was 72 h. Then, the solvent was evaporated, and the crude product was isolated and purified using various procedures.

Dimethyl aminophosphonates **2a–g** were isolated as follows: crude products were dissolved in DCM (30 mL), and the solution was washed 5 times with saturated aqueous sodium hydrogen carbonate. After the usual workup, DCM was evaporated, and the product was purified by column chromatography on silica gel using ethyl acetate-hexane in a 4:1 ratio.

Product **2h** was chromatographed after evaporating acetonitrile, eluted using ethyl acetate—hexane in a 3:2 ratio.

Dimethyl N-(2-methylphenyl)amino(2-thienyl)methylphosphonate (**2a**): Yield = 83% (2.58 g) yellow oil. ^{1}H-NMR ($CDCl_3$, 600 MHz): δ 7.27–7.25 (m, H_5^{thioph}, 1H), 7.21–7.19 (m, H_3^{thioph}, 1H), 7.11 (d, $^3J_{HH}$ = 7.4 Hz, *o*-C_6H_4, 1H), 7.07 (ddd, $^3J^{(1)}{}_{HH}$ = $^3J^{(2)}{}_{HH}$ = 7.4 Hz and $^4J_{HH}$ = 1.7 Hz, *o*-C_6H_4, 1H), 7.01 (ddd, $^3J_{HH}$ = 5.0 and 4.2 Hz and $^4J_{HH}$ = 0.7 Hz, H_4^{thioph}, 1H), 6.74 (ddd, $^3J^{(1)}{}_{HH}$ = $^3J^{(2)}{}_{HH}$ = 7.4 Hz and $^4J_{HH}$ = 1.0 Hz, *o*-C_6H_4, 1H), 6.65 (d, $^3J_{HH}$ = 7.4 Hz, *o*-C_6H_4, 1H), 5.16 (d, $^2J_{PH}$ = 23.7 Hz, CHP, 1H), 3.82 (d, $^3J_{PH}$ = 10.6 Hz, $POCH_3$, 3H), 3.67 (d, $^3J_{PH}$ = 10.6 Hz, $POCH_3$, 3H), 2.28 (s, CH_3, 3H). ^{13}C-NMR ($CDCl_3$, 150 MHz): δ 143.98 (d, $^2J_{CP}$ = 12.3 Hz, $C^2{}_{thioph}$), 139.81 (C_{arom}), 130.41 (C_{arom}), 127.19 (d, *J* = 2.5 Hz, C_{thioph}), 127.03 (C_{arom}), 126.19 (d, *J* = 7.2 Hz, C_{thioph}), 125.43 (d, *J* = 3.4 Hz, C_{thioph}), 123.39 (C_{arom}), 118.86 (C_{arom}), 111.48 (C_{arom}), 54.17 (d, $^2J_{CP}$ = 7.0 Hz, POC), 53.88 (d, $^2J_{CP}$ = 7.3 Hz, POC), 51.84 (d, $^1J_{CP}$ = 157.8 Hz, PC), 17.47 (ArC). ^{31}P-NMR (243 MHz, $CDCl_3$): δ 23.33. Elem. Anal. Calcd. for $C_{14}H_{18}NO_3PS$: C, 54.01, H, 5.83, N, 4.50. Found: C, 54.18, H, 5.97, N, 4.59.

Dimethyl N-(3-methylphenyl)amino(2-thienyl)methylphosphonate (**2b**): Yield = 94% (2.92 g) yellow crystals, m.p.: 81–83 °C. ^{1}H-NMR ($CDCl_3$, 600 MHz): δ 7.26–7.25 (m, H_5^{thioph}, 1H), 7.21–7.20 (m, H_3^{thioph}, 1H), 7.07 (dd, $^3J_{HH}$ = 7.2 and 7.0 Hz, *m*-C_6H_4, 1H), 7.01 (m, H_4^{thioph}, 1H), 6.62 (d, $^3J_{HH}$ = 7.0 Hz, *m*-C_6H_4, 1H), 6.55 (s, *m*-C_6H_4, 1H), 6.52 (dd, $^3J_{HH}$ = 7.2 Hz and $^4J_{HH}$ = 1.9 Hz, *m*-C_6H_4, 1H), 5.11 (d, $^2J_{PH}$ = 24.4 Hz, CHP, 1H), 3.82 (d, $^3J_{PH}$ = 10.4 Hz, $POCH_3$, 3H), 3.65 (d, $^3J_{PH}$ = 10.8 Hz, $POCH_3$, 3H), 2.28 (s, CH_3, 3H). ^{13}C-NMR ($CDCl_3$, 150 MHz): δ 145.94 (d, $^2J_{CP}$ = 13.1 Hz, $C^2{}_{thioph}$), 139.68 (C_{arom}), 139.10 (C_{arom}), 129.15 (C_{arom}), 127.17 (d, *J* = 3.1 Hz, C_{thioph}), 126.25 (d, *J* = 7.3 Hz, C_{thioph}), 125.41 (d, *J* = 4.0 Hz, C_{thioph}), 120.09 (C_{arom}), 114.94 (C_{arom}), 111.01 (C_{arom}), 54.13 (d, $^2J_{CP}$ = 6.7 Hz, POC), 53.83 (d, $^2J_{CP}$ = 7.4 Hz, POC), 51.73 (d, $^1J_{CP}$ = 158.4 Hz, PC), 21.54 (ArC). ^{31}P-NMR (243 MHz, $CDCl_3$): δ 23.20. *Elem. Anal.* Calcd. for $C_{14}H_{18}NO_3PS$: C, 54.01, H, 5.83, N, 4.50. Found: C, 54.17, H, 5.79, N, 4.51.

Dimethyl N-(4-methylphenyl)amino(2-thienyl)methylphosphonate (**2c**): Yield = 54% (1.68 g) yellow crystals, m.p.: 98–100 °C. ^{1}H-NMR ($CDCl_3$, 600 MHz): δ 7.25–7.24 (m, H_5^{thioph}, 1H), 7.20–7.19 (m, H_3^{thioph}, 1H), 7.05–6.98 (m, H_5^{thioph}, *p*-C_6H_4, 3H), 6.64 (part of AA'XX' system, $^3J_{HH}$ = 9.0 and $^4J_{HH}$ = 1.2 and 1.1 Hz, *p*-C_6H_4, 2H), 5.09 (d, $^2J_{PH}$ = 24.0 Hz, CHP, 1H), 3.82 (d, $^3J_{PH}$ = 10.8 Hz, $POCH_3$, 3H), 3.65 (d, $^3J_{PH}$ = 10.8 Hz, $POCH_3$, 3H), 2.24 (s, CH_3, 3H). ^{13}C-NMR ($CDCl_3$, 150 MHz): δ 143.63 (d, $^2J_{CP}$ = 13.4 Hz, $C^2{}_{thioph}$), 139.77 (C_{arom}), 129.79 (C_{arom}), 128.40 (C_{arom}), 127.15 (d, *J* = 2.4 Hz, C_{thioph}), 126.26 (d, *J* = 7.4 Hz, C_{thioph}), 125.41 (d, *J* = 3.4 Hz, C_{thioph}), 114.25 (C_{arom}), 54.13 (d, $^2J_{CP}$ = 7.1 Hz, POC), 53.81 (d, $^2J_{CP}$ = 7.1 Hz, POC), 52.12 (d, $^1J_{CP}$ = 157.9 Hz, PC), 20.40 (ArC). ^{31}P-NMR (243 MHz, $CDCl_3$): δ 23.27. *Elem. Anal.* Calcd. for $C_{14}H_{18}NO_3PS$: C, 54.01, H, 5.83, N, 4.50. Found: C, 54.13, H, 5.88, N, 4.63.

Dimethyl N-(3-methoxyphenyl)amino(2-thienyl)methylphosphonate (**2d**): Yield = 72% (2.35 g) yellow oil ^{1}H-NMR ($CDCl_3$, 600 MHz): δ 7.26–7.25 (m, H_5^{thioph}, 1H), 7.20–7.19 (m, H_3^{thioph}, 1H), 7.08 (dd, J_1 = J_2 = 7.8 Hz, *m*-C_6H_4, 1H), 7.01–6.99 (m, H_4^{thioph}, 1H), 6.35–6.34 (m, *m*-C_6H_4, 1H), 6.33–6.31 (m, *m*-C_6H_4, 1H), 6.28–6.26 (m, *m*-C_6H_4, 1H), 5.10 (d, $^2J_{PH}$ = 24.0 Hz, CHP, 1H), 3.81 (d, $^3J_{PH}$ = 10.8 Hz, $POCH_3$, 3H), 3.74 (s, OCH_3, 3H), 3.65 (d, $^3J_{PH}$ = 10.8 Hz, $POCH_3$, 3H). ^{13}C-NMR ($CDCl_3$, 150 MHz): δ 160.77 (C_{arom}), 147.34 (d, $^2J_{CP}$ = 13.5 Hz, $C^2{}_{thioph}$), 139.46 (C_{arom}), 130.12 (C_{arom}), 127.2 (d, *J* = 3.0 Hz, C_{thioph}), 126.3 (d, *J* = 6.9 Hz, C_{thioph}), 125.5 (d, *J* = 3.5 Hz, C_{thioph}), 106.91 (C_{arom}), 104.38 (C_{arom}), 100.23 (C_{arom}), 55.08 (ArOC), 54.12 (d, $^2J_{CP}$ = 6.7 Hz, POC), 53.90 (d, $^2J_{CP}$ = 6.7 Hz, POC), 51.69 (d, $^1J_{CP}$ = 158.4 Hz, PC).^{31}P-NMR (243 MHz, $CDCl_3$: δ 23.06. *Elem. Anal.* Calcd. for $C_{14}H_{18}NO_4PS \times {}^1/_5\ C_6H_{14}$: C, 52.98, H, 6.08, N, 4.06. Found: C, 52.97, H, 5.96, N, 4.38.

Dimethyl N-(4-methoxyphenyl)amino(2-thienyl)methylphosphonate (**2e**): Yield = 80% (2.62 g) yellow oil. ^{1}H-NMR ($CDCl_3$, 600 MHz): δ 7.25–7.23 (m, H_5^{thioph}, 1H), 7.18–7.16 (m, H_3^{thioph}, 1H), 6.99–6.98 (m, H_4^{thioph}, 1H), 6.75 and 6.66 (AA'XX' system, $^3J_{HH}$ = 9.0 and $^4J_{HH}$ = 1.2 and 1.1 Hz, *p*-C_6H_4, 4H), 4.94 (d, $^2J_{PH}$ = 23.4 Hz, CHP, 1H), 4.54 (d, *J* – 8.4 Hz, NH, 1H), 3.81 (d, $^3J_{PH}$ = 10.8 Hz, $POCH_3$, 3H), 3.72 (s, OCH_3, 3H), 3.64 (d, $^3J_{PH}$ = 10.8 Hz, $POCH_3$, 3H). ^{13}C-NMR ($CDCl_3$, 150 MHz): δ 153.26 (C_{arom}), 139.91 (d, $^2J_{CP}$ = 14.1 Hz, $C^2{}_{thioph}$), 139.73 (C_{arom}), 127.1 (d, *J* = 2.8 Hz, C_{thioph}), 126.3 (d, *J* = 6.9 Hz, C_{thioph}), 125.4 (d, *J* = 4.0 Hz, C_{thioph}), 115.69 (C_{arom}), 114.83 (C_{arom}), 55.63 (ArOC), 54.13 (d, $^2J_{CP}$ = 6.7 Hz, POC), 53.81 (d, $^2J_{CP}$ = 6.7 Hz, POC), 52.85 (d, $^1J_{CP}$ = 158.4 Hz, PC). ^{31}P-NMR (243 MHz, $CDCl_3$): δ 23.33. *Elem. Anal.* Calcd. for $C_{14}H_{18}NO_4PS \times {}^1/_6\ C_7H_8$: C, 53.16, H, 5.69, N, 4.09. Found: C, 53.21, H, 5.78, N, 4.26.

Dimethyl N-benzylamino(2-thienyl)methylphosphonate (**2f**): Yield = 57% (1.77 g) yellow oil. ^{1}H-NMR ($CDCl_3$, 600 MHz): 7.26–7.20 (m, PhH, H_5^{thioph}, 5H), 7.18–7.15 (m, PhH, 1H), 7.02–7.01 (m, H_3^{thioph}, 1H), 6.93 (dd, $^3J_{HH}$ = 4.8 and 3.6 Hz, H_4^{thioph}, 1H), 4.24 (d, $^2J_{PH}$ = 23.8 Hz, CHP, 1H), 3.82 (d, $^2J_{HH}$ = 13.8 Hz, CH_2Ph, 1H), 3.68 (d, $^3J_{PH}$ = 10.2 Hz, $POCH_3$, 3H), 3.57 (d, $^2J_{HH}$ = 13.8 Hz, CH_2Ph, 1H), 3.55 (d, $^3J_{PH}$ = 10.2 Hz, $POCH_3$, 3H). ^{13}C-NMR ($CDCl_3$, 150 MHz): δ 139.30 (d, $^2J_{CP}$ = 4.3 Hz, $C^2{}_{thioph}$), 138.99 (C_{arom}), 128.47 (d, *J* = 7.5 Hz, C_{thioph}), 128.38 (C_{arom}), 128.23 (C_{arom}), 127.28 (C_{arom}), 126.97 (d, *J* = 3.5 Hz, C_{thioph}), 125.68 (d, *J* = 3.3 Hz, C_{thioph}), 54.74 (d, $^1J_{CP}$ = 159.7 Hz, PC), 53.94 (d, $^2J_{CP}$ = 6.8 Hz, POC),

53.63 (d, $^2J_{CP}$ = 6.8 Hz, POC), 51.24 (d, $^3J_{CP}$ = 15.6 Hz, PCC). ^{31}P-NMR (243 MHz, $CDCl_3$): δ 23.98. *Elem. Anal.* Calcd. for $C_{14}H_{18}NO_3PS$: C, 54.01, H, 5.83, N, 4.50. Found: C, 53.77, H, 5.67, N, 4.46.

Dimethyl N-benzhydrylamino(2-thienyl)methylphosphonate (**2g**): Yield = 41% (1.59 g) yellow oil. ^{1}H-NMR ($CDCl_3$, 600 MHz): 7.44–7.34 (m, PhH, H_5^{thioph}, 6H), 7.32–7.22 (m, PhH, 5H), 7.07–7.06 (m, H_3^{thioph}, H_4^{thioph}, 2H), 4.91 (s, CH, 1H), 4.29 (d, $^2J_{PH}$ = 22.8 Hz, CHP, 1H), 3.91 (d, $^3J_{PH}$ = 10.8 Hz, $POCH_3$, 3H), 3.64 (d, $^3J_{PH}$ = 10.8 Hz, $POCH_3$, 3H), 2.57–2.48 (bs, NH, 1H). ^{13}C-NMR ($CDCl_3$, 150 MHz): δ 139.36 (d, $^2J_{CP}$ = 6.4 Hz, C^2_{thioph}), 139.73 (C_{arom}), 128.75 (C_{arom}), 128.56 (C_{arom}), 127.89 (C_{arom}), 127.17 (d, J = 7.9 Hz, C_{thioph}), 127.07 (d, J = 2.3 Hz, C_{thioph}), 125.72 (d, J = 3.4 Hz, C_{thioph}), 63.81 (d, $^3J_{CP}$ = 16.0 Hz, PCC), 54.12 (d, $^2J_{CP}$ = 6.7 Hz, POC), 53.53 (d, $^2J_{CP}$ = 6.7 Hz, POC), 53.19 (d, $^1J_{CP}$ = 161.5 Hz, PC). ^{31}P-NMR (243 MHz, $CDCl_3$): δ 24.32. *Elem. Anal.* Calcd. for $C_{20}H_{22}NO_3PS \times {}^1/_5\, C_7H_8$: C, 63.33, H, 5.86, N, 3.45. Found: C, 63.54, H, 5.69, N, 3.75.

Dibenzyl N-furfurylamino(2-thienyl)methylphosphonate (**2h**): Yield = 62% (2.81 g) yellow oil. ^{1}H-NMR ($CDCl_3$, 600 MHz): δ 7.36–7.32 (m, PhH, 10H), 7.25–7.23 (m, H_5^{thioph}, H_5^{fur}, 1H), 7.11–7.09 (m, H_3^{thioph}, 1H), 7.03–7.02 (m, H_4^{thioph}, 1H), 6.33–6.31 (m, H_3^{fur}, 1H), 6.15–6.14 (m, H_4^{fur}, 1H), 5.10–5.02 (Part AB of ABX system, $^3J_{PH}$ = 7.5 and 8.8 Hz, $^2J_{HH}$ = 11.7 Hz, $POCH_2Ph$, CHP, 2H), 5.00 (Part A of AMX system, $^3J_{PH}$ = 7.2 and $^2J_{HH}$ = 11.8 Hz, $POCH_2Ph$, 1H), 4.90 (Part M of AMX system, $^3J_{PH}$ = 8.2 and $^2J_{HH}$ = 11.8 Hz, $POCH_2Ph$, 1H), 4.43 (d, $^2J_{PH}$ = 19.3 Hz, CHP, 1H), 3.92 and 3.69 (2d, $^2J_{HH}$ = 14.6 Hz, CH_2Fur, 3H). ^{13}C-NMR (150 MHz, $CDCl_3$): δ 152.60 (C_{arom}), 142.12 (C_{arom}), 138.67 (d, J = 4.8 Hz, C_{arom}), 136.32 (d, $^3J_{CP}$ = 6.0 Hz, C_{arom}), 136.19 (d, $^3J_{CP}$ = 6.3 Hz, C_{arom}), 128.48 (d, J = 6.5 Hz, C_{arom}), 128.34 (C_{arom}), 128.28 (C_{arom}), 128.00 (C_{arom}), 127.84 (C_{arom}), 127.36 (C_{arom}), 127.30 (C_{arom}), 126.92 (d, J = 2.4 Hz, C_{arom}), 125.82 (d, J = 3.5 Hz, C_{arom}), 110.14 (C_{arom}), 107.92 (C_{arom}), 68.58 and 68.18 (2d, $^2J_{CP}$ = 40.6 Hz, POC), 55.11 (d, $^1J_{CP}$ = 159.8 Hz, PC), 43.72 (d, $^3J_{CP}$ = 16.4 Hz, NC). ^{31}P-NMR (243 MHz, $CDCl_3$): δ 22.19. *Elem. Anal.* Calcd. for $C_{24}H_{24}NO_4PS \times {}^1/_{10}\, CH_2Cl_2$: C, 62.66, H, 5.28, .N, 3.03. Found: C, 62.77, H, 5.03, N, 3.05.

3.2. Plant Growth Test of New Synthesized Compounds

The plant growth test of thiophene-derived aminophosphonates **2a–h** was performed in laboratory conditions following the OECD 208 Guideline Terrestrial Plants Growth Test for oat (*Avena sativa*) as a monocotyledonous plant and radish (*Raphanus sativus* L. subvar. radicula Pers.), a dicotyledonous plant.

According to the mentioned OECD 208 standard, the plant growth test of Compounds **2a–h** was carried out in sandy soil having the following parameters: granulometric composition of soil—77% sand, 14% dust and loam, organic carbon content of approx. 1.2%, pH (KCl) equal to 6.3.

Tests were carried out in polypropylene pots (diameter of 90 mm and capacity of 300 cm^3), which were filled with the control soil or with the soil mixed with the tested Compounds **2a–h** added at a specific concentration. Twenty identical seeds of each of the selected plant species were sown into the soil. Seeds originated from the same source. Plants were grown for 14 days under controlled conditions: a constant humidity content at the level required for the plants (70% field water capacity), light intensity (7000 lux), and temperature (20 ± 2 °C) were maintained. Then, seedlings were counted, and the dry as well as fresh weight of the plants was determined. The parts of plants above the soil surface were analyzed.

The performed plant growth was evaluated using a preliminary test; subsequently, the final test was conducted afterward based on the obtained screening results. Preliminary tests were performed to determine the range of concentrations of compounds affecting the soil quality; therefore, the used concentrations of Samples **2a–h** were as follows: 0 mg (control), 1 mg, 10 mg, 100 mg, and 1000 mg/kg of soil dry weight.

As already mentioned, based on the obtained results from preliminary test, the final, more precise test to find values of the no observed effect concentration (NOEC) and the lowest observed effect concentration (LOEC) of the compounds under study **2a–h** was performed.

The evaluation of phytotoxicity of the studied aminophosphonates **2a–h** at applied concentrations was made by comparing the germination and (dry and fresh) weight of control plant sprouts (seedlings) with germination and of (dry and fresh) plants sprouts grown in the soil with an admixture of given amounts of the tested compounds.

The dry weights of tested plants were measured after drying at 75 °C until the constant weight.

The visual evaluation of phytotoxicity of aminophosphonates **2a–h** at applied concentrations was performed by digital photography. Obtained pictures were analyzed in terms of any type of damage to tested plants, such as their growth inhibition, chlorosis, and necrosis. Tests were carried out three times for each sample.

The significance of the obtained results was evaluated using the analysis of variance (ANOVA). The least significant difference (LSD) values at a confidence level of 95% were computed using the Tukey test.

4. Conclusions

To conclude, the toxicity of *N*-methylphenyl aminophosphonates **2a–c** was diagnosed to be promising as a potential soil-applied herbicide, especially for their selectivity—they are evidently toxic for dicotyledonous radish and not so harmful for monocotyledonous oat. Moreover, their herbicidal efficiency is stronger for the *N*-methylphenyl aminophosphonates substituted with methyl group at position *ortho* (**2a**) and *meta* (**2b**) in the phenyl ring.

Dimethyl *N*-benzylamino(2-thienyl)methylphosphonate (**2f**) exhibited non-selective, harmful effects against both tested plants.

Among *N*-methoxyphenyl aminophosphonates **2d** and **2e**, the latter showed an inhibitive effect especially against dicotyldenous plants. Its impact, however, was significantly weaker than impact of **2a,b**. This is to point out that the sole structural difference between *N*-methoxyphenyl aminophosphonates **2d,e** and *N*-methylphenyl aminophosphonates **2a–c** is the substituent of the phenyl moiety, *i.e.*, the methoxy and methyl group, respectively, which indicates that the type and position of those groups in the phenyl ring play a key role in the toxicity effect against tested plants.

Among all tested samples, dibenzyl *N*-furfurylamino(2-thienyl)methylphosphonate (**2h**) was found to be the most toxic for radish and oat but without significant selectivity. The investigation of Compounds **2a,b**, **2f**, and **2h**, such as their potential applications and mechanistic approaches, will be continued.

Some thiophene-derived aminophosphonic derivatives were found to have various biological properties (*vide supra*); therefore, important microbiological, herbicidal, and cytotoxic properties of the studied aminophosphonic systems bearing thiophene moiety **2a–h** cannot be excluded. If so, the synthesis of such compounds for potential agrochemical or pharmacological application must be taken into consideration from the environmental protection point of view. Our results call attention to the necessity of further phytotoxicological investigation of any new synthesized aminophosphonic derivatives bearing thiophene moiety.

Acknowledgments: Studies were founded by the National Centre of Science of Polish State (NCN), grant no. 2014/13/B/NZ9/02418, where funds for publishing in open-access mode were scheduled. Special thanks are addressed to Paulina Tomczyk and Justyna Biernacka, former M.Sc. students at the University of Łódź, for their fruitful assistance in the preparation of the studied compounds.

Author Contributions: J.L. and P.R. conceived and designed the experiments; A.M., Z.M., M.M., and D.R. performed the experiments; J.L. and P.R. analyzed the data; J.L. and P.R. wrote the paper.

References

1. Andersen, S.M.; Hertz, P.B.; Holst, T.; Bossi, R.; Jacobsen, C.S. Mineralisation studies of C14-labelled metsulfuron-methyl, tribenuron-methyl, chlorsulfuron and thifensulfuronmethyl in one Danish soil and groundwater sediment profile. *Chemosphere* **2001**, *45*, 775–782. [CrossRef]
2. Polati, S.; Bottaro, M.; Frascarolo, P.; Gosetti, F.; Gianotti, V.; Gennaro, M.C. HPLC-UV and HPLC-MSn multiresidue determination of amidosulfuron, azimsulfuron, nicosulfuron, rimsulfuron, thifensulfuron methyl, tribenuron methyl and azoxystrobin in surface waters. *Anal. Chim. Acta* **2006**, *579*, 146–151. [CrossRef] [PubMed]
3. EFSA (European Food Safety Authority). Conclusion on the peer review of the pesticide risk assessment of the active substance thifensulfuron-methyl. *EFSA J.* **2015**, *13*, 4201. [CrossRef]
4. Rosenkrantz, R.Y.; Baun, A.; Kusk, O. Growth inhibition and recoveryof *Lemna gibba* after pulse exposure to sulfonylurea herbicides. *Ecotoxicol. Environ. Saf.* **2013**, *89*, 89–94. [CrossRef] [PubMed]
5. Beckie, H.J.; Tardif, F.J. Herbicide cross resistance in weeds. *Crop Prot.* **2012**, *35*, 15–28. [CrossRef]
6. Brown, D. *2,4-D and Dicamba-Resistant Crops and Their Implications for Susceptible Non-Target Crops*; Michigan State University Extension: East Lansing, MI, USA, 2013.
7. Filipe, O.M.S.; Santos, S.A.O.; Domingues, M.R.M.; Vidal, M.M.; Silvestre, A.J.D.; Neto, C.P.; Santos, E.B.H. Photodegradation of the fungicide thiram in aqueous solutions. Kinetic studies and identification of the photodegradation products by HPLC-MS/MS. *Chemosphere* **2013**, *91*, 993–1001. [CrossRef] [PubMed]
8. Sanchirico, R.; Pinto, G.; Pollio, A.; Cordella, M.; Cozzani, V. Thermal degradation of Fenitrothion: Identification and eco-toxicity of decomposition products. *J. Hazard. Mater.* **2012**, *199*, 390–400. [CrossRef] [PubMed]
9. Baghestani, M.A.; Zand, E.; Soufizadeh, S.; Jamali, M.; Mighany, F. Evaluation of sulfosulfuron for broadleaved and grass weed control in wheat (*Triticum aestivum* L.) in Iran. *Crop Prot.* **2007**, *26*, 1385–1389. [CrossRef]
10. Zhao, W.; Xu, L.; Li, D.; Li, X.; Wang, C.; Zheng, M.; Pan, C.; Qiu, L. Biodegradation of thifensulfuron-methyl by *Ochrobactrum* sp. in liquid medium and soil. *Biotechnol. Lett.* **2015**, *37*, 1385–1392. [CrossRef] [PubMed]
11. Cessna, A.J.; Donald, D.B.; Bailey, J.; Waiser, M. Persistence of the Sulfonylurea Herbicides Sulfosulfuron, Rimsulfuron, and Nicosulfuron in Farm Dugouts (Ponds). *J. Environ. Qual.* **2015**, *44*, 1948–1955. [CrossRef] [PubMed]
12. Elliott, J.A.; Cessna, A.J. Variability in the distribution and dissipation of the herbicide thifensulfuron-methyl in a prairie wetland. *J. Soil Water Conserv.* **2014**, *69*, 151–159. [CrossRef]
13. Aziz, S.; Dumas, S.; el Azzouzi, M.; Sarakham, M.; Chovelon, J.-M. Photophysical and photochemical studies of thifensulfuron-methyl herbicide in aqueous solution. *J. Photochem. Photobiol. A Chem.* **2010**, *209*, 210–218. [CrossRef]
14. Lazartigues, A.; Thomas, M.; Banas, D.; Brun-Bellut, J.; Cren-Olivé, C.; Feidt, C. Accumulation and half-lives of 13 pesticides in muscle tissue of freshwater fishes through food exposure. *Chemosphere* **2013**, *91*, 530–535. [CrossRef] [PubMed]
15. Brown, H.M.; Joshi, M.M.; van Ailien, T.; Carski, T.H.; Dulka, J.J.; Patrick, M.C.; Reiser, R.W.; Livingston, R.S.; Doughty, J. Degradation of thifensulfuron methyl in soil: Role of microbial carboxyesterase activity. *J. Agric. Food Chem.* **1997**, *45*, 955–961. [CrossRef]
16. Boduszek, B. Synthesis and Biological Activity of Heterocyclic Amino-phosphonates. *Phosphorus Sulfur Silicon Relat. Elem.* **1999**, *144*, 433–436. [CrossRef]
17. Janardhan Rao, A.; Visweswara Rao, P.; Koteswara Rao, V.; Mohan, C.; Naga Raju, C.; Suresh Reddy, C. Microwave Assisted One-pot Synthesis of Novel α-Amino-phosphonates and Their Biological Activity. *Bull. Korean Chem. Soc.* **2010**, *31*, 1863–1868.
18. Maheswara Rao, K.U.; Namkoong, S.; Yu, H.-C.; Park, J.; Chung, C.-M.; Oh, S.Y. Green Synthesis and Biological Evaluation of New Di-aminophosphonate Derivatives as Cytotoxic Agents. *Arch. Pharm. Chem. Life Sci.* **2013**, *346*, 851–859. [CrossRef] [PubMed]

19. Rao Devineni, S.; Doddaga, S.; Donka, R.; Raju Chamarthi, N. $CeCl_3 \cdot 7H_2O$-SiO_2: Catalyst promoted microwave assisted neat synthesis, antifungal and antioxidant activities of α-diaminophosphonates. *Chin. Chem. Lett.* **2013**, *24*, 759–763. [CrossRef]
20. Rasheed, S.; Venkataramana, K.; Chandrasekhar, K.; Fareeda, G.; Naga Raju, C. Ultrasound-Assisted Synthesis of Novel-Aminophosphonates and Their Biological Activity. *Arch. Pharm. Chem. Life Sci.* **2012**, *345*, 294–301. [CrossRef] [PubMed]
21. Banik, A.; Batta, S.; Bandyopadhyay, D.; Banik, B.K. A Highly Efficient Bismuth Salts-Catalyzed Route for the Synthesis of α-Aminophosphonates. *Molecules* **2010**, *15*, 8205–8213. [CrossRef] [PubMed]
22. Boduszek, B. An efficient synthesis of 1-aminophosphonic acids and esters bearing heterocyclic moiety. *Phosphorus Sulfur Silicon Relat. Elem.* **1995**, *104*, 63–70. [CrossRef]
23. Demir, A.S.; Tanyeli, C.; Sesenoglu, O.; Demic, S. A Simple Synthesis of 1-Amino-phosphonic Acids from 1-Hydroxyiminophosphonates with $NaBH_4$ in the Presence of Transition Metal Compounds. *Tetrahedron Lett.* **1996**, *37*, 407–410. [CrossRef]
24. Manjula, A.; Vittal Rao, B.; Neelakantan, P. One-Pot Synthesis of α-aminophosphonates: An Inexpensive Approach. *Synth. Commun.* **2003**, *33*, 2963–2969. [CrossRef]
25. Mu, X.-J.; Lei, M.-Y.; Zoua, J.-P.; Zhang, W. Microwave-assisted solvent-free and catalyst-free Kabachnik–Fields reactions for α-amino phosphonates. *Tetrahedron Lett.* **2006**, *47*, 1125–1127. [CrossRef]
26. Saito, B.; Egami, H.; Katsuki, T. Synthesis of an Optically Active Al(salalen) Complex and Its Application to Catalytic Hydrophosphonylation of Aldehydes and Aldimines. *J. Am. Chem. Soc.* **2007**, *129*, 1978–1986. [CrossRef] [PubMed]
27. Cherkasov, R.A.; Galkin, V.I. The Kabachnik–Fields reaction: Synthetic potential and the problem of the mechanism. *Russ. Chem. Rev.* **1998**, *67*, 857–882. [CrossRef]
28. Keglevich, G.; Bálint, E. The Kabachnik–Fields reaction: Mechanism and synthetic use. *Molecules* **2012**, *17*, 12821–12835. [CrossRef] [PubMed]
29. Kafarski, P.; Górniak, M.G.; Andrasiak, I. Kabachnik–Fields reaction under green conditions-A critical overview. *Curr. Green Chem.* **2015**, *5*, 218–222. [CrossRef]
30. Kafarski, P.; Lejczak, B. Biological Activity of Aminophosphonic Acids. *Phosphorus Sulfur Silicon Relat. Elem.* **1991**, *63*, 193–215. [CrossRef]
31. Forlani, G.; Berlicki, Ł.; Duò, M.; Dzięzdzioła, G.; Giberti, S.; Bertazzini, M.; Kafarski, P. Synthesis and Evaluation of Effective Inhibitors of Plant δ^1-Pyrroline-5-carboxylate Reductase. *J. Agric. Food Chem.* **2013**, *61*, 6792–6798. [CrossRef] [PubMed]
32. Occhipinti, A.; Berlicki, Ł.; Giberti, S.; Dzięzdzioła, G.; Kafarski, P.; Forlani, G. Effectiveness and mode of action of phosphonate inhibitors of plant glutamine synthetase. *Pest Manag. Sci.* **2010**, *66*, 51–58. [CrossRef] [PubMed]
33. Giberti, S.; Bertazzini, M.; Liboni, M.; Berlicki, Ł.; Kafarski, P.; Forlani, G. Phytotoxicity of aminobisphosphonates targeting both δ^1-pyrroline-5-carboxylate reductase and glutamine synthetase. *Pest Manag. Sci.* **2016**. [CrossRef] [PubMed]
34. Matusiak, A.; Lewkowski, J.; Rychter, P.; Biczak, R. Phytotoxicity of New Furan-derived Aminophosphonic Acids, *N*-Aryl Furaldimines and 5-Nitrofuraldimine. *J. Agric. Food Chem.* **2013**, *61*, 7673–7678. [CrossRef] [PubMed]
35. Klimczak, A.A.; Kuropatwa, A.; Lewkowski, J.; Szemraj, J. Synthesis of *N*-aryl, furan-derived aminophosponates and studies of their *in vitro* cytotoxicity against esophageal cancer cells. *Med. Chem. Res.* **2013**, *21*, 852–860. [CrossRef]

4

Practical and Efficient Synthesis of α-Aminophosphonic Acids Containing 1,2,3,4-Tetrahydroquinoline or 1,2,3,4-Tetrahydroisoquinoline Heterocycles

Mario Ordóñez [1,*], Alicia Arizpe [2], Fracisco J. Sayago [2], Ana I. Jiménez [2] and Carlos Cativiela [2,*]

[1] Centro de Investigaciones Químicas-IICBA, Universidad Autónoma del Estado de Morelos, Av. Universidad 1001, Cuernavaca 62209, Morelos, Mexico

[2] Departamento de Química Orgánica, Universidad de Zaragoza-CSIC, ISQCH, Zaragoza 50009, Spain; aliceiw@gmail.com (A.A.); jsayago@unizar.es (F.J.S.); anisjim@unizar.es (A.I.J.)

* Correspondence: palacios@uaem.mx (M.O.); cativiela@unizar.es (C.C.)

Academic Editor: György Keglevich

Abstract: We report here a practical and efficient synthesis of α-aminophosphonic acid incorporated into 1,2,3,4-tetrahydroquinoline and 1,2,3,4-tetrahydroisoquinoline heterocycles, which could be considered to be conformationally constrained analogues of pipecolic acid. The principal contribution of this synthesis is the introduction of the phosphonate group in the *N*-acyliminium ion intermediates, obtained from activation of the quinoline and isoquinoline heterocycles or from the appropriate δ-lactam with benzyl chloroformate. Finally, the hydrolysis of phosphonate moiety with simultaneous cleavage of the carbamate afforded the target compounds.

Keywords: α-aminophosphonic acids; *N*-acyliminium ions; conformationally constrained

1. Introduction

There is a continuously growing interest in the development of new peptidomimetics, compounds that mimic the bioactive conformation and action of therapeutic peptides while possessing greater bioavailability and stability and less undesirable effects. In this regard, the incorporation of rigid unusual secondary α-amino acids, where the nitrogen is involved in a ring, may result in significant consequences for the conformation of peptidomimetics as synthetic tools for drug discovery [1,2]. Some of the most important molecules are the 1,2,3,4-tetrahydroquinoline-2-carboxylic acid **1** [3,4] and 1,2,3,4-tetrahydroisoquinoline-1-carboxylic acid **2** [5,6], which are conformationally constrained analogues of unnatural pipecolic acid, and 1,2,3,4-tetrahydroisoquinoline-3-carboxylic acid (Tic) **3** [7,8], which is considered a conformationally constrained analogue of phenylalanine (Phe). These compounds are important unnatural α-amino acids, and they are used as key intermediates in organic synthesis for the preparation of biologically active compounds. Therefore, much effort has been dedicated to the preparation of these compounds.

On the other hand, the α-aminoalkylphosphonic acids are probably the most important analogues of the α-amino acids, obtained by isosteric substitution of the planar and less bulky carboxylic acid (CO_2H) by a tetrahedral phosphonic acid functionality (PO_3H_2). This class of compounds is currently attracting interest in organic and medicinal chemistry, due to their important biological and pharmacological properties [9–12]. The great importance of this type of compound has prompted organic chemists to report numerous procedures for their racemic or stereoselective synthesis [13–18], principally using the diastereoselective and enantioselective

Pudovik [19,20] and Kabachnik–Fields [21–25] reactions for acyclic α-aminoalkylphosphonates, and through *N*-acyl iminium ions for the synthesis of cyclic derivatives [26–34]. However, to the best of our knowledge, the synthesis of 1,2,3,4-tetrahydroquinoline-2-phosphonic acid **4** and 1,2,3,4-tetrahydroisoquinoline-1-phosphonic acid **5** analogues [35–38] has not yet been described in the literature, whereas the synthesis of 1,2,3,4-tetrahydroisoquinoline-3-phosphonic acid **6** has been recently described by our research group [39,40] (Figure 1).

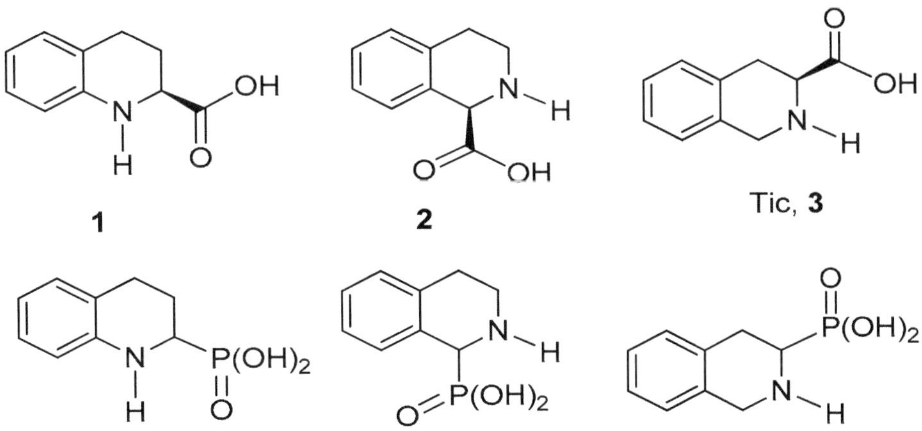

1 **2** Tic, **3**

4 **5** Tic^P, **6**

Figure 1. α-Amino acids characterized by a tetrahydroquinoline or tetrahydroisoquinoline heterocyles and their α-amino phosphonic analogues.

Considering the high value of these non-coded compounds in connection with our current research interest in the synthesis of novel conformationally restricted α-aminophosphonic acids [26–28,41–43], we now report herein the practical and efficient synthesis of α-aminophosphonic acids incorporating 1,2,3,4-tetrahydroquinoline **4**, and 1,2,3,4-tetrahydroisoquinoline rings **5** and **6**, which could be considered to be conformationally constrained analogues of pipecolic acid. The principal contribution of this work is the regioselective introduction of the phosphonate group in the *N*-acyliminium ions derivatives obtained from activation of the quinoline and isoquinoline nucleous or from the appropriate δ-lactams with benzyl chloroformate.

2. Results and Discussion

For the synthesis of 1,2,3,4-tetrahydroquinoline-2-phosphonic acid **4**, we proposed the *N*-acyliminium ions as suitable precursors, considering that the well-known reduction of the carbonyl group of lactams such as the 3,4-dihydro-2(1*H*)-quinolinone **7**, and their subsequent transformation into a *N*-acyliminium ion, is one of the methods that allows the incorporation of the phosphonate functionality into the α position of the nitrogen atom such as we have previously described [26–28]. These proposals prompted us to explore further application of the *N*-acyliminium strategy for the synthesis of the target compound. In this context, the commercially available 3,4-dihydro-2(1*H*)-quinolinone **7** was reacted with lithium bis(trimethylsilyl)amide (LiHMDS) as a base and benzyl chloroformate in THF at −78 °C, to obtain the *N*-Cbz-3,4-dihydro-2(1*H*)-quinolinone **8** in 91% yield. Reduction of the carbonyl group in **8** with diisobutylaluminium hydride (DIBAL-H) and subsequent reaction with methanol and catalytic amounts of pyridinium *p*-toluenesulfonate (PPTS), gave the methoxyaminal **9**, which was treated immediately with trimethyl phosphite in the presence of $BF_3{\cdot}OEt_2$, obtaining the dimethyl *N*-Cbz-1,2,3,4-tetrahydroquinoline-2-phosphonate **11** in 62% via the *N*-acyliminium ion **10** in 62% yield from **8**. Treatment of this *N*-Cbz-protected phosphonate with a 33% solution of hydrogen bromide in acetic acid, afforded the 1,2,3,4-tetrahydroquinoline-2-phosphonic acid **4** as hydrobromide in quantitative yield (Scheme 1).

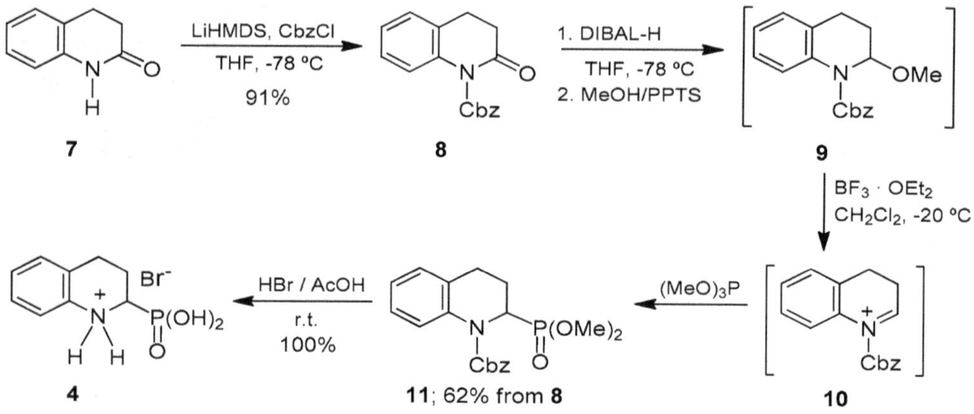

Scheme 1. Synthesis of 1,2,3,4-tetrahydroquinoline-2-phosphonic acid **4** from lactam **7**.

The literature has described the hydrophosphonylation of quinoline derivatives to obtain the 1,2-dihydroquinolin-2-ylphosphonate and 2,4-diphosphono-1,2,3,4-tetrahydroquinoline derivatives using activating agents [44–48]. These reactions proceed via quinolinium ions that can be viewed as counterparts of the above-mentioned iminium species. Thus, the addition of an acyl chloride to quinoline can be considered as a way to generate the iminium ion necessary for subsequent regioselective incorporation of the phosphonate functionality α to the nitrogen atom.

Based on this precedent, we decided to compare the efficiency of the *N*-acylquinolinium salts as intermediates in the synthesis of α-aminophosphonic acid **4**. For this purpose, benzyl chloroformate was added to quinoline and the *N*-Cbz-quinolinium chloride formed **12** was reacted with trimethyl phosphite in acetonitrile at 50 °C, obtaining the dimethyl *N*-Cbz-1,2-dihydroquinoline-2-phosphonate derivative **13** in 74% yield from quinolone. The diphosphonylation products at the 2- and 4-positions of the quinoline ring were not detected. The regioselectivity observed in the addition of trimethyl phosphite to the *N*-acylquinolinium ion **12** may be attributed to the electron-withdrawing character of the benzyloxycarbonyl group, which increases the electrophilic nature of the carbon adjacent to the nitrogen atom. Additionally, the benzyloxycarbonyl group was selected for its easy incorporation and removal under mild conditions. Thus, the catalytic hydrogenation resulted in simultaneous removal of the Cbz group and reduction of the double bond at positions 3,4 to afford the α-aminophosphonate **14**, which, by treatment with hydrogen bromide in acetic acid, gave the 1,2,3,4-tetrahydroquinoline-2-phosphonic acid **4** as hydrobromide. The latter two steps proceeded quantitatively and obviated the need for purification (Scheme 2).

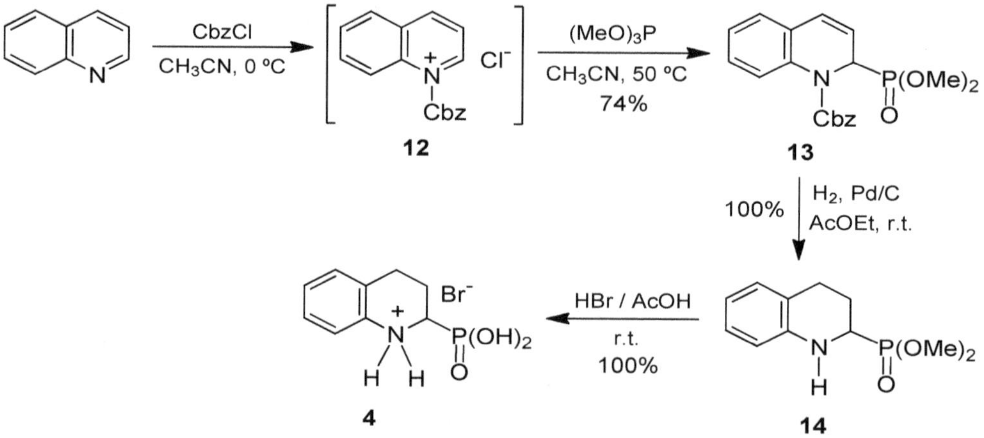

Scheme 2. Synthesis of 1,2,3,4-tetrahydroquinoline-2-phosphonic acid **4** from quinoline.

The 1,2,3,4-tetrahydroquinoline-2-phosphonic acid **4** was obtained in 74% overall yield from quinoline following the synthetic strategy in Scheme 2, which compares favorably with the 56% global yield achieved in Scheme 1, when starting from quinolinone **1**. Accordingly, the superior global yield, together with the much lower price of quinoline in comparison with 3,4-dihydro-2(1*H*)-quinolinone **7**, makes the methodology in Scheme 2 more advantageous than the lactam-based one for the preparation of α-aminophosphonic acid **4**. From an operational viewpoint, both routes required few purification steps—by column chromatography—of intermediate compounds.

We next addressed the preparation of 1,2,3,4-tetrahydroisoquinoline-1-phosphonic acid **5**. Similarly to that described above for the quinoline counterpart, the preparation of *N*-substituted-1,2-dihydroisoquinoline-1-phosphonates through the isoquinolinium salts formed by reaction of isoquinoline with acyl chlorides and subsequent addition of trialkyl phosphites has been described [49,50]. Thus, the synthesis of dimethyl *N*-Cbz-1,2-dihydroisoquinoline-1-phosphonate should be straightforward, as shown above for the preparation of dimethyl *N*-Cbz-1,2-dihydroquinoline-2-phosphonate **13**.

On the other hand, the lactam to be used as a starting product for the preparation of the target amino acid **5** is not easily available from commercial sources in this case. This fact together with the advantageous preparation of amino acid **4** via the formation of quinolinium salt prompted us to undertake the synthesis of 1,2,3,4-tetrahydroisoquinoline-1-phosphonic acid following an analogous synthetic route.

Thus, using the reaction conditions described above, the isoquinoline was reacted with benzyl chloroformate followed by the addition of trimethyl phosphite in acetonitrile at 50 °C. However, under these reaction conditions, the desired compound **16** was not obtained. It is noteworthy that during the addition of benzyl chloroformate to isoquinoline, a yellow solid was formed, which was later identified as the isoquinolinium salt **15**. This compound remained unaltered upon addition of trimethyl phosphite so that the formation of the expected 1,2-dihydroisoquinoline-1-phosphonate **16** did not take place. We reasoned that if the trimethyl phosphite was present in the reaction medium before benzyl chloroformate was added, the isoquinolinium ion formed **16** would immediately react with it, thus preventing precipitation. To our delight, when the order of addition of these reagents was exchanged (that is, trimethyl phosphite prior to benzyl chloroformate), the dimethyl *N*-Cbz-1,2-dihydroisoquinoline-1-phosphonate **16** was obtained at 92% yield. In the next step, we carried out the catalytic hydrogenation of the double bond in **16** under an atmospheric pressure of hydrogen gas and using Pd/C as a catalyst, to generate the tetrahydroisoquinoline moiety. However, the Cbz group in **16** was readily eliminated at room temperature, whereas the double bond at positions 3,4 was only partially hydrogenated even after long reaction times. As a consequence, mixtures of the desired tetrahydroisoquinoline derivative **17** and the analogous dihydroisoquinoline **18** were obtained. Attempts to improve this result by changing the solvent as well as by increasing the reaction temperature or hydrogen gas pressure proved unsuccessful and a mixture of **17** and **18** was always obtained (Scheme 3).

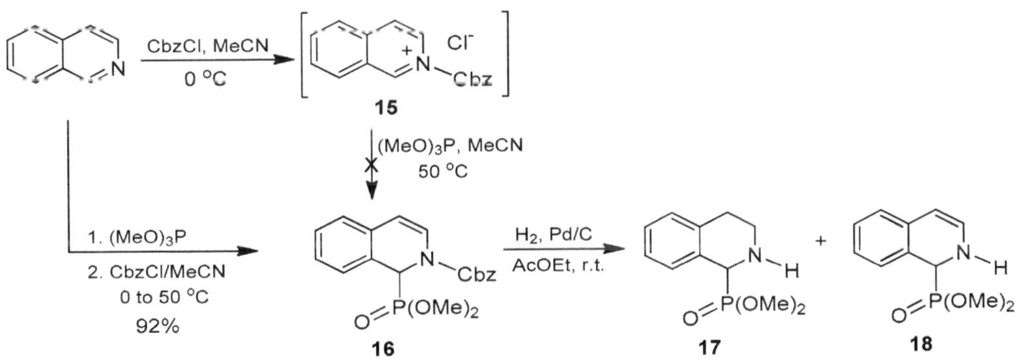

Scheme 3. First attempt to synthetize 1,2,3,4-tetrahydroisoquinoline-1-phosphonic acids derivatives.

To circumvent this problem, we synthesized the compound analogous to **16** that bears the less acid-sensitive diethyl phosphonate group **19**. A procedure identical to that established for the dimethyl derivative but using triethyl phosphite as the phosphorus source provided the desired diethyl *N*-Cbz-1,2-dihydroisoquinoline-1-phosphonate **19** in 98% yield from isoquinoline. Taking into account the fact that compound **19** can be considered not only as an *N*-acyl-α-aminophosphonate, but also as an enamide, we decided to perform the reduction of the double bond using trifluoroacetic acid and triethylsilane as the reducing agent, following the conditions described by Jacobsen et al. [51] for cyclic enamides. In this case, the diethyl *N*-Cbz-1,2,3,4-tetrahydroisoquinoline-1-phosphonate **20** was readily obtained in quantitative yield by reaction with triethylsilane and trifluoroacetic acid. Subsequent cleavage of the protecting groups in **20** by treatment with hydrogen bromide in acetic acid, afforded the 1,2,3,4-tetrahydroisoquinoline-1-phosphonic acid **5** as hydrobromide in quantitative yield. This compound was isolated in 97% overall yield from isoquinoline (Scheme 4).

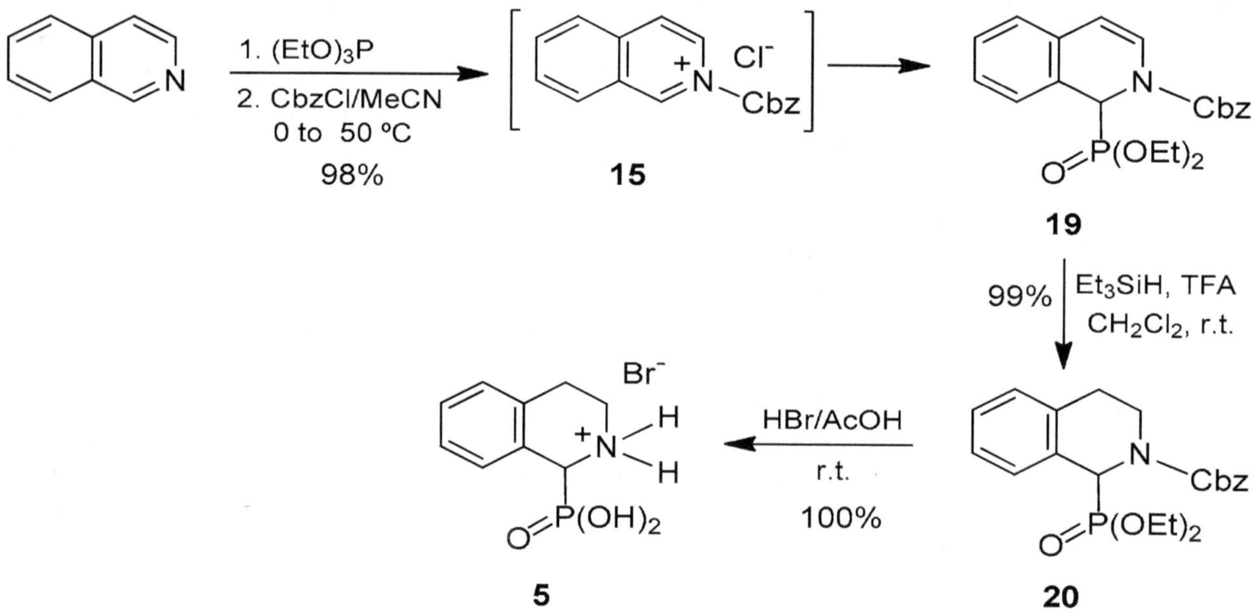

Scheme 4. Synthesis of 1,2,3,4-tetrahydroisoquinoline-1-phosphonic acid **5** from isoquinoline.

Finally, we focused on the preparation of 1,2,3,4-tetrahydroisoquinoline-3-phosphonic acid **6**, that is, the phosphonic analogue of Tic (Tic^P). This compound exhibits a tetrahydroisoquinoline core, as does the aminophosphonic acid **5**. However, in the case of Tic^P, the phosphonate group is at the 3-position. However, the isoquinoline selectively provides the phosphonate group at the 1-position of the ring. Therefore, we believe that the generation of a suitable iminium ion for the introduction of the phosphonate moiety at the desired 3-position, the lactam precursor was the only route to consider in this case. It should be noted that, following this strategy, the position of the carbonyl group in the starting lactam determines with complete regiocontrol the introduction of the phosphonate substituent, which is a distinctive advantage of this methodology. The adequate lactam **21** used in the synthesis of the target amino acid was easily prepared following reported procedures that involved reaction of 2-phenylacetyl chloride with aqueous ammonia to give 2-phenylacetamide [52] and subsequent condensation with formaldehyde [53]. Lactam **21** was readily transformed into 1,2,3,4-tetrahydroisoquinoline-3-phosphonic acid (Tic^P, **6**) following the synthetic route strictly similar to that reported above for the preparation of the α-aminophosphonic acid **4** from lactam **7** (Scheme 1). The 1,2,3,4-tetrahydroisoquinoline-3-phosphonic acid **6** was thus obtained as hydrobromide in 52% global yield from the starting substrate **21**, with isolation and purification of only two synthetic intermediates **22, 25** (Scheme 5).

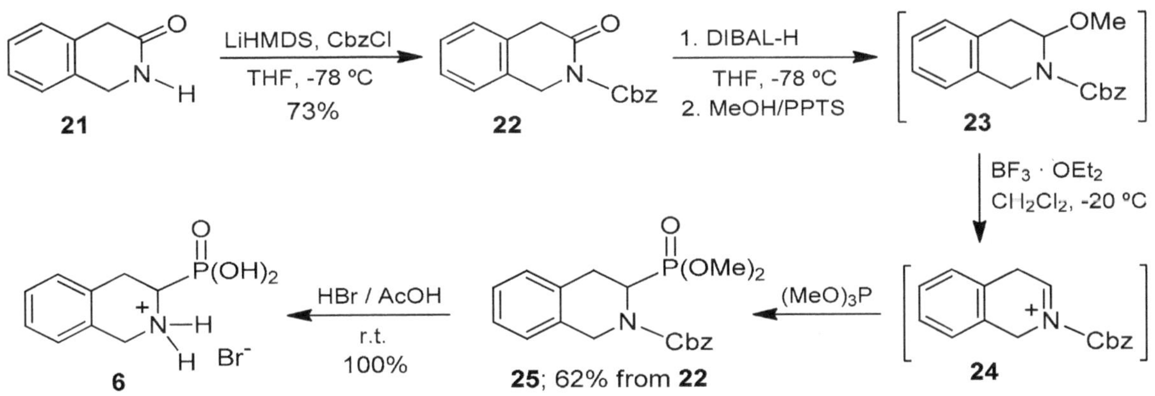

Scheme 5. Synthesis of 1,2,3,4-tetrahydroisoquinoline-3-phosphonic acid, Tic^P **6**.

The results obtained show the successful use of the lactam **21** to generate the key intermediate *N*-acyliminium ion, which by phosphonylation, afforded the desired diethyl *N*-Cbz-1,2,3,4-tetrahydroisoquinoline-3-phosphonate **25**, demonstrating thus the versatility of this methodology in the preparation of α-aminophosphonic acids structurally related to pipecolic acid.

3. Materials and Methods

3.1. General

All reagents were used as received from commercial suppliers (Sigma-Aldrich Chemie GmbH, Buchs, Switzerland) without further purification. Thin-layer chromatography (TLC) was performed on Macherey-Nagel Polygram® SIL G/UV$_{254}$ (Macherey-Nagel, Duren, Germany) precoated silica gel polyester plates. The products were visualized by exposure to UV light (254 nm), iodine vapour or ethanolic solution of phosphomolybdic acid. Column chromatography was performed using 60 M (0.04–0.063 mm) silica gel from Macherey-Nagel. Melting points were determined on a Gallenkamp apparatus (Weiss Gallenkamp, Loughborough, UK). IR spectra were registered on a Nicolet Avatar 360 FTIR spectrophotometer (Thermo Electron Corporation, Madison, WI, USA); ν_{max} is given for the main absorption bands. ^{1}H-, ^{13}C- and ^{31}P-NMR spectra were recorded on Bruker AV-400 or AV-300 instruments (Bruker BioSpin GmbH, Rheinstetten, Germany) at room temperature, unless otherwise indicated, using the residual solvent signal as the internal standard; chemical shifts (δ) are expressed in ppm and coupling constants (*J*) in Hertz. High-resolution mass spectra were obtained on a Bruker Microtof-Q spectrometer. Compound **22** was prepared by reaction of 2-phenylacetamide [52] with formaldehyde [53]. ^{1}H-, ^{13}C- and ^{31}P- NMR spectra of all final compounds are showed in the supplementary material (Figures S1–S34).

3.2. Synthesis of N-Benzyloxycarbonyl-3,4-dihydro-2-quinolinone **8**

A 1 M solution of lithium bis(trimethylsilyl)amide in tetrahydrofuran (2.80 mL, 2.80 mmol) was slowly added to a solution of 3,4-dihydro-2(1*H*)-quinolinone **7** (400 mg, 2.72 mmol) in anhydrous tetrahydrofuran (10 mL) kept at −78 °C under argon. After 30 min, benzyl chloroformate (0.39 mL, 464 mg, 2.72 mmol) was added dropwise and stirring was continued for additional 3 h. The reaction mixture was then treated with saturated aqueous ammonium chloride (10 mL) and allowed to warm to room temperature. The two layers were separated and the aqueous phase was extracted with dichloromethane (2 × 20 mL). The combined organic extracts were dried, filtered, and concentrated. Purification by column chromatography (eluent:hexanes/ethyl acetate 4:1) afforded **8** as a colourless oil (694 mg, 2.47 mmol, 91% yield). IR (neat) ν_{max} 1773, 1699 cm^{-1}. ^{1}H-NMR (400 MHz, $CDCl_3$): δ = 7.48–7.33 (m, 5H, Ar), 7.20–7.13 (m, 2H, Ar), 7.10–7.04 (m, 1H, Ar), 6.92–6.88 (m, 1H, Ar), 5.41 (s, 2H, CH_2Ph), 2.98–2.92 (m, 2H, H-4), 2.72–2.67 (m, 2H, H-3) ppm. ^{13}C-NMR (100 MHz, $CDCl_3$): δ = 169.90 (CO), 153.46 (COO), 136.94 (Ar), 134.52 (Ar), 128.85 (Ar), 128.76 (Ar), 128.75 (Ar), 127.95 (Ar),

127.39 (Ar), 126.95 (Ar), 124.78 (Ar), 118.65 (Ar), 70.05 (CH_2Ph), 33.05 (C-3), 25.55 (C-4) ppm. HRMS (ESI): calcd. for $C_{17}H_{15}NNaO_3$ $[M + Na]^+$ 304.0944; found 304.0947.

3.3. *Synthesis of Dimethyl N-benzyloxycarbonyl-1,2,3,4-tetrahydroquinoline-2-phosphonate* **11**

A 1 M solution of diisobutylaluminium hydride in hexanes (2.40 mL, 2.40 mmol) was slowly added to a solution of **8** (448 mg, 1.59 mmol) in anhydrous tetrahydrofuran (8 mL) kept at −78 °C under argon. After stirring at this temperature for 2 h, the reaction was treated with saturated aqueous sodium acetate (5 mL) and warmed to room temperature. A 3:1 mixture of diethyl ether and saturated aqueous ammonium chloride (16 mL) was then added and the resulting mixture was stirred at room temperature until a suspension was formed. The solid was filtered off under reduced pressure and washed with diethyl ether (2 × 10 mL). The organic layer was separated and the aqueous phase was extracted with diethyl ether (2 × 20 mL). The combined organic extracts were washed with brine (20 mL), dried, filtered, and evaporated to provide the hemiaminal as an oil. It was dissolved in methanol (6 mL) and treated with pyridinium *p*-toluenesulfonate (40 mg, 0.16 mmol). After stirring at room temperature for 2 h, triethylamine (0.10 mL, 74 mg, 0.73 mmol) was added. The solvent was evaporated and the crude methoxyaminal **9** thus obtained was dissolved in anhydrous dichloromethane (7 mL) and kept under argon. Trimethyl phosphite (0.19 mL, 198 mg, 1.59 mmol) was added and the resulting solution was cooled to −20 °C. Boron trifluoride-diethyl ether (0.20 mL, 226 mg, 1.59 mmol) was added dropwise and the reaction mixture was slowly warmed to room temperature and stirred for 12 h. After quenching with water (2 mL), the two layers were separated and the aqueous phase was extracted with dichloromethane (2 × 10 mL). The combined organic extracts were dried, filtered, and concentrated. Purification by column chromatography (eluent:ethyl acetate/hexanes 4:1) afforded **11** as a colourless oil (372 mg, 0.99 mmol, 62% yield). IR (neat) ν_{max} 1702, 1279, 1029 cm^{-1}. ^1H-NMR (400 MHz, $CDCl_3$): δ = 7.53–7.46 (m, 1H, Ar), 7.37–7.28 (m, 5H, Ar), 7.22–7.16 (m, 1H, Ar), 7.14–7.06 (m, 2H, Ar), 5.31 (d, *J* = 12.5 Hz, 1H, CH_2Ph), 5.16 (d, *J* = 12.5 Hz, 1H, CH_2Ph), 5.04 (ddd, *J* = 13.4, 8.7, 7.3 Hz, 1H, H-2), 3.63 (d, *J* = 10.6 Hz, 3H, OMe), 3.51 (d, *J* = 10.6 Hz, 3H, OMe), 2.85–2.75 (m, 1H, H-4), 2.67–2.57 (m, 1H, H-4′), 2.54–2.40 (m, 1H, H-3), 2.17–2.03 (m, 1H, H-3′) ppm. ^{13}C-NMR (100 MHz, $CDCl_3$): δ = 154.68 (CO), 136.84 (Ar), 136.08 (Ar), 132.78 (Ar), 128.54 (Ar), 128.16 (Ar), 127.88 (Ar), 127.56 (Ar), 126.36 (Ar), 125.97 (Ar), 125.22 (Ar), 68.11 (CH_2Ph), 53.07 (d, *J* = 6.2 Hz, OMe), 52.82 (d, *J* = 7.2 Hz, OMe), 49.33 (d, *J* = 158.2 Hz, C-2), 25.75 (C-3), 25.66 (C-4) ppm. ^{31}P-NMR (162 MHz, $CDCl_3$): δ = 26.39 ppm. HRMS (ESI): calcd. for $C_{19}H_{22}NNaO_5P$ $[M + Na]^+$ 398.1128; found 398.1145.

3.4. *Synthesis of Dimethyl N-Benzyloxycarbonyl-1,2-dihydroquinoline-2-phosphonate* **13**

Benzyl chloroformate (0.61 mL, 726 mg, 4.26 mmol) was added dropwise to a solution of quinoline (0.46 mL, 500 mg, 3.87 mmol) in anhydrous acetonitrile (6 mL) kept at 0 °C under argon. After stirring at this temperature for 10 min, trimethyl phosphite (0.50 mL, 528 mg, 4.26 mmol) was slowly added followed by sodium iodide (853 mg, 5.69 mmol). The mixture was heated at 50 °C for 10 min. The solvent was evaporated and the crude product was partitioned between dichloromethane (15 mL) and saturated aqueous sodium bicarbonate (15 mL). The organic phase was separated and the aqueous layer was further extracted with dichloromethane (3 × 15 mL). The combined organic extracts were washed with brine (10 mL), dried, filtered, and concentrated. Purification by column chromatography (eluent:ethyl acetate/hexanes 4:1) afforded **13** as a colourless oil (1.07 g, 2.87 mmol, 74% yield). IR (neat) ν_{max} 1706, 1653, 1603, 1263, 1125, 1026 cm^{-1}. ^1H-NMR (400 MHz, DMSO-d_6, 80 °C): δ = 7.58–7.52 (m, 1H, Ar), 7.44–7.30 (m, 5H, Ar), 7.25–7.07 (m, 3H, Ar), 6.71–6.65 (m, 1H, H-4), 6.07 (ddd, *J* = 9.5, 6.3, 4.5 Hz, 1H, H-3), 5.61 (ddd, *J* = 21.4, 6.3, 1.2 Hz, 1H, H-2), 5.27 (s, 2H, CH_2Ph), 3.47 (d, *J* = 10.6 Hz, 6H, OMe) ppm. ^{13}C-NMR (100 MHz, DMSO-d_6, 80 °C): δ = 152.90 (d, *J* = 4.7 Hz, CO), 135.68 (Ar), 134.66 (Ar), 128.05 (Ar), 127.70 (Ar), 127.40 (Ar), 127.35 (Ar), 127.01 (d, *J* = 9.7 Hz, C-4), 126.77 (d, *J* = 4.0 Hz, Ar), 126.06 (d, *J* = 1.5 Hz, Ar), 124.35 (Ar), 123.66 (Ar), 122.55 (d, *J* = 3.0 Hz, C-3), 67.44 (CH_2Ph), 52.70 (d, *J* = 7.0 Hz, OMe), 52.39 (d, *J* = 6.6 Hz, OMe), 50.84 (d, *J* = 151.5 Hz, C-2) ppm.

^{31}P-NMR (162 MHz, DMSO-d_6, 80 °C): δ = 20.41 ppm. HRMS (ESI): calcd. for $C_{19}H_{20}NNaO_5P$ [M + Na]$^+$ 396.0971; found 396.0991.

3.5. Synthesis of Dimethyl 1,2,3,4-Tetrahydroquinoline-2-phosphonate **14**

A mixture of **13** (200 mg, 0.54 mmol) and 10% Pd/C (20 mg) in ethyl acetate (10 mL) was stirred at room temperature under an atmospheric pressure of hydrogen gas for 12 h. Filtration of the catalyst and evaporation of the solvent provided pure **14** as a colourless oil (130 mg, 0.54 mmol, 100% yield). IR (neat) ν_{max} 3316, 1231, 1057, 1029 cm^{-1}. ^{1}H-NMR (400 MHz, $CDCl_3$): δ = 7.02–6.93 (m, 2H, Ar), 6.70–6.63 (m, 1H, Ar), 6.58–6.53 (m, 1H, Ar), 4.16 (br s, 1H, NH), 3.82 (d, *J* = 10.3 Hz, 3H, OMe), 3.81 (d, *J* = 10.5 Hz, 3H, OMe), 3.71 (ddd, *J* = 10.1, 6.7, 3.4 Hz, 1H, H-2), 2.88–2.75 (m, 2H, H-4), 2.29–2.19 (m, 1H, H-3), 2.12–1.98 (m, 1H, H-3′) ppm. ^{13}C-NMR (100 MHz, $CDCl_3$): δ = 143.08 (d, *J* = 10.9 Hz, Ar), 129.40 (Ar), 127.11 (Ar), 121.05 (Ar), 118.26 (Ar), 114.97 (Ar), 53.83 (d, *J* = 6.7 Hz, OMe), 53.08 (d, *J* = 7.3 Hz, OMe), 49.05 (d, *J* = 161.5 Hz, C-2), 26.09 (d, *J* = 13.7 Hz, C-4), 22.39 (d, *J* = 5.2 Hz, C-3) ppm. ^{31}P-NMR (162 MHz, $CDCl_3$): δ = 27.39 ppm. HRMS (ESI): calcd. for $C_{11}H_{16}NNaO_3P$ [M + Na]$^+$ 264.0760; found 264.0752.

3.6. Synthesis of 1,2,3,4-Tetrahydroquinoline-2-phosphonic Acid Hydrobromide **4**

From **11**: A 33% solution of hydrogen bromide in acetic acid (2 mL) was added to compound **11** (110 mg, 0.29 mmol) and the reaction mixture was stirred at room temperature for 3 h. The solvent was evaporated and the residue was taken up in water and lyophilised to afford **4** as a white solid (86 mg, 0.29 mmol, 100% yield). M.p. 94–96 °C (dec.). IR (nujol) ν_{max} 3381, 1171, 1077, 1057 cm^{-1}. ^{1}H-NMR (400 MHz, CD_3OD): δ = 7.40–7.28 (m, 4H, Ar), 3.78 (ddd, *J* = 13.1, 12.1, 2.5 Hz, 1H, H-2), 3.08–3.01 (m, 2H, H-4), 2.52–2.43 (m, 1H, H-3), 2.19–2.06 (m, 1H, H-3′) ppm. ^{13}C-NMR (100 MHz, CD_3OD): δ = 132.21 (d, *J* = 0.7 Hz, Ar), 131.87 (Ar), 131.78 (Ar), 130.25 (Ar), 128.64 (Ar), 124.49 (Ar), 52.52 (d, *J* = 153.2 Hz, C-2), 25.94 (d, *J* = 12.4 Hz, C-4), 22.37 (d, *J* = 2.2 Hz, C-3) ppm. ^{31}P-NMR (162 MHz, CD_3OD): δ = 14.61 ppm. HRMS (ESI): calcd. for $C_9H_{13}NO_3P$ [M − Br]$^+$ 214.0628; found 214.0632.

From **14**: A 33% solution of hydrogen bromide in acetic acid (2 mL) was added to compound **14** (130 mg, 0.54 mmol) and the reaction mixture was stirred at room temperature for 3 h. The solvent was evaporated and the residue was taken up in water and lyophilised to afford **4** as a white solid (158 mg, 0.54 mmol, 100% yield). Spectroscopic data were identical to those described above.

3.7. Synthesis of Dimethyl N-Benzyloxycarbonyl-1,2-dihydroisoquinoline-1-phosphonate **16**

Benzyl chloroformate (0.61 mL, 726 mg, 4.26 mmol) was added dropwise to a solution of isoquinoline (500 mg, 3.87 mmol) and trimethyl phosphite (0.50 mL, 528 mg, 4.26 mmol) in anhydrous acetonitrile (6 mL) kept at 0 °C under argon. Sodium iodide (853 mg, 5.69 mmol) was slowly added and the mixture was heated at 50 °C for 10 min. The solvent was evaporated and the crude product was partitioned between dichloromethane (15 mL) and saturated aqueous sodium bicarbonate (15 mL). The organic phase was separated and the aqueous layer was further extracted with dichloromethane (3 × 15 mL). The combined organic extracts were washed with brine (10 mL), dried, filtered, and concentrated. Purification by column chromatography (eluent: ethyl acetate/hexanes 4:1) afforded **16** as a colourless oil (1.33 g, 3.56 mmol, 92% yield). IR (neat) ν_{max} 1715, 1342, 1295, 1238, 1121, 1028 cm^{-1}. ^{1}H-NMR (300 MHz, DMSO-d_6, 50 °C): δ = 7.51–7.11 (m, 9H, Ar), 6.91 (d, *J* = 7.7 Hz, 1H, H-3), 6.08–5.94 (m, 1H, H-4), 5.83 (d, *J* = 15.8 Hz, 1H, H-1), 5.27 (s, 2H, CH_2Ph), 3.57–3.38 (m, 6H, OMe) ppm. ^{13}C-NMR (100 MHz, DMSO-d_6): δ = (duplicate signals are observed for some carbons; asterisks indicate those corresponding to the minor rotamer) 152.37* (CO), 151.88 (CO), 136.01 (Ar), 135.91* (Ar), 131.03* (d, *J* = 3.6 Hz, Ar), 130.84 (d, *J* = 3.6 Hz, Ar), 128.75* (d *J* = 3.0 Hz, Ar), 128.68 (d, *J* = 3.1 Hz, Ar), 128.54 (Ar), 128.47* (Ar), 128.26 (Ar), 128.11* (Ar), 127.92 (Ar), 127.55 (d, *J* = 5.2 Hz, Ar), 127.30 (d, *J* = 2.3 Hz, Ar), 127.18 (d, *J* = 2.0 Hz, Ar), 125.43 (Ar), 125.35 (Ar), 124.86 (d, *J* = 2.9 Hz, C-3), 124.79* (C-3), 109.86 (C-4), 67.86* (CH_2Ph), 67.77 (CH_2Ph), 53.40* (d, *J* = 148.4 Hz, C-1), 53.16* (d, *J* = 6.1 Hz,

OMe), 53.12 (d, J = 7.0 Hz, OMe), 52.64 (d, J = 149.6 Hz, C-1) ppm. ^{31}P-NMR (122 MHz, DMSO-d_6, 50 °C): δ = 21.34 ppm. HRMS (ESI): calcd. for $C_{19}H_{20}NNaO_5P$ [M + Na]$^+$ 396.0971; found 396.0982.

3.8. *Synthesis of Diethyl N-Benzyloxycarbonyl-1,2-dihydroisoquinoline-1-phosphonate* **19**

Benzyl chloroformate (0.61 mL, 726 mg, 4.26 mmol) was added dropwise to a solution of isoquinoline (500 mg, 3.87 mmol) and triethyl phosphite (0.73 mL, 708 mg, 4.26 mmol) in anhydrous acetonitrile (6 mL) kept at 0 °C under argon. Sodium iodide (853 mg, 5.69 mmol) was slowly added and the mixture was heated at 50 °C for 10 min. The solvent was evaporated and the crude product was partitioned between dichloromethane (15 mL) and saturated aqueous sodium bicarbonate (15 mL). The organic phase was separated and the aqueous layer was further extracted with dichloromethane (3 × 15 mL). The combined organic extracts were washed with brine (10 mL), dried, filtered, and concentrated. Purification by column chromatography (eluent:ethyl acetate/hexanes 3:2) afforded **19** as a colourless oil (1.52 g, 3.79 mmol, 98% yield). IR (neat) ν_{max} 1715, 1294, 1253, 1120, 1023 cm^{-1}. ^{1}H-NMR (300 MHz, DMSO-d_6, 70 °C): δ = 7.50–7.10 (m, 9H, Ar), 6.91 (d, J = 7.8 Hz, 1H, H-3), 5.98 (d, J = 7.8 Hz, 1H, H-4), 5.77 (d, J = 15.9 Hz, 1H, H-1), 5.26 (s, 2H, CH_2Ph), 3.95–3.69 (m, 4H, OCH_2), 1.08 (t, J = 7.0 Hz, 3H, Me), 1.07 (t, J = 7.0 Hz, 3H, Me) ppm. ^{13}C-NMR (100 MHz, DMSO-d_6): δ = (duplicate signals are observed for some carbons; asterisks indicate those corresponding to the minor rotamer) 152.48* (CO), 151.91 (CO), 136.05 (Ar), 135.88* (Ar), 131.15* (d, J = 3.6 Hz, Ar), 130.98 (d, J = 3.7 Hz, Ar), 128.69* (d, J = 3.3 Hz, Ar), 128.61 (d, J = 3.3 Hz, Ar), 128.56 (Ar), 128.47* (Ar), 128.28 (Ar), 128.25* (Ar), 128.13* (Ar), 127.97 (Ar), 127.53 (d, J = 5.2 Hz, Ar), 127.22 (d, J = 2.6 Hz, Ar), 127.11* (d, J = 2.7 Hz, Ar), 125.59 (d, J = 1.8 Hz, Ar), 125.56* (d, J = 2.4 Hz, Ar), 125.42 (Ar), 124.84* (C-3), 124.81 (C-3), 109.95 (C-4), 67.84* (CH_2Ph), 67.71 (CH_2Ph), 62.54 (d, J = 7.1 Hz, OCH_2), 62.42* (d, J = 6.6 Hz, OCH_2), 62.38* (d, J = 7.1 Hz, OCH_2), 53.97* (d, J = 148.9 Hz, C-1), 53.15 (d, J = 150.2 Hz, C-1), 16.17 (d, J = 5.3 Hz, Me), 16.15* (d, J = 5.7 Hz, Me) ppm. ^{31}P-NMR (122 MHz, DMSO-d_6, 70 °C): δ = 18.77 ppm. HRMS (ESI): calcd. for $C_{21}H_{25}NO_5P$ [M + H]$^+$ 402.1465; found 402.1466.

3.9. *Synthesis of Diethyl N-Benzyloxycarbonyl-1,2,3,4-tetrahydroisoquinoline-1-phosphonate* **20**

Triethylsilane (1.0 mL, 730 mg, 6.28 mmol) and trifluoroacetic acid (0.48 mL, 716 mg, 6.28 mmol) were added to a solution of **19** (300 mg, 0.75 mmol) in anhydrous dichloromethane (15 mL) kept at 0 °C under argon. The solution was allowed to warm to room temperature and stirred for 18 h. Evaporation of the solvent followed by column chromatography (eluent:hexanes/ethyl acetate 1:1) afforded **20** as a colourless oil (298 mg, 0.74 mmol, 99% yield). IR (neat) ν_{max} 1701, 1294, 1249, 1230, 1051, 1022 cm^{-1}. ^{1}H-NMR (300 MHz, DMSO-d_6, 70 °C): δ = 7.44–7.16 (m, 9H, Ar), 5.53 (d, J = 20.4 Hz, 1H, H-1), 5.17 (s, 2H, CH_2Ph), 4.14–3.77 (m, 5H, H-3, OCH_2), 3.72–3.52 (m, 1H, H-3′), 2.98–2.78 (m, 2H, H-4), 1.17 (t, J = 7.0 Hz, 3H, Me), 1.08 (t, J = 6.9 Hz, 3H, Me) ppm. ^{13}C-NMR (100 MHz, DMSO-d_6): δ = (duplicate signals are observed for some carbons; asterisks indicate those corresponding to the minor rotamer) 154.60 (d, J = 3.9 Hz, CO), 154.18* (d, J = 2.4 Hz, CO), 136.66 (Ar), 136.46* (Ar), 134.73 (d, J = 5.6 Hz, Ar), 134.63* (d, J = 5.6 Hz, Ar), 129.22 (Ar), 129.13* (d, J = 2.2 Hz, Ar), 128.95 (d, J = 2.3 Hz, Ar), 128.85 (Ar), 128.38 (Ar), 128.33* (Ar), 127.94* (Ar), 127.91 (Ar), 127.86* (Ar), 127.67 (d, J = 4.0 Hz, Ar), 127.58 (Ar), 127.38 (Ar), 125.88 (d, J = 2.8 Hz, Ar), 125.83* (d, J = 2.8 Hz, Ar), 66.92* (CH_2Ph), 66.82 (CH_2Ph), 62.58 (d, J = 7.2 Hz, OCH_2), 62.19 (d, J = 7.1 Hz, OCH_2), 52.76* (d, J = 149.9 Hz, C-1), 52.34 (d, J = 152.0 Hz, C-1), 39.05* (C-3), 38.84 (C-3), 27.43 (C-4), 27.15* (C-4), 16.08* (d, J = 5.6 Hz, Me), 16.05 (d, J = 5.3 Hz, Me). ^{31}P-NMR (122 MHz, DMSO-d_6, 70 °C): δ = 20.75. HRMS (ESI): calcd. for $C_{21}H_{27}NO_5P$ [M + H]$^+$ 404.1621; found 404.1611.

3.10. *Synthesis of 1,2,3,4-Tetrahydroisoquinoline-1-phosphonic Acid Hydrobromide* **5**

A 33% solution of hydrogen bromide in acetic acid (2 mL) was added to **20** (160 mg, 0.40 mmol) and the reaction mixture was stirred at room temperature for 3 h. The solvent was evaporated and the residue was taken up in water and lyophilised to afford **5** as a white solid (117 mg, 0.40 mmol, 100% yield). M.p. 85–87 °C (dec.). IR (nujol) ν_{max} 3421, 1212, 1118, 1019 cm^{-1}. ^{1}H-NMR (400 MHz,

D_2O): δ = 7.45–7.41 (m, 1H, Ar), 7.37–7.27 (m, 3H, Ar), 4.70 (d, *J* = 17.6 Hz, 1H, H-1), 3.78 (ddd, *J* = 12.7, 9.2, 6.5 Hz, 1H, H-3), 3.58–3.51 (m, 1H, H-3′), 3.18–3.12 (m, 2H, H-4) ppm. ^{13}C-NMR (100 MHz, CD_3OD): δ = 132.88 (d, *J* = 5.4 Hz, Ar), 130.40 (d, *J* = 2.2 Hz, Ar), 129.47 (d, *J* = 2.7 Hz, Ar), 128.94 (d, *J* = 3.6 Hz, Ar), 127.92 (d, *J* = 2.6 Hz, Ar), 127.40 (d, *J* = 5.1 Hz, Ar), 54.00 (d, *J* = 147.3 Hz, C-1), 41.15 (d, *J* = 2.1 Hz, C-3), 25.88 (C-4) ppm. ^{31}P-NMR (162 MHz, D_2O): δ = 10.02 ppm. HRMS (ESI): calcd. for $C_9H_{13}NO_3P$ [M − Br]$^+$ 214.0628; found 214.0633.

3.11. *Synthesis of N-Benzyloxycarbonyl-1,4-dihydro-3-isoquinolinone* **22**

A 1 M solution of lithium bis(trimethylsilyl)amide in tetrahydrofuran (0.83 mL, 0.83 mmol) was slowly added to a solution of 1,4-dihydro-3(2*H*)-isoquinolinone **21** (122 mg, 0.83 mmol) in anhydrous tetrahydrofuran (5 mL) kept at −78 °C under argon. After 30 min, benzyl chloroformate (0.12 mL, 142 mg, 0.83 mmol) was added dropwise and stirring was continued for additional 3 h. The reaction was then treated with saturated aqueous ammonium chloride (10 mL) and allowed to warm to room temperature. The two layers were separated and the aqueous phase was extracted with dichloromethane (2 × 10 mL). The combined organic extracts were dried, filtered, and concentrated. Purification by column chromatography (eluent:hexanes/ethyl acetate 7:3) afforded **22** as a white solid (172 mg, 0.61 mmol, 73% yield). M.p. 63–65 °C. IR (nujol) ν_{max} 1692 cm^{-1}. ^{1}H-NMR (400 MHz, $CDCl_3$): δ = 7.49–7.44 (m, 2H, Ar), 7.40–7.25 (m, 6H, Ar), 7.23–7.19 (m, 1H, Ar), 5.34 (s, 2H, CH_2Ph), 4.92 (s, 2H, H-1), 3.73 (s, 2H, H-4) ppm. ^{13}C-NMR (100 MHz, $CDCl_3$): δ = 169.43 (CO), 153.62 (COO), 135.38 (Ar), 132.25 (Ar), 132.13 (Ar), 128.74 (Ar), 128.50 (Ar), 128.46 (Ar), 128.22 (Ar), 127.35 (Ar), 126.99 (Ar), 125.81 (Ar), 68.97 (CH_2Ph), 48.75 (C-1), 41.53 (C-4) ppm. HRMS (ESI): calcd. for $C_{17}H_{15}NNaO_3$ [M + Na]$^+$ 304.0944; found 304.0943.

3.12. *Synthesis of Dimethyl N-Benzyloxycarbonyl-1,2,3,4-tetrahydroisoquinoline-3-phosphonate* **25**

A 1 M solution of diisobutylaluminium hydride in hexanes (0.78 mL, 0.78 mmol) was slowly added to a solution of **22** (146 mg, 0.52 mmol) in anhydrous tetrahydrofuran (5 mL) kept at −78 °C under argon. After stirring at this temperature for 2 h, the reaction was treated with saturated aqueous sodium acetate (10 mL) and allowed to warm to room temperature. A 3:1 mixture of diethyl ether and saturated aqueous ammonium chloride (16 mL) was then added and the resulting mixture was stirred at room temperature until a suspension was formed. The solid was filtered off under reduced pressure and washed with diethyl ether (2 × 10 mL). The organic layer was separated and the aqueous phase was extracted with diethyl ether (2 × 10 mL). The combined organic extracts were washed with brine (10 mL), dried, filtered, and evaporated to provide the hemiaminal as an oil. It was dissolved in methanol (5 mL) and treated with pyridinium *p*-toluenesulfonate (13 mg, 0.05 mmol). After stirring at room temperature for 2 h, triethylamine (31 μL, 22 mg, 0.22 mmol) was added. The solvent was evaporated and the crude methoxyaminal obtained **23** was dissolved in anhydrous dichloromethane (5 mL) and kept under argon. Trimethyl phosphite (61 μL, 65 mg, 0.52 mmol) was added and the resulting solution was cooled to −20 °C. Boron trifluoride-diethyl ether (65 μL, 74 mg, 0.52 mmol) was added and the reaction mixture was slowly warmed to room temperature and stirred for 12 h. After quenching with water (10 mL), the two layers were separated and the aqueous phase was extracted with dichloromethane (2 × 10 mL). The combined organic extracts were dried, filtered, and concentrated. Purification by column chromatography (eluent:ethyl acetate/hexanes 9:1) afforded **25** as a colourless oil (138 mg, 0.37 mmol, 71% yield). IR (neat) ν_{max} 1703, 1410, 1246, 1055, 1031 cm^{-1}. ^{1}H-NMR (400 MHz, $CDCl_3$): δ = (duplicate signals are observed for some protons; asterisks indicate those corresponding to the minor rotamer) 7.42–7.30 (m, 5H, Ar), 7.23–7.04 (m, 4H, Ar), 5.31–5.12 (m, 2H, CH_2Ph), 5.11–5.03 (m, 1H, H-3), 5.00–4.86 (m, 1H, H-1) overlapped with 4.97*–4.89* (m, 1H, H-3), 4.52 (d, *J* = 16.6 Hz, 1H, H-1′), 4.45* (d, *J* = 16.6 Hz, 1H, H-1′), 3.65 (d, *J* = 10.7 Hz, 3H, OMe), 3.51* (d, *J* = 10.7 Hz, 3H, OMe), 3.35 (d, *J* = 10.7 Hz, 3H, OMe), 3.33–3.16 (m, 2H, H-4) overlapped with 3.24* (d, *J* = 10.7 Hz, 3H, OMe) ppm. ^{13}C-NMR (100 MHz, $CDCl_3$): δ = (duplicate signals are observed for some carbons; asterisks indicate those corresponding to the minor rotamer)

155.72 (CO), 155.05* (CO), 136.37 (Ar), 136.18* (Ar), 132.61* (Ar), 132.46 (Ar), 131.41 (Ar), 131.00* (Ar), 128.64 (Ar), 128.53 (Ar), 128.45* (Ar), 128.24 (Ar), 128.17* (Ar), 127.95 (Ar), 126.70 (Ar), 126.64* (Ar), 126.60 (Ar), 126.09* (Ar), 125.86 (Ar), 67.79 (CH_2Ph), 52.74 (d, J = 6.4 Hz, OMe), 52.58 (d, J = 6.6 Hz, OMe), 52.34* (d, J = 6.9 Hz, OMe), 47.07* (d, J = 154.2 Hz, C-3), 46.29 (d, J = 154.1 Hz, C-3), 44.37 (C-1), 28.56* (C-4), 28.32 (C-4) ppm. ^{31}P-NMR (162 MHz, $CDCl_3$): δ = (a duplicate signal is observed; an asterisk indicates that corresponding to the minor rotamer) 27.67, 27.07* ppm. HRMS (ESI): calcd. for $C_{19}H_{22}NNaO_5P$ $[M + Na]^+$ 398.1128; found 398.1144.

3.13. Synthesis of 1,2,3,4-Tetrahydroisoquinoline-3-phosphonic Acid Hydrobromide **6**

A 33% solution of hydrogen bromide in acetic acid (2 mL) was added to **25** (100 mg, 0.27 mmol) and the reaction mixture was stirred at room temperature for 3 h. The solvent was evaporated and the residue was taken up in water and lyophilised to afford **6** as a white solid (79 mg, 0.27 mmol, 100% yield). M.p. 102–103 °C (dec.). IR (nujol) ν_{max} 3404, 1232, 1010 cm^{-1}. ^{1}H-NMR (400 MHz, CD_3OD): δ = 7.35–7.22 (m, 4H, Ar), 4.52 (d, J = 15.7 Hz, 1H, H-1), 4.42 (dd, J = 15.7, 3.1 Hz, 1H, H-1′), 3.94–3.83 (m, 1H, H-3), 3.34–3.25 (m, 2H, H-4) ppm. ^{13}C-NMR (100 MHz, CD_3OD): δ = 131.71 (d, J = 12.2 Hz, Ar), 129.97 (Ar), 129.27 (Ar), 128.69 (Ar), 128.32 (Ar), 127.77 (Ar), 51.34 (d, J = 153.1 Hz, C-3), 47.01 (d, J = 7.7 Hz, C-1), 27.44 (C-4) ppm. ^{31}P-NMR (162 MHz, CD_3OD): δ = 14.06 ppm. HRMS (ESI): calcd. for $C_9H_{13}NO_3P$ $[M - Br]^+$ 214.0628; found 214.0631.

4. Conclusions

The synthesis of α-aminophosphonic acids characterized by a tetrahydroquinoline or tetrahydroisoquinoline heterocycles is described. Specifically, 1,2,3,4-tetrahydroquinoline-2-phosphonic acid **4**, 1,2,3,4-tetrahydroisoquinoline-1-phosphonic acid **5**, and 1,2,3,4-tetrahydro-isoquinoline-3-phosphonic acid **6** have been prepared in high overall yields using efficient methods that make use of easily available substrates. The compounds obtained can be viewed as higher homologues of phosphopipecolic acid that bear a benzene ring fused at different positions of the six-membered piperidine cycle. In particular, the latter is the phosphonic surrogate of Tic, an amino acid of extraordinary value in the design of pharmacologically useful peptides. The synthetic routes developed rely on the addition of trialkyl phosphites to iminium ions generated from quinoline/isoquinoline or from the appropriate δ-lactam bearing a fused benzene ring. Thus, lactams have proven to be versatile starting materials for the efficient and selective synthesis of α-aminophosphonic acids containing an azacyclic skeleton. The presence of the carbonyl group in the starting lactam determines, with full regiocontrol, the position at which the phosphonate substituent is incorporated. Formation of the intermediate *N*-acyliminium ion with the adequate regiochemistry has allowed the synthesis of the phosphonic counterpart of Tic (Tic^P). Importantly, this α-aminophosphonic acid is not accessible through other methodologies traditionally used to generate iminium ions.

Acknowledgments: Financial support to this work was provided by Consejo Nacional de Ciencia y Tecnología (CONACYT-Mexico, grants 181816 and 248868). Ministerio Economia y Competitividad (grant CTQ2013-40855-R), and Gobierno de Aragón-Fondo Social Europeo (research group E40). Consejo Superior de Investigaciones Científicas (grant 2008MX0044; JAE predoctoral fellowship to A.A).

Author Contributions: M.O., C.C. and A.I.J. provided the concepts of the work, interpreted the results and prepared the manuscript, A.A., F.J.S., carried out the experimental work, interpreted the results and prepared the manuscript. All authors read and approved the final manuscript.

References

1. Ko, E.; Liu, J.; Perez, L.M.; Lu, G.; Schaefer, A.; Burgess, K. Universal peptidomimetics. *J. Am. Chem. Soc.* **2011**, *133*, 462–477. [CrossRef] [PubMed]

2. Hruby, V.J. Designing peptide receptor agonists and antagonists. *Nat. Rev. Drug. Discov.* **2002**, *1*, 847–858. [CrossRef] [PubMed]
3. Brindisi, M.; Butini, S.; Franceschini, S.; Brogi, S.; Trotta, F.; Ros, S.; Cagnotto, A.; Salmona, M.; Casagni, A.; Andreassi, M.; et al. Targeting Dopamine D_3 and Serotonin 5-HT_{1A} and 5-HT_{2A} Receptors for developing effective antipsychotics: Synthesis, biological characterization, and behavioral studies. *J. Med. Chem.* **2014**, *57*, 9578–9597. [CrossRef] [PubMed]
4. Mitsunaga, S.; Ohbayashi, T.; Sugiyama, S.; Saitou, T.; Tadokoro, M.; Satoh, T. Asymmetric synthesis of cyclic α-amino acid derivatives by the intramolecular reaction of magnesium carbenoid with an *N*-magnesio arylamine. *Tetrahedron: Asymmetry* **2009**, *20*, 1697–1708. [CrossRef]
5. Kurata, K.; Inoue, K.; Nishimura, K.; Hoshiya, N.; Kawai, N.; Uenishi, J. Synthesis of optically pure (*R*)- and (*S*)-tetrahydroisoquinoline-1- and -3-carboxylic acids. *Synthesis* **2015**, *47*, 1238–1244.
6. Hu, L.; Magesh, S.; Chen, L.; Wang, L.; Lewis, T.A.; Chen, Y.; Khodier, C.; Inoyama, D.; Beamer, L.J.; Emge, T.J.; et al. Discovery of a small-molecule inhibitor and cellular probe of Keap1-Nrf2 protein-protein interaction. *Bioorg. Med. Chem. Lett.* **2013**, *23*, 3039–3043. [CrossRef] [PubMed]
7. Kotha, S.; Deodhar, D.; Khedkar, P. Diversity-oriented synthesis of medicinally important 1,2,3,4-tetrahydroisoquinoline-3-carboxylic acid (Tic) derivatives and higher analogs. *Org. Biomol. Chem.* **2014**, *12*, 9054–9091. [CrossRef] [PubMed]
8. Yolacan, C.; Mavis, M.E.; Aydogan, F. Evaluation of mono- and dipeptides as organocatalysts for enantioselective aldol reaction. *Tetrahedron* **2014**, *70*, 3707–3713. [CrossRef]
9. Mucha, A.; Kafarski, P.; Berlicki, Ł. Remarkable Potential of the α-aminophosphonate/phosphinate structural motif in medicinal chemistry. *J. Med. Chem.* **2011**, *54*, 5955–5980. [CrossRef] [PubMed]
10. Orsini, F.; Sello, G.; Sisti, M. Aminophosphonic acids and derivatives. Synthesis and biological applications. *Curr. Med. Chem.* **2010**, *17*, 264–289. [CrossRef] [PubMed]
11. Naydenova, E.D.; Todorov, P.T.; Troev, K.D. Recent synthesis of aminophosphonic acids as potential biological importance. *Amino Acids* **2010**, *38*, 23–30. [CrossRef] [PubMed]
12. Lejczak, B.; Kafarski, P. Biological activity of aminophosphonic acids and their short peptides. *Top. Heterocycl. Chem.* **2009**, *20*, 31–63.
13. Ordóñez, M.; Viveros-Ceballos, J.L.; Cativiela, C.; Sayago, F.J. An update on the stereoselective synthesis of α-aminophosphonic acids and derivatives. *Tetrahedron* **2015**, *71*, 1745–1784. [CrossRef]
14. Ordóñez, M.; Sayago, F.J.; Cativiela, C. Synthesis of quaternary α-aminophosphonic acids. *Tetrahedron* **2012**, *68*, 6369–6412. [CrossRef]
15. Ordóñez, M.; Viveros-Ceballos, J.L.; Cativiela, C.; Arizpe, A. Stereoselective synthesis of α-aminophosphonic acids analogs of the 20 proteinogenic α-amino acids. *Curr. Org. Synth.* **2012**, *9*, 310–341. [CrossRef]
16. Kudzin, Z.H.; Kudzin, M.H.; Drabowicz, J.; Stevens, C.V. Aminophosphonic acids-phosphorus analogues of natural Amino Acids. Part 1: Syntheses of α-aminophosphonic acids. *Curr. Org. Chem.* **2011**, *15*, 2015–2071. [CrossRef]
17. Gulyukina, N.S.; Makukhin, N.N.; Beletskaya, I.P. Synthesis methods of (1-aminocyclopropyl)phosphonic acids. *Russ. J. Org. Chem.* **2011**, *47*, 633–649. [CrossRef]
18. Ordóñez, M.; Rojas-Cabrera, H.; Cativiela, C. An overview of stereoselective synthesis of α-aminophosphonic acids and derivatives. *Tetrahedron* **2009**, *65*, 17–49. [CrossRef] [PubMed]
19. Pudovik, A.N. Addition of dialkyl phosphites to imines. New method of synthesis of esters of amino phosphonic acids. *Dokl. Akad. Nauk. SSSR* **1952**, *83*, 865–869.
20. Alfonsov, V.A. Diastereoselective Synthesis of Enantiopure α-Aminophosphonic Acid Derivatives: Pudovik Reaction in Stereoselective Synthesis (Dedicated to A.N. Pudovik, 1916–2006). *Phosphorus. Sulfur. Silicon Relat. Elem.* **2008**, *183*, 2637–2644. [CrossRef]
21. Kabachnik, M.I.; Medved, T.Y. New synthesis of aminophosphonic acids. *Dokl. Akad. Nauk SSSR* **1952**, *83*, 689–692.
22. Fields, E.K. The synthesis of esters of substituted amino phosphonic acids. *J. Am. Chem. Soc.* **1952**, *74*, 1528–1531. [CrossRef]
23. Zefirov, N.S.; Matveeva, E.D. Catalytic Kabachnik-Fields reaction: New horizons for old reaction. *Arkivoc* **2008**, *2008*. [CrossRef]
24. Kafarski, P.; Gorny vel Gorniak, M.; Andrasiak, I. Kabachnik-Fields Reaction under Green Conditions—A Critical Overview. *Curr. Green Chem.* **2015**, *2*, 218–222. [CrossRef]

25. Keglevich, G.; Bálint, E. The Kabachnik–Fields Reaction: Mechanism and Synthetic Use. *Molecules* **2012**, *17*, 12821–12835. [CrossRef] [PubMed]
26. Bonilla-Landa, I.; Viveros-Ceballos, J.L.; Ordóñez, M. Diastereoselective synthesis of novel 5-substituted morpholine-3-phosphonic acids: Further exploitation of *N*-acyliminium intermediates. *Tetrahedron: Asymmetry* **2014**, *25*, 485–487. [CrossRef]
27. Arizpe, A.; Sayago, F.J.; Jiménez, A.I.; Ordóñez, M.; Cativiela, C. Synthesis of phosphoproline derivatives with an octahydroisoindole structure. *Eur. J. Org. Chem.* **2011**, *2001*, 6732–6738. [CrossRef]
28. Arizpe, A.; Sayago, F.J.; Jiménez, A.I.; Ordóñez, M.; Cativiela, C. Stereodivergent synthesis of two novel α-aminophosphonic acids characterised by a *cis*-fused octahydroindole system. *Eur. J. Org. Chem.* **2011**, *2001*, 3074–3081. [CrossRef]
29. Boto, A.; Gallardo, J.A.; Hernández, R.; Saavedra, C.J. One-pot synthesis of α-amino phosphonates from α-amino acids and β-amino alcohols. *Tetrahedron Lett.* **2005**, *46*, 7807–7811. [CrossRef]
30. Shono, T.; Matsumura, Y.; Tsubata, K. A new carbon—Phosphorous bond forming reaction and synthesis of aminoalkylphosphonic acid derivatives. *Tetrahedron Lett.* **1981**, *22*, 3249–3252. [CrossRef]
31. Maury, C.; Wang, Q.; Gharbaoui, T.; Chiadmi, M.; Tomas, A.; Royer, J.; Husson, H.-P. Asymmetric synthesis of (*R*)- and (*S*)-piperidin-2-yl-phosphonic acid by diastereoselective addition of trialkyl phosphite onto potential iminium salt. *Tetrahedron* **1997**, *53*, 3627–3636. [CrossRef]
32. Katritzky, A.R.; Qiu, G.; He, H.-Y.; Yang, B. Novel Syntheses of Hexahydro-1*H*-pyrrolo[1,2-*a*]imidazoles and Octahydroimidazo[1,2-*a*]pyridines. *J. Org. Chem.* **2000**, *65*, 3683–3689. [CrossRef] [PubMed]
33. Kaname, M.; Mashige, H.; Yoshifuji, S. Chemical Conversion of Cyclic α-Amino Acids to Cyclic α-Aminophosphonic Acids. *Chem. Pharm. Bull.* **2001**, *49*, 531–536. [CrossRef] [PubMed]
34. Yang, D.; Zhao, D.; Mao, L.; Wang, L.; Wang, R. Copper/DIPEA-Catalyzed, Aldehyde-Induced Tandem Decarboxylation—Coupling of Natural α-Amino Acids and Phosphites or Secondary Phosphine Oxides. *J. Org. Chem.* **2011**, *76*, 6426–6431. [CrossRef] [PubMed]
35. Han, W.; Mayer, P.; Ofial, A.R. Iron-Catalyzed Oxidative Mono- and Bis-Phosphonation of *N,N*-Dialkylanilines. *Adv. Synth. Catal.* **2010**, *352*, 1667–1676. [CrossRef]
36. Rueping, M.; Zhu, S.; Koenigs, R.M. Photoredox catalyzed C-P bond forming reactions-visible light mediated oxidative phosphonylations of amines. *Chem. Commun.* **2011**, *47*, 8679–8681. [CrossRef] [PubMed]
37. Wang, H.; Li, X.; Wu, F.; Wan, B. Direct oxidative phosphonylation of amines under metal-free conditions. *Tetrahedron Lett.* **2012**, *53*, 681–683. [CrossRef]
38. Yoo, W.-J.; Kobayashi, S. Efficient visible light-mediated crossdehydrogenative coupling reactions of tertiary amines catalyzed by a polymer-immobilized iridium-based photocatalyst. *Green Chem.* **2014**, *16*, 2438–2442. [CrossRef]
39. Viveros-Ceballos, J.L.; Ordóñez, M.; Sayago, F.J.; Jimenez, A.I.; Cativiela, C. First synthesis of (*R*)- and (*S*)-1,2,3,4-tetrahydroisoquinoline-3-phosphonic acid (Tic^P) using a Pictet-Spengler reaction. *Eur. J. Org. Chem.* **2016**, 2711–2719. [CrossRef]
40. Ramirez-Marroquin, O.A.; Romero-Estudillo, I.; Viveros-Ceballos, J.L.; Cativiela, C.; Ordóñez, M. Convenient synthesis of cyclic α-aminophosphonates by alkylation-cyclization reaction of iminophosphoglycinates using phase-transfer catalysis. *Eur. J. Org. Chem.* **2016**, *2016*, 308–313. [CrossRef]
41. López-Iglesias, M.; Arizpe, A.; Sayago, F.J.; Gotor, V.; Cativiela, C.; Gotor-Fernández, V. Lipase-catalyzed dynamic kinetic resolution of dimethyl (1,3-dihydro-2*H*-isoindol-1-yl)phosphonate. *Tetrahedron* **2016**. [CrossRef]
42. Arizpe, A.; Rodríguez-Mata, M.; Sayago, F.J.; Pueyo, M.J.; Gotor, V.; Jiménez, A.I.; Gotor-Fernández, V.; Cativiela, C. Enzymatic and chromatographic resolution procedures applied to the synthesis of the phosphoproline enantiomers. *Tetrahedron: Asymmetry* **2015**, *26*, 1469–1477. [CrossRef]
43. Viveros-Ceballos, J.L.; Sayago, F.J.; Cativiela, C.; Ordóñez, M. First practical and efficient synthesis of 3-phosphorylated β-carboline derivatives using the Pictet-Spengler reaction. *Eur. J. Org. Chem.* **2015**, *2015*, 1084–1091. [CrossRef]
44. Liu, K.; Liu, L.-L.; Gu, C.-Z.; Dai, B.; He, L. Aryne-induced dearomatized phosphonylation of electron-deficient azaarenes. *RSC Adv.* **2016**, *6*, 33606–33610. [CrossRef]
45. Zhang, Q.; Wei, D.; Cui, X.; Zhang, D.; Wang, H.; Wu, Y. Direct diphosphonylation of quinolines with *H*-phosphonates under metal-free conditions. *Tetrahedron* **2015**, *71*, 6087–6093. [CrossRef]

46. Shaabani, A.; Sarvary, A.; Mousavi-Faraz, S.; Ng, S.W. Synthesis of 1,2-dihydroquinolin-2-ylphosphonates and 1,2-dihydroisoquinolin-1-ylphosphonates via three-component reactions. *Monatsh. Chem.* **2012**, *143*, 1061–1065. [CrossRef]
47. De Blieck, A.; Masschelein, K.G.R.; Dhaene, F.; Rozycka-Sokolowska, E.; Marciniak, B.; Drabowicz, J.; Stevens, C.V. One-pot tandem 1,4–1,2-addition of phosphites to quinolines. *Chem. Commun.* **2010**, *46*, 258–260. [CrossRef] [PubMed]
48. De Blieck, A.; Catak, S.; Debrouwer, W.; Drabowicz, J.; Hemelsoet, K.; Verstraelen, T.; Waroquier, M.; van Speybroeck, V.; Stevens, C.V. Diphosphonylation of Aromatic Diazaheterocycles and Theoretical Rationalization of Product Yields. *Eur. J. Org. Chem.* **2013**, *2013*, 1058–1067. [CrossRef]
49. Hamada, Y.; Sugiura, M.; Asai, K.; Hamada, Y. Simple synthetic method of dialkyl 1,2-dihydro(iso)quinoline (1 or 2)-phosphonates. *Heterocycles* **1996**, *43*, 953–958. [CrossRef]
50. Akiba, K.; Negishi, Y.; Inamoto, N. A facile general synthesis of dimethyl 1-acyl-1,2-dihydroquinoline-2-phosphonates and 2-acyl-1,2-dihydroisoquinoline-1-phosphonates. *Synthesis* **1979**, 55–56. [CrossRef]
51. Taylor, M.S.; Tokunaga, N.; Jacobsen, E.N. Enantioselective thiourea-catalyzed acyl-Mannich reactions of isoquinolines. *Angew. Chem. Int. Ed.* **2005**, *44*, 6700–6704. [CrossRef] [PubMed]
52. Ilayaraja, N.; Noel, M. A Comparative study of anodic fluorination of *N*-alkyl and *N,N*-dialkyl phenylacetamides in $Et_3N{\cdot}4HF$ medium. *J. Electroanal. Chem.* **2009**, *632*, 45–54. [CrossRef]
53. Kamochi, Y.; Watanabe, Y. Studies on the synthesis of benzolactam rings. II. Synthesis of 1,4-dihydro-3(2*H*)-isoquinolinone derivatives. *Heterocycles* **1987**, *26*. [CrossRef]

5

1-(Acylamino)alkylphosphonic Acids—Alkaline Deacylation

Marek Cypryk [1], Jozef Drabowicz [1,2,*], Bartlomiej Gostynski [1], Marcin H. Kudzin [3], Zbigniew H. Kudzin [4,*] and Pawel Urbaniak [4]

1 Centre of Molecular and Macromolecular Studies, Polish Academy of Sciences, Sienkiewicza 120a, Lodz 90-363, Poland; mcypryk@cbmm.lodz.pl (M.C.); bgostyns@cbmm.lodz.pl (B.G.)
2 Department of Chemistry and Environment Protection, Jan Dlugosz University, Czestochowa 42-200, Poland
3 Textile Research Institute, Brzezinska 5/15, Lodz 92-103, Poland; kudzin@iw.lodz.pl
4 Faculty of Chemistry, University of Lodz, Tamka 12, Lodz 91-403, Poland; pawelurb@chemia.uni.lodz.pl
* Correspondence: drabowicz@gmail.com (J.D.); zhkudzin@uni.lodz.pl (Z.H.K.)

Academic Editor: György Keglevich

Abstract: The alkaline deacylation of a representative series of 1-(acylamino)alkylphosphonic acids [(AC)-AA^P: (AC) = Ac, TFA, Bz; AA^P = Gly^P, Ala^P, Val^P, Pgl^P and Phe^P] in an aqueous solution of KOH (2M) was investigated. The results suggested a two-stage reaction mechanism with a quick interaction of the hydroxyl ion on the carbonyl function of the amide R-C(O)-N(H)- group in the first stage, which leads to instant formation of the intermediary acyl-hydroxyl adducts of R-C$(O^-)_2$-N(H)-, visible in the ^{31}P NMR spectra. In the second stage, these intermediates decompose slowly by splitting of the RC$(O^-)_2$-N(H)- function with the subsequent formation of 1-aminoalkylphosphonate and carboxylate ions.

Keywords: amino acids; aminophosphonic acids; 1-aminoalkylphosphonic acids; 1-(acylamino) alkylphosphonic acids; ^{31}P NMR spectra of intermediates; hydrolytic deacylation

1. Introduction

Aminoalkylphosphonic acids (AA^P) are structural analogues of amino acids (AA^C) [1], some of which are of natural origin [2]. Due to the structural analogy, they present similar biological properties to the class and are important inhibitors of enzymes of the amino acids metabolism [1,3] (Table 1).

Table 1. Structures of aminophosphonic acids (AA^P)—phosphonic analogs of amino acids (AA^C) and 1-(acylamino)alkylphosphonic acids (AC)-AA^P and phosphonopeptides AA^C-AA^P and AA^P-AA^C [4–14].

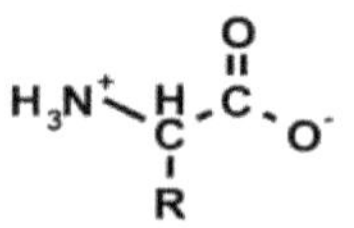	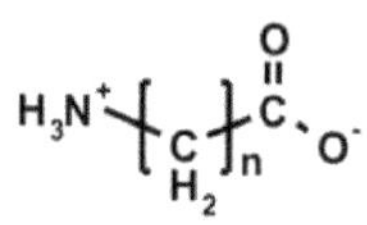	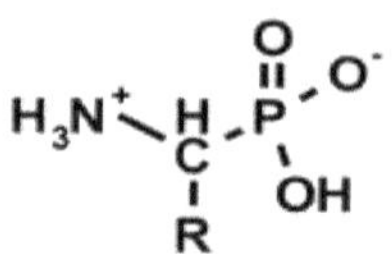	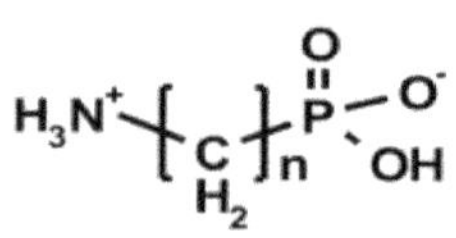	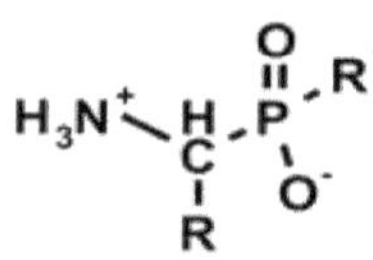
AA^C	ω-AA^C	AA^P	ω-AA^P	AA^{P-R}
Naturally occurring aminophosphonic acids.				

Table 1. *Cont.*

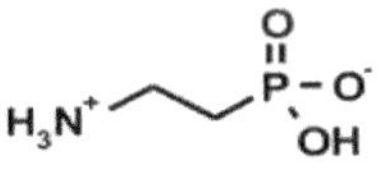	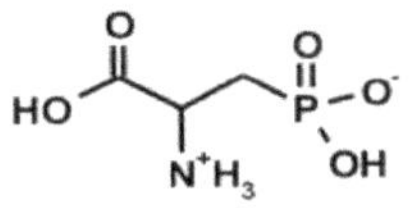	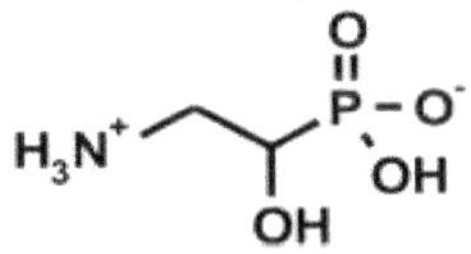	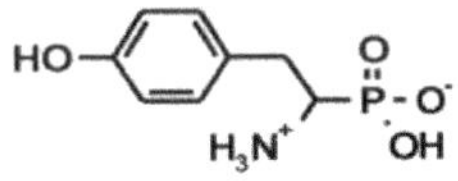	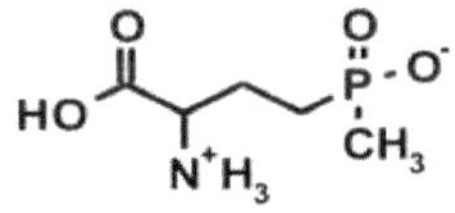
β-Ala^P [4]	Asp$^{\beta-P}$ [5]	Iser^P [6]	Tyr^P [7]	Glu$^{-\gamma-P(Me)}$ [8]
Representative, biologically active aminophosphonates.				
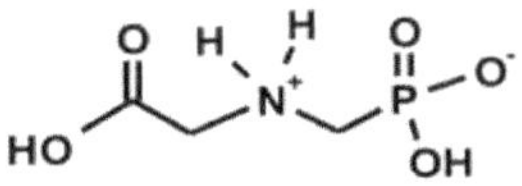	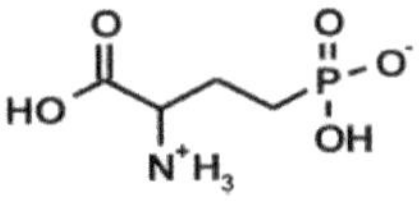	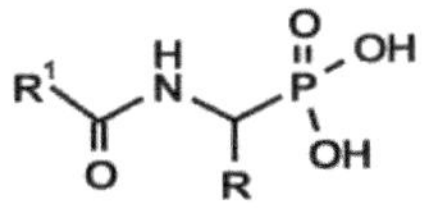	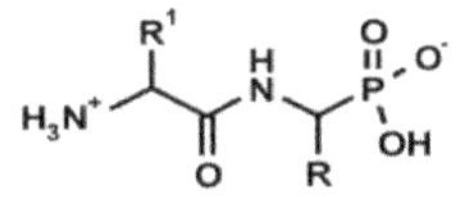	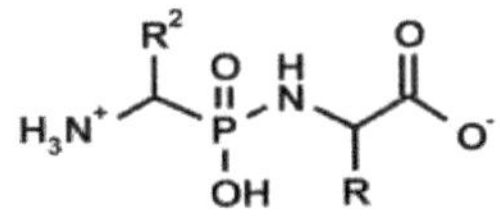
PMG [9]	Glu$^{\gamma-P}$ [10]	(AC)-AA^P[R=H, (AC)-Gly^P] [11]	AA^C-AA^P(R=R^1=Me, Ala-Ala^P) [12]	AA^P-AA^C [13,14]

Abbreviation of AA^P and (AC)-AA^P follow the general rules elaborated by Kudzin et al. [15].

Several papers reflected the complexing abilities of this class of compounds [16–18] and their pharmacological [19–22], agro-chemical [23] and industrial [11,24,25] applications. Therefore, the studies on the synthesis [26–30] and physico-chemical properties [31–40] of AA^P and their derivatives constitute an important topic in chemistry and biochemistry also in addition to material science (e.g., Self-assembled monolayers agents, SAMs [41]).

The 1-(acylamino)alkylphosphonic acids (AC)-AA^P belong to the interesting group of compounds that is of potential pharmacological importance since the corresponding 1-*N*-aminoacylamino derivatives of AA^C-AA^P (mixed *P*-terminal phosphono-dipeptides, e.g., Alaphosphaline) exhibit antibacterial activity [1,23] (Table 1).

1-(Acylamino)alkylphosphonic acids [36,37] and esters [42] are easily formed from starting aminoalkylphosphonic acids, the latter being substrates for selective deacylation to *O*,*O*-dialkyl aminoalkylphosphonates [43]. Both of these groups are substrates for stereoselective enzymatic hydrolysis [44–48] (Figure 1). 1-(Acylamino)alkylphosphonic acids are also formed in so-called Engelman-Pikl-Oleksyszyn method synthesis of 1-aminoalkylphosphonic and 1-aminoalkylphosphonic acids, which starts from appropriate amides and phosphorus chlorides [10,27,49,50] (Figure 2). As some secondary 1-aminoalkanephosphonates are unstable in the process of acidic degradation [40], alkaline hydrolysis can present an alternative method in such procedures.

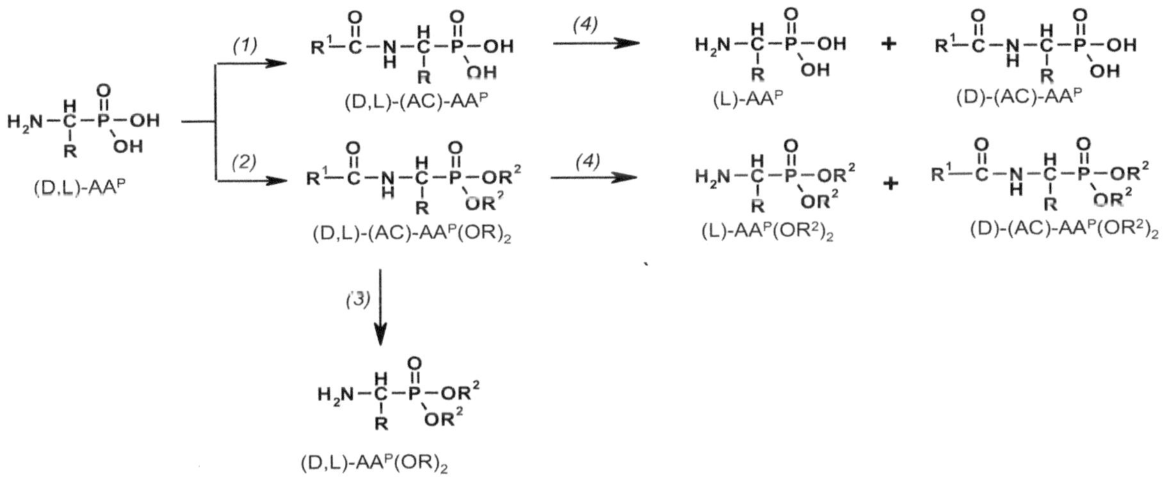

Figure 1. Transformation of 1-aminoalkylphosphonic acids via *N*-acyl-derivatives [(*1*) R^1-C(O)OH/[R^1-C(O)]$_2$O [36–38]; (*2*) R^1-C(O)OH/[R^1-C(O)]$_2$O/HC(OR^2)$_3$ [42]; (*3*) NaBH$_4$/MeOH [43]; (*4*) enzymatic hydrolysis of racemic (D,L)-(AC)-AA^P and (D,L)-(AC)-AA^P(OR)$_2$ [44–48]].

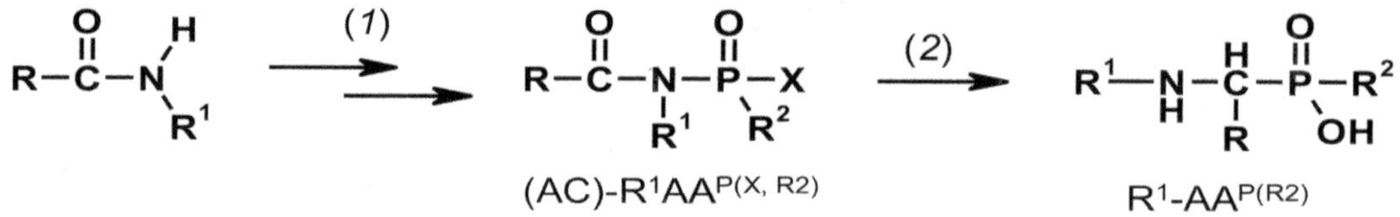

Figure 2. Scheme of Engelmann-Pikl-Oleksyszyn methods for synthesis of secondary 1-aminoalkylphosphonic (R^2=OH) and 1-aminoalkylphosphinic (R^2=H, alkyl) acids (*1*) aldehyde, R-PCl_2/AcOH; and (2) 5M HCl-H_2O, Δ8–10 h.

The susceptibility of (AC)-AA^P to hydrolysis/solvolysis constitutes an important factor influencing their biological activity, especially during penetration through cell membranes. Therefore, the detailed studies on deacylation of these compounds should be helpful in obtaining a deeper understanding of this phenomenon. To tackle this problem, we have decided to carry out studies devoted to deacylation of (AC)-AA^P under different reaction conditions. Our preliminary results of experiments were carried out in aqueous media in the pH range of 0–6.5, which occurs namely in an aqueous 2M HCl and buffer solutions. These are briefly described in a recently published article [51]. As an obvious and necessary continuation of this topic, this paper presents our results on the deacylation of representative types of 1-(acylamino)alkylphosphonic acids (AC)-AA^P in aqueous 2M KOH solution. These include 1-(acetylamino)alkylphosphonic Ac-AA^P, 1-(trifluoroacetylamino)alkylphosphonic acids TFA-AA^P and 1-(benzoylamino)alkylphosphonic Bz-AA^P, derived from representative 1-aminoalkylphosphonic acids AA^P (Gly^P, Ala^P, Val^P, Pgl^P and Phe^P).

2. Results and Discussion

It is generally known that amides can be hydrolyzed with either acidic or basic catalysis, with products being the free acid and the ammonium/substituted ammonium ions or the salts of the acid and ammonia/amine, respectively. Both the acid- and base-catalyzed hydrolyses are essentially irreversible, since salts are formed in both cases [52–55]. Water alone is not sufficient to hydrolyze most of amides [56]. The very low rate of amide hydrolysis by water has been measured by Kahne & Still [57].

A kinetic study has been conducted on the alkaline hydrolyses of several types of amides, including formanilides [58,59], *N*-methylformanilides and *N*-acetanilides [60], 1,8-bis(trifluoroacetylamino)-naphthalene [61] also in addition to *N*-methylacetamide, *N*-methylbenzamide and acetanilide [62]. A mild protocol for the alkaline hydrolysis of secondary and tertiary amides in non-aqueous conditions at room temperature or under reflux has been recently described [63].

The structural analogy of 1-(acylamino)alkylphosphonic acids (AC)-AA^P and 1-(acylamino)-alkanoic acids (AC)-AA^C or amides causes the mechanism of hydrolytic scission of the amide linkage R-C(O)-N in common amides and in 1-(acylamino)alkylphosphonic acids. These should be ruled by similar if not the same mechanisms.

2.1. *Investigations on the Deacylation Course of 1-(acylamino)alkylphosphonic Acids*

The (AC)-AA^P contain amidic functions, which has a hydrolytic sensitivity that should be dependent on the structure of their acyl moieties and on the type of applied hydrolytic medium [49]. For conducting an inquiry into the hydrolytic stability of (AC)-AA^P, we undertook deacylation investigations of these compounds in aqueous 2M KOH solutions.

The application of ^{31}P NMR monitoring of the reaction course enabled identification and quantification of phosphoric components of the reaction mixtures formed. Essentially, monitoring of the deacylation course of Ac-Gly^P in 2M KOH solution reveals the formation of Gly^P (Figure 3). The results of ^{31}P NMR monitoring of the deacylation of the various types of (AC)-AA^P (Ac-AA^P, Bz-AA^P and TFA-AA^P) at a temperature of 25 °C and exposition period up to 240 h are presented in Figure 4 (Ac-AA^P), Figure 5 (TFA-AA^P) and Figure 6 (TFA-AA^P), respectively.

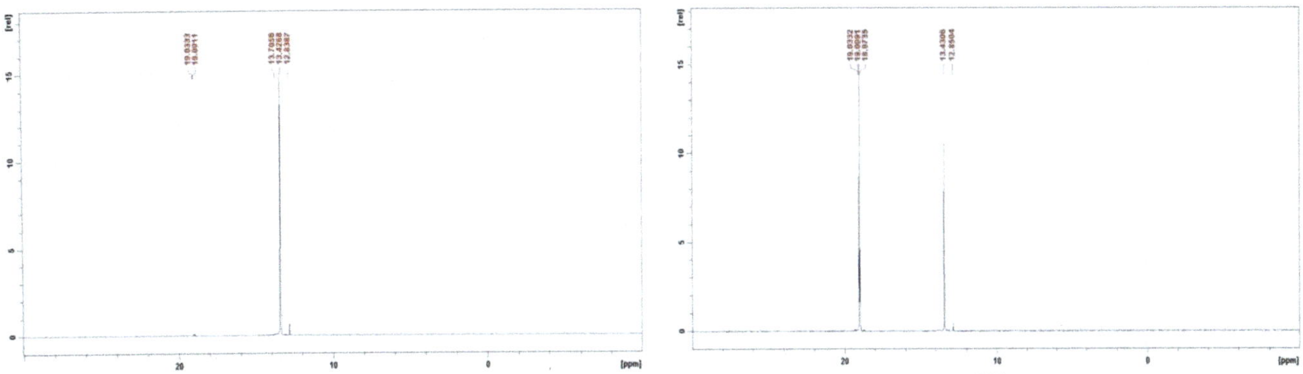

(A) (B)

Figure 3. ^{31}P NMR spectra of Ac-Gly^P recorded in 2 M KOH. (**A**) Immediately after mixing the reaction mixture, and (**B**) After 192 h of exposition of the reaction mixture. [Ac-Gly^P: 13.4 ppm; Gly^P: 19.0 ppm].

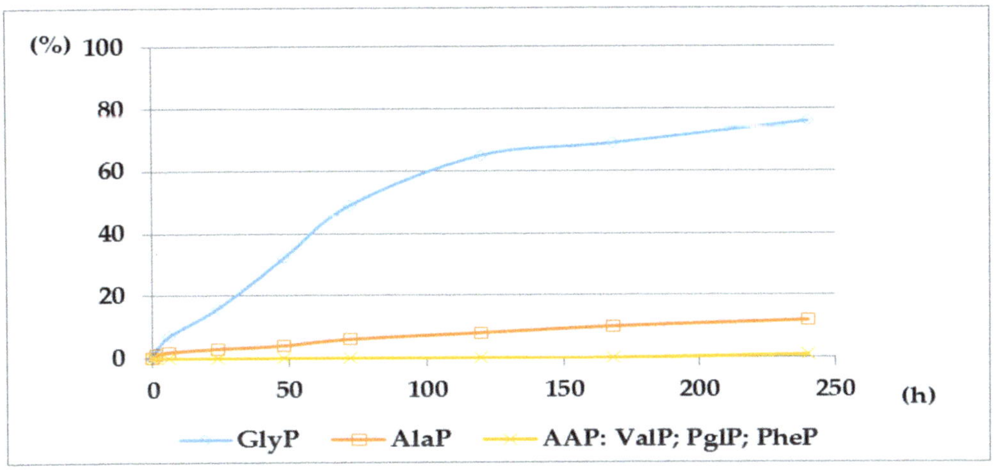

Figure 4. A solution of 0.1 M 1-(acetylamino)alkylphosphonic acids Ac-AA^P in 2 M KOH reacted at 25 °C, which was monitored by ^{31}P NMR [DD (%) vs. time (h)].

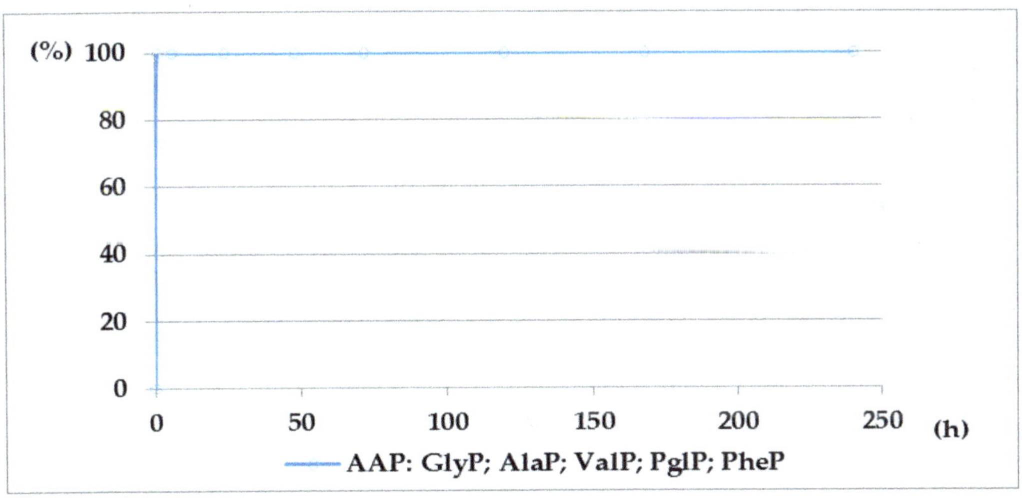

Figure 5. A solution of 0.1 M 1-(trifluoroacetylamino)alkylphosphonic acids TFA-AA^P in 2 M KOH reacted at 25 °C, which was monitored by ^{31}P NMR [DD (%) vs. time (h)].

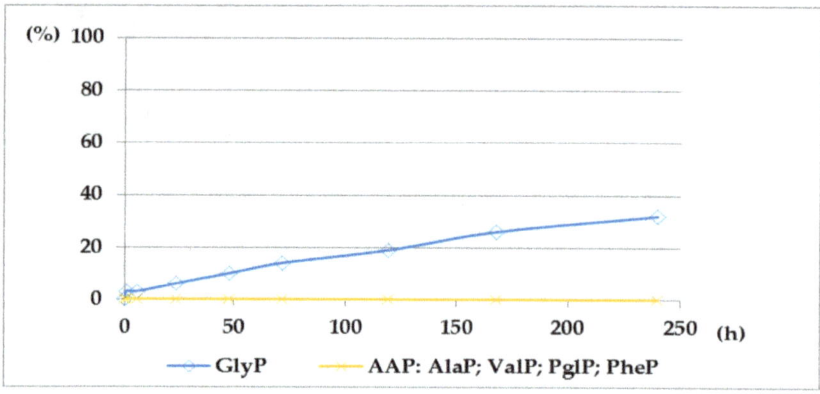

Figure 6. A solution of 0.1 M 1-(benzoylamino)alkylphosphonic acids Bz-AA^P in 2 M KOH reacted at 25 °C, which was monitored by ^{31}P NMR [DD (%) vs. time (h)]. Thus, the presented graphs illustrate much higher stability of benzoyl-derivatives Bz-AA^P (Bz-Gly^P: DD_{200h} ~30%; for Bz-Ala^P and Bz-Pgl^P: DD_{200h} < 3%) in comparison with acetyl-derivatives Ac-AA^P (Ac-Gly^P: DD_{200h} ~70%; for Ac-Ala^P: DD_{200h} ~20%; and Ac-Pgl^P: DD_{200h} < 7%).

The comparison of hydrolytic stability of TFA-Gly^P, Ac-Gly^P and Bz-Gly^P in 2M KOH is presented in Figure 7.

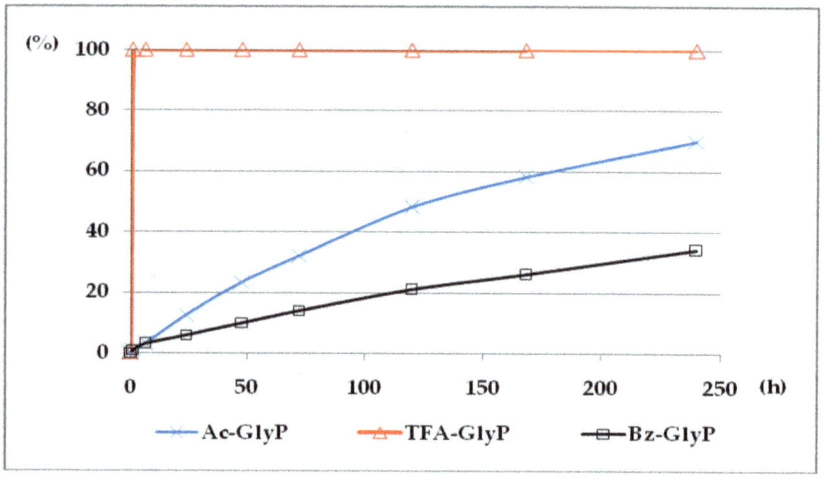

Figure 7. A solution of 0.1 M 1-(acylamino)methylphosphonic acids (AC)-Gly^P (AC: Ac, TFA, Bz) in 2M KOH reacted at 25 °C, which was monitored by ^{31}P NMR [DD (%) vs. time (h)]. Points on the plots corresponds to Gly^P released from the parent (AC)-Gly^P, which means Gly^P from Ac-Gly^P, Gly^P from Bz-Gly^P and Gly^P from TFA-Gly^P.

Since no products of the scission of the C-P bond in all investigated cases were observed, the deacylation mechanism of (AC)-AA^P in alkaline solutions can be illustrated by an analogy to the corresponding basic deacylation of 1-(acylamino)alkanoic acids (Scheme 1) [54,62,63].

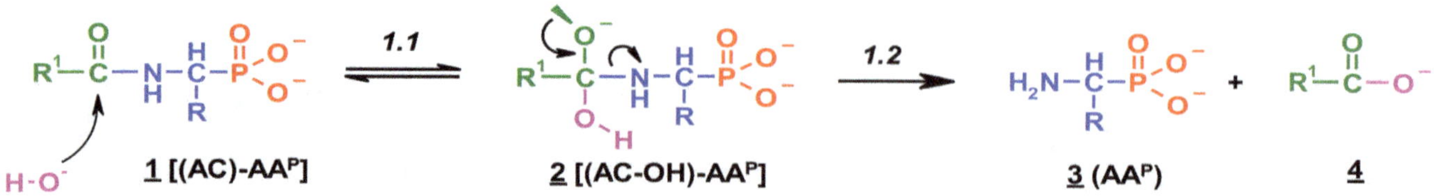

Scheme 1. Hypothetical mechanism of deacylation of (AC)-AA^P in 2 M KOH solution.

This includes the prior interaction of hydroxyl ion on the carbonyl carbon of 1, which occurs with the formation of the tetrahedral intermediate state 2 (Scheme 1). These intermediates were observed on ^{31}P NMR spectra as signals down-shifted in comparison with the parent signals of AC-AA^P (e.g., Figure 2). The ^{31}P NMR chemical shifts of (AC)-AA^P (1) and the corresponding adduct compounds AC(OH)-AA^P (2) formed are listed in Table 2.

Table 2. ^{31}P NMR chemical shifts of (AC)-AA^P and the corresponding adduct compounds (AC)(OH)-AA^P formed immediately after mixing the reaction mixture. (**A**) ^{31}P NMR chemical shifts of Ac-AA^P and the corresponding adduct compounds Ac(OH)-AA^P formed. (**B**) ^{31}P NMR chemical shifts of (AC)-Gly^P and the corresponding adduct compounds [AC(OH)]-Gly^P formed.

(A)

	Ac-Gly^P		**Ac-Ala^P**		**Ac-Nva^P**		**Ac-Val^P**		**Ac-Pgl^P**	
	Ac-Gly^P	Ac(OH)-Gly^P	Ac-Ala^P	Ac(OH)-Ala^P	Ac-Nva^P	Ac(OH)-Nva^P	Ac-Val^P	Ac(OH)-Val^P	Ac-Pgl^P	Ac(OH)-Pgl^P
δ	13.4	*12.8*	17.0	*16.5*	15.9	*15.2*	16.8	*16.1*	13.3	*12.7*
%	97.1	*2.9*	97.1	*2.9*	96.3	*3.7*	97.0	*3.0*	97.0	*3.0*

(B)

	Ac-Gly^P			**Prp-Gly^P**			**Bz-Gly^P**		
	Gly^P	Ac-Gly^P	Ac(OH)-Gly^P	Gly^P	Prp-Gly^P	Prp(OH)-Gly^P	Gly^P	Bz-Gly^P	Bz(OH)-Gly^P
δ	19.1	13.4	12.8	19.3	13.4	12.8	19.3	13.5	13.2
%	-	97.1	2.9	-	98.3	1.7	-	99.5	<0.5

These compounds [AC(OH)]-AA^P] slowly disappeared during the deacylation progress (Figure 8).

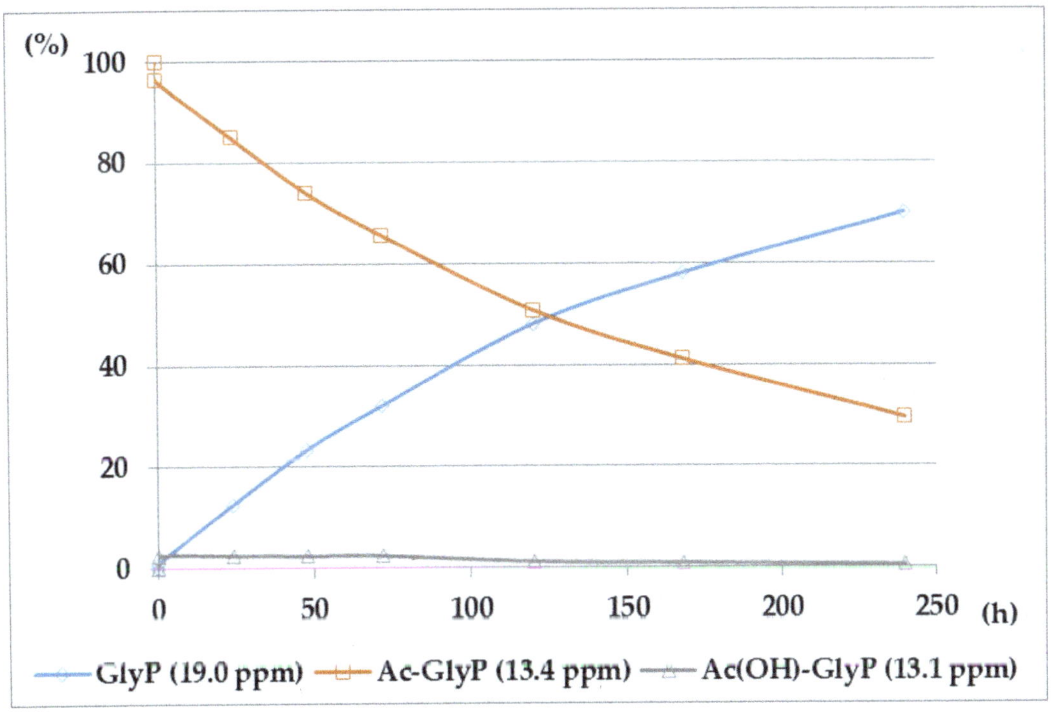

Figure 8. Ac-Gly hydrolysis in 2M KOH [DD (%) vs. time (h)].

The rehybridization (sp^3→sp^2) of the carbonyl carbon in 2 enforces the splitting of the amide bond R-C(O)-N and the subsequent formation of the aminophosphonate 3.

The comparison of (AC)-AA^P stability in aqueous 2M KOH and 2M HCl solutions is given in Figure 9.

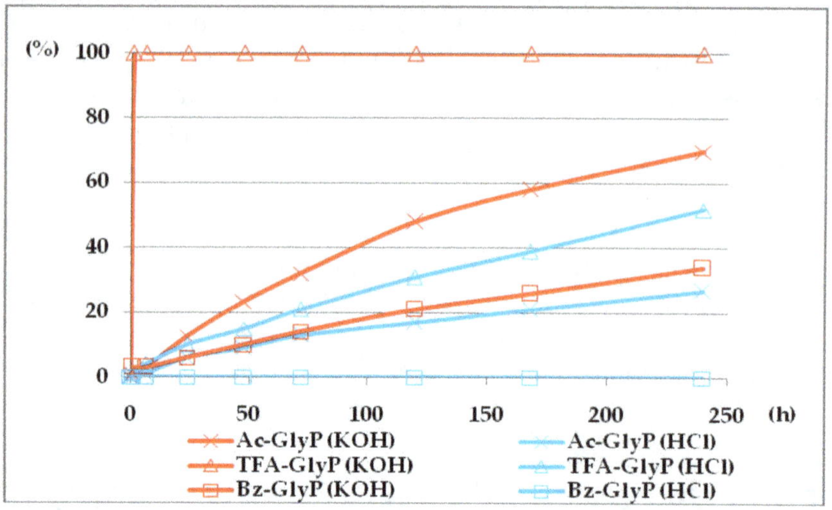

Figure 9. (AC)-Gly^P(Ac-Gly^P, TFA-Gly^P, TFA-Gly^P) hydrolysis in aqueous solutions of 2M KOH and 2M HCl [DD (%) vs. time (h)].

2.2. Quantum Chemical Calculations for Deacylation of (AC)-AA^P in Basic Conditions

The reaction illustrated in Scheme 1 was divided into two stages (reactions *1.1* and *1.2*, respectively), which were studied separately. Stage *1.1* concerns the addition of hydroxide to amidophosphonate anion 1 with the formation of trianion 2 (the example structure of the addition product 2 for Ac-Ala^P (R=R^1=Me) is given in Figure 10). Stage *1.2* involves dissociation of the trianion 2 into aminophosphonate dianion 3 and carboxylate anion 4. We assumed that the transfer of proton from OH to the amide nitrogen in 2 proceeds simultaneously with dissociation. The thermodynamics of this two-stage process was studied by density functional theory (DFT) methods in the gas phase and in aqueous solution. Gas phase calculations were performed only for comparison with the aqueous system, which was the real reaction system to illustrate the essential role of the solvation effect.

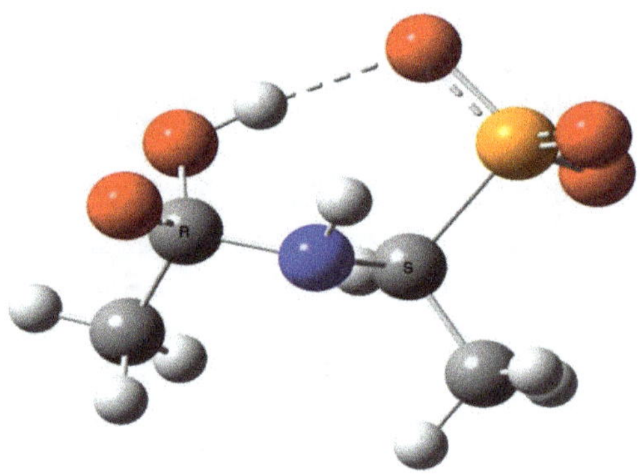

Figure 10. The structure of the addition product **2** (R=R^1=Me, Ac(OH)-Ala^P) with intramolecular hydrogen bond between OH and phosphoryl group (orange—phosphorus, blue—nitrogen, red—oxygen).

Stage 1: Hydroxide anion attack on the carbonyl moiety of (AC)-AA^P dianion 1 (Scheme 1, r. *1.1*).

This reaction stage presumably proceeds by the attack of a lone pair of negatively charged oxygen atoms on sp^2 carbon in the carbonyl group. This process is highly unfavorable since it involves the reaction of two anions with the formation of the molecule with a total charge of −3. The energy of this reaction in the gas phase is very high, so this process seems to be thermodynamically very unlikely (Table 3).

Table 3. Gibbs free energies ΔG (298 K) for reaction *1.1* in gas phase (kcal/mol).

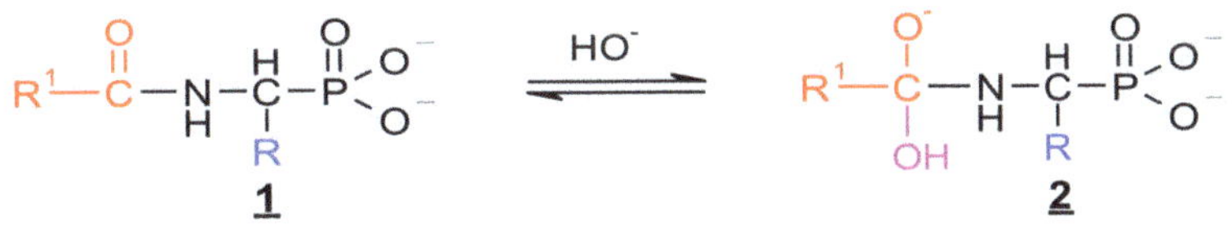

(AC)-AA^P (1Xx)	R	*R^1-C(O)-* (1Xx)		
		CH_3- (1Ax)	*CF_3-* (1Bx)	*Ph-* (1Cx)
(AC)-Gly^P (1Xa)	H-	139.43	123.16	126.87
(AC)-Ala^P (1Xb)	CH_3-	135.19	119.26	124.82
(AC)-Val^P (1Xc)	$(CH_3)_2CH$-	133.61	117.48	132.28
(AC)-Pgl^P (1Xd)	Ph-	125.63	121.43	117.08
(AC)-Phe^P (1Xe)	$PhCH_2$-	139.31	122.06	129.57

However, when carried out in aqueous solutions, this reaction is much less unfavorable due to the stabilization of ionic structures by the polar solvent (Table 2). Similarly, this reaction should also be facilitated by counterions, although these were omitted in order to keep the model system simple. The values of the Gibbs free energy ΔG^{298} (kcal/mol) for reaction *1.1* in gas phase and in aqueous phase are listed in Tables 3 and 4, respectively.

Table 4. Gibbs free energies ΔG (298 K) for reaction *1.1* in aqueous phase (kcal/mol).

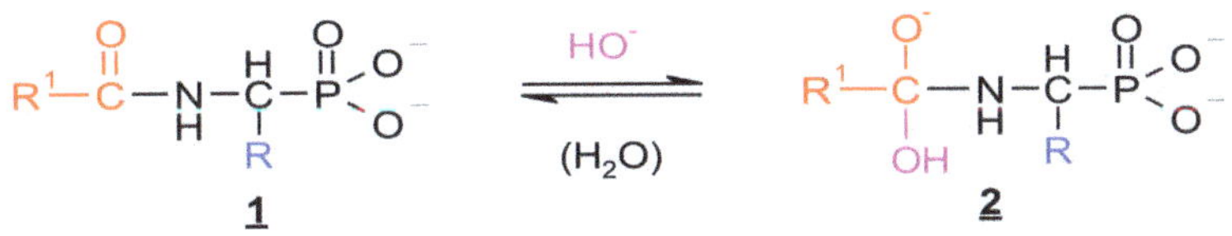

(AC)-AA^P (1Xx)	R	*R^1-C(O)-* (1Xx)		
		CH_3- (1Ax)	*CF_3-* (1Bx)	*Ph-* (1Cx)
(AC)-Gly^P (1Xa)	H-	26.70	11.77	25.70
(AC)-Ala^P (1Xb)	CH_3-	26.26	10.54	25.45
(AC)-Val^P (1Xc)	$(CH_3)_2CH$-	26.15	10.90	26.57
(AC)-Pgl^P (1Xd)	Ph-	25.47	12.27	24.96
(AC)-Phe^P (1Xe)	$PhCH_2$-	24.22	7.73	24.53

The free energy barriers, $\Delta G^{\ddagger}$ (298 K), calculated for reaction *1.1* in a water solution are shown in Table 5. The lowest energy barriers for OH^- addition (Scheme 1, r. *1.1*) occur when $R^1 = CF_3$, which results from stabilization of the negative charge by the electron-withdrawing CF_3 group.

Table 5. Free energy barriers ($\Delta G^{\ddagger}$ (298 K), kcal/mol) for OH^- addition (reaction *1.1*) in water.

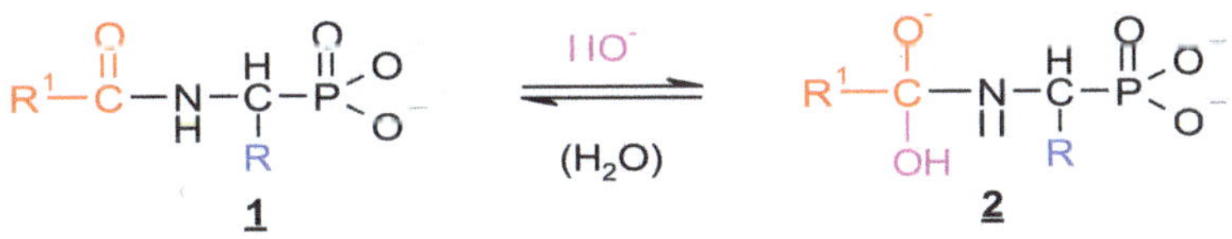

(AC)-AA^P (1Xx)	R	*R^1-C(O)-* (1Xx)		
		CH_3- (1Ax)	*CF_3-* (1Bx)	*Ph-* (1Cx)
(AC)-Gly^P (1Xa)	H-	29.99	19.51	25.70
(AC)-Ala^P (1Xb)	CH_3-	30.85	19.12	26.75
(AC)-Val^P (1Xc)	$(CH_3)_2CH$-	31.40	18.47	29.68
(AC)-Pgl^P (1Xd)	Ph-	28.57	17.60	28.44
(AC)-Phe^P (1Xe)	$PhCH_2$-	28.85	18.66	30.23

Stage 2: Dissociation of trianion 2 into aminophosphonate dianion 3 and carboxylate anion 4 (Scheme 1, reaction *1.2*).

This reaction stage presumably involves the proton transfer from the geminal hydroxyl group to nitrogen with dissociation of the C-N bond. This hypothesis is supported by the transition structures found for this reaction. The example of this structure for the proton transfer/dissociation of the C-N bond in Ac-Ala$^{P(3-)}$ is shown in Figure 11. Vibrational analysis indicates that the imaginary vibration is associated with proton transfer from OH to NH. In the structure of the transition state shown in Figure 11, the geminal C-O bond distance is 1.284 Å, which corresponds well to C=O double bond. Furthermore, the C-N distance is 1.802 Å compared to 1.503 Å in the starting amide, which indicates that the bond breaking in the transition state is significantly advanced.

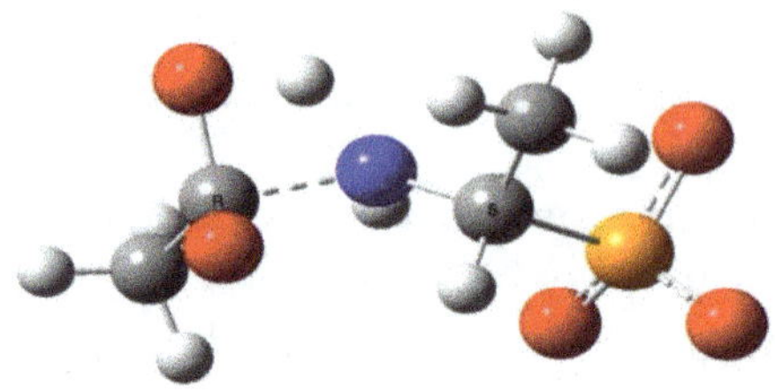

Figure 11. The transition state for proton transfer/dissociation of the C-N bond in Ac(OH)-Ala$^{P(3-)}$.

The Gibbs free energies for the dissociation reaction *1.2* in the gas phase and in water are presented in Tables 6 and 7, respectively.

Table 6. Gibbs free energy ΔG (298 K) for the reaction *1.2* in gas phase (kcal/mol).

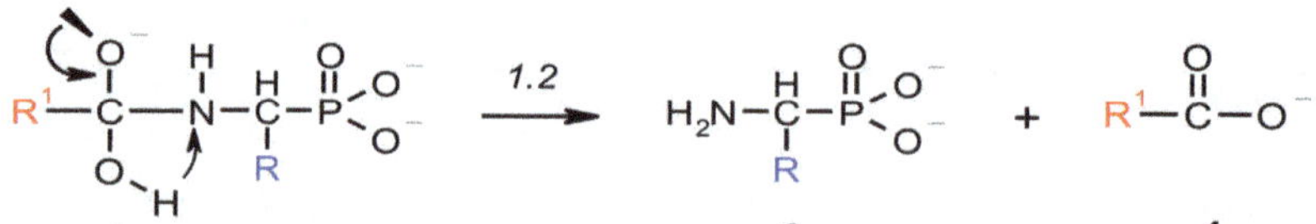

2 3 4

(AC-OH)-AA^P (2-Xx)	R	R^1-C(O⁻)(OH)- (2Xx)		
		CH_3- (2Ax)	CF_3- (2Bx)	*Ph*- (2Cx)
(AC-OH)-Gly^P (2Xa)	H-	−165.22	−153.75	−152.91
(AC-OH)-Ala^P (2Xb)	CH_3-	−159.50	−149.37	−149.31
(AC-OH)-Val^P (2Xc)	$(CH_3)_2CH$-	−162.42	−152.74	−161.46
(AC-OH)-Pgl^P (2Xd)	Ph-	−157.24	−160.78	−150.14
(AC-OH)-Phe^P (2Xe)	$PhCH_2$-	−168.00	−158.21	−159.56

Table 7. Gibbs free energy ΔG (298 K) for the reaction *1.2* in an aqueous phase (kcal/mol).

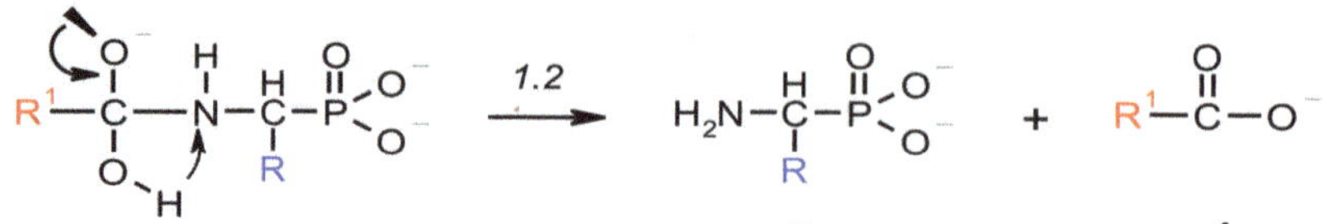

2 3 4

(AC-OH)-AA^P (2-Xx)	R	R^1-C(O⁻)(OH)- (2Xx)		
		CH_3- (2Ax)	CF_3- (2Bx)	*Ph*- (2Cx)
(AC-OH)-Gly^P (2Xa)	H-	-46.41	-40.32	-48.51
(AC-OH)-Ala^P (2Xb)	CH_3-	-46.62	-38.58	-48.56
(AC-OH)-Val^P (2Xc)	$(CH_3)_2CH$-	-44.45	-37.28	-47.29
(AC-OH)-Pgl^P (2Xd)	Ph-	-45.60	-40.90	-47.44
(AC-OH)-Phe^P (2Xe)	$PhCH_2$-	-47.74	-39.80	-50.89

Comparing the thermodynamic quantities in Tables 3 and 6, it is evident that the total free energy of the reaction *1* is negative in all cases by approximately −30 kcal/mol and thus, the entire process *1* is thermodynamically feasible. However, the very high energy barrier for the first step of reaction *1.1* (Table 3) prevents this reaction from proceeding in the gas phase.

Analogously, the reaction 1 in water is equally thermodynamically favorable (ΔG = −20 to −30 kcal/mol, Tables 4 and 7), but the energy barrier associated with the first step *1.1* is much lower in this case (Table 4). Table 8 presents the results of free energy barrier calculations for reaction *1.2* in water.

Table 8. Free energy barriers (ΔG‡ (298 K), kcal/mol) for proton transfer/dissociation of the C-N bond in water (reaction *1.2*).

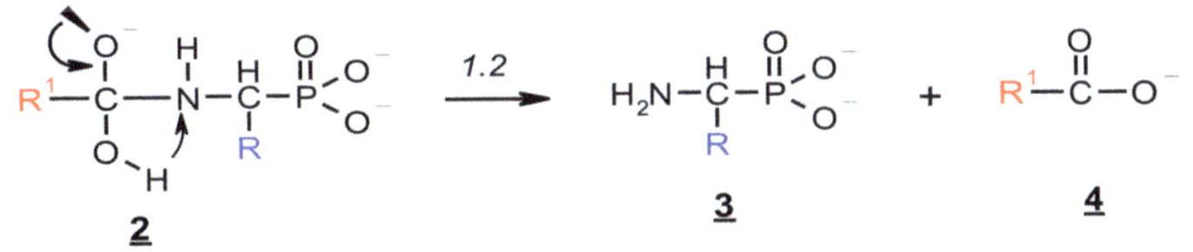

(AC-OH)-AA^P (2-Xx)	R	*R*¹-*C(O⁻)(OH)*- (2Xx)		
		CH₃- (2Ax)	*CF₃*- (2Bx)	*Ph*- (2Cx)
(AC-OH)-Gly^P (2Xa)	H-	20.76	21.54	16.91
(AC-OH)-Ala^P (2Xb)	CH_3-	21.47	23.43	19.00
(AC-OH)-Val^P (2Xc)	$(CH_3)_2CH$-	23.73	24.91	20.56
(AC-OH)-Pgl^P (2Xd)	Ph-	20.05	21.74	18.51
(AC-OH)-Phe^P (2Xe)	$PhCH_2$-	22.50	25.00	18.99

The total Gibbs free energy profiles of entire process *1* for two selected pairs of 1-(acylamino)alkylphosphonic acids [Ac-Ala$^{P(2-)}$, TFA-Ala$^{P(2-)}$, Bz-Pgl$^{P(2-)}$ and TFA-Pgl$^{P(2-)}$] in water solution are shown in Figure 12.

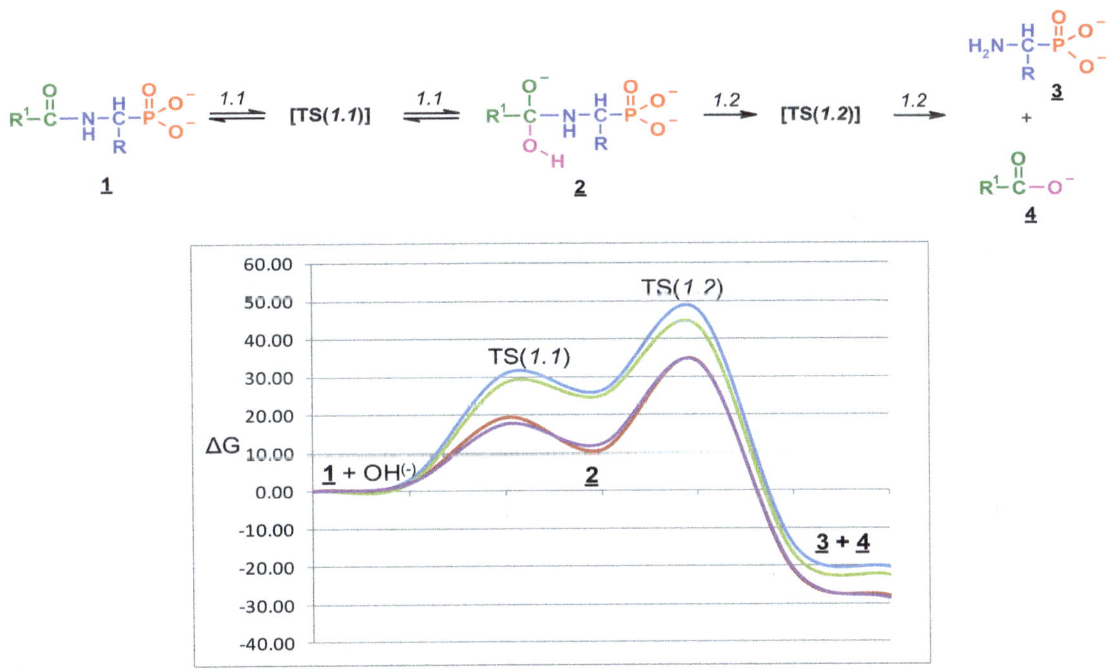

Figure 12. Free energy profiles ΔG^{298} (kcal/mol) for the reaction of Ac-Ala$^{P(2-)}$ (blue), TFA-Ala$^{P(2-)}$ (red), Bz-Pgl$^{P(2-)}$ (green), and TFA-Pgl$^{P(2-)}$ (purple) with hydroxyl ions in water solution (the general reaction scheme is given above).

The total Gibbs free energies and reaction barriers for the entire process *1* are shown in Table 9. Moreover, Figure 12 reveals that the energy profiles for both TFA-substituted phosphonic acids are essentially identical and all stationary points are significantly lower in energy than Ac-substituted phosphonic acids.

Table 9. Total Gibbs free energies$\Delta G^{298}(1.1) + \Delta G^{298}(1.2)$ for the overall deacylation process in the aqueous phase and total Gibbs free energy barriers ($\Delta G^{298}(1.1) + \Delta G^{\ddagger}(1.2)$, in parentheses) (kcal/mol).

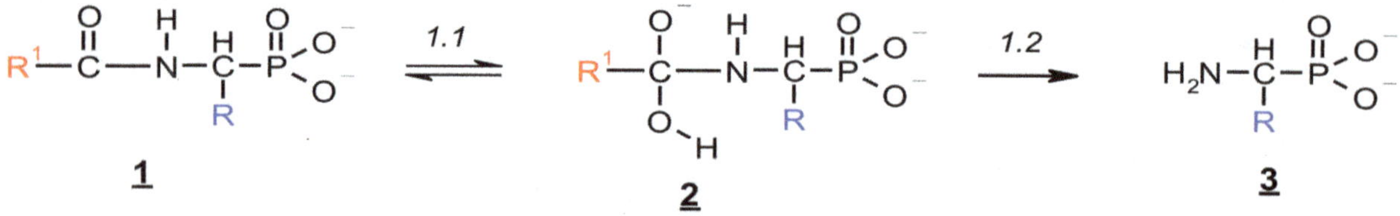

(AC)-AA^P (1Xx)	R	R¹-C(O)- (1Xx)		
		CH_3- (1Ax)	CF_3- (1Bx)	Ph- (1Cx)
(AC)-Gly^P (1Xa)	H-	−19.72 (47.5)	−28.56 (31.3)	−22.82 (42.6)
(AC)-Ala^P (1Xb)	CH_3-	−20.36 (47.7)	−28.04 (34.0)	−23.11 (44.5)
(AC)-Val^P (1Xc)	$(CH_3)_2CH$-	−18.30 (49.9)	−26.38 (35.8)	−20.72 (47.1)
(AC)-Pgl^P (1Xd)	Ph-	−20.13 (45.5)	−28.63 (34.0)	−22.48 (43.5)
(AC)-Phe^P (1Xe)	$PhCH_2$-	−23.53 (46.7)	−32.08 (32.7)	−26.36 (43.5)

Therefore, the presence of a TFA group is the only factor determining the free energy barrier of the reaction. From the energy profile, it is obvious that the overall energy barrier in water $\Delta G^{\ddagger}$ is G^{298}[TS (*1.2*)] − (G^{298}(1) + $G^{298}(OH^-)$), while the net Gibbs energy effect for the reaction in water solution can be calculated as $\Delta G_r^{298} = G^{298}(\underline{3}) + G^{298}(\underline{4}) - [(G^{298}(\underline{1}) + G^{298}(OH^-)] = \Delta G^{298}(1.1) + \Delta G^{\ddagger}(1.2)$.

The total energetic effect of the process *1* in water is negative, which means that the overall reaction is thermodynamically feasible (Table 8). The free energy of the process when R^1=CF_3 is also approximately 8 kcal/mol more favorable than for R^1=Me. The comparison of values in Table 9 also reveals that the transformation *1* of substrates with R^1=Ph are thermodynamically more favored by approximately 3 kcal/mol than R^1=Me. Thus, the thermodynamic effect of reaction *1* depends on the R^1 substituents in the order of: CH_3 < Ph < CF_3. Table 9 also leads to the conclusion that the substrates with a trifluoromethyl group in the R^1 position systematically show free energy barriers that are approximately 10–15 kcal/mol lower than those for other substituents R^1, due to the additional stabilization of the trianion 2 by the inductive effect of the strongly electron-withdrawing CF_3 group. This suggests that trifluoromethyl-substituted anions should be more reactive than the other derivatives.

3. Materials and Methods

3.1. General Information

The ^{31}P NMR spectra were recorded on a Bruker AV 200 spectrometer (Rheinstetten, Germany) operating at 81.01 MHz and on a Bruker Avance III 600 spectrometer (Bruker BioSpin, Rheinstetten, Germany) operating at 242.9 MHz. The1H NMR spectra were recorded on a Bruker Avance III 600 spectrometer operating at 600 MHz. The positive chemical shift values of ^{31}P were reported for compounds that were absorbing at lower fields than H_3PO_4. The purity of the 1-aminoalkylphosphonic acids and corresponding 1-(acylamino)alkylphosphonic acids were determined by pH-metric titration using a computer aided automatic titrator connected to the EMU-meter (Wroclaw University of Science and Technology, Wrocław, Poland), which was fitted with a combined glass microelectrode Crison 5028.

3.2. *Investigated Compounds*

Phosphonoglycine (Gly^{P}) was obtained according to a previous study [64]. The 1-Aminoalkylphosphonic acids of phosphonoalanine (Ala^{P}); phosphonovaline (Val^{P}); phosphonophenylglycine (Pgl^{P}) and phosphonophenylalanine (Phe^{P}) have been prepared according to a previous study [65]. 1-(Acetylamino)alkylphosphonic acids (Ac-AA^{P}: Ac-Gly^{P}; Ac-Ala^{P}; Ac-Val^{P}; Ac-Pgl^{P} and Ac-Phe^{P}) and 1-(benzoylamino)alkylphosphonic acids (Bz-AA^{P}: Bz-Gly^{P}; Bz-Ala^{P}; Bz-Val^{P}; Bz-Pgl^{P} and Bz-Phe^{P}) were synthesized according to a previous study [36]. 1-(Trifluoroacetylamino)alkylphosphonic acids (TFA-AA^{P}: TFA-Gly^{P}; TFA-Ala^{P}; TFA-Val^{P}; TFA-Pgl^{P} and TFA-Phe^{P}) were synthesized according to a previous study [37]. (Structures, names and abbreviations of aminophosphonic acids and corresponding 1-(acylamino)phosphonic acids discussed in this work are given in Table 10). Purity of all (AC)-AA^{P} synthesized was determined by ^{31}P NMR and ^{1}H NMR in addition to potentiometric titration. Analytical standards (fixanals) and other reagents were purchased from Aldrich–Sigma (Poznań, Poland).

Table 10. Structures, names and abbreviations of aminophosphonic acids and corresponding 1-(acylamino)phosphonic acids discussed in this work [a].

		Name	Trivial Name	Abbr./(No)	Ref.
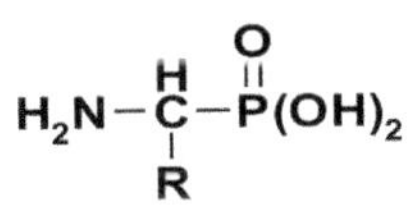		Aminoalkyl-phosphonic acid	Phosphono-Amino Acid	AA^{P} (<u>3</u>)	
	R				
	H	Aminomethyl-phosphonic acid	Phosphono-glycine	Gly^{P} (<u>3a</u>)	[64]
	Me	1-Aminoethyl-phosphonic acid	Phosphono-alanine	Ala^{P} (<u>3b</u>)	[65]
	iPr	1-Amino-1-methylethyl-phosphonic acid	Phosphono-valine	Val^{P} (<u>3c</u>)	
	Ph	1-Amino-1-phenylmethyl-phosphonic acid	Phosphono-phenyl-glycine	Pgl^{P} (<u>3d</u>)	
	$PhCH_2$	1-Amino-1-phenylethyl-phosphonic acid	Phosphono-phenyl-alanine	Phe^{P} (<u>3e</u>)	
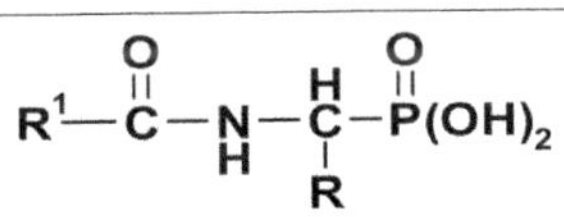		1-(Acylamino)alkyl-phosphonic acid	Acylamino-Phosphono Acid	(AC)-AA^{P} (<u>2</u>)	
R^{1}	R				
CH_3	H, Me, iPr, Ph, $PhCH_2$	1-(Acetylamino)alkyl -phosphonic acid	Acetylamino-Phosphono Acid	Ac-AA^{P} (<u>2A</u>)	[36]
CH_3	H	1-(Acetylamino)methyl-phosphonic acid	Phosphono-Acetylamino Glycine	Ac-Gly^{P} (<u>2Aa</u>)	
CH_3	Ph	1-(Acetylamino)-1-phenyl-methyl-phosphonic acid	Phosphono-Acetyl-amino-PhenylGlycine	Ac-Pgl^{P} (<u>2Ad</u>)	
CF_3	H, Me, iPr, Ph, $PhCH_2$	1-(Trifluoroacetylamino)-alkyl-phosphonic acid	Phosphono-Trifluoroacetyl-amino Acid	TFA-AA^{P} (<u>2B</u>)	[37]
CF_3	iPr	1-(Trifluoroacetylamino)-2-methylethyl-phosphonic acid	Phosphono-Trifluoroacetyl-Valine	TFA-Val^{P} (<u>2Bc</u>)	
Ph	H, Me, iPr, Ph, $PhCH_2$	1-(Benzoylamino)alkyl-phosphonic acid	Phosphono-Benzoylamino Acid	Bz-AA^{P} (<u>2C</u>)	[36]
Ph	$PhCH_2$	1-(Benzoylamino)-2-phenyl-ethyl-phosphonic acid	Phosphono-Benzoyl-Phenyl-alanine	Bz-Phe^{P} (<u>2Ce</u>)	

[a] Applied names were in accordance with the IUPAC rules, and the abbreviations were in agreement with the general rules elaborated by Kudzin et al. [15].

3.3. *Reaction of Deacylation of 1-(acylamino)alkylphosphonic Acids*

A sample of (AC)-AA^P (0.5 mmol) was placed into 5-mL V-vials, before being dissolved in 5 mL of appropriate solution fortified with D_2O (2 M KOH). The vials were maintained at 25 °C ± 0.2 °C. At various time intervals, the vials were removed from the baths and the samples of 0.5 mL were taken for ^{31}P NMR analysis. The concentration of the substrate and formed products were determined from the integration of its NMR signal.

3.4. *Computational Methods*

All quantum mechanical calculations were performed using the Gaussian 09 suite of programs [66]. Geometries of model compounds in the gas phase and in water solution were optimized using the B3LYP hybrid functional containing three-parameter Becke (B3) exchange and Lee-Yang-Parr (LYP) correlation functional and the 6–31+G(d) basis set. All stationary points were identified as stable minima by frequency calculations. The vibrational analysis also provided the thermal enthalpy and entropy corrections at 298 K within the rigid rotor/harmonic oscillator/ideal gas approximation. Computed frequencies were scaled by a factor of 0.98. Transition states were verified by the intrinsic reaction path (IRC) procedure to confirm if they did link the substrates and products along the reaction path.

Free energies of the reaction in water were calculated within a continuum (implicit) solvent approximation using the Conductor-like Polarizable Continuum Model (CPCM) with UFF (Universal Force Field) cavities (SCRF = CPCM option as defined in Gaussian 09 program) [67]. This approximation assumes that molecule of the solute is placed in a cavity within the solvent, which is treated as a dielectric continuum (self-consistent reaction field). Thermodynamic functions in solution were calculated based on vibrational analysis, which was described above.

4. Conclusions

The analysis of ^{31}P NMR results presented above leads to the following conclusions:

- The alkaline deacylation of (AC)-AA^P occurs through the hydroxyl adduct intermediates, which was observed on ^{31}P NMR spectra of the reaction mixtures;
- The deacylation ability of investigated (AC)-AA^P derivatives exhibited strong dependence of electron-acceptor character of the acyl group, with CF_3-C(O)- > CH_3-C(O)-;
- The 1-(acylamino)alkylphosphonic acids are present in aqueous solutions with substantial stability at ambient temperatures, which decreases with temperature elevation;
- The deacylation of (AC)-AA^P increases substantially in basic (2 M KOH) solutions;
- The lowest deacylation ability (highest stability) was found for the 1-(benzoylamino acids Bz-AA^P;
- For the same type of 1-(acylamino)-derivatives (AC)-AA^P, the highest stability was found in(AC)-Val^P and (AC)-Phe^P; lower stability in (AC)-Ala^P; and the lowest in (AC)-Gly^P and (AC)-Pgl^P;
- All examined ^{31}P NMR spectra of deacylation mixtures did not reveal any trace of H_3PO_3 and/orH_3PO_4, which represent the products of dephosphonylation or oxidative dephosphonylation of (AC)-AA^P.

The theoretical calculations lead to the following conclusions:

The substrates with the trifluoromethyl group in R^1 position are predicted to be significantly more reactive than the other phosphonic amino acids studied, which is consistent with the experimental results. The intermediates and products when R^1=CF_3 are also more thermodynamically favored than the other derivatives. This difference in reactivity is due to additional stabilization of the trianion **2** by the inductive effect of the strongly electron-withdrawing CF_3 group. The calculations did not provide unequivocal information about relative reactivities of other AA derivatives as the calculated reaction energy barriers are similar (within a computational error), which is shown in

Table 9 and Figure 12. The calculation results are consistent with those reported for acidic deacylation of 1-(acylamino)alkylphosphonic acids [51]. As CF_3 is strongly electron-withdrawing, this group plays an activating role in basic media, increasing the acidity of carbonyl moiety. In contrast, it has a deactivating effect in acidic media, decreasing the basicity of carbonyl oxygen.

Acknowledgments: The studies at the Textile Research Institute, Łódź, in 2017 were partly financed by the Polish Ministry of Science and Higher Education within statutory research founds (for M.H.K.). DFT calculations were supported by the PL-Grid infrastructure. We would like to thank the National Science Center (Krakow, Poland) for the support in the frame of the grant UMO 2014/15/B/ST5/05329 (for J.D.). Dedicated to the memory of Professor Andrzej Okruszek.

Author Contributions: J.D. and Z.H.K. designed the research study and contributed to the data interpretation and to the manuscript drafting and revisions. M.C. and B.G. carried out the Quantum Chemical Calculations and contributed to writing the manuscript. M.H.K. performed the synthesis and purification of investigated 1-(acylamino)alkylphosphonic acids and was involved in the conception of the research study, analyzed the data, and contributed to writing the manuscript. P.U. recorded NMR spectra and analyzed the experimental data.

References

1. Kukhar, V.P.; Hudson, H.R. (Eds.) *Aminophosphonic and Aminophosphinic Acids: Chemistry and Biological Activity*; J. Wiley & Sons, Ltd.: Chichester, UK; New York, NY, USA; Weinheim, Germany; Brisbane, Australia; Singapore; Toronto, ON, Canada, 2000; ISBN 0-471-89149-5.
2. Kittredge, J.S.; Roberts, E. A carbon-phosphorus compounds in nature. *Science* **1969**, *164*, 37–42. [CrossRef] [PubMed]
3. Kafarski, P.; Mastalerz, P. *Aminophosphonates, Natural Occurrence, Biochemistry and Biological Properties*; Beiträge zur Wirkstofforschung; Institut für Wirkstofforschung: Berlin, Germany, 1984; Volume 24, pp. 1–110.
4. Horiguchi, M.; Kandatsu, M. Isolation of 2-aminoethyl phosphonic amid from rumen protozoa. *Nature* **1959**, *184*, 901–902. [CrossRef] [PubMed]
5. Kittredge, J.S.; Hughes, R.R. Occurence of α-amino-β-phosphono-propionic acid in the zoanthid, zoanthussociatus, and the ciliate, Tetrahymena pyrifornis. *Biochemistry* **1964**, *3*, 991–996. [CrossRef] [PubMed]
6. Korn, E.D.; Deaborn, D.G.; Falles, H.M.; Sokoloski, E.A. A major polysaccharide constituents of the amoeba plasma membrane contains 2-aminoethylphosphonic acid and 1-hydroxy-2-aminoethyl-phosphonic acid. *J. Biol. Chem.* **1973**, *248*, 2257–2259. [PubMed]
7. Kido, Y.; Hamakado, T.; Anno, M.; Miyagawa, E.; Motoki, Y.; Wakamiya, T.; Shiba, T. Isolation and characterization of 15112, a new phosphorus containing inhibitor of angiotensin I converting enzyme produced by *Actinomadura* sp. *J. Antibiot.* **1984**, *37*, 965–969. [CrossRef] [PubMed]
8. Bayer, E.; Gugel, K.H.; Haegele, K.; Hagenmaier, H.; Jessipov, S.; Koenig, W.A.; Zaehner, H. Metabolic products of microorganisms. 98. Phosphinothricine and phosphinothricylo-alanylo-alanine. *Helv. Chim. Acta* **1972**, *55*, 224–239. [CrossRef] [PubMed]
9. Franz, J.E. Herbicidal compositions and methods employing esters of *N*-phosphonoglycine. U.S. Patent US 3997860, 14 December 1976.
10. Mastalerz, P. Inhibition of glutamine synthetase by phosphonic analogs of glutamic acid. *Arch. Immun. Ter. Dośw.* **1959**, *7*, 201–210.
11. Engelmann, M.; Pikl, J. Phosphonic Acids Derived from Organic Acylamidomethyl Compounds. U.S. Patent 2304156, 8 December 1942.
12. Allen, J.G.; Atherton, F.R.; Hall, M.J.; Hassal, C.H.; Holmes, S.W.; Lambert, R.W.; Nisbet, L.J.; Ringrose, P.S. Phosphonopeptides, a new class of synthetic antibacterial agents. *Nature* **1978**, *272*, 56–58. [CrossRef] [PubMed]
13. Jacobsen, N.E.; Bartlett, P.A. A phosphonamidate dipeptide analog as an inhibitor of carboxypeptidase A. *J. Am. Chem. Soc.* **1981**, *103*, 654–657. [CrossRef]
14. McLeod, D.A.; Brinkworth, R.; Ashley, J.A.; Janda, K.D.; Wirsching, P. Phosphonamidates and phosphonamidate esters as HIV-1 protease inhibitors. *Bioorg. Med. Chem. Lett.* **1991** *1*, 653–658. [CrossRef]
15. Drabowicz, J.; Jakubowski, H.; Kudzin, M.H.; Kudzin, Z.H. The nomenclature of 1-aminoalkylphosphonic acids and derivatives: Evolution of the code system. *Acta Biochim. Pol.* **2015**, *62*, 139–150. [CrossRef] [PubMed]

16. Kabachnik, M.I.; Medved, T.Y.; Dyatlova, N.M.; Archipova, O.G.; Rudomino, M.W. Phosphoroorganic complexones. *Usp. Khim.* **1974**, *43*, 1554–1574. [CrossRef]
17. Rizkalla, E.N. Metal chelates of phosphonate-containing ligands. *Rev. Inorg. Chem.* **1983**, *5*, 223–304.
18. Nowack, B. Environmental chemistry of phosphonates. *Water Res.* **2003**, *37*, 2533–2546. [CrossRef]
19. Kafarski, P.; Lejczak, B. Aminophosphonic acids of potential medical importance. *Curr. Med. Chem. Anti-Cancer Agents* **2001**, *1*, 301–312. [CrossRef] [PubMed]
20. Lejczak, B.; Kafarski, P. Biological activity of aminophosphonic acids and their short peptides. *Top. Heterocycl. Chem.* **2009**, *20*, 31–63. [CrossRef]
21. Orsini, F.; Sello, G.; Sisti, M. Aminophosphonic acids and derivatives. Synthesis and biological applications. *Curr. Med. Chem.* **2010**, *17*, 264–289. [CrossRef] [PubMed]
22. Mucha, A.; Kafarski, P.; Berlicki, L. Remarkable potential of the α-amino-phosphonate/phosphinate structural motif in medicinal chemistry. *J. Med. Chem.* **2011**, *54*, 5955–5980. [CrossRef] [PubMed]
23. Sikorski, J.A.; Logush, E.W. Aliphatic carbon-phosphorus compound as herbicides. In *Handbook in Organophosphorus Chemistry*; Engel, R., Ed.; Marcel Dekker Inc.: New York, NY, USA, 1988; Volume Chapter 15, pp. 737–806.
24. Maier, L. Phosphoroorganic detergents. *Chimia* **1969**, *23*, 323–330.
25. Petrov, K.A.; Chauzov, V.A.; Erokhina, T.E. Aminoalkyl organo-phosphorus compounds. *Usp. Khim.* **1974**, *43*, 2045–2087. [CrossRef]
26. Palacios, F.; Alonso, C.; de los Santos, J.M. Synthesis of β-amino-phosphonates and -phosphinates. *Chem. Rev.* **2005**, *105*, 899–931. [CrossRef] [PubMed]
27. Kudzin, Z.H.; Kudzin, M.H.; Drabowicz, J.; Stevens, C. Aminophosphonic acids—Phosphorus analogues of natural amino acids. P. 1: Syntheses of α-amino-phosphonic acids. *Curr. Org. Chem.* **2011**, *15*, 2015–2071. [CrossRef]
28. Kudzin, Z.H.; Kudzin, M.H.; Drabowicz, J. Thioureidoalkylphosphonates in the synthesis of 1-aminoalkylphosphonic acids. The Ptc-aminophosphonate method. *Arkivoc* **2011**, *6*, 227–269. [CrossRef]
29. Ma, J.-N. Catalytic asymmetric synthesis of α- and β-amino phosphonic acid derivatives. *Chem. Soc. Rev.* **2006**, *35*, 630–636. [CrossRef] [PubMed]
30. Ordonez, M.; Rojas-Cabrera, H.; Cativiela, C. An overview of stereoselective synthesis of α-aminophosphonic acids and derivatives. *Tetrahedron* **2009**, *65*, 17–49. [CrossRef] [PubMed]
31. Kukhar, V.P.; Solodenko, V.A. The phosphorus analogs of aminocarboxylic acids. *Usp. Khim.* **1987**, *56*, 1504–1532. [CrossRef]
32. Kudzin, Z.H.; Mokrzan, J.; Skowroński, R. Long chain 1-aminothia-alkanephosphonates, their sulphinyl and sulphonyl derivatives. A new class of complexane type surfactants. *Phosphorus Sulfur Silicon Relat. Elem.* **1989**, *42*, 41–46. [CrossRef]
33. Kudzin, Z.H.; Skrzypek, S.W.; Skowroński, R.; Ciesielski, W.; Cristau, H.J.; Plenat, F. Polarographic investigations of functionalized alkanephosphonic acids. Part II. *Phosphorus Sulfur Silicon Relat. Elem.* **1996**, *119*, 201–207. [CrossRef]
34. Chęcińska, L.; Kudzin, Z.H.; Małecka, M.; Nazarski, R.B.; Okruszek, A. [(Diphenoxyphosphinyl) methylidene]triphenylphosphorane—The double P^{+}-stabilisedcarboanion: A crystallographic, computational and solution NMR comperative study on the ylidic bonding. *Tetrahedron* **2003**, *59*, 7681–7693. [CrossRef]
35. Kudzin, Z.H.; Saganiak, M.; Andrijewski, G.; Drabowicz, J. Oxidation of phosphonocysteine and phosphonohomocysteine. Synthesis of phosphonocysteic and phosphonohomocysteic acids. *Pol. J. Chem.* **2005**, *79*, 529–539.
36. Kudzin, Z.H.; Depczyński, R.; Andrijewski, G.; Drabowicz, J.; Łuczak, J. 1-(*N*-acylamino)alkanephosphonates. IV. *N*-acylation of 1-aminoalkanephosphonic acids. *Pol. J. Chem.* **2005**, *79*, 499–513.
37. Kudzin, Z.H.; Depczyński, R.; Kudzin, M.H.; Drabowicz, J.; Łuczak, J. 1-(*N*-Trifluoroacetylamino) alkylphosphonic acids. Synthesis and properties. *Amino Acids* **2007**, *33*, 663–667. [CrossRef] [PubMed]
38. Kudzin, Z.H.; Depczyński, R.; Kudzin, M.H.; Drabowicz, J. 1-(*N*-Chloroacetylamino)alkylphosphonic acids—Synthetic precursors of glycylophosphono-peptides and related compounds. *Amino Acids* **2008**, *34*, 163–168. [CrossRef] [PubMed]

39. Drabowicz, J.; Jordan, F.; Kudzin, M.H; Kudzin, Z.H.; Urbaniak, P.; Stevens, C. Reactivity of aminophosphonic acids. oxidative dephosphonylation of 1-aminoalkylphosphonic acids by aqueous halogens. *Dalton Trans.* **2016**, *45*, 2308–2317. [CrossRef] [PubMed]
40. Gancarz, R.; Kiersnowska, A.D. Associative vs. dissociative mechanism of P–C bond breaking in α-aminophosphonates leading to phosphoric acid [P(V)] derivatives. *Arkivoc* **2017**, *2*, 285–302. [CrossRef]
41. Aldrich-Sigma Catalog 2016. Aminophosphonic Self-Assembled Monolayers Agents (e.g., 795798, SIK 7701-10; SIK 7701-101). Available online: https://www.sigmaaldrich.com/china-mainland.html/ (accessed on 9 April 2018).
42. Kudzin, Z.H.; Łuczak, J. A facile conversion of aminoalkanephosphonic acids into *O,O*-dialkyl 1-(*N*-acylamino)alkanephosphonate derivatives. *Synthesis* **1995**, 509–511. [CrossRef]
43. Kudzin, Z.H.; Łyżwa, P.; Łuczak, J.; Andrijewski, G. Aminoalkanephosphonates. P.II. Facile conversion of 1-aminoalkanephosphonic acids their *O,O*-diester derivatives. *Synthesis* **1997**, 44–47. [CrossRef]
44. Svedas, V.K.; Kozlova, E.V.; Mironenko, D.A.; Kukhar, V.P.; Kasheva, T.N.; Solodenko, V.A.; Belozersky, A.N. Enzymatic hydrolysis of *N*-acylated 1-aminophosphonic acids. *Phosphorus Sulfur Silicon Relat. Elem.* **1990**, *51*, 411. [CrossRef]
45. Solodenko, V.; Kasheva, T.; Kukhar, V.; Kozlova, E.; Mironenko, D.; Svedas, V. Preparation of optically active 1-aminoalkylphosphonic acids by stereoselective enzymatic hydrolysis of racemic *N*-acylated 1-aminoalkyl-phosphonic acids. *Tetrahedron* **1991**, *47*, 3989–3998. [CrossRef]
46. Solodenko, V.; Belik, M.; Galushko, S.; Kukhar, V.; Kozlova, E.; Mironenko, D.; Svedas, V. Enzymatic preparation of both L- and D-enantiomers of phosphonic and phosphonous analogues of alanine using penicillin acylase. *Tetrahedron Asymmetry* **1993**, *4*, 1965–1968. [CrossRef]
47. Zielinska, K.; Mazurkiewicz, R.; Szymanska, K.; Jarzebski, A.; Magierac, S.; Erfurt, K. Penicillin G acylase-mediated kinetic resolution of racemic 1-(*N*-acylamino)alkylphosphonic and 1-(*N*-acylamino)alkylphosphinic acids and their esters. *J. Mol. Catal. B Enzym.* **2016**, *132*, 31–40. [CrossRef]
48. Zielinska, K.; Szymanska, K.; Mazurkiewicz, R.; Jarzebski, A. Batch and in-flow kinetic resolution of racemic 1-(*N*-acylamino)alkylphosphonic and 1-(*N*-acylamino)alkylphosphinic acids and their esters using immobilized penicillin G acylase. *Tetrahedron Asymmetry* **2017**, *28*, 146–152. [CrossRef]
49. Oleksyszyn, J. An amidoalkylation of trivalent phosphorous compounds with P(O)H functions including acetic acid solutions of PCl_3, $RPCl_2$ or R_2PCl, diesters of phosphorous acid and phosphorous-III-acids. *J. Prakt. Chem.* **1987**, *329*, 19–29. [CrossRef]
50. Soroka, M. The synthesis of 1-aminoalkylphosphonic acids. A revised mechanism of the reaction of phosphorus trichloride, amides and aldehydes or ketones in acetic acid solution (Oleksyszyn reaction). *Liebigs Ann. Chem.* **1990**, *4*, 331–334. [CrossRef]
51. Cypryk, M.; Drabowicz, J.; Gostynski, B.; Kudzin, M.H.; Kudzin, Z.H.; Urbaniak, P. 1-(*N*-Acylamino)-alkylphosphonic acids. Deacylation in aqueous solutions. *Phosphorus Sulfur Silicon Relat. Elem.* **2017**, *192*, 651–658. [CrossRef]
52. Smith, M.B.; March, J. *March's Advanced Organic Chemistry: Reactions, Mechanisms, and Structure*, 6th ed.; John Wiley & Sons, Inc.: Hoboken, NJ, USA, 2007; Volume Chapters 16–60, pp. 1408–1411. ISBN 13-978-0-471-72091-1
53. O'Connor, C. Acidic and basic amide hydrolysis. *Q. Rev. Chem. Soc.* **1970**, *24*, 553–564. [CrossRef]
54. Brown, R.S.; Bennet, A.J.; Slebocka-Tilk, H. Recent perspectives concerning the mechanism of H_3O^+ and OH^- promoted amide hydrolysis. *Acc. Chem. Res.* **1992**, *25*, 481–488. [CrossRef]
55. Bagno, A.; Lovato, G.; Scorrano, G. Thermodynamics of protonation and hydration of aliphatic amides. *J. Chem. Soc. Perkin Trans.* **1993**, *2*, 1091–1098. [CrossRef]
56. Zahn, D. On the role of water in amide hydrolysis. *Eur. J. Org. Chem.* **2004**, 4020–4023. [CrossRef]
57. Kahne, D.; Still, W.C. Hydrolysis of a peptide bond in neutral water. *J. Am. Chem. Soc.* **1988**, *110*, 7529–7534. [CrossRef]
58. DeWolfe, R.H.; Newcombe, R.C. Hydrolysis of formanilides in alkaline solutions. *J. Org. Chem.* **1971**, *36*, 3870–3878. [CrossRef]
59. Desai, S.D.; Kirsch, L.E. The ortho effect on the acidic and alkaline hydrolysis of substituted formanilides. *Int. J. Chem. Kinet.* **2015**, *47*, 471–488. [CrossRef]

60. Bowden, K.; Bromley, K. Reactions of carbonyl compounds in basic solutions. Part 14. The alkaline hydrolysis of substituted *N*-methylformanilides, *N*-methylacetanilides, 1-phenylazetidin-2-ones, 1-phenyl-2-pyrrolidones, and 1-phenyl-2-piperidones. *J. Chem. Soc. Perkin Trans.* **1990**, *2*, 2103–2109. [CrossRef]
61. Hibbert, F.; Malana, M.A. Kinetics of the alkaline hydrolysis of 1,8-bis(trifluoroacetylamino)-naphthalene to 1-amino-8-trifluoroacetylaminonaphthalene in 70%, 80% and 90%(v/v) $Me_2SO–H_2O$. *J. Chem. Soc. Perkin Trans.* **1992**, *2*, 755–759. [CrossRef]
62. Cheshmedzhieva, D.; Ilieva, S.; Hadjieva, B.; Galabov, B. The mechanism of alkaline hydrolysis of amides: A comparative computational and experimental study of the hydrolysis of *N*-methylacetamide, *N*-methylbenzamide, and acetanilide. *J. Phys. Org. Chem.* **2009**, *22*, 619–631. [CrossRef]
63. Theodorou, V.; Paraskevopoulos, G.; Skobridis, K. A mild alkaline hydrolysis of *N*- and *N,N*-substituted amides and nitriles. *Arkivoc* **2015**, *7*, 101–112. [CrossRef]
64. Soroka, M. Comments on the synthesis of aminomethylphosphonic acid. *Synthesis* **1989**, 547–548. [CrossRef]
65. Kudzin, Z.H.; Stec, W.J. Synthesis of 1-aminoalkanephosphonic acids via thioureidoalkanephosphonates. *Synthesis* **1978**, 469–472. [CrossRef]
66. Frisch, M.J.; Trucks, G.W.; Schlegel, H.B.; Scuseria, G.E.; Robb, M.A.; Cheeseman, J.R.; Scalmani, G.; Barone, V.; Mennucci, B.; Petersson, G.A.; et al. *Gaussian 09, Revision D.01*; Gaussian, Inc.: Wallingford, CT, USA, 2009.
67. Cossi, M.; Rega, N.; Scalmani, G.; Barone, V. Energies, structures, and electronic properties of molecules in solution with the C-PCM solvation model. *J. Comput. Chem.* **2003**, *24*, 669–681. [CrossRef] [PubMed]

6

1-(*N*-Acylamino)alkyltriarylphosphonium Salts with Weakened C_α-P^+ Bond Strength—Synthetic Applications

Jakub Adamek [1,2,*], Anna Węgrzyk [1,2], Justyna Kończewicz [1,2], Krzysztof Walczak [1] and Karol Erfurt [3]

1 Department of Organic Chemistry, Bioorganic Chemistry and Biotechnology, Silesian University of Technology, B. Krzywoustego 4, 44-100 Gliwice, Poland; anna.wegrzyk@polsl.pl (A.W.); j.konczewicz@gmail.com (J.K.); krzysztof.walczak@polsl.pl (K.W.)

2 Biotechnology Center of Silesian University of Technology, B. Krzywoustego 8, 44-100 Gliwice, Poland

3 Department of Chemical Organic Technology and Petrochemistry, Silesian University of Technology, B. Krzywoustego 4, 44-100 Gliwice, Poland; karol.erfurt@polsl.pl

* Correspondence: jakub.adamek@polsl.pl

Abstract: The α-amidoalkylating properties of 1-(*N*-acylamino)alkyltriarylphosphonium salts with weakened C_α-P^+ bond strength are discussed and examined. It is demonstrated that such type of phosphonium salts reacts smoothly with a diverse array of carbon- and heteroatom-based nucleophiles, including 1-morpholinocyclohexene, 1,3-dicarbonyl compounds, benzotriazole sodium salt, *p*-toluenesulfinate sodium salt, benzylamine, triarylphosphines, and other P-nucleophiles. Reactions are conducted at room temperature, in a short time (5–15 min) and mostly without catalysts. Simple work-up procedures result in good or very good yields of products. The structures of known compounds were established by spectroscopic methods and all new compounds have been fully characterized using ^{1}H-, ^{13}C-, ^{31}P-NMR, IR spectroscopy, and high-resolution mass spectrometry. Mechanistic aspects of described transformations are also performed and discussed. It was demonstrated that unique properties make 1-(*N*-acylamino)alkyl-triarylphosphonium salts with weakened C_α-P^+ bond strength interesting building blocks with great potential, especially in α-amidoalkylation reactions.

Keywords: organophosphorus chemistry; phosphonium salts; α-amidoalkylating agents; *N*-acyliminium cation; *N*-acylimine

---

1. Introduction

Specific structural features of 1-(*N*-acylamino)alkylphosphonium salts make them very interesting reagents. The presence of a positively charged nucleofugal phosphonium moiety in the close surroundings of the *N*-acyl group determines its unique chemical properties such as high reactivity in α-amidoalkylations [1–10]. This type of reactions has enjoyed unflagging interest for years as a synthetic method with great potential, especially valuable for C-C and C-heteroatom bond formation [1,11–37].

In general, the reactivity of α-amidoalkylating agents **1** depends on the efficiency of the generation of *N*-acyliminium cation **2** or *N*-acylimine **3** from its precursor and the equilibrium constant of this reaction. Of course, the reactivity of *N*-acyliminium cation **2** or *N*-acylimine **3** toward a nucleophile is also significant [1–3]. To produce the proper α-amidoalkylating agents, for instance *N*-acyliminium cation **2** or *N*-acylimine **3** from the most popular precursors such as α-amido sulfones (Z = SO_2Ar), *N*-(1-benzotriazolil)alkylamides (Z = Bt), and *N*-(1-alkoxyalkyl)amides (Z = OR), it is necessary to use catalysts, mainly Lewis acids (route **a**, Scheme 1 (I)) [18–25,29–37]. In contrast, 1-(*N*-acylamino)alkyltriphenylphosphonium salts **1** ($Z = Ph_3P^+ X^-$) do not require the use of acidic catalysts because of the permanent positive charge on the phosphonium group. However, the relatively high stability of C_α-P^+ bond forces the use of a base catalyst (e.g., Hünig's base, DBU, TBD; route **b**, Scheme 1 (I)) and sometimes microwave radiation [1,4–10].

Recently, we have proven that the use of 1-(*N*-acylamino)alkyltriarylphosphonium salts **4** derived from the EWG-substituted triarylphosphines facilitates the cleavage of C_α-P^+ bond and thereby the generation of *N*-acyliminium cation (Scheme 1 (II)) [2,3]. This phenomenon significantly increases the reactivity of 1-(*N*-acylamino)alkyltriarylphosphonium salts with weakened C_α-P^+ bond strength and allows us to conduct α-amidoalkylation without the need for any catalyst. In this work, we demonstrate that the abovementioned phosphonium salts **4** react smoothly with various nucleophiles in a short time under mild conditions and create new carbon-carbon or carbon-heteroatom bonds with good or very good yields (Scheme 1 (III)).

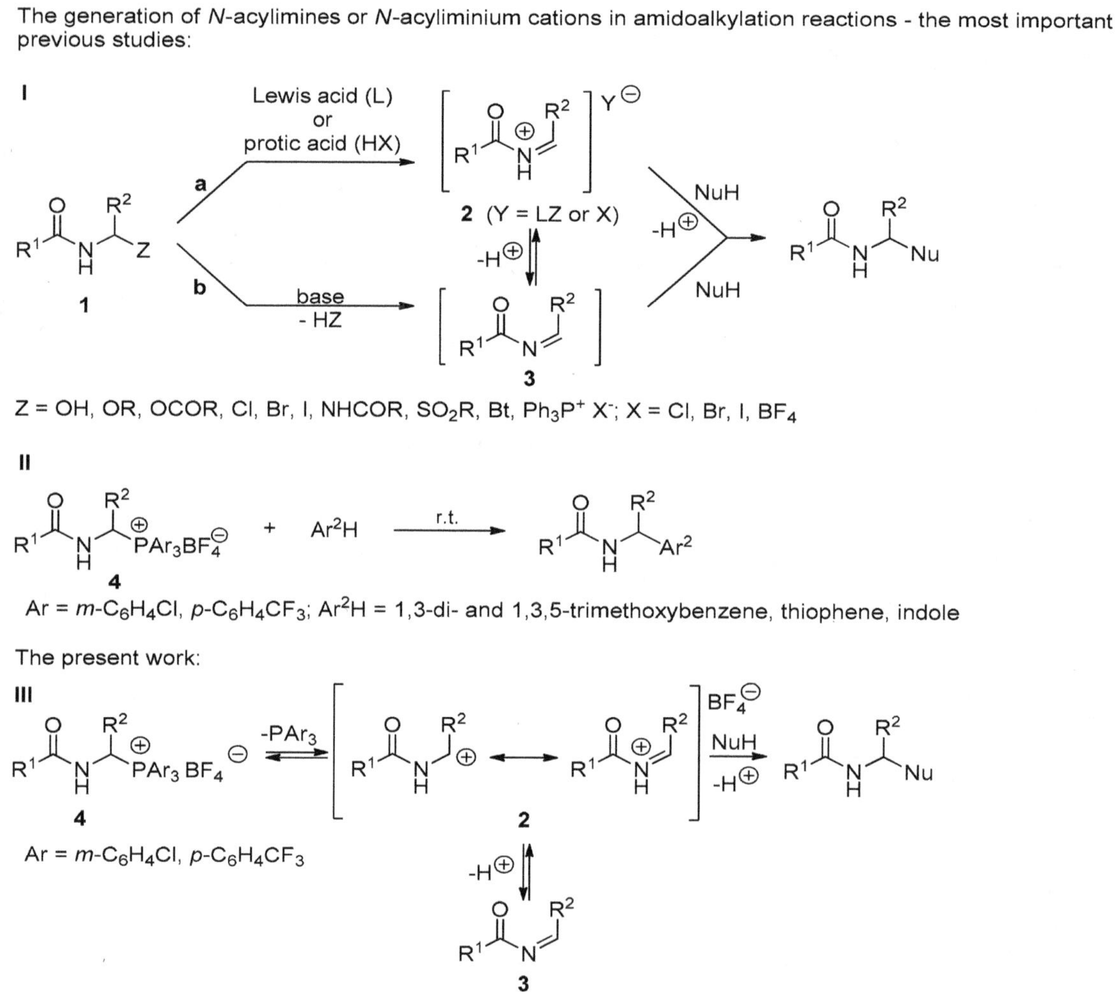

Scheme 1. α-Amidoalkylation reactions—various synthetic routes.

2. Results and Discussion

In this section, we focus our studies on the reactivity of phosphonium salts **4** with weakened C_α-P^+ bond strength in α-amidoalkylation of various types of carbon- and heteronucleophiles, as shown in Scheme 2. 1-Morpholinocyclohexene and 1,3-dicarbonyl compounds, such as dimethyl malonate, diethyl malonate, and ethyl acetoacetate, are used as carbon nucleophiles. In the case of heteronucleophiles, the reaction toward benzotriazole sodium salt, *p*-toluenesulfinate sodium salt, benzylamine, triarylphosphines, trimethyl phosphite, dimethyl phenylphosphonite, and methyl diphenylphosphinite is examined.

The synthesis of 1-(*N*-acylamino)alkyltriarylphosphonium salts **4** from α-amino acids was performed according to our previously described procedure, for which the electrochemical decarboxylative α-methoxylation of *N*-acyl-α-amino acids and substitution of the methoxy group by triarylphosphonium moiety are crucial steps (see Scheme 2) [2,9].

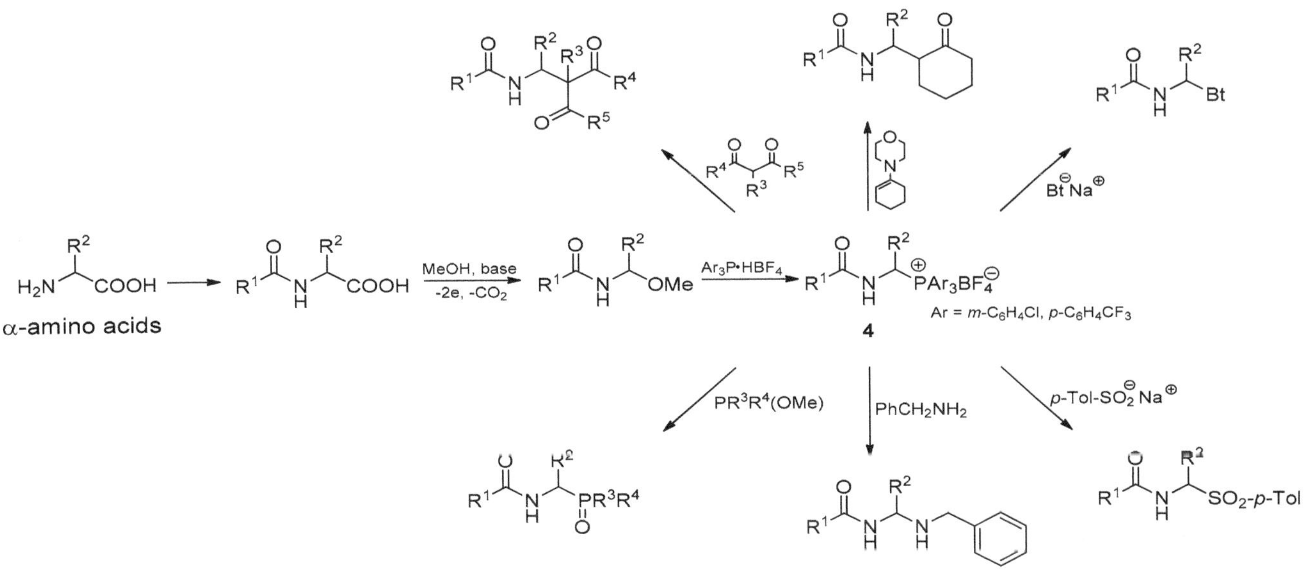

Scheme 2. 1-(*N*-Acylamino)alkyltriarylphosphonium salts with weakened C_{α}-P^{+} bond strength **4**—preparation and synthetic application.

2.1. 1-(N-Acylamino)alkyltriarylphosphonium Salts with Weakened C_{α}-P^{+} Bond Strength in the Selected Reaction of C-C Bond Formation

One of the crucial issues of organic synthesis is the formation of new C-C bonds. The possibility of an effective extension of the carbon skeleton is important in many fields, especially in medicinal chemistry, agrochemical synthesis, or in the synthesis of natural products.

Several years ago, we had proved that 1-(*N*-acylamino)alkyltriphenylphosphonium salts **4** (Ar = Ph) reacted quite easily with 1,3-dicarbonyl compounds in the presence of a base such as DBU (1,8-diazabicyclo[5.4.0]undec-7-ene), DBN (1,5-diazabicyclo[4.3.0]non-5-ene), or TBD (1,5,7-triaza-bicyclo[4.4.0]dec-5-ene) under microwave irradiation at 60 °C. However, the use of organic bases (DBU, DBN, TBD) complicates the course of α-amidoalkylation reaction due to the formation of amidinium and guanidinium salts [10]. To avoid this, in the current work, we used lithium diisopropylamide (LDA) as a base to produce enolate anions from the corresponding 1,3-dicarbonyl compounds. In our protocol, α-amidoalkylation was conducted under argon, in THF, at room temperature. A THF solution of enolate anions generated using LDA was introduced into the THF solution of the phosphonium salt **4**. The reaction was performed for 15 min, and the use of microwave irradiation was not necessary. In addition, it was found that the most favorable molar ratio of phosphonium salt **4**: 1,3 dicarbonyl compound: LDA was 1:8:1 (compare experiments 1 and 2, Table 1).

Under the abovementioned conditions, phosphonium salts with weakened C_{α}-P_{+} bond strength **4** react with 1,3-dicarbonyl compounds, including diethyl and dimethyl malonate and ethyl acetoacetate, to give the corresponding products **6** with good yields, regardless of whether tris(4-trifluoromethylphenyl)- or tris(3-chlorophenyl)phosphonium salts were used as substrates. Only in the case of phosphonium salts with the benzyloxy carbamate protective group, the yields were significantly lower and did not exceed 30% (entry 8, Table 1).

Table 1. Conditions and yields for α-amidoalkylation of 1,3-dicarbonyl compounds by 1-(*N*-acylamino) alkyltriarylphosphonium salts **4**.

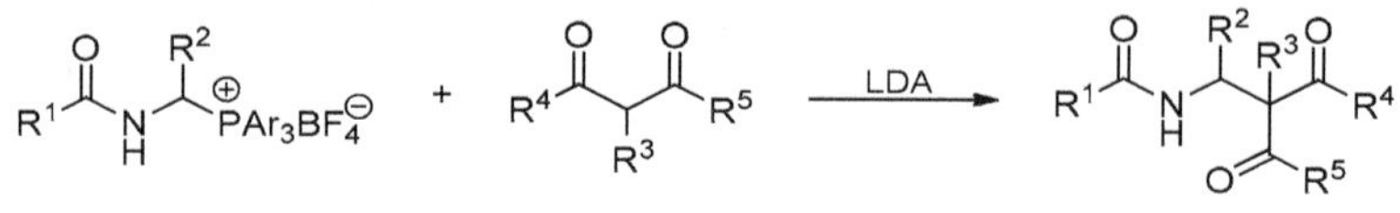

4 **5** **6**

Entry	Phosphonium Salt 4			1,3-Dicarbonyl Compound, 5	Molar Ratio of 4:5:LDA	6	Yield, %
	R^1	R^2	Ar				
1	*t*-Bu	Me	*m*-C_6H_4Cl	diethyl malonate	1:2:1	**6a**	40
2	*t*-Bu	Me	*m*-C_6H_4Cl	diethyl malonate	1:8:1	**6a**	65
3	*t*-Bu	Me	*p*-$C_6H_4CF_3$	diethyl malonate	1:8:1	**6a**	67
4	*t*-Bu	Me	*m*-C_6H_4Cl	ethyl acetoacetate	1:8:1	**6b**	62 [a]
5	Bn	*i*-Bu	*m*-C_6H_4Cl	ethyl acetoacetate	1:8:1	**6c**	83 [b]
6	Bn	*i*-Bu	*m*-C_6H_4Cl	dimethyl malonate	1:8:1	**6d**	63
7	Bn	*i*-Bu	*m*-C_6H_4Cl	diethyl malonate	1:8:1	**6e**	52
8	BnO	Bn	*m*-C_6H_4Cl	ethyl acetoacetate	1:8:1	**6f**	30 [c]

[a] A mixture of diastereoisomers in a molar ratio of 1.4:1. [b] A mixture of diastereoisomers in a molar ratio of 1.25:1. [c] Attempts to isolate an analytically pure sample failed.

α-Amidoalkylation of carbon nucleophiles was successfully extended to enamines. Based on the recently described protocol [8], we have demonstrated that 1-(*N*-acylamino)alkyltriarylphosphonium salts **4** with weakened C_α-P^+ bond strength react with 1-morpholinocyclohexene in Stork-type enamination to give the expected products **8** with good yields. We have proved that this reaction can be conducted in acetonitrile, at room temperature and without using any base catalysts. The optimized reaction time was 60 min, and the molar ratio of phosphonium salt to 1-morpholinocyclohexene was 1:2. We were able to separate the major diastereoisomer from the mixture using column chromatography and crystallization technique. Furthermore, the use of phosphonium salts with the benzyloxy carbamate protective group resulted in the decrease in reaction efficiency, as already observed for reactions with 1,3-dicarbonyl compounds (see Table 2).

Table 2. Conditions and yields for α-amidoalkylation of 1-morpholinocyclohexene by 1-(*N*-acylamino) alkyltriarylphosphonium salts **4**.

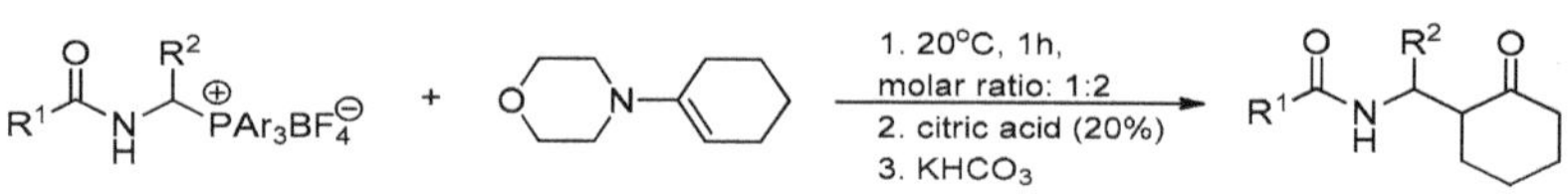

4 **7** **8**

Entry	Phosphonium Salt 4			8	Yield, %
	R^1	R^2	Ar		
1	*t*-Bu	Me	*m*-C_6H_4Cl	**8a**	63 [a]
2	Bn	*i*-Bu	*m*-C_6H_4Cl	**8b**	62 [b]
3	BnO	*i*-Bu	*m*-C_6H_4Cl	**8c**	40 [c]
4	BnO	Bn	*m*-C_6H_4Cl	**8d**	33 [d]
5	BnO	CH_2O*t*-Bu	*p*-$C_6H_4CF_3$	**8e**	31 [c]

[a] A mixture of diastereoisomers in a molar ratio of 3.5:1. [b] A mixture of diastereoisomers in a molar ratio of 3.2:1. [c] Only one diastereoisomer was detected and isolated. [d] A mixture of diastereoisomers in a molar ratio of 13.4:1.

2.2. 1-(N-Acylamino)alkyltriarylphosphonium Salts with Weakened C_α-P^+ Bond Strength in α-Amidoalkylation of Selected Heteronucleophiles

Applications of 1-(*N*-acylamino)alkyltriphenylphosphonium salts **4** (Ar = Ph) in the α-amidoalkylation of heteronucleophiles have been reported many times in the literature [5–8,10]. Usually in these types of reactions, it was necessary to use base catalysts, elevated temperature, and microwave irradiation. The effect of the use of 1-(*N*-acylamino)alkyltriarylphosphonium salts with weakened C_α-P^+ bond strength **4** (R = *m*-C_6H_4Cl, *p*-$C_6H_4CF_3$) on α-amidoalkylation has not been investigated so far. Therefore, we selected several heteronucleophiles such as benzotriazole sodium salt, *p*-toluenesulfinate sodium salt, benzylamine, and triphenylphosphine, and examined their α-amidoalkylation by phosphonium salts with weakened C_α-P^+ bond strength. As it was expected, reactions of salts with a modified structure of the phosphonium group occurred much faster and under mild conditions. In none of the described examples was it necessary to use a base catalyst. The reaction time was only 5–15 min. α-Amidoalkylation of benzotriazole sodium salt and sodium *p*-toluenesulfinate (entries 1–3, Table 3) was conducted at room temperature in $CHCl_3$. Similar conditions were used for reaction with benzylamine (entries 4 and 5, Table 3). The obtained aminal derivatives **10d** and **10e** exhibited limited stability. Therefore, the use of excess amount of the nucleophile was required to make the alkaline environment safe for aminals.

Table 3. Conditions and yields for α-amidoalkylation of selected heteronucleophiles by 1-(*N*-acylamino) alkyltriarylphosphonium salts **4**.

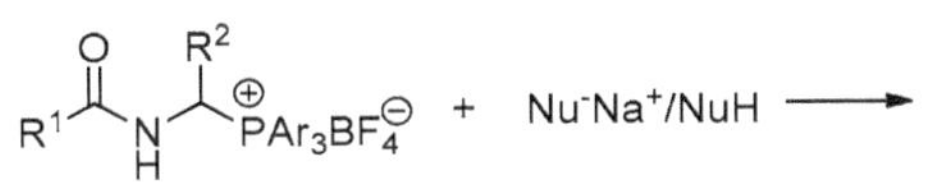

Entry	Phosphonium Salt 4 R¹	R²	Ar	Nu⁻Na⁺/NuH	Temp., °C	Time, Min.	Molar Ratio of 4:9	10	Yield, %
1	*t*-Bu	Me	*p*-$C_6H_4CF_3$	Bt^-Na^+	20	15	1:1	**10a**	99
2	BnO	Bn	*m*-C_6H_4Cl	Bt^-Na^+	20	15	1:1	**10b**	70
3	*t*-Bu	Me	*p*-$C_6H_4CF_3$	$TolSO_2^-Na^+$	20	15	1:1	**10c**	88
4	*t*-Bu	Me	*m*-C_6H_4Cl	Bn-NH_2	20	5	1:4	**10d**	91
5	Bn	*i*-Bu	*p*-$C_6H_4CF_3$	Bn-NH_2	20	5	1:4	**10e**	55

1-(*N*-Acylamino)alkyltriarylphosphonium salts **4** react also with triarylphosphines. For example, 1-(*N*-pivaloylamino)ethyltris(4-trifluoromethylphenyl)phosphonium tetrafluoroborate (**4a**) is completely transformed into 1-(*N*-pivaloylamino)ethyltriphenylphosphonium tetrafluoroborate (**4b**) during the reaction with triphenylphosphine. On the other hand, using tris(3 chlorophenyl)- phosphine as the nucleophile, we obtain a reaction mixture in which both 1-(*N*-pivaloylamino)ethyltris(4-trifluoromethylphenyl)phosphonium tetrafluoroborate (**4a**) and 1-(*N*-pivaloylamino)ethyltris(3-chlorophenyl)phosphonium tetrafluoroborate (**4c**) are present in a molar ratio of 1:3. A mixture of the same composition can also be obtained by the reaction of salt **4c** with tris(4-trifluoromethylphenyl)phosphine. These observations may suggest the existence of some equilibrium, as described by the equations in Scheme 3. The equilibrium of the reaction is shifted toward more stable and less reactive phosphonium salts, which is evident for the reaction with triphenylphosphine (equation a, Scheme 3).

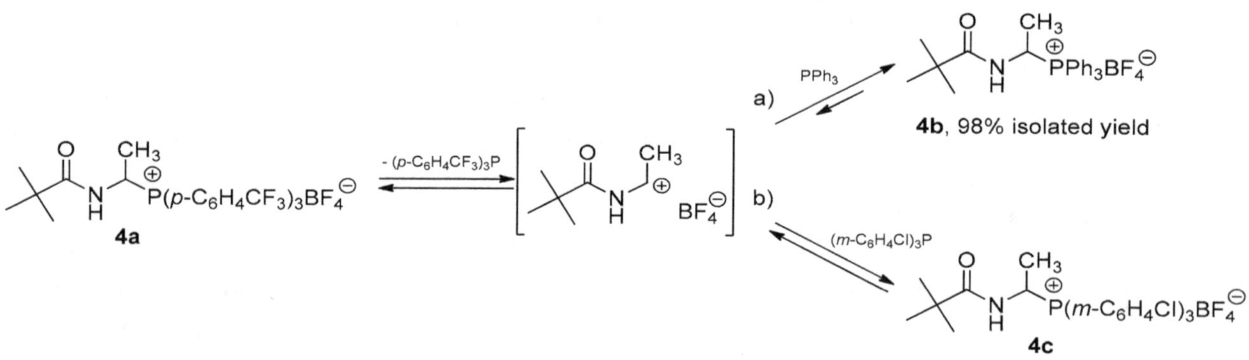

Scheme 3. 1-(*N*-Acylamino)alkyltriarylphosphonium salts with weakened C_α-P^+ bond strength **4**—reactions with triarylphosphines.

The results described above have encouraged us to extend the range of P-nucleophiles by trimethyl phosphite, dimethyl phenylphosphonite, and methyl diphenylphosphinite. It was expected that reactions with these types of phosphorus nucleophiles may occur quickly at room temperature in a non-catalytic environment, as for earlier tested nucleophiles. Surprisingly, the first experiments have shown that reactions are much slower. The analysis of reaction kinetics allowed us to explain these observations. In 2013, we proposed a plausible reaction mechanism wherein the *N*-acyliminium cation or *N*-acylimine, both generated from the 1-(*N*-acylamino)alkyltriphenylphosphonium salt **4** (Ar = Ph), reacts with phosphorus nucleophile to form alkoxyphosphonium salt **13**—the characteristic intermediate of the Michaelis–Arbuzov reaction. The final step of the reaction is the dealkylation of the alkoxyphosphonium salt **13** and may occur directly with triphenylphosphine (see Scheme 4) [7]. Although we have not yet been able to isolate or even observe the formation of postulated intermediate product, we assume that 1-(*N*-acylamino)alkyltriarylphosphonium salts with weakened C_α-P^+ bond strength react with phosphorus nucleophiles in an analogous manner. To prove this, we performed the reaction of 1-(*N*-pivaloylamino)ethyltris(3-chlorophenyl)phosphonium tetrafluoroborate (**4c**) with trimethyl phosphite. The reaction progress was monitored using NMR spectroscopy. In the reaction mixture, besides substrates, two products were also detected. Changes in their concentration as a function of time were measured, as shown in Figure 1. The concentration of the first product quickly reaches its maximum and then rapidly decreases, while the concentration of the second one at the beginning of the reaction is low and then increases. The induction period, which is characteristic for the formation of the final product in the consecutive-type reaction, is very clearly visible. Detailed NMR (^{1}H- and ^{31}P-NMR) and HR-MS analysis confirmed that the first, fast-growing product was 1-(*N*-acylamino)alkyltrimethoxyphosphonium tetrafluoroborate (**13a**) —the postulated intermediate of the Michaelis–Arbuzov reaction. As a result of its demethylation, dimethyl 1-(*N*-pivaloylamino)ethanephosphonate (**12a**) is formed with a much slower reaction. It seems that the second step determines the overall rate of the process.

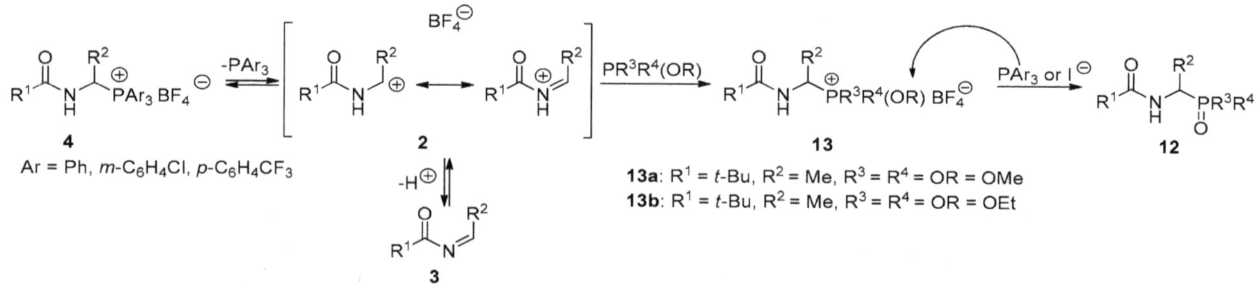

Scheme 4. α-Amidoalkylation of P-nucleophiles by 1-(*N*-acylamino)alkyltriarylphosphonium salts **4**—a plausible mechanism.

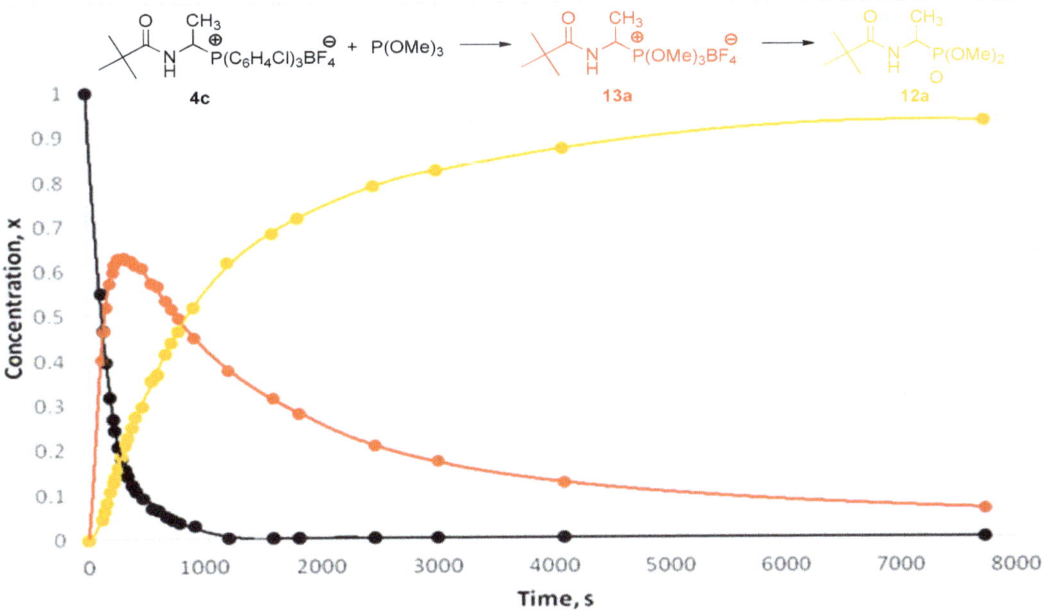

Figure 1. Concentration of the substrate **4c**, intermediate **13a**, and product **12a** as a function of time for the reaction of 1-(*N*-pivaloylamino)ethyltris(3-chlorophenyl)phosphonium tetrafluoroborate (**4c**) with trimethyl phosphite at 26 °C.

A plausible mechanism that explains the described kinetic facts is shown in Scheme 4 and is consistent with the mechanism proposed earlier in 2013. The only difference is that the generation of *N*-acyliminium cation from 1-(*N*-acylamino)alkyltriphenylphosphonium salt **4** (Ar = Ph) is more difficult. On the other hand, the highnucleophilicity of triphenylphosphine facilitates dealkylation, which makes the first step crucial for the course of the entire reaction. In the case of 1-(*N*-acylamino)alkyltriarylphosphonium salts with weakened C_{α}-P^+ bond strength **4** (R = *m*-C_6H_4Cl, *p*-$C_6H_4CF_3$), the first step is easier, which facilitates the formation of alkoxyphosphonium salts **13**. However, due to the lower nucleophilicity of phosphines with electron-withdrawing substituents [P(*m*-$C_6H_4Cl)_3$, P(*p*-$C_6H_4CF_3)_3$], the rate of dealkylation drops. To overcome this drawback, we decided to add to the reaction mixture the substoichiometric amounts of methyltriphenylphosphonium iodide as the dealkylating agent (molar ratio of 1:0.25). This protocol allowed us to obtain a series of phosphorus analogs of α-amino acids **12** (*N*-acyl-α-aminoalkanephosphonates **12a–c**, *N*-acyl-α-aminoalkanephosphinate (**12d**) and *N*-acyl-α-aminoalkylphosphine oxide (**12e**)) with good or very good yields. Usually, reactions occur efficiently at room temperature. Only in the reaction with dimethyl phenylphosphonite, it was necessary to raise the temperature to 60 °C. At room temperature, the reaction is very slow even after the addition of methyltriphenylphosphonium iodide (see Table 4).

Finally, we tried to isolate and fully characterize the intermediate of the Michaelis Arbuzov reaction **13**. To this end, we conducted the reaction of 1-(*N*-pivaloylamino)ethyltris(3-chlorophenyl)phosphonium tetrafluoroborate (**4c**) with trimethyl phosphite. Unfortunately, the expected 1-(*N*-pivaloylamino) ethyltrimethoxyphosphonium tetrafluoroborate (**13a**, Scheme 4) was too reactive and attempts to isolate the analytically pure sample failed. To increase the stability of the intermediate **13**, we used triethyl phosphite as the phosphorus nucleophile. This allowed us to isolate 1-(*N*-pivaloylamino)ethyltriethoxyphosphonium tetrafluoroborate (**13b**) and determine its basic physicochemical properties.

Table 4. Conditions and yields for α-amidoalkylation of P-nucleophiles by 1-(*N*-acylamino) alkyltriarylphosphonium salts **4**.

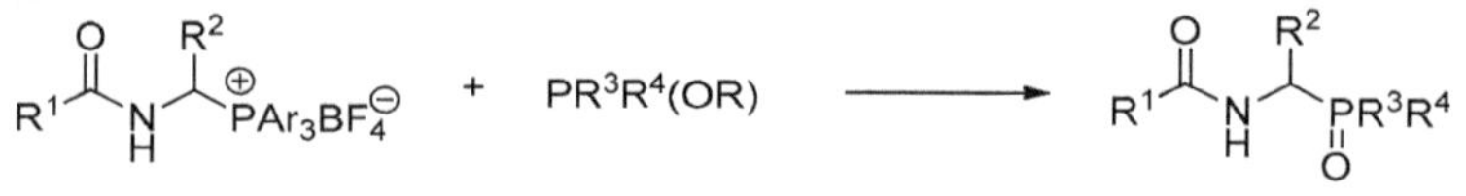

4 **11** **12**

Entry	Phosphonium Salt 4				P-Nu 11		Temp., °C	Time, h	Molar Ratio of 4:11	12	Yield, %
	R^1	R^2	Ar	R	R^3	R^4					
1	*t*-Bu	Me	*m*-C_6H_4Cl	Me	OMe	OMe	20	3	1:1.5	**12a**	47
2	*t*-Bu	Me	*m*-C_6H_4Cl	Me	OMe	OMe	20	3	1:1.5 [a]	**12a**	85
3	Bn	*i*-Bu	*m*-C_6H_4Cl	Me	OMe	OMe	20	3	1:1.5 [a]	**12b**	77
4	BnO	CH_2O*t*-Bu	*p*-$C_6H_4CF_3$	Me	OMe	OMe	20	3	1:1.5 [a]	**12c**	61
5	*t*-Bu	Me	*m*-C_6H_4Cl	Me	Ph	OMe	60	2	1:1.5 [a]	**12d**	69
6	BnO	*i*-Bu	*p*-$C_6H_4CF_3$	Me	Ph	Ph	20	3	1:1.5 [a]	**12e**	83

[a] Substoichiometric amounts of methyltriphenylphosphonium iodide as a dealkylating agent (molar ratio of 1:0.25) was used.

3. Experimental Section

3.1. General Information

Melting points were determined in capillaries and were uncorrected. IR spectra were measured on an FT-IR spectrophotometer (ATR method). ^{1}H- and ^{13}C-NMR were recorded at operating frequencies of 400 and 100 MHz, respectively, using TMS as the resonance shift standard. ^{31}P-NMR spectra were recorded at an operating frequency of 161.9 MHz without the resonance shift standard, with respect to H_3PO_4 as zero ppm. All chemical shifts (δ) are reported in ppm and coupling constants (*J*) in Hz. High-resolution mass spectrometry (HR-MS) analyses were performed on a Xevo G2 Q-TOF mass spectrometer (Waters, Milford, MA, USA) equipped with an ESI source operating in the positive ion mode. The accurate mass and composition of molecular ion adducts were calculated using the MassLynx software incorporated within the instrument.

^{1}H, ^{13}C, and ^{31}P NMR spectra of all new compounds **6**, **8**, **10**, **12**, **13** as well as the summary table in which we compare conditions and yields for reactions of 1-(*N*-acylamino)alkyltriphenylphosphonium salts (former studies) and 1-(*N*-acylamino)alkyltriarylphosphonium salts (the current work) with selected nucleophiles, are placed in the Supplementary data.

3.2. Syntheses

3.2.1. Substrate Synthesis

Electrochemical decarboxylative α-methoxylation of N-acyl-α-amino acids [2,9]

The electrolysis was conducted in an undivided glass electrolyzer (85 cm^3) equipped with a thermostatic jacket, a magnetic stirrer, a concentrically arranged, cylindrical Pt mesh anode (47 cm^2) and cathode (44 cm^2). To the thus prepared electrolyzer, *N*-acyl-α-amino acid (3.0 mmol), SiO_2-Pip (200 mg), and methanol (30 cm^3) were added. The electrochemical decarboxylative α-methoxylation was executed while stirring, at a constant current of 0.15 A, at 10 °C until a 3.75 F/mol charge had passed. Then, SiO_2-Pip was filtered off, and methanol was evaporated under reduced pressure to obtain *N*-(1-methoxyalkyl)amide, which was used in the next reaction without further purification.

Transformation of N-(1-methoxyalkyl)amides to 1-(N-acylamino)alkyltriarylphosphonium salts **4** *[2,10]*

To a solution of triarylphosphine (1 mmol) in DCM (2 mL), a tetrafluoroboric acid diethyl ether complex ($HBF_4 \cdot Et_2O$, 136 μl, 161.9 mg, 1 mmol) was added at 0 °C. The reaction mixture was stirred for 2 h at room temperature. Thereafter, *N*-(1-methoxyalkyl)amide (1 mmol) was added and stirring was continued for 15 min. Evaporation of the solvent yielded the crude 1-(*N*-acylamino) alkyltriarylphosphonium salt **4**, which was used in the next reaction without further purification.

3.2.2. Reactivity of 1-(*N*-acylamino)alkyltriarylphosphonium Salts **4** toward Carbon Nucleophiles

Reaction of 1-(N-acylamino)alkyltriarylphosphonium salts **4** *with C-H acids—general procedure*

A solution of a nucleophile (1.6 mmol), THF (0.5 mmol), and base LDA (2.0 M solution in THF, 0.1 cm^3) was stirred under argon atmosphere. After a few minutes, this mixture was relegated to a solution of 1-(*N*-acylamino)alkyltriarylphosphonium salt (0.2 mmol) in THF (0.5 cm^3) under argon atmosphere. The resulting reaction mixture was stirred for 15 min at room temperature, and then it was evaporated under reduced pressure. The product was separated using column chromatography [toluene/AcOEt 5:1 *v*/*v*].

Diethyl 1-(pivaloylamino)etylpropanedioate (**6a**) [10]. Colorless oil (38.5 mg, 67% yield). ^{1}H-NMR ($CDCl_3$) δ 6.80 (d, *J* = 7.5 Hz, 1H, NH), 4.74–4.62 (m, 1H, $C_\alpha H$), 4.29–4.21 (m, 2H, OCH_2), 4.21–4.12 (m, 2H, OCH_2), 3.57 (d, *J* = 3.4 Hz, 1H, CH), 1.32–1.25 (m, 9H, CH_3 and 2 × OCH_2CH_3), 1.17 (s, 9H, *t*-Bu) ppm; ^{13}C-NMR ($CDCl_3$) δ 177.6 (NHC = O), 168.7 (C = O), 167.7 (C = O), 61.7 ($\underline{O}CH_2$), 61.5 (OCH_2), 55.6 ($C_\alpha H$), 44.2 (CH), 38.6 ($\underline{C}(CH_3)_3$), 27.4 ($C(\underline{C}H_3)_3$), 19.0 (CH_3), 14.0 ($OCH_2\underline{C}H_3$) ppm; IR (ATR) 2977, 1731, 1647, 1510, 1175 cm^{-1}.

Ethyl 2-acetyl-3-(pivaloylamino)butanoate (**6b**). A mixture of two disatereoisomers in a molar ratio of 1.4:1 (31.9 mg, 62% yield). ^{1}H-NMR ($CDCl_3$) δ 6.92 (d, *J* = 8.0 Hz, 1H, NH)[a], 6.41 (d, *J* = 7.8 Hz, 1H, NH)[a], 4.85–4.74 (m, 1H, $C_\alpha H$)[a], 4.68–4.57 (m, 1H, $C_\alpha H$)[a], 4.33–4.10 (m, 2H, OCH_2)[b], 3.72 (d, *J* = 4.9 Hz, 1H, CH)[a], 3.65 (d, *J* = 3.6 Hz, 1H, CH)[a], 2.26 (s, 3H, CH_3CO)[a], 2.25 (s, 3H, CH_3CO)[a], 1.35–1.20 (m, 6H, CH_3 and $OCH_2\underline{CH_3}$)[b], 1.16 (s, 9H, *t*-Bu)[a], 1.15 (s, 9H, *t*-Bu)[a] ppm; ^{13}C-NMR ($CDCl_3$) δ 203.4[a] ($CH_3\underline{C}$ = O), 202.1[a] ($CH_3\underline{C}$ = O), 178.0[a] (NHC = O), 177.7[a] (NHC = O), 169.5[a] (C = O), 168.2[a] (C = O), 63.4[a] ($C_\alpha H$), 62.2[a] ($C_\alpha H$), 61.6[a] (OCH_2), 61.4[a] (OCH_2), 44.3[a] (CH), 43.4[a] (CH), 38.6[a] ($\underline{C}(CH_3)_3$), 38.6[a] ($\underline{C}(CH_3)_3$), 30.4[a] ($\underline{C}H_3CO$), 29.0[a] ($\underline{C}H_3CO$), 27.4[a] ($C(\underline{C}H_3)_3$), 27.3[a] ($C(\underline{C}H_3)_3$), 19.7[a] (CH_3), 18.7[a] (CH_3), 14.1[a] ($OCH_2\underline{C}H_3$), 14.1[a] ($OCH_2\underline{C}H_3$) ppm; IR (ATR) 3318, 2971, 1730, 1635, 1534, 1301, 1190 cm^{-1}. HRMS (TOF-ESI) calcd for $C_{13}H_{24}NO_4$ $[M + H]^+$ 258.1705, found 258.1701. [a] Separate signals from both diastereoisomers. [b] Overlapping signals of both diastereoisomers.

Ethyl 2-acetyl-5-methyl-3-(phenylacetylamino)hexanoate (**6c**). A mixture of two disatereoisomers in a molar ratio of 1.25:1 (55.4 mg, 83% yield). ^{1}H-NMR ($CDCl_3$) δ 7.39–7.19 (m, 5H, Ph)[a], 6.43 (d, *J* = 9.7 Hz, 1H, NH)[b], 6.02 (d, *J* = 9.5 Hz, 1H, NH)[b], 4.77–4.68 (m, 1H, $C_\alpha H$)[b], 4.63–4.54 (m, 1H, $C_\alpha H$)[b], 4.21–4.07 (m, 2H, OCH_2)[a], 3.64 (d, *J* = 4.9 Hz, 1H, CH)[b], 3.60 (d, *J* = 3.7 Hz, 1H, CH)[b], 3.51 (s, 2H, $Ph\underline{CH_2}$)[a], 2.23 (s, 3H, CH_3CO)[b], 2.19 (s, 3H, CH_3CO)[b], 1.56–1.37 (m, 2H, CH_2)[a], 1.28–1.14 (m, 4H, CH and $OCH_2\underline{CH_3}$)[a], 0.95–0.80 (m, 6H, 2 × CH_3)[a] ppm; ^{13}C-NMR ($CDCl_3$) δ 203.2[b] ($CH_3\underline{C}$ = O), 202.2[b] ($CH_3\underline{C}$ = O), 170.6[b] (NHC = O), 170.4[b] (NHC = O), 169.1[b] (C = O), 168.1[b] (C = O), aromatic carbons: 134.7[b], 134.6[b], 129.3[b], 129.3[b], 128.9[b], 128.8[b], 127.2[b], 127.2[b], 62.4[b] ($C_\alpha H$), 61.8[b] ($C_\alpha H$), 61.6[b] (OCH_2), 61.3[b] (OCH_2), 47.1[b] ($Ph\underline{C}H_2$), 46.2[b] ($Ph\underline{C}H_2$), 43.9[b] (CH), 43.8[b] (CH), 42.7[b] (CH_2), 41.9[b] (CH_2), 30.3[b] ($\underline{C}H_3CO$), 29.1[b] ($\underline{C}H_3CO$), 25.1[b] (CH), 25.1[b] (CH), 23.1[b](CH_3), 22.7[b] (CH_3), 22.1[b] (CH_3), 21.7[b] (CH_3), 14.0[b] ($OCH_2\underline{C}H_3$), 14.0[b] ($OCH_2\underline{C}H_3$) ppm; IR (ATR) 3275, 2956, 1740, 1648, 1496, 1260, 1144 cm^{-1}. HRMS (TOF-ESI) calcd for $C_{19}H_{28}NO_4$ $[M + H]^+$ 334.2018, found 334.2009. [a] Overlapping signals of both diastereoisomers. [b] Separate signals from both diastereoisomers.

Dimethyl 3-methyl-1-(phenylacetylamino)butylpropanedioate (**6d**). Colorless crystals (42.3 mg, 63% yield), mp 78–80 °C. ^{1}H-NMR ($CDCl_3$) δ 7.41–7.18 (m, 5H, Ph), 6.25 (d, *J* = 9.6 Hz, 1H, NH), 4.68–4.59 (m, 1H, $C_\alpha H$), 3.69 (s, 3H, OCH_3), 3.62 (s, 3H, OCH_3), 3.55 (d, *J* = 3.9 Hz, 1H, CH), 3.52 (s, 2H, $Ph\underline{CH_2}$), 1.54–1.39 (m, 2H, CH_2), 1.28–1.19 (m, 1H, CH), 0.90 (d, *J* = 6.5 Hz, 3H, CH_3), 0.86 (d, *J* = 6.6 Hz, 3H, CH_3) ppm; ^{13}C-NMR ($CDCl_3$) δ 170.3 (NHC = O), 168.6 (C = O), 168.0 (C = O), aromatic carbons: 134.8, 129.3, 128.8, 127.1, 54.6 ($C_\alpha H$), 52.6 (OCH_3), 52.3 (OCH_3), 46.8 ($Ph\underline{C}H_2$), 43.9 (CH), 42.1 (CH_2), 25.0 (CH), 22.9 (CH_3), 21.9 (CH_3) ppm; IR (ATR) 3283, 2954, 1736, 1648, 1454, 1263 cm^{-1}. HRMS (TOF-ESI) calcd for $C_{18}H_{26}NO_5$ $[M + H]^+$ 336.1811, found 336.1805.

Diethyl 3-methyl-1-(phenylacetylamino)butylpropanedioate (**6e**). Colorless crystals (37.8 mg, 52% yield), mp 93.5–95.5 °C. ^{1}H-NMR ($CDCl_3$) δ 7.42–7.19 (m, 5H, Ph), 6.30 (d, *J* = 9.6 Hz, 1H, NH), 4.68–4.60 (m, 1H, $C_\alpha H$), 4.30–4.00 (m, 4H, 2 × OCH_2), 3.52 (s, 2H, $PhC\underline{H_2}$), 3.52 (d, *J* = 2.8 Hz, 1H, CH), 1.57–1.37 (m, 2H, CH_2), 1.36–1.06 (m, 7H, CH and 2 × OCH_2CH_3), 0.90 (d, *J* = 6.5 Hz, 3H, CH_3), 0.86 (d, *J* = 6.6 Hz, 3H, CH_3) ppm; ^{13}C-NMR ($CDCl_3$) δ 170.2 (NHC = O), 168.3 (C = O), 167.7 (C = O), aromatic carbons: 134.9, 129.3, 128.8, 127.1, 61.7 (OCH_2), 61.4 (OCH_2), 54.9 ($C_\alpha H$), 46.8 ($Ph\underline{C}H_2$), 43.9 (CH), 42.2 (CH_2), 25.0 (CH), 22.9 (CH_3), 22.0 (CH_3), 14.0 ($OCH_2\underline{C}H_3$), 13.9 ($OCH_2\underline{C}H_3$) ppm; IR (ATR) 3362, 2964, 1745, 1656, 1533, 1345, 1144 cm^{-1}. HRMS (TOF-ESI) calcd for $C_{20}H_{30}NO_5$ $[M + H]^+$ 364.2124, found 364.2122.

Reaction of 1-(N-acylamino)alkyltriarylphosphonium salts **4** *with 1-morpholinocyclohexene—general procedure*

To a solution of 1-(*N*-acylamino)alkyltriarylphosphonium salt **4** (0.25 mmol) in MeCN (1 cm^3), 1-morpholinocyclohexene (84.1 μL, 83.7 mg, 0.5 mmol) was added. After 1 h of stirring at room temperature, an aqueous solution of citric acid (20%) (1.125 cm^3) was added. Stirring was continued for 45 min, and then a saturated solution of $KHCO_3$ was added. Thereafter, the mixture was extracted with DCM (5 × 3 cm^3), and the organic layer was combined and dried over $MgSO_4$. Then, the solvent was evaporated under reduced pressure and the crude product was purified using column chromatography [toluene/EtOAc 2:1 *v*/*v* (**8a**), 5:1 (**8b**, **8c**, **8e**) or 10:1 *v*/*v* (**8d**)].

N-[1-(2-oxocyclohexyl)ethyl]pivalamide (**8a**). A mixture of two disatereoisomers in a molar ratio of 3.5:1 (35.5 mg, 63% yield). The major diastereoisomers was isolated using column chromatography [toluene/EtOAc 2:1] and crystallization from toluene. Colorless crystals, mp 124–126 °C. ^{1}H-NMR ($CDCl_3$) δ 6.51 (d, *J* = 8.1 Hz, 1H, NH), 4.19–4.10 (m, 1H, $C_\alpha H$), 2.56–2.48 (m, 1H, CH), 2.39–2.30 (m, 2H, CH_2), 2.11–2.01 (m, 2H, CH_2), 1.91–1.83 (m, 1H, $\underline{C}HH$), 1.74–1.57 (m, 3H, $CH\underline{H}$ and CH_2), 1.22 (d, *J* = 7.0 Hz, 3H, CH_3), 1.17 (s, 9H, *t*-Bu) ppm; ^{13}C-NMR ($CDCl_3$) δ 214.2 (C = O), 178.0 (NHC = O), 55.4 ($C_\alpha H$), 45.6 (CH), 43.1 (CH_2), 38.7 ($\underline{C}(CH_3)_3$), 32.7 (CH_2), 28.4 (CH_2), 27.6 ($C(\underline{C}H_3)_3$), 24.8 (CH_2), 20.0 (CH_3) ppm; IR (ATR) 3339, 2961, 2869, 1706, 1627, 1526, 1305, 1209, 1117 cm^{-1}. HRMS (TOF-ESI) calcd for $C_{13}H_{24}NO_2$ $[M + H]^+$ 226.1807, found 226.1800.

N-[1-(2-oxocyclohexyl)-3-methylbutyl]phenylacetamide (**8b**). A mixture of two disatereoisomers in a molar ratio of 3.2:1 (46.7 mg, 62% yield). The major diastereoisomers was isolated using column chromatography [toluene/EtOAc 5:1] and crystallization from toluene. Colorless crystals, mp 115.5–117.5 °C. ^{1}H-NMR ($CDCl_3$) δ 7.42–7.09 (m, 5H, Ph), 6.13 (d, *J* = 9.7 Hz, 1H, NH), 4.10–3.99 (m, 1H, $C_\alpha H$), 3.51 (s, 2H, $PhC\underline{H_2}$), 2.49–2.40 (m, 1H, CH), 2.31–2.18 (m, 2H, CH_2), 2.07–1.96 (m, 2H, CH_2), 1.86–1.75 (m, 1H, CH), 1.65–1.37 (m, 5H, CH and 2 × CH_2), 1.21–1.09 (m, 1H, CH), 0.86 (d, *J* = 6.5 Hz, 3H, CH_3), 0.83 (d, *J* = 6.6 Hz, 3H, CH_3) ppm; ^{13}C-NMR ($CDCl_3$) δ 213.5 (C = O), 170.7 (NHC = O), aromatic carbons: 135.3, 129.1, 128.8, 127.1, 54.8 ($C_\alpha H$), 48.7 ($Ph\underline{C}H_2$), 44.2 (CH), 43.1 (CH_2), 43.1 (CH_2), 32.6 (CH_2), 28.3 (CH_2), 25.2 (CH), 25.0 (CH_2), 23.1 (CH_3), 22.0 (CH_3) ppm; IR (ATR) 3273, 2954, 2865, 1702, 1633, 1552, 1455, 1315, 1138 cm^{-1}. HRMS (TOF-ESI) calcd for $C_{19}H_{28}NO_2$ $[M + H]^+$ 302.2120, found 302.2115.

Benzyl N-[1-(2-oxocyclohexyl)-3-methylbutyl]carbamate (**8c**). Only one disatereoisomer was detected and isolated. Colorless oil (31.7 mg, 40% yield). ^{1}H-NMR ($CDCl_3$) δ 7.42–7.27 (m, 5H, Ph), 5.42 (d, *J* = 10.1 Hz, 1H, NH), 5.09 (d, *J* = 12.5 Hz, 1H, $PhC\underline{H}HO$), 5.06 (d, *J* = 12.4 Hz, 1H, $PhCH\underline{H}O$), 3.83–3.69 (m, 1H, $C_\alpha H$), 2.54–2.44 (m, 1H, CH), 2.38–2.23 (m, 2H, CH_2), 2.15–1.99 (m, 2H, CH_2), 1.93–1.82 (m, 1H, $\underline{C}HH$), 1.74–1.55 (m, 5H, $CH\underline{H}$ and 2 × CH_2), 1.28–1.16 (m, 1H, CH), 0.91 (d, *J* = 6.4 Hz, 3H, CH_3), 0.89 (d, *J* = 6.5 Hz, 3H, CH_3) ppm; ^{13}C-NMR ($CDCl_3$) δ 213.1 (C = O), 156.5 (NHC = O), aromatic carbons: 136.8, 128.4, 127.9, 127.9, 66.5 ($Ph\underline{C}H_2O$), 54.7 ($C_\alpha H$), 50.8 (CH), 43.4 (CH_2), 43.2 (CH_2), 32.3 (CH_2), 28.1 (CH_2), 25.1 (CH_2), 25.1 (CH), 23.2 (CH_3), 21.9 (CH_3) ppm; IR (ATR) 3439, 3333, 2952, 2866, 1699, 1499, 1213, 1052 cm^{-1}. HRMS (TOF-ESI) calcd for $C_{19}H_{28}NO_3$ $[M + H]^+$ 318.2069, found 318.2061.

Benzyl N-[1-(2-oxocyclohexyl)-2-phenylethyl]carbamate (**8d**). A mixture of two disatereoisomers in a molar ratio of 13.4:1 (29.0 mg, 33% yield). The major diastereoisomers was isolated using column chromatography [toluene/EtOAc 10:1]. Colorless crystals, mp 99–101 °C. ^{1}H-NMR ($CDCl_3$) δ 7.42–7.06 (m, 10H, 2 × Ph), 5.65 (d, J = 10.0 Hz, 1H, NH), 5.07 (d, J = 12.3 Hz, 1H, PhC$\underline{H}$HO), 5.00 (d, J = 12.4 Hz, 1H, PhCH$\underline{H}$O), 3.95–3.87 (m, 1H, C$_\alpha$H), 2.99 (dd, J_1 = 13.5, J_2 = 7.3 Hz, 1H, PhC$\underline{H}$H), 2.92 (dd, J_1 = 13.5, J_2 = 8.5 Hz, 1H, PhCH$\underline{H}$), 2.51–2.41 (m, 1H, CH), 2.40–2.31 (m, 1H, CH), 2.29–2.17 (m, 1H, CH), 2.09–1.92 (m, 2H, CH_2), 1.87–1.78 (m, 1H, CH), 1.77–1.51 (m, 3H, CH and CH_2) ppm; ^{13}C-NMR ($CDCl_3$) δ 213.5 (C = O), 156.4 (NHC = O), aromatic carbons: 138.6, 136.7, 129.1, 128.5, 128.4, 127.9, 127.8, 126.4, 66.4 (Ph$\underline{C}H_2$O), 54.2 (C$_\alpha$H), 51.9 (CH), 43.2 (CH_2), 40.3 (Ph$\underline{C}H_2$), 32.6 (CH_2), 28.1 (CH_2), 25.0 (CH_2) ppm; IR (ATR) 3308, 2929, 1721, 1697, 1541, 1341, 1242, 1094, 1051 cm^{-1}. HRMS (TOF-ESI) calcd for $C_{22}H_{26}NO_3$ $[M + H]^+$ 352.1913, found 352.1913.

Benzyl N-[1-(2-oxocyclohexyl)-2-tert-butoxyethyl]carbamate (**8e**). Only one disatereoisomer was detected and isolated. Colorless oil (26.9 mg, 31% yield). ^{1}H-NMR ($CDCl_3$) δ 7.38–7.28 (m, 5H, Ph), 5.57 (d, J = 9.7 Hz, 1H, NH), 5.11 (d, J = 12.3 Hz, 1H, PhC$\underline{H}$HO), 5.06 (d, J = 12.3 Hz, 1H, PhCH$\underline{H}$O), 3.89–3.79 (m, 1H, C$_\alpha$H), 3.44 (d, J = 7.1 Hz, 2H, $\underline{CH_2}$Ot-Bu), 2.90–2.82 (m, 1H, CH), 2.38–2.27 (m, 2H, CH_2), 2.09–2.00 (m, 2H, CH_2), 1.92–1.85 (m, 1H, $\underline{C}$$\underline{H}$H), 1.78–1.64 (m, 3H, CH$\underline{H}$ and CH_2), 1.12 (s, 9H, t-Bu) ppm; ^{13}C-NMR ($CDCl_3$) δ 213.8 (C = O), 156.4 (NHC = O), aromatic carbons: 136.6, 128.5, 128.0, 73.1 (O$\underline{C}$(CH_3)$_3$), 66.6 (Ph$\underline{C}H_2$O), 62.1 ($\underline{C}H_2$Ot-Bu), 52.5 (C$_\alpha$H), 50.0 (CH), 42.8 (CH_2), 31.9 (CH_2), 28.1 (CH_2), 27.5 (OC(CH_3)$_3$), 24.9 (CH_2) ppm; IR (ATR) 3432, 2972, 2935, 2867, 1701, 1499, 1363, 1195, 1057 cm^{-1}. HRMS (TOF-ESI) calcd for $C_{20}H_{30}NO_4$ $[M + H]^+$ 348.2175, found 348.2173.

3.2.3. Reactivity of 1-(*N*-Acylamino)alkyltriarylphosphonium Salts 4 toward Heteronucleophiles

Reaction of 1-(N-acylamino)alkyltriarylphosphonium salts **4** *with benzotriazole sodium salt—general procedure*

To a solution of 1-(*N*-acylamino)alkyltriarylphosphonium salt **4** (0.5 mmol) in $CHCl_3$ (1 cm^3), benzotriazole sodium salt (0.5 mmol) was added. The reaction mixture was stirred at room temperature for 15 min. Thereafter, it was filtered through a fluted filter and the filtrate was evaporated under reduced pressure. The residue was crystallized from toluene.

N-[1-(Benzotriazol-1-yl)ethyl]pivalamide (**10a**) [6]. Colorless crystals (121.9 mg, 99% yield), mp 142–144 °C. ^{1}H-NMR ($CDCl_3$) δ 8.06–8.02 (m, 1H, aromatic), 7.85–7.79 (m, 1H, aromatic), 7.54–7.48 (m, 1H, aromatic), 7.41–7.35 (m, 1H, aromatic), 6.88 (dq, J_1 = 9.1, J_2 = 6.7 Hz, 1H, C$_\alpha$H), 6.63 (d, J = 8.8 Hz, 1H, NH), 2.04 (d, J = 6.7 Hz, 3H, CH_3), 1.15 (s, 9H, t-Bu) ppm; ^{13}C-NMR (100 MHz, $CDCl_3$) δ 178.2 (C = O), aromatic carbons: 145.5, 132.4, 127.7, 124.4, 119.5, 110.4, 58.7 (C$_\alpha$H), 38.7 ($\underline{C}$(CH_3)$_3$), 27.2 (C($\underline{C}H_3$)$_3$), 20.7 (CH_3) ppm; IR (ATR) 3346, 2969, 1668, 1512, 1193, 1152, 1065 cm^{-1}.

Benzyl N-[1-(benzotriazol-1-yl)-2-phenylethyl]carbamate (**10b**) [6]. Colorless crystals (130.4 mg, 70% yield), mp 117.5–119.5 °C. ^{1}H-NMR ($CDCl_3$) δ 8.00 (d, J = 8.4 Hz, 1H, aromatic), 7.54 (d, J = 7.5 Hz, 1H, aromatic), 7.42–7.35 (m, 1H, aromatic), 7.33–7.27 (m, 4H, aromatic), 7.27–7.20 (m, 2H, aromatic), 7.19–7.12 (m, 3H, aromatic), 7.11–7.04 (m, 2H, aromatic), 6.71–6.58 (m, 1H, C$_\alpha$H), 6.02 (d, J = 8.9 Hz, 1H, NH), 5.10 (d, J = 12.4 Hz, 1H, PhC$\underline{H}$HO), 4.97 (d, J – 12.2 Hz, 1H, PhCH$\underline{H}$O), 3.80–3.68 (m, 1H, PhC$\underline{H}$H), 3.63 (dd, J_1 – 13.8, J_2 = 6.6 Hz, 1H, PhCHH) ppm; ^{13}C-NMR ($CDCl_3$) δ 160.3 (C = O), aromatic carbons: 145.5, 136.0, 134.7, 131.9, 129.1, 128.7, 128.5, 128.4, 128.1, 127.7, 127.3, 124.1, 119.7, 109.7, 67.5 (Ph$\underline{C}H_2$O), 65.8 (C$_\alpha$H), 41.1 (PH$\underline{C}H_2$) ppm; IR (ATR) 3177, 3008, 1712, 1548, 1280, 1261, 1244, 1195, 1046, 1022 cm^{-1}.

Reaction of 1-(N-pivaloylamino)ethyltris(4-trifluoromethylphenyl)phosphonium tetrafluoroborate **4a** *with sodium p-toluenesulfinate*

To a solution of 1-(*N*-pivaloylamino)ethyltris(4-trifluoromethylphenyl)phosphonium tetrafluoroborate **4a** (340.7 mg, 0.5 mmol) in $CHCl_3$ (1 cm^3), sodium p-toluenesulfinate (89.1 mg, 0.5 mmol) was added.

The reaction mixture was stirred at room temperature for 15 min. Thereafter, it was filtered through a fluted filter and the filtrate was evaporated under reduced pressure. The residue was crystallized from toluene.

N-[1-(p-Toluenesulfonyl)ethyl]pivalamide (**10c**) [5]. Colorless crystals (124.7 mg, 88% yield), mp 143–145 °C. ^{1}H-NMR ($CDCl_3$) δ 7.77 (d, *J* = 8.3 Hz, 2H, aromatic), 7.33 (d, *J* = 7.9 Hz, 2H, aromatic), 5.98 (d, *J* = 10.2 Hz, 1H, NH), 5.41 (dq, J_1 = 10.2, J_2 = 7.0 Hz, 1H, $C_\alpha H$), 2.42 (s, 3H, CH_3), 1.62 (d, *J* = 7.0 Hz, 3H, CH_3), 1.01 (s, 9H, *t*-Bu) ppm; ^{13}C-NMR ($CDCl_3$) δ 176.9 (C = O), aromatic carbons: 145.1, 133.5, 129.6, 129.1, 64.3 ($C_\alpha H$), 38.7 ($\underline{C}(CH_3)_3$), 27.2 ($C(\underline{C}H_3)_3$), 21.6 (CH_3), 13.2 (CH_3) ppm; IR (ATR) 3372, 2973, 1677, 1516, 1308, 1287, 1135, 1083, 1016, 724 cm^{-1}.

Reaction of 1-(N-acylamino)alkyltriarylphosphonium salts **4** *with benzylamine—general procedure*

To a stirred solution of 1-(*N*-acylamino)alkyltriarylphosphonium salt **4** (0.25 mmol) in DCM (1 cm^3), benzylamine (109.4 μl, 110 mg, 1 mmol) was added dropwise. Stirring was continued for 5 min at room temperature. Thereafter, the mixture was evaporated under reduced pressure and dried. The residue was purified using column chromatography [DCM/MeOH/Et_3N, 5:1:0.2 *v*/*v*/*v*].

N-[1-(Benzylamino)ethyl]pivalamide (**10d**). Yellow crystals (53.3 mg, 91% yield), mp 56.5–58.5 °C. ^{1}H-NMR ($CDCl_3$) δ 7.35–7.29 (m, 4H, aromatic), 7.26–7.21 (m, 1H, aromatic), 5.67 (d, *J* = 6.7 Hz, 1H, NH), 4.90 (qd, J_1 = 6.2, J_2 = 1.6 Hz, 1H, $C_\alpha H$), 3.77 (s, 2H, $PhC\underline{H_2}$), 1.92 (br s, 1H, NH), 1.32 (d, *J* = 6.2 Hz, 3H, CH_3), 1.17 (s, 9H, *t*-Bu) ppm; ^{13}C-NMR ($CDCl_3$) δ 178.1 (C = O), aromatic carbons: 140.1, 128.4, 127.9, 126.9, 60.9 ($C_\alpha H$), 49.8 ($Ph\underline{C}H_2$), 38.6 ($\underline{C}(CH_3)_3$), 27.4 ($C(\underline{C}H_3)_3$), 21.6 ppm (CH_3); IR (ATR) 3328, 2971, 1622, 1527, 1475, 1144, 1097 cm^{-1}. HRMS (TOF-ESI) calcd for $C_{14}H_{23}N_2O$ $[M + H]^+$ 235.1810, found 235.1802.

N-[1-(Benzylamino)-3-methylbutyl]phenylacetamide (**10e**). Colorless crystals (42.7 mg, 55% yield), mp 93–95 °C. ^{1}H-NMR ($CDCl_3$) δ 7.59–6.96 (m, 10H, 2 × Ph), 5.33 (d, *J* = 8.3 Hz, 1H, NH), 4.82–4.75 (m, 1H, $C_\alpha H$), 3.68 (s, 2H, $PhC\underline{H_2}NH$), 3.54 (s, 2H, $PhC\underline{H_2}$), 1.66 (br s, 1H, NH), 1.64–1.53 (m, 1H, CH), 1.45–1.28 (m, 2H, CH_2), 0.87 (d, *J* = 6.6 Hz, 3H, CH_3), 0.84 (d, *J* = 6.6 Hz, 3H, CH_3) ppm; ^{13}C-NMR ($CDCl_3$) δ 170.9 (C = O), aromatic carbons: 140.3, 135.0, 129.3, 129.0, 128.4, 128.0, 127.3, 126.9, 63.3 ($C_\alpha H$), 49.6 ($Ph\underline{C}H_2$), 44.5 ($Ph\underline{C}H_2$), 44.1 (CH_2), 24.9 (CH), 22.6 (CH_3), 22.5 (CH_3) ppm; IR (ATR) 3306, 2962, 2946, 1632, 1518, 1144, 1009 cm^{-1}. HRMS (TOF-ESI) calcd for $C_{20}H_{27}N_2O$ $[M + H]^+$ 311.2123, found 311.2121.

Reaction of 1-(N-pivaloylamino)ethyltris(4-trifluoromethylphenyl)phosphonium tetrafluoroborate **4a** *with triphenylphosphine*

To a solution of 1-(*N*-pivaloylamino)ethyltris(4-triflurometylphenyl)phosphonium tetrafluoroborate **4a** (102.2 mg, 0.15 mmol) in DCM (1 cm^3), triphenylphosphine (39.3 mg, 0.15 mmol) was added. The homogeneous mixture was allowed to react at room temperature for 5 min, and 1-(*N*-pivaloylamino)ethyltriphenylphosphonium tetrafluoroborate (**4b**) [9] was precipitated with Et_2O, separated by decantation, and dried under reduced pressure.

1-(*N*-Pivaloylamino)ethyltriphenylphosphonium tetrafluoroborate (**4b**). Colorless crystals (70.2 mg, 98% yield), mp 160.5–162.5 °C. ^{1}H-NMR ($CDCl_3$) δ 7.89–7.64 (m, 16H, 3 × Ph and NH), 5.82–5.73 (m, 1H, $C_\alpha H$), 1.72 (dd, J_1 = 17.8, J_2 = 7.4 Hz, 3H, CH_3), 0.91 (s, 9H, *t*-Bu) ppm; ^{13}C-NMR ($CDCl_3$) δ 179.7 (d, *J* = 2.3 Hz, C = O), aromatic carbons: 134.8 (d, *J* = 3.0 Hz), 134.5 (d, *J* = 9.2 Hz), 130.0 (d, *J* = 12.3 Hz), 118.4 (d, *J* = 82.5 Hz), 45.0 (d, *J* = 53.4 Hz, $C_\alpha H$), 38.5 ($\underline{C}(CH_3)_3$), 26.8 ($C(\underline{C}H_3)_3$), 17.4 (d, *J* = 4.6 Hz, CH_3) ppm; ^{31}P NMR (161.9 MHz, $CDCl_3$) δ 29.2 ppm; IR (ATR) 3373, 1684, 1516, 1447, 1136, 1040 cm^{-1}.

Reaction of 1-(N-acylamino)alkyltriarylphosphonium salts **4** *with triarylphosphines—NMR scale*

To a solution of 1-(*N*-acylamino)alkyltriarylphosphonium salt (0.025 mmol) in $CDCl_3$ (0.65 cm^3), the corresponding triarylphosphine (0.025 mmol) was added. Reactions were conducted in NMR tubes at 26 °C, and their course was monitored using NMR spectroscopy.

Reaction of 1-(N-acylamino)alkyltriarylphosphonium salts **4** *with phosphorus nucleophiles in Michaelis–Arbuzov type reaction—general procedure*

To a stirred solution of 1-(*N*-acylamino)alkyltriarylphosphonium salt **4** (0.25 mmol) and methyltriphenylphosphonium iodide (25.3 mg, 0.0625 mmol) in DCM (2 cm^3), phosphorus nucleophile (0.375 mmol) was added, and the mixture was stirred for 3 h at room temperature (**12a–c** and **12e**) or for 2 h at 60 °C (**12d**). The mixture was evaporated under reduced pressure and the residue was extracted with toluene (3 × 1 cm^3) at 50 °C. The solvent was evaporated under reduced pressure, and the product was isolated using column chromatography [DCM/MeOH 20:1 *v*/*v*].

Dimethyl 1-(N-pivaloylamino)ethanephosphonate (**12a**) [7]. Colorless crystals (50.4 mg, 85% yield), mp 125.5–127.5 °C. ^{1}H-NMR ($CDCl_3$) δ 5.96 (d, J = 8.1 Hz, 1H, NH), 4.65–4.49 (m, 1H, C_αH), 3.78 (d, J = 4.4 Hz, 3H, OCH_3), 3.75 (d, J = 4.3 Hz, 3H, OCH_3), 1.38 (dd, J_1 = 16.8, J_2 = 7.4 Hz, 3H, CH_3), 1.21 (s, 9H, *t*-Bu) ppm; ^{13}C-NMR ($CDCl_3$) δ 177.6 (d, J = 5.1 Hz, C = O), 53.1 (d, J = 7.0 Hz, OCH_3), 52.9 (d, J = 6.5 Hz, OCH_3), 40.1 (d, J = 156.4 Hz, C_αH), 38.6 ($\underline{C}(CH_3)_3$), 27.3 ($C(\underline{C}H_3)_3$), 15.5 (CH_3) ppm; ^{31}P-NMR ($CDCl_3$) δ 28.5 ppm; IR (ATR) 3281, 2956, 1659, 1529, 1224, 1205, 1057, 1026, 1007 cm^{-1}.

Dimethyl 1-(N-phenylcetylamino)-3-methylbutanephosphonate (**12b**) [4]. Colorless crystals (60.3 mg, 77% yield), mp 117–119 °C. ^{1}H-NMR ($CDCl_3$) δ 7.39–7.23 (m, 5H, Ph), 5.72 (d, J = 9.9 Hz, 1H, NH), 4.62–4.48 (m, 1H, C_αH), 3.73 (d, J = 10.6 Hz, 3H, OCH_3), 3.63 (d, J = 10.6 Hz, 3H, OCH_3), 3.60 (s, 2H, $Ph\underline{CH_2}$), 1.61–1.44 (m, 3H, CH_2 and CH), 0.88 (d, J = 5.9 Hz, 6H, 2 × CH_3) ppm; ^{13}C-NMR ($CDCl_3$) δ 170.5 (d, J = 4.3 Hz, C = O), aromatic carbons: 134.6, 129.2, 128.9, 127.4, 53.1 (d, J = 7.2 Hz, OCH_3), 53.0 (d, J = 6.6 Hz, OCH_3), 43.7 ($Ph\underline{C}H_2$), 43.2 (d, J = 155.5 Hz, C_αH), 38.1 (d, J = 1.8 Hz, CH_2), 24.5 (d, J = 13.4 Hz, CH), 23.2 (CH_3), 21.1 (CH_3) ppm; ^{31}P-NMR ($CDCl_3$) δ 27.6 ppm; IR (ATR) 3243, 2959, 1673, 1540, 1217, 1027, 836, 742 cm^{-1}.

Dimethyl 1-(N-benzyloxycarbonylamino)-2-tert-butoxyethanephosphonate (**12c**) [4]. Colorless crystals (54.8 mg, 61% yield), mp 59.5–61.5 °C. ^{1}H-NMR ($CDCl_3$) δ 7.41–7.28 (m, 5H, Ph), 5.29 (d, J = 8.8 Hz, 1H, NH), 5.13 (s, 2H, $Ph\underline{CH_2}O$), 4.33–4.19 (m, 1H, C_αH), 3.76 (d, J = 10.6 Hz, 3H, OCH_3), 3.75 (d, J = 10.3 Hz, 3H, OCH_3), 3.61 (dd, J_1 = 9.4, J_2 = 3.8 Hz, 1H, C$\underline{H}$HO*t*-Bu), 3.54 (dd, J_1 = 9.4, J_2 = 3.8 Hz, 1H, CH$\underline{H}$O*t*-Bu), 1.18 (s, 9H, *t*-Bu) ppm; ^{13}C-NMR ($CDCl_3$) δ 155.7 (C = O), aromatic carbons: 136.2, 128.5, 128.2, 128.1, 73.7 ($O\underline{C}(CH_3)_3$), 67.2 ($Ph\underline{C}H_2O$), 60.6 ($\underline{C}H_2O$*t*-Bu), 53.3 (d, J = 6.0 Hz, OCH_3), 52.6 (d, J = 5.8 Hz, OCH_3), 48.3 (d, J = 156.0 Hz, C_αH), 27.3 ($OC(\underline{C}H_3)_3$) ppm; ^{31}P-NMR ($CDCl_3$) δ 26.1 ppm; IR (ATR) 3293, 2974, 1699, 1532, 1278, 1235, 1039, 1023, 756, 697 cm^{-1}.

Methyl phenyl(1-pivaloylaminoethyl)phosphinate (**12d**). Colorless crystals (48.9 mg, 69% yield), mp 131–133 °C. ^{1}H-NMR ($CDCl_3$) δ 7.84–7.75 (m, 2H, aromatic), 7.59–7.43 (m, 3H, aromatic), 5.76 (d, J = 9.6 Hz, 1H, NH), 4.79–4.67 (m, 1H, C_αH), 3.67 (d, J = 10.8 Hz, 3H, OCH_3), 1.45 (dd, J_1 = 14.6, J_2 = 7.3 Hz, 3H, CH_3), 0.92 (s, 9H, *t*-Bu) ppm; ^{13}C-NMR ($CDCl_3$) δ 177.3 (d, J = 4.7 Hz, C = O), aromatic carbons: 132.8 (d, J = 2.8 Hz), 132.4 (d, J = 9.4 Hz), 128.5 (d, J = 12.5 Hz), 127.5 (d, J = 124.2 Hz), 51.8 (d, J = 7.1 Hz, OCH_3), 42.3 (d, J = 115.5 Hz, C_αH), 38.5 ($\underline{C}(CH_3)_3$), 27.1 ($C(\underline{C}H_3)_3$), 14.4 (CH_3) ppm; ^{31}P-NMR ($CDCl_3$) δ 42.9 ppm. IR (ATR) 3264, 2954, 1660, 1531, 1198, 1139, 1027, 799 cm^{-1}; HRMS (TOF-ESI) calcd for $C_{14}H_{23}NO_3P$ $[M + H]^+$ 284.1416, found 284.1405.

Diphenyl 1-(N-benzyloxycarbonylamino)-3-methylbutylphosphine oxide (**12e**) [8]. Colorless crystals (87.5 mg, 83% yield), mp 173.5–175.5 °C. ^{1}H-NMR ($CDCl_3$) δ 7.87–7.09 (m, 15H, 3 × Ph), 5.40 (d, J = 10.6 Hz, 1H, NH), 5.02 (d, J = 12.5 Hz, 1H, PhC$\underline{H}$HO), 4.89 (d, J = 12.5 Hz, 1H, PhCH$\underline{H}$O), 4.83–4.72 (m, 1H, C_αH), 1.83–1.66 (m, 2H, CH_2), 1.36–1.21 (m, 1H, CH), 0.90 (d, J = 6.5 Hz, 3H, CH_3), 0.86 (d, J = 6.6 Hz, 3H, CH_3) ppm; ^{13}C-NMR ($CDCl_3$) δ 156.1 (d, J = 4.5 Hz, C = O), aromatic carbons: 136.4, 132.0 (d,

J = 2.8 Hz), 131.9 (d, J = 2.8 Hz), 131.5, 131.3, 131.1 (d, J = 9.2 Hz), 130.9 (d, J = 9.0 Hz), 128.8 (d, J = 11.4 Hz), 128.4 (d, J = 11.6 Hz), 128.4, 127.9, 127.6, 66.8 (Ph$\underline{C}H_2$O), 48.1 (d, J = 79.5 Hz, C_αH), 37.5 (d, J = 3.5 Hz, CH_2), 24.5 (d, J = 11.0 Hz, CH), 23.4 (CH_3), 21.0 (CH_3) ppm; ^{31}P-NMR ($CDCl_3$) δ 33.5 ppm; IR (ATR) 3181, 2954, 1702, 1545, 1436, 1263, 1187, 1120 cm^{-1}.

Reaction of 1-(N-pivaloylamino)ethyltris(3-chlorophenyl)phosphonium tetrafluoroborate **4c** *with trimethyl phosphite—measurement of changes in concentrations of the substrate* **4c**, *intermediate* **13a** *and product* **12a** *by NMR*

To a solution of 1-(*N*-pivaloylamino)ethyltris(3-chlorophenyl)phosphonium tetrafluoroborate **4c** (29.0 mg, 0.05 mmol) in $CDCl_3$ (0.65 cm^3), trimethyl phosphite (8.9 μL, 9.3 mg, 0.075 mmol) was added. Dimethyldiphenylsilane (5 mg) was used as the internal standard. The reaction mixture was placed directly into the NMR tube. Changes in the concentrations of substrate **4c**, intermediate **13a**, and product **12a** were monitored using ^{1}H-NMR spectroscopy.

Reaction of 1-(N-pivaloylamino)ethyltris(3-chlorophenyl)phosphonium tetrafluoroborate **4c** *with triethyl phosphite–synthesis of 1-(N-pivaloylamino)ethyltriethoxyphosphonium tetrafluoroborate* **13b**

To a stirred solution of 1-(*N*-pivaloylamino)ethyltris(3-chlorophenyl)phosphonium tetrafluoroborate (**4c**, 145.2 mg, 0.25 mmol) in DCM (1 cm^3), triethyl phosphite (64.6 μL, 62.3 mg, 0.375 mmol) was added. Stirring was continued for 30 min at room temperature. Thereafter the intermediate product **13b** was precipitated with Et_2O, separated by decantation, and dried under reduced pressure. In order to obtain the triethoxyphosphonium salt **13** with a higher purity, the reaction should be carried out under argon atmosphere.

1-(N-Pivaloylamino)ethyltriethoxyphosphonium tetrafluoroborate (**13b**). Colorless oil (47.6 mg, 50% yield). ^{1}H-NMR ($CDCl_3$) δ 7.53 (t, J = 8.2 Hz, 1H, NH), 4.69 (dq, J_1 = 19.7, J_2 = 7.0 Hz, 1H, C_αH), 4.62–4.46 (m, 6H, 3 × OCH_2), 1.56 (dd, J_1 = 20.2, J_2 = 7.3 Hz, 3H, CH_3), 1.48 (td, J_1 = 7.0, J_2 = 0.9 Hz, 9H, 3 × $OCH_2\underline{CH_3}$), 1.23 (s, 9H, *t*-Bu) ppm; ^{13}C-NMR ($CDCl_3$) δ 179.9 (C = O), 70.2 (d, J = 9.1 Hz, OCH_2), 42.3 (d, J = 148.6 Hz, C_αH), 38.5 ($\underline{C}(CH_3)_3$), 27.1 ($C(\underline{C}H_3)_3$), 15.9 (d, J = 6.0 Hz, $OCH_2\underline{C}H_3$), 14.0 (CH_3) ppm; ^{31}P-NMR ($CDCl_3$) δ 37.0 ppm; IR (ATR) 3375, 2973, 1664, 1515, 1030 cm^{-1}. HRMS (TOF-ESI) calcd for $C_{13}H_{29}NO_4P$ $[M]^+$ 294.1834, found 294.1829.

4. Conclusions

Modification of the phosphonium group by introducing electron-withdrawing substituents results in the weakening of the C_α-P^+ bond and makes it susceptible to cleavage. This phenomenon is the cause of the high reactivity of 1-(*N*-acylamino)alkyltriarylphosphonium salts with weakened C_α-P^+ bond strength. As we have demonstrated, these types of phosphonium salts react smoothly, usually at room temperature with various types of nucleophiles such as 1-morpholinocyclohexene, 1,3-dicarbonyl compounds, benzotriazole sodium salt, *p*-toluenesulfinate sodium salt, benzylamine, triarylphosphines, and other P-nucleophiles. Only in the case of 1,3-dicarbonyl compounds, it was necessary to use a strong base to generate enolate anions. Reactions with 1-morpholinocyclohexene, benzotriazole sodium salt, *p*-toluenesulfinate sodium salt, benzylamine, and triarylphosphines do not require the use of any catalysts and occur quite fast (5–60 min). Other examined P-nucleophiles also react efficiently with 1-(*N*-acylamino)alkyltriarylphosphonium salts. However, the quickly formed intermediate, in the absence of any dealkylating agent, slowly transforms into a final product. Therefore, to facilitate the reaction, we used a substoichiometric amount of the dealkylating agent in the form of methyltriphenylphosphonium iodide.

The use of 1-(*N*-acylamino)alkyltriarylphosphonium salts with weakened C_α-P^+ bond strength allowed us to discover interesting mechanistic aspects of the examined reactions. Detection, isolation, and characterization of 1-(*N*-acylamino)alkyltrialkoxyphosphonium salt **13b**—the reactive intermediate in the Michaelis–Arbuzov type reaction, were particularly important.

Futher studies on expanding the range of nucleophiles, which can be used in α-amidoalkylation by 1-(*N*-acylamino)alkyltriarylphosphonium salts with weakened C_α-P^+ bond strength, are in progress.

Author Contributions: Conceptualization, J.A.; Formal analysis, J.A., A.W., J.K. and K.E.; Investigation, J.A., A.W., J.K. and K.E.; Methodology, J.A. and A.W.; Supervision, J.A. and K.W.; Writing—original draft, J.A.; Writing—review & editing, J.A. and A.W.

Acknowledgments: This work received financial support from the National Science Center, Poland (NCN) under Grant No. 2015/19/D/ST5/00733.

References

1. Mazurkiewicz, R.; Październiok-Holewa, A.; Adamek, J.; Zielińska, K. α-Amidoalkylating agents: Structure, synthesis, reactivity and application. *Adv. Heterocycl. Chem.* **2014**, *111*, 43–94. [CrossRef]
2. Adamek, J.; Węgrzyk, A.; Krawczyk, M.; Erfurt, K. Catalyst-free Friedel-Crafts reaction of 1-(*N*-acylamino)alkyltriarylphosphonium salts with electron-rich arenes. *Tetrahedron* **2018**, *74*, 2575–2583. [CrossRef]
3. Adamek, J.; Mazurkiewicz, R.; Węgrzyk, A.; Erfurt, K. 1-Imidoalkylphosphonium salts with modulated C_α-P^+ bond strength: Synthesis and application as new active α-imidoalkylating agents. *Beilstein J. Org. Chem* **2017**, *13*, 1446–1455. [CrossRef] [PubMed]
4. Zielińska, K.; Mazurkiewicz, R.; Szymańska, K.; Jarzębski, A.; Magiera, S.; Erfurt, K. Penicillin G acylase-mediated kinetic resolution of racemic 1-(*N*-acylamino)alkylphosphonic and 1-(*N*-acylamino)alkylphosphinic acids and their esters. *J. Mol. Catal. B Enzym.* **2016**, *132*, 31–40. [CrossRef]
5. Adamek, J.; Mazurkiewicz, R.; Październiok-Holewa, A.; Grymel, M.; Kuźnik, A.; Zielińska, K. 1-(*N*-Acylamino)alkyl Sulfones from *N*-Acyl-α-amino Acids or *N*-Alkylamides. *J. Org. Chem.* **2014**, *79*, 2765–2770. [CrossRef] [PubMed]
6. Adamek, J.; Mazurkiewicz, R.; Październiok-Holewa, A.; Kuźnik, A.; Grymel, M.; Zielińska, K.; Simka, W. *N*-[1-(Benzotriazol-1-yl)alkyl] amides from *N*-acyl-α-amino acids or *N*-alkylamides. *Tetrahedron* **2014**, *70*, 5725–5729. [CrossRef]
7. Adamek, J.; Październiok-Holewa, A.; Zielińska, K.; Mazurkiewicz, R. Comparative Studies on the Amidoalkylating Properties of *N*-(1-Methoxyalkyl)Amides and 1-(*N*-Acylamino) Alkyltriphenylphosphonium Salts in the Michaelis–Arbuzov-Like Reaction: A New One-Pot Transformation of *N*-(1-Methoxyalkyl)Amides into Phosphonic or Phosphinic Analogs of *N*-Acyl-α-Amino Acids. *Phosphorus Sulfur Silicon Relat. Elem.* **2013**, *188*, 967–980. [CrossRef]
8. Październiok-Holewa, A.; Adamek, J.; Mazurkiewicz, R.; Zielińska, K. Amidoalkylating Properties of 1-(*N*-Acylamino)Alkyltriphenylphosphonium Salts. *Phosphorus Sulfur Silicon Relat. Elem.* **2013**, *188*, 205–212. [CrossRef]
9. Mazurkiewicz, R.; Adamek, J.; Październiok-Holewa, A.; Zielińska, K.; Simka, W.; Gajos, A.; Szymura, K. α-Amidoalkylating Agents from *N*-Acyl-α-amino Acids: 1-(*N*-Acylamino)alkyltriphenylphosphonium Salts. *J. Org. Chem.* **2012**, *77*, 1952–1960. [CrossRef] [PubMed]
10. Październiok-Holewa, A.; Adamek, J.; Zielińska, K.; Piernikarczyk, K.; Mazurkiewicz, R. *N*-(1 acyloaminoalkyl) amidinium salts derived from DBU or related bases as reactive intermediates in α-amidoalkylation reactions. *Arkivoc* **2012**, 314–329. [CrossRef]
11. Zaugg, H.E. Recent Synthetic Methods Involving Intermolecular alpha-Amidoalkylation at Carbon. *Synthesis* **1970**, 49–73. [CrossRef]
12. Zaugh, H.E. α-Amidoalkylation at Carbon: Recent Advances–Part I. *Synthesis* **1984**, 85–110. [CrossRef]
13. Zaugg, H.E. α-Amidoalkylation at Carbon: Recent Advances–Part II. *Synthesis* **1984**, 181–212. [CrossRef]
14. Speckamp, W.N.; Hiemstra, H. Intramolecular reactions of *N*-acyliminium intermediates. *Tetrahedron* **1985** *41*, 4367–4416. [CrossRef]
15. Hiemstra, H.; Speckamp, W.N. *N*-Acyliminium Ions as Intermediates in Alkaloid Synthesis. *Alkaloids* **1988** *32*, 271–339. [CrossRef]

16. Hiemstra, H.; Speckamp, W.N. *Comprehensive Organic Synthesis*; Throst, B.M., Flemming, I., Eds.; Oxford: Pergamon, Turkey, 1991; Volume 2, pp. 1047–1082. [CrossRef]
17. Kleinman, E.F. The Bimolecular Aliphatic Mannich and Related Reactions. In *Comprehensive Organic Synthesis*; Trost, B.M., Ed.; Pergamon Press: Oxford, UK, 1991; Volume 2, pp. 893–951.
18. Katritzky, A.R.; Lan, X.; Yang, J.Z.; Denisko, O.V. Properties and Synthetic Utility of *N*-Substituted Benzotriazoles. *Chem. Rev.* **1998**, *98*, 409–548. [CrossRef] [PubMed]
19. Speckamp, W.N.; Moolenaar, M.J. New developments in the chemistry of *N*-acyliminium ions and related intermediates. *Tetrahedron* **2000**, *56*, 3817–3856. [CrossRef]
20. Mecozzi, T.; Petrini, M.; Profeta, R. Reactivity of chiral exocyclic *N*-acyliminium ions with aromatic derivatives. *Tetrahedron Asymmetry* **2003**, *14*, 1171–1178. [CrossRef]
21. Maryanoff, B.E.; Zhang, H.C.; Cohen, J.H.; Turchi, I.J.; Maryanoff, C.A. Cyclizations of *N*-acyliminium ions. *Chem. Rev.* **2004**, *104*, 1431–1628. [CrossRef] [PubMed]
22. Katritzky, A.R.; Manju, K.; Singh, S.K.; Meher, N.K. Benzotriazole mediated amino-, amido-, alkoxy- and alkylthioalkylation. *Tetrahedron* **2005**, *61*, 2555–2581. [CrossRef]
23. Petrini, M. α-Amido Sulfones as Stable Precursors of Reactive *N*-Acylimino Derivatives. *Chem. Rev.* **2005**, *105*, 3949–3977. [CrossRef] [PubMed]
24. Ollevier, T.; Li, Z. The first catalytic Sakurai reaction of *N*-alkoxycarbonylamino sulfones with allyltrimethylsilane. *Org. Biomol. Chem.* **2006**, *4*, 4440–4443. [CrossRef] [PubMed]
25. Ballini, R.; Palmieri, A.; Petrini, M.; Torregiani, E. Solventless Clay-Promoted Friedel−Crafts Reaction of Indoles with α-Amido Sulfones: Unexpected Synthesis of 3-(1-Arylsulfonylalkyl) Indoles. *Org. Lett.* **2006**, *8*, 4093–4096. [CrossRef] [PubMed]
26. Marianacci, O.; Micheletti, G.; Bernardi, L.; Fini, F.; Fochi, M.; Pettersen, D.; Sgarzani, V.; Ricci, A. Organocatalytic Asymmetric Mannich Reactions with *N*-Boc and *N*-Cbz Protected α-Amido Sulfones. *Chem. Eur. J.* **2007**, *13*, 8338–8351. [CrossRef] [PubMed]
27. Yazici, A.; Pyne, S.G. Intermolecular addition reactions of *N*-acyliminium ions (Part I). *Synthesis* **2009**, 339–368. [CrossRef]
28. Yazici, A.; Pyne, S.G. Intermolecular addition reactions of *N*-acyliminium ions (Part II). *Synthesis* **2009**, 513–541. [CrossRef]
29. Thirupathi, P.; Kim, S.S. $InBr_3$: A Versatile Catalyst for the Different Types of Friedel−Crafts Reactions. *J. Org. Chem.* **2009**, *74*, 7755–7761. [CrossRef] [PubMed]
30. Das, B.; Damodar, K.; Bhunia, N. A Simple and Efficient Access to α-Amino Phosphonates from *N*-Benzyloxycarbonylamino Sulfones Using Indium(III) Chloride. *J. Org. Chem.* **2009**, *74*, 5607–5609. [CrossRef] [PubMed]
31. Martínez-Estibalez, U.; Gómez-SanJuan, A.; García-Calvo, O.; Aranzamendi, E.; Lete, E.; Sotomayor, N. Strategies based on aryllithium and *N*-acyliminium ion cyclizations for the stereocontrolled synthesis of alkaloids and related systems. *Eur. J. Org. Chem.* **2011**, 3610–3633. [CrossRef]
32. Schneider, A.E.; Manolikakes, G. $Bi(OTf)_3$-Catalyzed Multicomponent α-Amidoalkylation Reactions. *J. Org. Chem.* **2015**, *80*, 6193–6212. [CrossRef] [PubMed]
33. Aranzamendi, E.; Arrasate, S.; Sotomayor, N.; González-Díaz, H.; Lete, E. Chiral Brønsted Acid Catalyzed Enantioselective α-Amidoalkylation Reactions: A Joint Experimental and Predictive Study. *ChemistryOpen* **2016**, *5*, 540–549. [CrossRef] [PubMed]
34. Touati, B.; El Bouakher, A.; Azizi, M.S.; Taillier, C.; Othman, R.B.; Trabelsi-Ayadi, M.; Antoniotti, S.; Dunach, E.; Dalla, V. Enolizable Carbonyls and N,O-Acetals: A Rational Approach for Room-Temperature Lewis Superacid-Catalyzed Direct α-Amidoalkylation of Ketones and Aldehydes. *Chem. Eur. J.* **2016**, *22*, 6012–6022. [CrossRef] [PubMed]
35. Vinogradov, M.G.; Turova, O.V.; Zlotin, S.G. The progress in the chemistry of *N*-acyliminium ions and their use in stereoselective organic synthesis. *Russ. Chem. Rev.* **2017**, *86*, 1–17. [CrossRef]
36. Aranzamendi, E.; Sotomayor, N.; González-Díaz, H.; Lete, E. Phenolic Activation in Chiral Brønsted Acid-Catalyzed Intramolecular α-Amidoalkylation Reactions for the Synthesis of Fused Isoquinolines. *ACS Omega* **2017**, *2*, 2706–2718. [CrossRef]
37. Touati, B.; El Bouakher, A.; Azizi, M.S.; Taillier, C.; Othman, R.B.; Trabelsi-Ayadi, M.; Antoniotti, S.; Dunach, E.; Dalla, V. Atom-Economic Catalytic Direct Substitution of N,O-Acetals with Simple Ketones. *Eur. J. Org. Chem.* **2017**, 4445–4460. [CrossRef]

7

Synthetic Strategies for Dinucleotides Synthesis

Lucie Appy, Crystalle Chardet, Suzanne Peyrottes and Béatrice Roy *

Nucleosides & Phosphorylated Effectors, Institut des Biomolécules Max Mousseron (IBMM), UMR 5247 CNRS, Université de Montpellier, ENSCM, Campus Triolet, cc 1705, Place Eugène Bataillon, CEDEX 5, 34095 Montpellier, France; lucie.appy@gmail.com (L.A.); crystalle.chardet@gmail.com (C.C.); suzanne.peyrottes@umontpellier.fr (S.P.)

* Correspondence: beatrice.roy@umontpellier.fr

Academic Editor: György Keglevich

Abstract: Dinucleoside 5′,5′-polyphosphates (DNPs) are endogenous substances that play important intra- and extracellular roles in various biological processes, such as cell proliferation, regulation of enzymes, neurotransmission, platelet disaggregation and modulation of vascular tone. Various methodologies have been developed over the past fifty years to access these compounds, involving enzymatic processes or chemical procedures based either on P(III) or P(V) chemistry. Both solution-phase and solid-support strategies have been developed and are reported here. Recently, green chemistry approaches have emerged, offering attracting alternatives. This review outlines the main synthetic pathways for the preparation of dinucleoside 5′,5′-polyphosphates, focusing on pharmacologically relevant compounds, and highlighting recent advances.

Keywords: phosphorylation; dinucleotides; organophosphorus chemistry; mechanochemistry; dry eye syndrome; diquafosol; denufosol

1. Introduction

Dinucleoside 5′,5′-polyphosphates (DNPs), commonly abbreviated as Np$_n$Ns, are essential to human biological systems [1,2]. They contain two ribonucleosides, which are linked at the 5′-position of their sugar moiety through n phosphate groups (Figure 1).

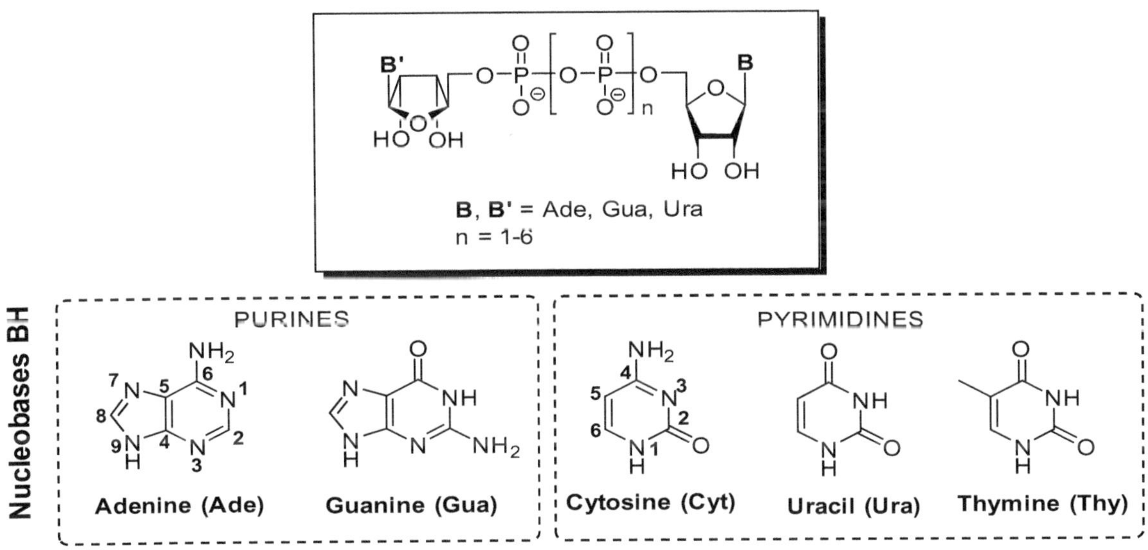

Figure 1. General structure of natural dinucleoside 5′,5-polyphosphates and canonical nucleobases.

Historically, dinucleoside 5′,5′-polyphosphates (i.e., Ap_2A and Up_2U) were first isolated by chemists [3]. In 1976, E. Rapaport and P.C. Zamecnik brought to light the presence of P^1,P^4-diadenosine-5′-tetraphosphate (Ap_4A) in mammals [4]. The authors assumed that this dinucleotide could have a role in cellular communication as a "signal nucleotide". Later, Up_4A was the first endogenous dinucleotide, possessing both purine and pyrimidine moieties, to be identified in living organisms [5,6]. Currently, 17 symmetrical or mixed Np_nNs have been isolated from human tissues and characterized [2]. DNPs are released into the extracellular space from different types of cells, such as platelets and endothelial cells, where they stimulate several cell-surface purinergic receptors. They have a strong physiological and pathophysiological impact on the cardiovascular system [2,7], and may also interact with enzymes, acting as inhibitors (e.g., kinases) or substrates [8–11]. Three classes of enzymes are known to catalyze the degradation of dinucleotides [10–12]: symmetrically cleaving dinucleoside polyphosphate hydrolases, asymmetrically cleaving dinucleoside polyphosphate hydrolases, and dinucleoside polyphosphate phosphorylases. Over the years, DNPs have gained increased attention and their therapeutic potential has been revealed. Likewise, structural analogues have been developed and some of them are drug candidates.

Herein, we first describe purinergic signalling, i.e., the role of nucleotides as extracellular signal molecules, with a focus on the P2Y receptor subtypes that interact with dinucleotides. Then, we present the synthetic strategies to obtain dinucleotides. Pyrophosphate bond formation has been addressed in some recent reviews and book chapters [1,13–15]. The current review focuses on methods reported for the synthesis of dinucleoside 5′,5′-polyphosphates, especially those of biological and pharmacological interest. We also highlight recent advances in the field, such as green chemistry approaches. It should be noted that methods for preparing cyclic dinucleotides [16] are beyond the scope of this review. Furthermore, the synthesis of nicotinamide adenine dinucleotides [17] and mRNA cap analogues such m^7Gp_nN (see reviews [18–20]) are not presented.

2. Interaction of DNPs with Purine and Pyrimidine Receptors

Purinergic receptors are extracellular receptors, currently consisting of four subtypes of P1 (or adenosine) receptors, seven subtypes of P2X ion channel receptors, and eight subtypes of P2Y receptors (Figure 2). Specifically, P2Y receptors (P2YRs) are G-protein-coupled receptors (GPCRs) activated by extracellular nucleotides, which are divided into two groups on the basis of sequence homology and the type of G protein they are primarily coupled to [21–23].

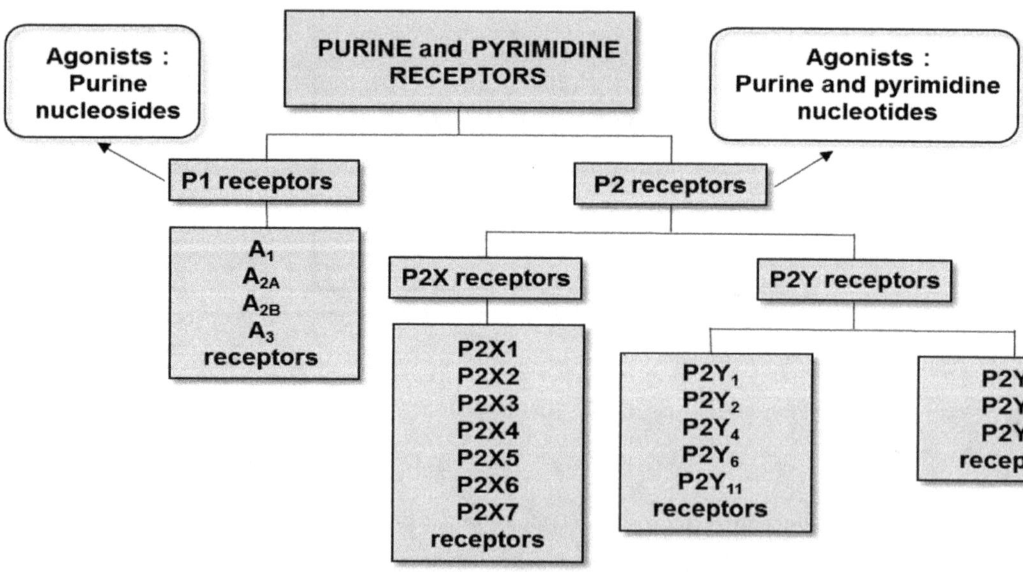

Figure 2. The classification of purine and pyrimidine receptors and their main physiological agonists.

Purinergic receptors are expressed in nearly all cell types, and have been found to play crucial roles in many biological processes, e.g., neurotransmission, neuroprotection in hypoxia and ischemia, regulation of cardiovascular function, platelet aggregation, smooth muscle contraction, secretion of hormones, modulation of immune response, control of cell proliferation, differentiation, and apoptosis. In this regard, different aspects of purinergic signaling such as the pathophysiology and the therapeutic potential have been extensively reviewed [21–28]. Several purinergic compounds are already on the market to treat thrombosis and stroke, and one dinucleotide is used to treat dry eye disease (see Section 2.1).

DNPs are strong endogenous agonists of the purinergic system [2]. In particular, they interact with the P2YRs, a family of high therapeutic relevance [2,23,29,30]. The $P2Y_1$, $P2Y_{11}$, $P2Y_{12}$, and $P2Y_{13}$ receptors are activated by adenosine 5′-di- or triphosphates (ADP and ATP), while uracil nucleotides (UDP and UTP) are the endogenous agonists for the $P2Y_4$, $P2Y_6$, and $P2Y_{14}$ receptors. On the other hand, the $P2Y_2$ receptor responds to both ATP and UTP. Natural dinucleoside tetraphosphates Ap_4A and Up_4U have agonist potencies comparable to those of ATP and UTP at $P2Y_2$ receptors [31–34]. It should be noted that dinucleotides are more resistant to hydrolysis than their parent mononucleotides, but are generally less potent, and lack selectivity in many cases [9,23]. Ligand development for the class of uracil nucleotide-activated P2Y receptors has been extensively reviewed by Rahefi and Müller in 2018 [23]. They have compiled existing data on, inter alia, dinucleotides such as Ap_nA (n = 2–6), Up_nU (n = 3–4) and structural analogues, as agonists for the uracil activated P2YRs. Given the significant clinical potential of the P2YRs, substantial research efforts directed towards developing P2YR ligands for use as pharmacological tools and drugs have led to the discovery and development of a significant number of agonists but, so far, only a moderate number of antagonists. Two relevant dinucleotides, namely Up_4U and Up_4dC, have undergone clinical development to cure several diseases involving the $P2Y_2$ receptor (Figure 3).

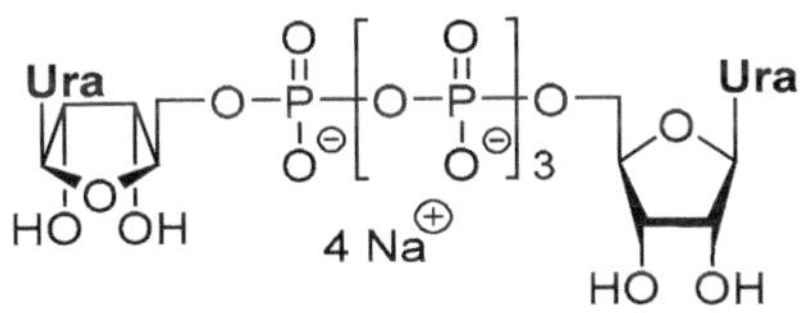

Diquafosol tetrasodium
INS365, Up_4U

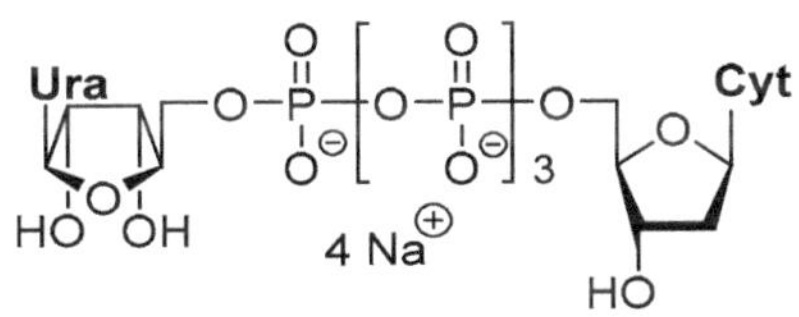

Denufosol tetrasodium
INS37217, Up_4dC

Figure 3. Dinucleotides with clinical developments.

2.1. *Up_4U for the Treatment of Dry Eye Disease (DED)*

Diquafosol tetrasodium (*P^1, P^4*-diuridine 5′,5′-tetraphosphate, Up_4U, NS365, *Diquas*®) is the first agonist of the $P2Y_2$ receptor subtype that has been approved in Japan as 3% ophthalmic solution (Santen Pharmaceutical Co, Ltd., Osaka, Japan) for the management of dry eye disease (see reviews by Lau et al. [35], and Keating [36]). This pathology commonly causes symptoms including dryness, irritation, itching, and light sensitivity. It is associated with tear film instability, increased tear film osmolarity, and ocular surface inflammation. It impacts the daily life of patients and has a prevalence that varies from 5% to 35%. DED results from either decreased aqueous tear production (aqueous tear-deficient dry eye) or increased tear evaporation (evaporative dry eye), or both. Several studies demonstrated the presence of P2X and P2Y receptors in ocular tissues (retina, ciliary body, and lens), and indicated that $P2Y_2$ receptors may be the main subtype of purinergic receptor located at the ocular surface [33]. Clinical data show that diquafosol tetrasodium improves ocular surface staining and may improve tear film volume and stability.

2.2. *Up_4dC for the Treatment of Cystic Fibrosis*

Denufosol tetrasodium (*P^1*-uridine-*P^4*-2′-deoxycytidine 5′,5′-tetraphosphate, Up_4dC, dCp_4U, INS37217) is a mixed dinucleoside tetraphosphate. Along with Ap_4A and Up_4U, dCp_4U acts as a $P2Y_2$ agonist [29,37]. It has been developed by Inspire Pharmaceuticals (now part of Merck) as an inhaled drug for the treatment of cystic fibrosis [38]. This recessive genetic disease is characterized by pulmonary and reproductive tract dysfunctions, involving abnormal ion transport and defective mucociliary clearance. Denufosol acts on P2Y receptors expressed on the surface of airway epithelia to stimulate chloride secretion independent of the chloride channel, which is dysfunctional in cystic fibrosis. It was shown to significantly enhance tracheal mucus transport in an animal model [39,40]. The first clinical trials established a good safety profile. It was evaluated in two phase 3 clinical trials, TIGER-1 and TIGER-2, as a therapy for cystic fibrosis patients [41,42]. In the first trial (TIGER-1), it was found to significantly improve lung function in cystic fibrosis patients with normal to mildly impaired lung function. Unfortunately, less than 3 weeks after publication of the TIGER-1 data, Inspire Pharmaceuticals announced that the 466-patients, 48-week placebo-controlled phase 3 TIGER-2 clinical trial, had failed to demonstrate any benefit [41,42].

3. Synthesis Based on P(V) Chemistry

To date, the most widely used methods to access dinucleotides involve P(V) chemistry. They are based on the activation of a 5′-nucleotide (i.e., nucleoside 5′-mono, 5′-di and 5′-triphosphate) and the reaction of the corresponding intermediate with a second 5′-nucleotide or inorganic pyrophosphate to form dinucleotides containing up to six bridging phosphate groups. These steps are usually performed in dry, aprotic solvents (mostly *N*,*N*-dimethylformamide, DMF), and, therefore, require the use of nucleotides, as well as inorganic phosphate or pyrophosphate, in their tri- or tetra-*n*-butylammonium forms due to solubility issues. Divalent cations, especially Mg^{2+} and Zn^{2+}, are sometimes added as catalysts for anhydride bond formation. It is assumed that the metal ion could serve two roles, namely, the activation of the electrophilic P(V) center and templating the incoming phosphate nucleophile and the P(V) electrophilic center [13].

3.1. *Synthesis via a Phosphoromorpholidate Intermediate*

The first description of a chemical synthesis of dinucleotides was reported in the mid-60s by Moffatt and Khorana [43,44]. This strategy relies on the conversion of nucleoside 5′-monophosphates (NMPs) into their phosphoromorpholidate derivatives, followed by reaction with a phosphate salt, i.e., orthophosphate or pyrophosphate (Scheme 1). Briefly, the acidic form of the nucleotides were activated with N,N′-dicyclohexylcarbodiimide (DCC) in the presence of morpholine, leading to the nucleoside 5′-phosphoromorpholidates as 4-morpholine N,N′-dicyclohexylcarboxamide salts **1–3** (Scheme 1) [45]. In the second step, addition of pyrophosphate afforded the symmetrical dinucleoside tetraphosphates. Similarly, dinucleoside triphosphates were obtained when pyrophosphate was replaced by orthophosphate.

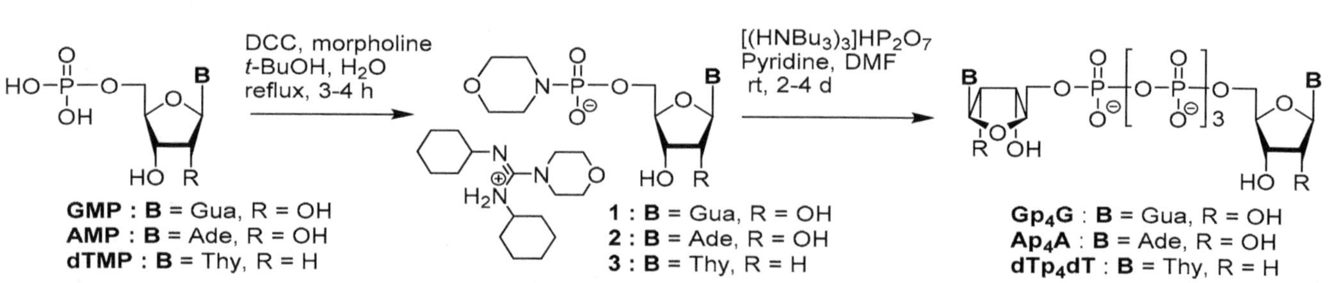

Scheme 1. Synthesis via phosphoromorpholidate intermediates [44].

Since then, this approach has been adapted by several research groups to access Np$_2$N, Np$_3$N, as well as some analogues. For example, Ap$_2$A was obtained in 90% yield, through the condensation of the bis-(tri-*n*-butylammonium) salt of AMP and the adenosine-5′-phosphoromorpholidate-4-morpholine-N,N′-dicyclohexylcarboxamide salt in anhydrous pyridine [46]. Mixed dinucleoside triphosphates were obtained in 10–30% yields, by reacting nucleoside 5′-phosphoromorpholidates with nucleoside 5′ diphosphates (NDPs), possibly in the presence of tetrazole [31,47,48]. However, due to long reaction times and generally low yields, this method has lost its relevance.

3.2. *Synthesis via a Phosphorimidazolide Intermediate*

Phosphorimidazolide derivatives exhibit high reactivity toward various nucleophiles. Therefore, they have been extensively used as intermediates for pyrophosphate bond formation. Typically, activation of a 5′-nucleotide (NMP, NDP or NTP) with *N,N′*-carbonyldiimidazole (CDI) in DMF, followed by the in situ condensation with a second 5′-nucleotide or pyrophosphate, affords dinucleotides containing two to four bridging phosphate groups in 10–60% yields [47–50]. Accordingly, the tri-*n*-butylammonium salt of UMP was activated with CDI to form a phosphorimidazolide intermediate **4**, which reacted with the remaining nucleotide to form Up$_2$U (Scheme 2) [47,49]. Slight modifications of this protocol allowed access to mixed dinucleotides [50]. First, activation of the bis(tri-*n*-butylammonium) salt of CMP was performed using ≈ 2 equiv of CDI in DMF for 2 h at rt, followed by the addition of dry methanol to quench the remaining CDI. Treatment with the bis(tri-*n*-butylammonium) salt of GMP for 2 h at rt afforded Gp$_2$C in 54% yield. Addition of GDP instead of GMP in the second step afforded Gp$_3$C, albeit in only 3% yield. Similarly, activation of NDPs with CDI followed by reaction with a NMP allowed the isolation of dinucleoside triphosphates in low yields [48].

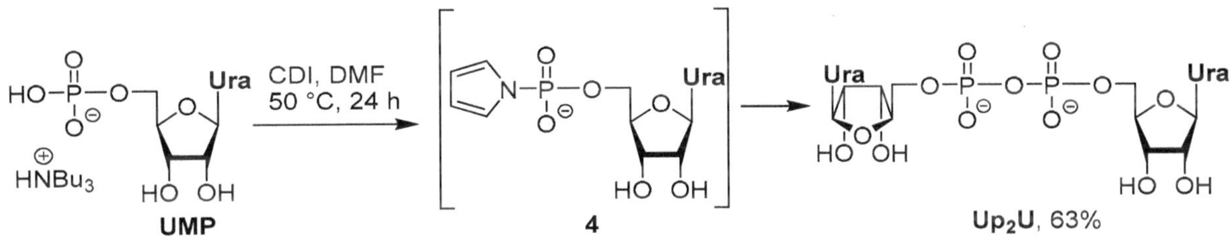

Scheme 2. Activation of UMP by CDI to access Up$_2$U [47,49].

Alternatively, symmetrical dinucleoside tetraphosphate Up$_4$U was obtained, either by activation of UMP with CDI followed by coupling with bis(tri-*n*-butylammmonium) pyrophosphate [47,49] or dimerization of UDP in the presence of CDI [51] (Scheme 3).

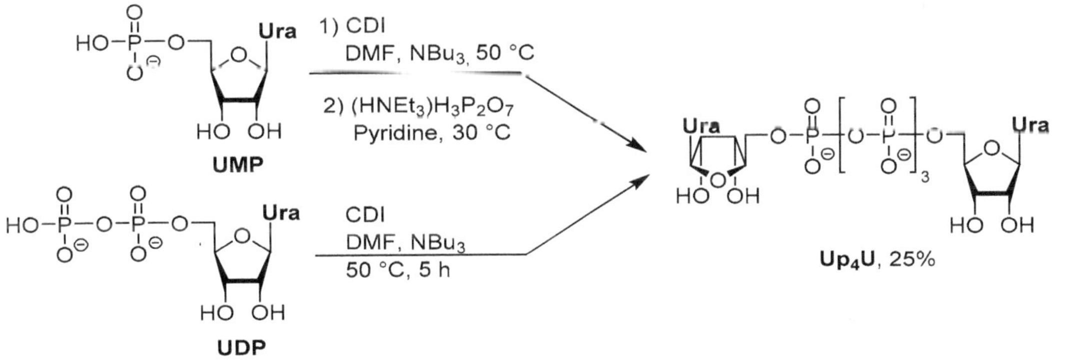

Scheme 3. Dimerization of nucleotides to access symmetrical Up$_4$U [49,51].

Activation of ribonucleotides in the presence of an excess of CDI usually results in the carbonation of the ribose moiety (Scheme 4). In ^{1}H-decoupled ^{31}P NMR spectroscopy, adenosine 5′-phosphorimidazolide is characterized by a singlet at −7.56 ppm, whereas phosphorimidazolide **5** exhibits a singlet at −7.76 ppm [52]. However, in some of the above-mentioned publications [47,49], carbonation was not observed, and may be due to the experimental conditions (number of equivalents of CDI, reaction time, temperature, adventitious presence of water). Nonetheless, the removal of the 2′,3′-carbonate protecting group can be easily performed under basic conditions [52,53].

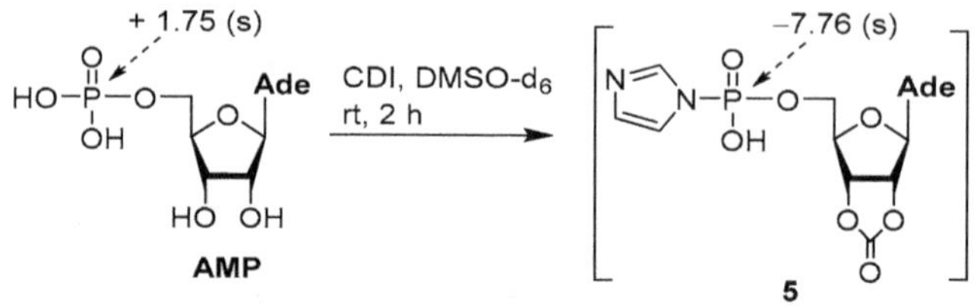

Scheme 4. Reaction of AMP with 4 equiv CDI [52]. ^{31}P chemical shifts are indicated in ppm together with their multiplicity.

In 2011, a variant of the CDI method was developed by Yanachkov and co-workers [54], where the activation of pyrophosphate by CDI gave rise to P^1,P^2-di(1-imidazolyl)pyrophosphate (Scheme 5). Using ^{13}C-labeled CDI and monitoring the reaction by ^{31}P NMR (^{13}C-^{31}P and ^{1}H-^{31}P couplings), the authors proposed a reaction mechanism involving the fast formation of mixed anhydrides, which react slowly with imidazole to give diimidazolyl pyrophosphate. This activated pyrophosphate was isolated as its disodium salt **6** by precipitation with sodium perchlorate in acetone, and then reacted with AMP or UMP to afford the symmetrical 5′,5′-dinucleoside tetraphosphates Ap_4A and Up_4U in good yields (Scheme 5). Noteworthy, the use of 1*H*-tetrazole or zinc chloride during the second step allowed the reaction time to be shortened from 24–48 h to 3–16 h.

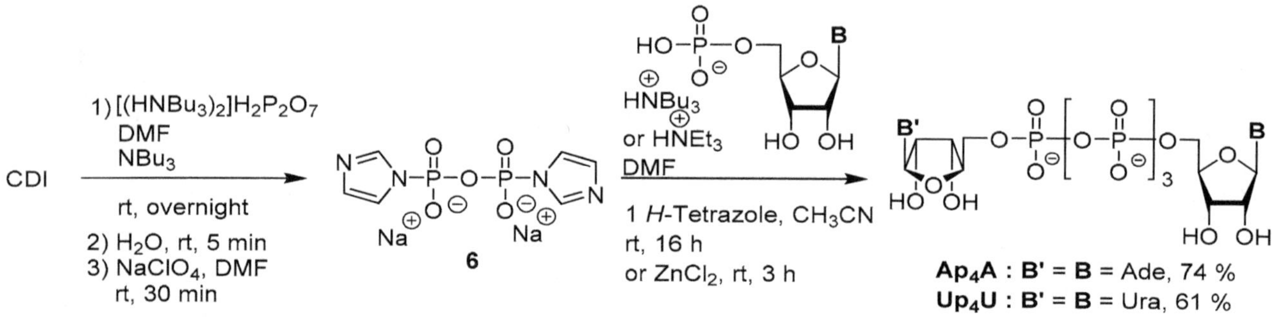

Scheme 5. Np_4Ns synthesis using P^1,P^2-di(1-imidazolyl)pyrophosphate **6 [54].**

Interestingly, this approach allows the preparation of dinucleotide analogues, starting from modified pyrophosphate (the central oxygen being replaced with monofluoro, monochloro or dichloromethylene). These DNPs analogues were obtained in 70–80% yields.

While these methods involving phosphoromidazolides are straightforward and rapid, they all require dry conditions to prevent side-product formation and most often, use trialkylammonium salts. To avoid this drawback, our research group has recently developed a one-pot synthesis of dinucleotides, either in aqueous solution or mechanochemically, through the use of phosphorimidazolides intermediates [55,56]. These strategies will be detailed in Section 5 and are related to green chemistry approaches.

3.3. *Synthesis via a Cyclic Trimetaphosphate Intermediate*

The initial synthesis of trimetaphosphate esters from ATP was described by Khorana [57]. Later, Ng and Orgel reported that adenosine 5′-polyphosphates can react with 1-ethyl-3-(3-dim ethylaminopropyl)carbodidimide (EDC) in water to form diadenosine 5′,5′-polyphosphates as the main products [58]. Yields were improved by performing the reaction in organic solvents such as DMF or DMSO, in the presence of DCC and using the tris or tetrakis(tri-*n*-butylammonium) salts of nucleotides [31,47]. The activation of nucleoside 5′-triphosphates (NTPs) with DCC proceeds via a cyclic nucleoside 5′-trimetaphosphate intermediates (**7–8**), which can further react with another NMP to produce the corresponding dinucleoside tetraphosphates (Scheme 6). Accordingly, Up_4U and Up_4A were obtained in low yields starting from UTP and ATP, respectively. One should note that *N,N′*-dicyclohexylurea (DCU) may be removed as a precipitate either after the first or the second step of the reaction.

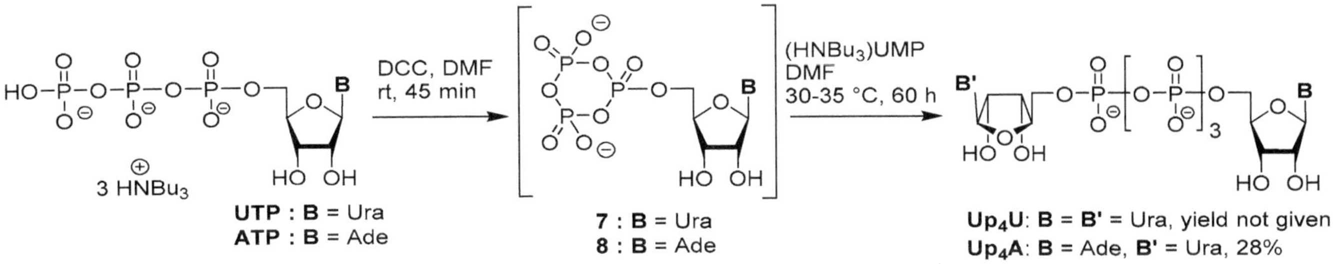

Scheme 6. Synthesis of Np_4Ns starting from nucleoside 5′-triphosphates (NTPs) [47].

Similarly, the activation of UTP or ATP with DCC in anhydrous DMF, followed by the addition of the tris(tri-*n*-butylammonium) salt of UDP or ADP in the second step of the reaction, afforded Up_5U and Ap_5A, in 10–60% yields [8,49,50]. Reactions must be performed under strictly anhydrous conditions and long reaction times are required.

In 2013, the Taylor research group reported a high-yielding synthesis of dinucleoside pentaphosphates (Np_5N) using the tri(tetra-*n*-butylammonium) salt of cyclic trimetaphosphate **9** as a phosphorylating agent [59]. The latter was prepared in almost quantitative yield by conversion of the trisodium salt of trimetaphosphate **10** into its pyridinium form, and then by titration of the solution to pH 7.0 with a dilute solution of tetrabutylammonium hydroxide and freeze-drying (Scheme 7).

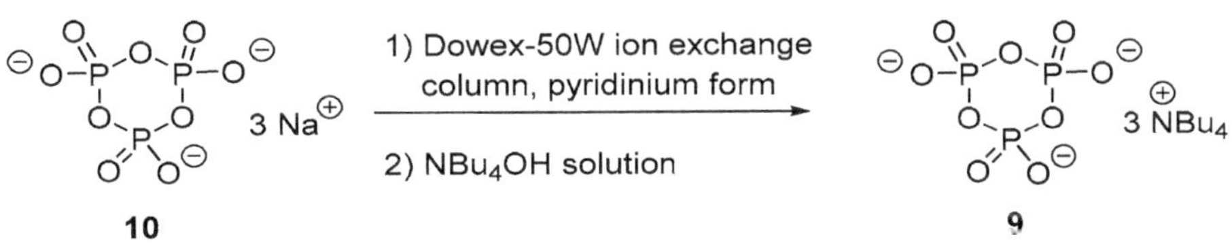

Scheme 7. Preparation of the cyclic trimetaphosphate **9** [59].

Then, the electrophilic reagent **9** was functionalized by a sulfonyl leaving group (LG) using *N*-methylimidazole (NMI) in the presence of either 2-mesitylenesulfonyl chloride **11** or 1-benzenesulfonyl-3-methyl-imidazolium triflate **12** in DMF (Scheme 8). Compound **11** was commercially available, whereas **12** was prepared in two steps starting from benzenesulfonyl chloride and imidazole [60]. Based on ^{31}P NMR monitoring of the reaction, the authors proposed that the activation proceeds via a mixed anhydride **13**, which reacts with NMI to give rise to intermediate **14**.

Addition of sub-stœchiometric amounts of NMPs to the activated cyclic trimetaphosphate **9** allowed their conversion to intermediates **15–18**, as suggested by ^{31}P NMR (Scheme 9). Addition of slight excess of the second NMP and anhydrous $MgCl_2$ gave rise to the corresponding dinucleoside pentaphosphates. In the absence of Mg^{2+} ions, the reaction was extremely slow.

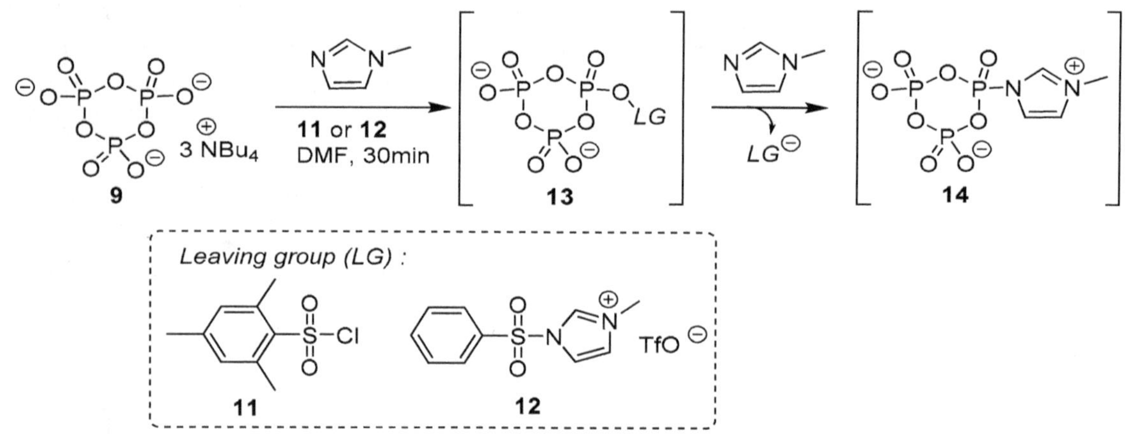

Scheme 8. Activation of the cyclic trimetaphosphate **9** [59].

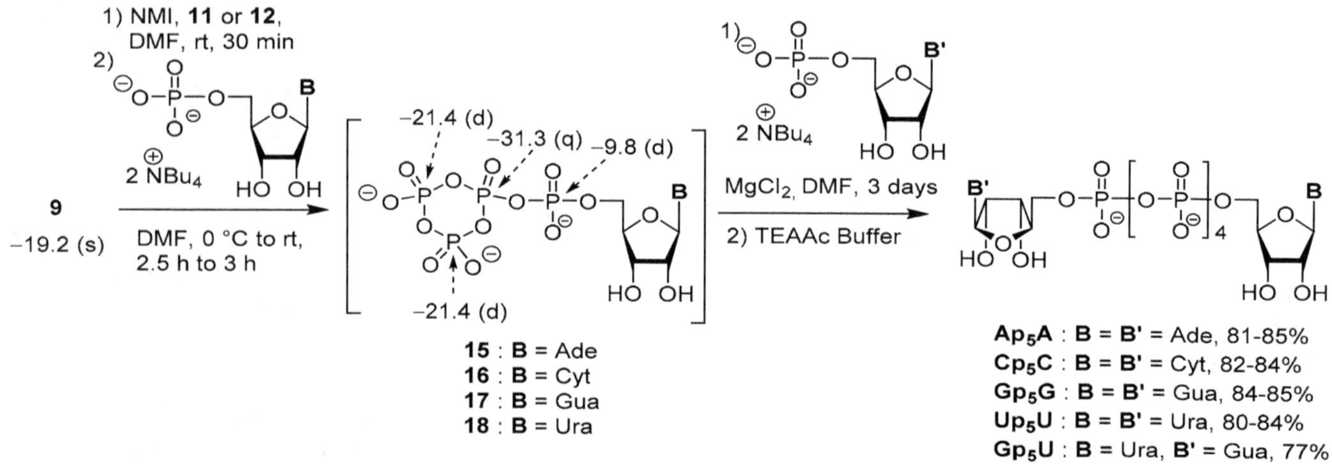

Scheme 9. Synthesis of dinucleoside pentaphosphates via an activated cyclic trimetaphosphate [59]. ^{31}P chemical shifts are indicated in ppm, together with their multiplicity.

This method constitutes an interesting approach as it allows formation of symmetric as well as mixed Np_5Ns in high yields, and uses inexpensive reagents (sodium cyclic trimetaphosphate, mesitylene chloride, NMI). It has recently been applied for the synthesis of dicaptides, used as substrates for polymerases [61]. However, it requires the preparation of the tetra-*n*-butylammonium salts of the cyclic trimetaphosphate and those of the nucleotides, reaction times are long (> 3 days), and must be performed under strictly anhydrous conditions.

3.4. Synthesis via a Phosphoromethylimidazolium Intermediate

Reacting the free acids of canonical nucleoside 5′-monophosphates with a large excess of NMI in the presence of triphenylphosphine/2,2′-dipyridyl sulfide as coupling agents and triethylamine in DMF or DMSO allowed dimerization to Np_2N in about 85% yields [62]. Unsymmetrical dinucleotides could also be obtained, albeit in much lower yields, by using a equimolar amount of the two NMPs to be coupled.

Another methodology to provide symmetrical and mixed dinucleoside polyphosphates Np_nN ($n = 2–4$) was inspired by the Bogachev procedure for the synthesis of deoxynucleoside 5′-triphosphates (dNTPs) [63]. It involves the reaction of nucleoside 5′-monophosphates-N-methylimidazolium as donors with an NMP, NDP or NTP to obtain dinucleoside di, tri and tetraphosphates (Scheme 10) [64]. These donors were prepared by treatment of the NMPs (disodium salts dihydrate or free acid hydrate) with a 16-fold excess of trifluoroacetic anhydride (TFAA) in acetonitrile (ACN) in the presence of excess triethylamine. This step resulted in the temporary protection of the hydroxyl and the amino (if present) groups and formation of a mixed anhydride, as shown by ^{31}P NMR ($\delta = 2$ ppm for the mixed anhydride formed within a few minutes). It offered the additional benefit of drying the starting materials without deleterious effects on the reaction outcome. However, in the case of GMP, these conditions resulted in the complete decomposition of the starting material. This limitation could be circumvented by using GMP in its tri-*n*-butylammonium form, together with reducing the number of equivalent of TFAA.

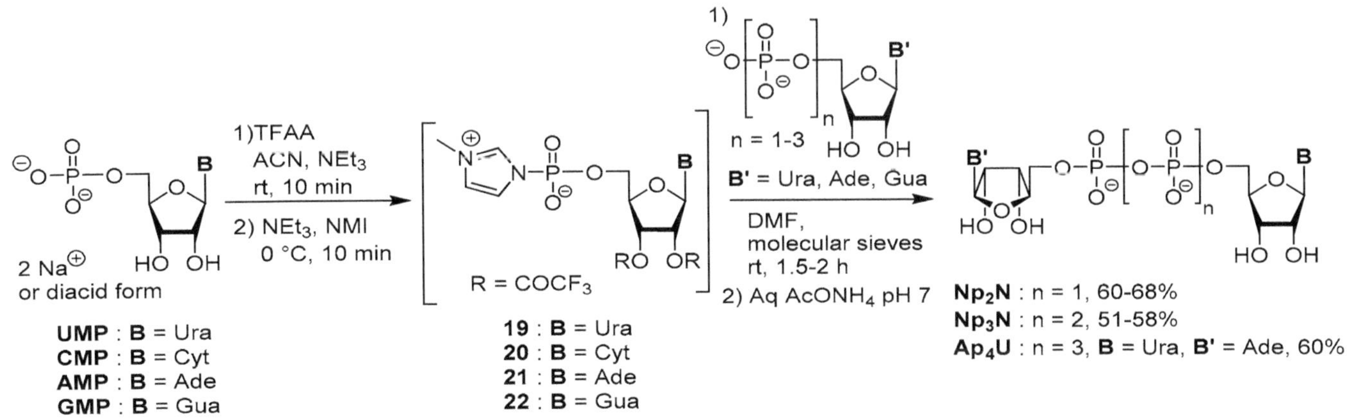

Scheme 10. Synthesis of dinucleotides via phosphoromethylimidazolium intermediates [64].

After removal of unreacted TFAA, the mixed anhydride was reacted with excess NMI in the presence of triethylamine to quantatively afford the highly reactive N-methylimidazolium salt donors **19–22** within 10 min. Formation of the N-methylimidazolium salts could also be easily followed by ^{31}P NMR spectroscopy as these species exhibit chemical shifts of approximately −9 to −10 ppm. The crude acetonitrile solutions were used directly in the next step. Thus, addition of the donors solutions to the bis(tri-*n*-butylammonium) salts of the nucleotide acceptors in DMF (in acetonitrile precipitation of the acceptors was observed) led to partially protected products. The latter were treated with aqueous ammonium acetate to remove the protecting groups, providing the desired compounds. For dinucleotides involving guanine nucleotides, N,N-dimethylaniline was found to be beneficial. Using this procedure, up to nine dinucleotides were obtained in yields of 51–68%. It should be noted that this method cannot be used to activate nucleoside 5′-di- or triphosphates. While the reactions were performed under argon, the authors pointed out that vigorous drying of the donor precursors was not always necessary when preparing the dinucleoside di- and triphosphates. Advantages of this one-pot method are short reaction time and good yields.

The same group also reported the synthesis of symmetrical di- and tetraphosphates by dimerization of nucleotides, using sulfonyl imidazolium salt **12** as a reagent (Scheme 11) [60,65]. The tetra-*n*-butylammonium salts of nucleoside 5′-mono or 5′-diphosphates were activated in DMF with 0.6–0.75 equiv of **12** in presence of 1–3 equiv of NMI and $MgCl_2$ (Scheme 11). Dimerization of the nucleotides led to Np_2N and Np_4N in high yields.

^{31}P NMR analyses showed that the reaction proceeds via the formation of the N-methylimidazolium salt intermediates **23–27** (Figure 4, ^{31}P $\delta = -9.2$ ppm), which are able to react with the remaining nucleotide to form the corresponding dimer.

Mixed dinucleotides could also be obtained in high yields, by introducing minor modifications into the protocol. Accordingly, activation was performed in the presence of a small excess of sulfonyl imidazolium salt **12**, and NMI was replaced by diisopropylethylamine (DIPEA) to prevent dimerization (Scheme 12). Moreover, the time of reaction of the first step was reduced to 1 min. Cp_2A and Ap_3U were obtained in similar yields starting from AMP and UMP, respectively.

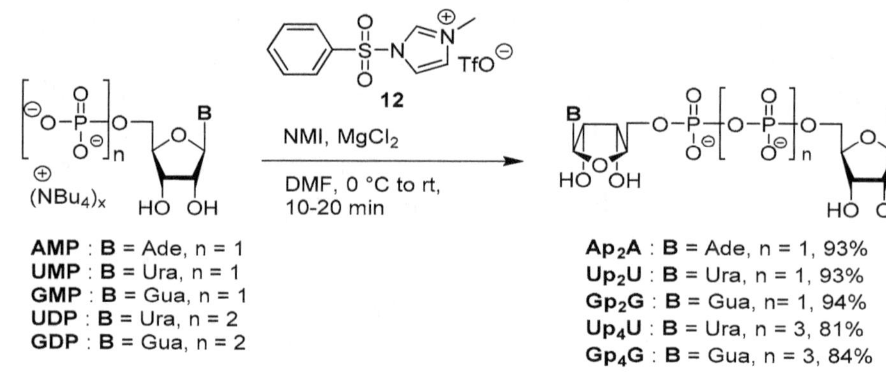

Scheme 11. Synthesis of symmetrical Np_2N and Np_4N [60].

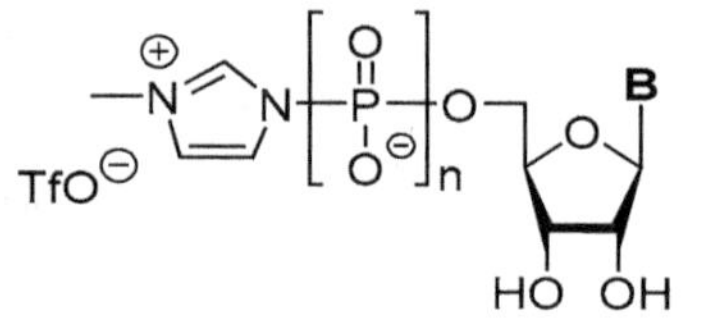

23 : **B** = Ade, n = 1
24 : **B** = Ura, n = 1
25 : **B** = Gua, n = 1
26 : **B** = Ura, n = 2
27 : **B** = Gua, n = 2

Figure 4. Phosphoromethylimidazolium intermediates [60].

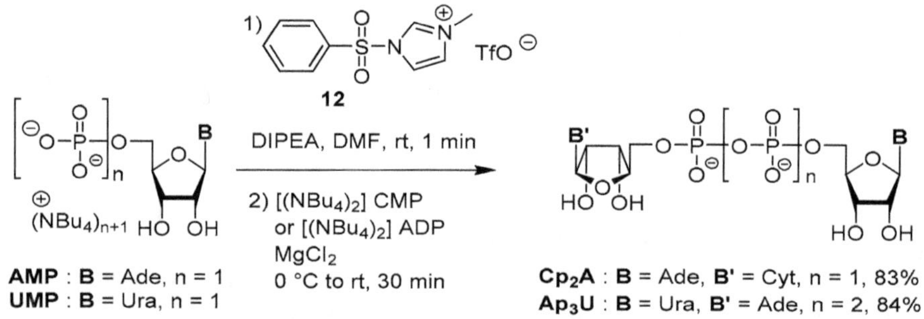

Scheme 12. Synthesis of mixed dinucleotides [60].

Finally, mixed adenine dinucleotides were obtained in high yields by activating the tetra-*n*-butylammonium salt of ATP with sulfonyl imidazolium salt **28**, followed by reaction with a second nucleotide (Scheme 13). Investigation of the first step by ^{31}P NMR revealed that the reaction proceeds via a cyclic adenosine trimetaphosphate intermediate **8**.

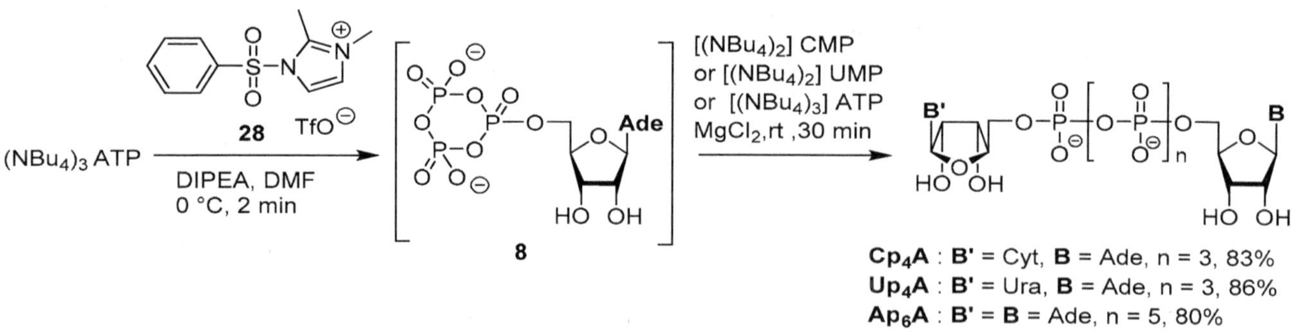

Scheme 13. Synthesis of mixed dinucleotides starting from ATP [60].

The advantages of this method using sulfonylimidazolium salts as key reagents, include short reaction times, access to a wide range of N_pNs (n = 2–6) and high yields. However, careful attention must be paid to maintain anhydrous conditions.

3.5. Synthesis via a Phosphoropiperidate Intermediate

In 2014, the Sun research group developed an approach based on the activation of nucleoside 5′-phosphoropiperidates with 4,5-dicyanoimidazole (DCI) as activator. The phosphoropiperidates **29–32** were obtained beforehand from unprotected NMPs via a redox condensation involving triphenylphosphine and 2,2′-dithiodianiline (Scheme 14) [66].

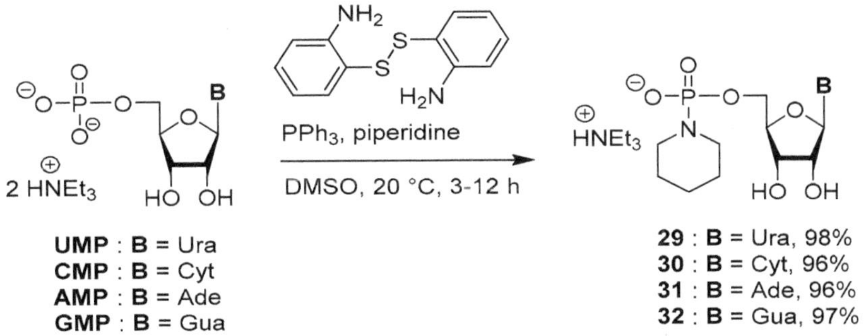

Scheme 14. Synthesis of nucleoside 5′-phosphoropiperidates [66].

As shown in Scheme 15, the reaction of nucleoside 5′-phosphoropiperidates with the tri-*n*-butylammonium salts of nucleotides (NMPs, NDPs, NTPs) in the presence of 2.5–3 equiv of DCI, led to the formation of more than 20 symmetrical or mixed dinucleoside di-, tri- and tetraphosphates [67,68]. Yields of Np_nN's dropped significantly along with the increase of the polyphosphate chain.

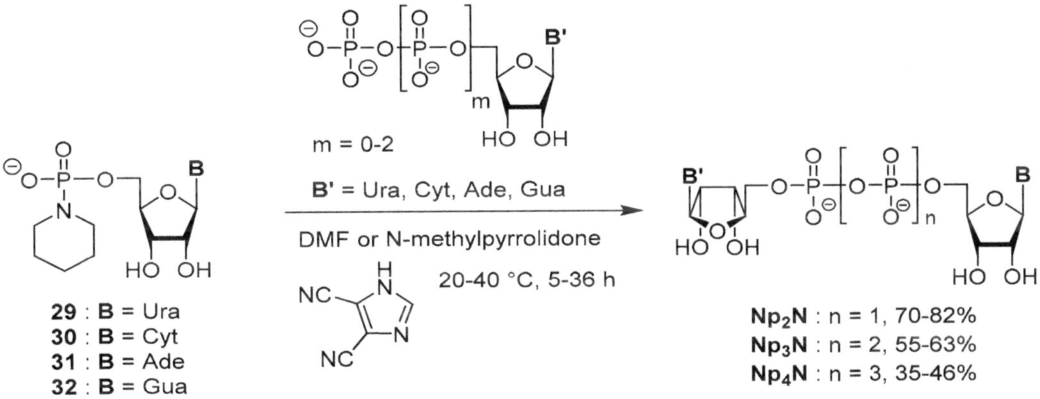

Scheme 15. Synthesis of dinucleotides from nucleoside 5′-phosphoropiperidates [67,68].

Among the polar aprotic solvents, *N*-methylpyrrolidone exhibited the best solvency for the tri-*n*-butylammonium salts of both NDPs and NTPs, while DMF was preferred for NMPs. Considering the cost of nucleoside 5′-polyphosphates and their similar chromatography properties with the products, only 0.5 equiv of NDPs or NTPs were used. This P(V) method is an efficient one, allowing access to a large range of Np_nNs (n = 2–4) in short reaction times and good to high isolated yields.

Symmetrical Np_nN (n = 3–5) could also be obtained by reacting nucleoside 5′-phosphoropiperidates with sub-stœchiometric amounts (0.35–0.4 equiv) of inorganic phosphorylating agents in the presence of excess DCI [69]. In this alternative protocol, bis(tetra-*n*-butylammonium) hydrogen phosphate, tris(tetra-*n*-butylammonium) hydrogen pyrophosphate or tris(tetra-*n*-butyl-ammonium) dihydrogen triphosphate were used instead of a nucleotide (Scheme 16). ^{31}P NMR tracking experiments showed that pyrophosphate reacts first with the starting phosphoropiperidate to form an NTP, and then, with the remaining phosphoropiperidate to give rise to Np_4N. This methodology was also adapted to the synthesis of nucleotides with a higher number of phosphate groups (such as Np_5N), using inorganic triphosphate instead of pyrophosphate.

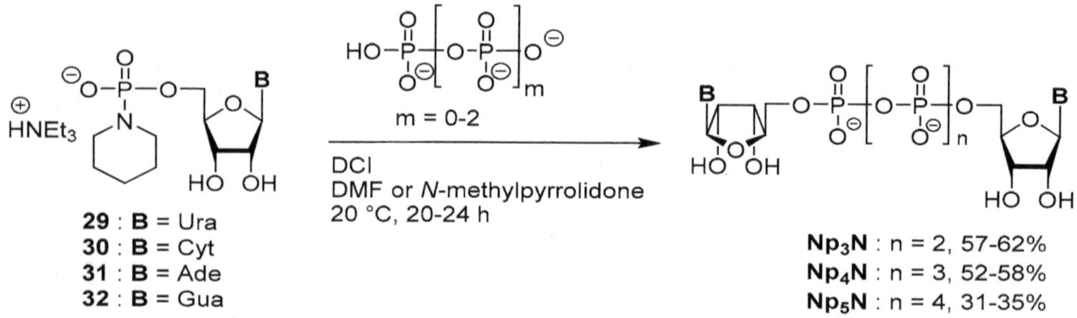

Scheme 16. Synthesis of symmetrical NpN by dimerization [67,68].

Interestingly, this strategy can be adapted to the synthesis of nucleotide analogues such as dinucleoside P^2,P^3-(dihalo)methylene tetraphosphates, by replacing pyrophosphate with bisphosphonate reagents.

4. Synthesis Based on P(III) Chemistry

Several methods involve P(III) reagents to provide phosphite or phosphoramidite intermediates, which are further oxidized to P(V). Initially developped for the synthesis of oligonucleotides, they have been adapted to dinucleotides by introducing minor modifications in the last steps of the reaction sequences. Because of the high reactivity of P(III) species, most of these approaches require the use of N-, and O-protecting groups on the starting nucleos(t)ides and strictly anhydrous conditions.

4.1. Synthesis via a Salicylphosphite Intermediate

This strategy is based on the well-known Ludwig–Eckstein procedure for NTP synthesis [70], and allows the one-pot preparation of symmetric dinucleoside tetra- and pentaphosphates starting from protected nucleosides [71]. Phosphitylation of triacetyl adenosine **33** with 2-chloro-4*H*-1,3,2-benzo-dioxaphosphorin-4-one (salicylchlorophosphite) led to intermediate **34**, then addition of bis(tri-*n*-butylammonium) pyrophosphate gave rise to the cyclic intermediate P(III)–P(V) **35** (Scheme 17). Oxidation to the cyclic trimetaphosphate, intermediate **36**, followed by reaction with adenosine 5′-monophosphate (AMP) or adenosine 5′-diphosphate (ADP) in dry DMF in the presence of $ZnCl_2$ as catalyst, afforded **37** and **38**, respectively. Removal of the acetyl groups under basic conditions, ion exchange to remove the Zn^{2+} cations and purification led to Ap_4A and Ap_5A as ammonium salts. It should be noted that symmetrical Ap_4A was obtained in much better yield than Ap_5A.

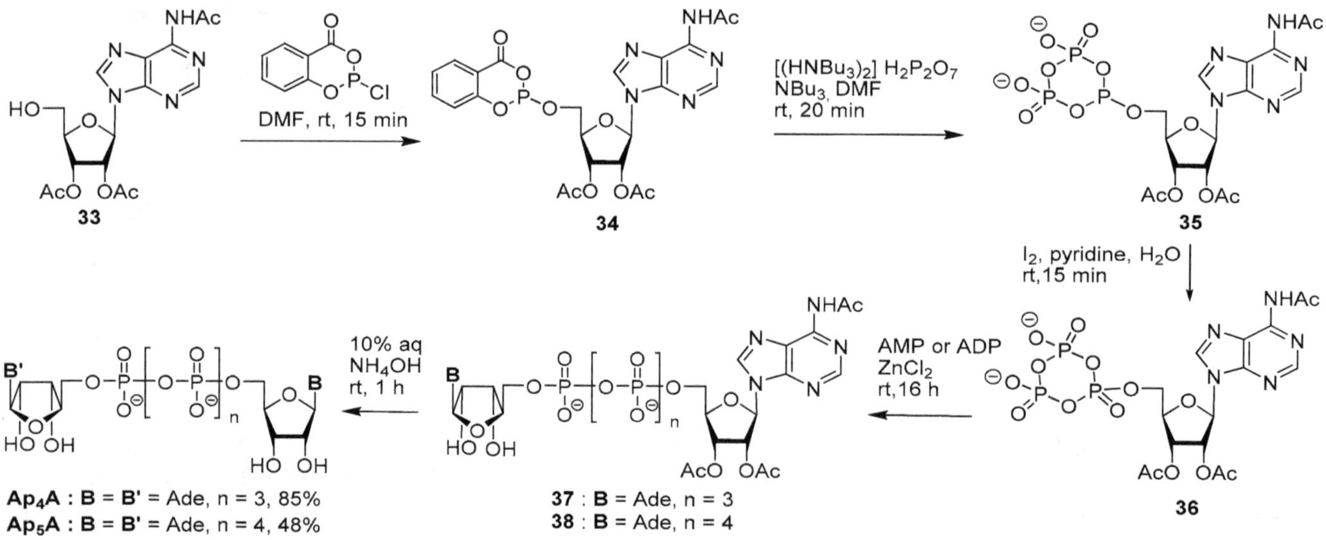

Scheme 17. Synthesis of the symmetrical dinucleotides Ap_4A and Ap_5A by phosphitylation [71].

The guanosine derivatives Gp_4G and Ap_4G were prepared in a similar manner, starting from 2′,3′-*O*-2-*N*-triacetylguanosine **39** (Scheme 18).

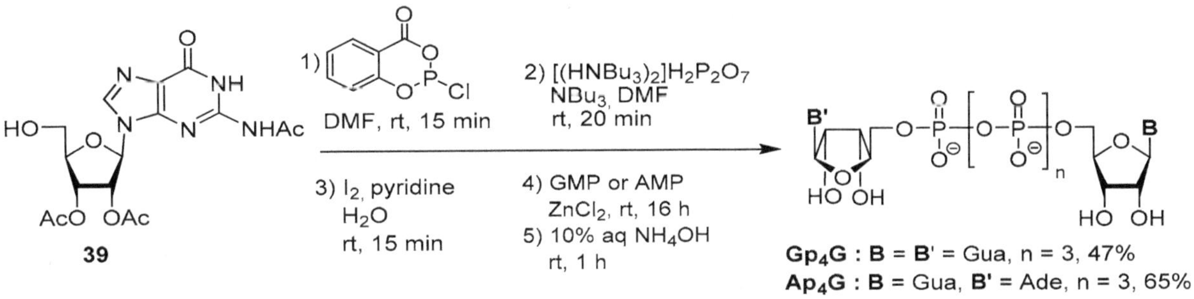

Scheme 18. Synthesis of dinucleotides Gp_4G and Ap_4G by phosphitylation [71].

4.2. Synthesis via a Cyclosaligenyl Phosphite Intermediate

This strategy initially developed by Meier and co-workers for the synthesis of prodrugs of nucleotides, called *cyclo*Sal (for *cyclo*saligenyl) pronucleotides [72,73], has been applied to the synthesis of nucleoside 5′-polyphosphates, sugar nucleotides and mixed dinucleotides [74,75]. Phosphochloridite **40** was prepared from the corresponding salicyl alcohol and phosphorus trichloride (Scheme 19). Reaction with 3′-*O*-acetylthymidine followed by oxidation with oxone (2 $KHSO_5.KHSO_4.K_2SO_4$) afforded 5-nitro-*cyclo*Sal-3′-*O*-acetylthymidine phosphotriester **41**.

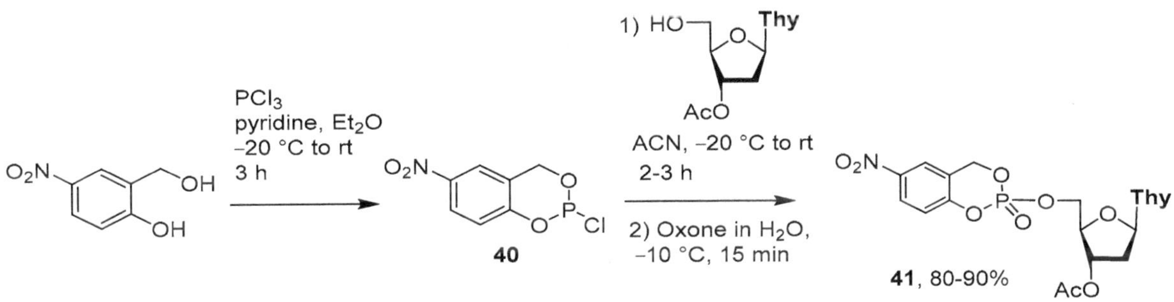

Scheme 19. Synthesis of a *cyclo*Sal phosphotriester derivative of thymidine [74].

This phosphotriester was too sensitive to moisture to be purified by chromatography. Therefore, the crude material (obtained after liquid-liquid extraction) was directly engaged in the next step, i.e., reaction with rigorously dried tetra-*n*-butylammonium UMP and tris(tetra-*n*-butylammonium) ATP in DMF (Scheme 20). After removal of the acetyl group and ion-exchange (n-NBu_4^+ to NH_4^+), Up_2T and Ap_4T were obtained in their ammonium form.

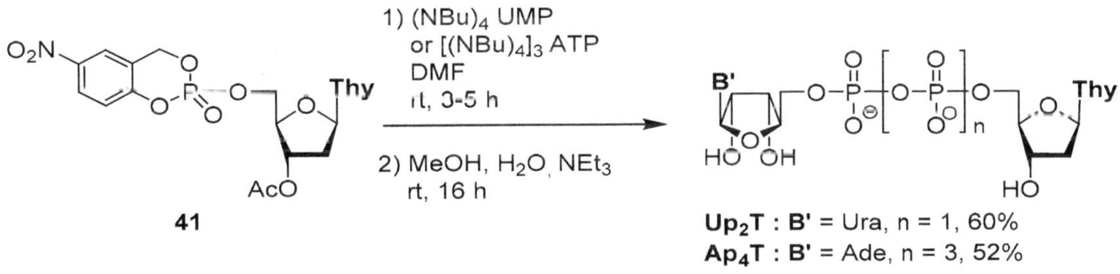

Scheme 20. Synthesis of the dinucleotides Up_2T and Ap_4T [74].

This approach has been adapted to solid phase [76]. The key intermediates of the synthesis are the 5′-substituted *cyclo*Sal nucleotides linked to a polystyrene support through the 3′-OH group and a succinyl linker (**42–44**, Figure 5).

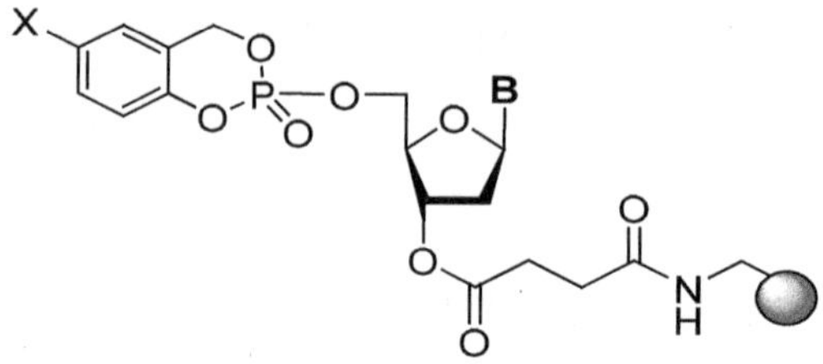

42 : X = Cl, **B** = Thy
43 : X = Cl, **B** = Ade
44 : X = $MeSO_2$, **B** = Thy

Figure 5. Structure of the supported nucleoside phosphotriesters [76].

Preparation of these intermediates started with the anchoring of the succinyl linker to the 5′-*O*-(4,4′-dimethoxytrityl)-protected nucleosides, and removal of the protecting group with trifluoroacetic acid (TFA). The 3′-*O*-succinylnucleosides were then treated with the appropriate chlorophosphites, followed by oxidation with oxone, to give rise to compounds **45–47** (Scheme 21). Coupling with the aminomethyl polystyrene support was achieved using 1-hydroxybenzotriazole (HOBt) and *N,N*′-diisopropylcarbodiimide (DIC) as coupling agents. Alternatively, 2-(1*H*-benzotriazole-1-yl)-1,1,3,3-tetramethyluronium tetrafluoroborate (TBTU) and 4-ethylmorpholine may be used. Conversion to Up_2T and Up_2dA was carried out by adding a large excess of tetra-*n*-butylammonium UMP, and then cleavage under basic conditions (70–78% purity, no overall yield was given). Compared to the solution phase synthesis, this supported approach is less straightforward and involves multiple steps.

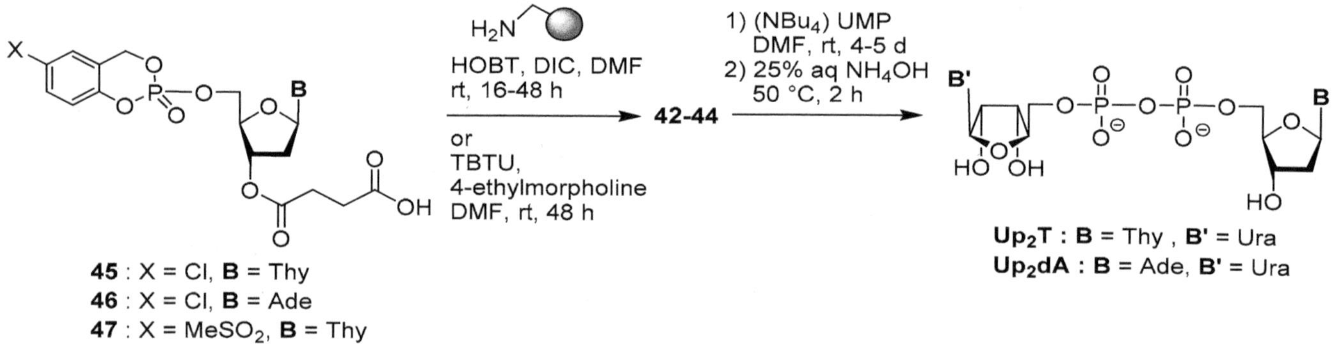

Scheme 21. Solid-phase synthesis of dinucleoside 5′,5′-diphosphates [76].

4.3. Synthesis via a Supported Phosphite Intermediate

Another P(III) strategy combined with a solid-phase approach was reported by Ahmadibeni and Parang [77]. Using these polymer-bound phosphitylating reagents, symmetrical 5′,5′-dinucleoside mono, di, tri, and tetraphosphates were prepared. Compounds **48–51** were obtained in 2 to 5 steps starting from phosphorus trichloride (Figure 6). Coupling with aminomethyl polystyrene resin-bound *p*-acetoxybenzyl alcohol yielded four classes of supported-phosphitylating reagents **52–55** (Figure 6).

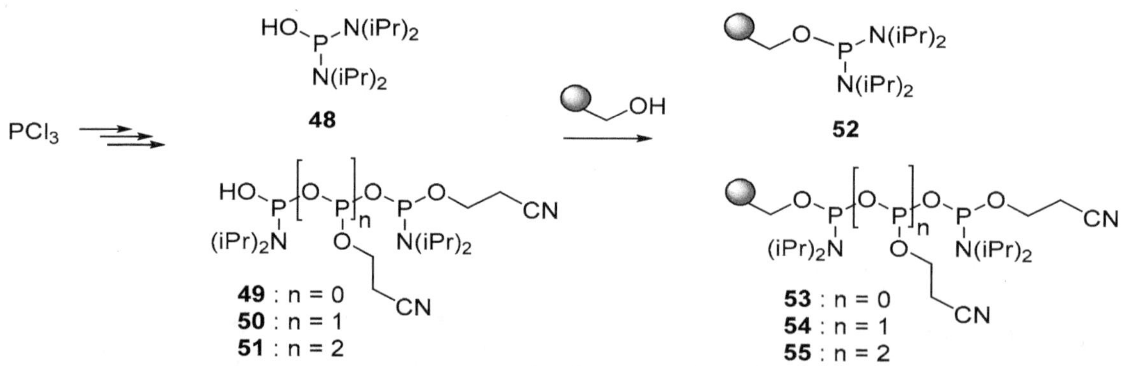

Figure 6. Synthesis of the immobilized phosphitylating reagents [77].

These supported reagents were reacted with unprotected nucleosides (e.g., thymidine, adenosine, cytidine, inosine or the analogue 3′-azido-3′-deoxythymidine) in the presence of 5-(ethylthio)-1*H*-tetrazole to afford compounds **56–59** (Scheme 22). Polymer-bound nucleotides underwent oxidation with *tert*-butyl hydroperoxide to **60–63**, then removal of the cyanoethoxy groups with 1,8-diazabicyclo[5.4.0]undec-7-en (DBU) was performed. Acidic treatment of **60** and **64–66** gave rise to the corresponding symmetrical dinucleotides in 59–78% yields starting from the supported reagents **52–55**.

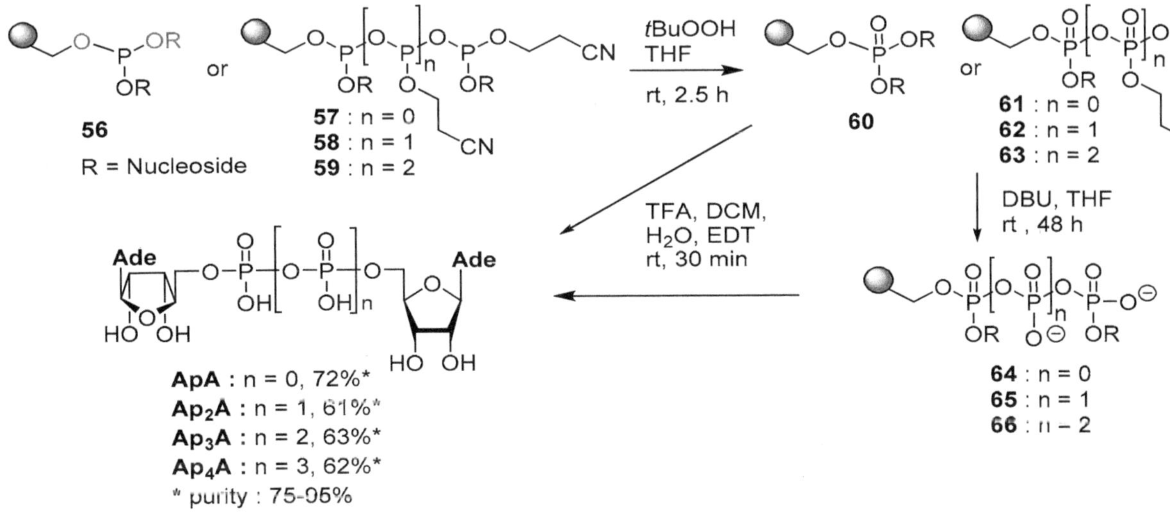

Scheme 22. Synthesis of symmetrical dinucleotides using polymer-bound phosphitylating reagents. Yields are given for the adenine derivatives [77].

This strategy allowed the synthesis of symmetrical dinucleotides starting from unprotected nucleosides without the need to purify the intermediates. It also offers the advantage of facile recovery of final products by filtration (trapped linkers on the resins). However, it requires the multistep synthesis of the supported polyphosphite reagents. It should be highlighted that part of this work could not be reproduced by others [78].

4.4. Synthesis via a 5′-H-Phosphonate Intermediate

This one-pot strategy developed by Sun et al. [79], and based on their previous work on the synthesis of nucleoside triphosphates [80], involves the formation of a pyridinium phosphoramidate intermediate starting from a nucleoside 5′-*H*-phosphonate monoester. Nucleoside 5′-*H*-phosphonate monoesters **67–70** were obtained from 2′,3′-*O*-isopropylidene ribonucleosides **71–74** using diphenylphosphite as a reagent (Scheme 23). Then, removal of the isopropylidene group with aqueous TFA afforded 5′-*H*-phosphonate monoesters **75–78** in yields ranging from 68% to 82% over two steps.

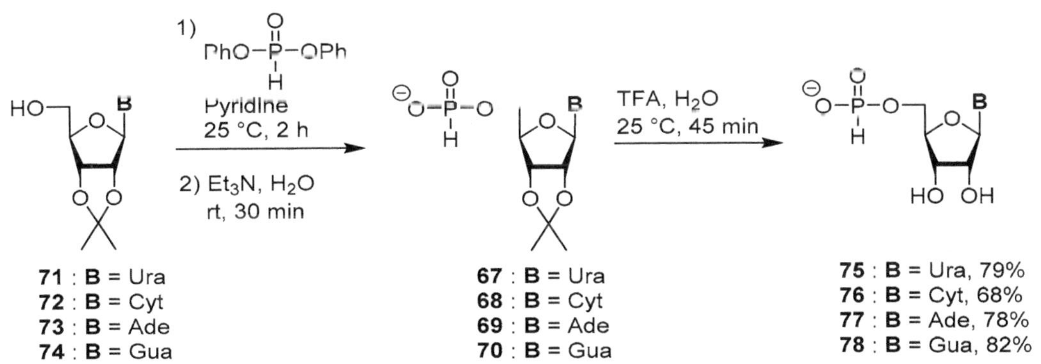

Scheme 23. Synthesis of nucleoside 5′-*H*-phosphonates from the 2′,3′-protected nucleosides [79].

H-phosphonates were silylated with trimethylsilyl chloride (TMSCl) in pyridine, and in situ oxidized with iodine to generate the highly reactive zwitterionic pyridinium phosphoramidate intermediates **79–82** (Scheme 24). Indeed, even trace amounts of water may result in the formation of polyphosphate by-products. The authors hypothesized that if 50% of the phosphoramidate intermediate could be hydrolyzed with precise control, NMP generated in situ would react with the remaining phosphoramidate to afford Np_2N as a major product. Accordingly, hydrolysis and coupling steps were optimized using ^{31}P NMR tracking experiments and accomplished by two sequential additions of H_2O in DMF. After purification and ion exchange, symmetrical dinucleoside diphosphates sodium salts were obtained in 68–91% yields starting from the nucleoside 5′-*H*-phosphonates. Advantages of this method are the short reaction time, simple purification procedure and high yields. However, it requires the preliminary synthesis of the *H*-phosphonate starting materials.

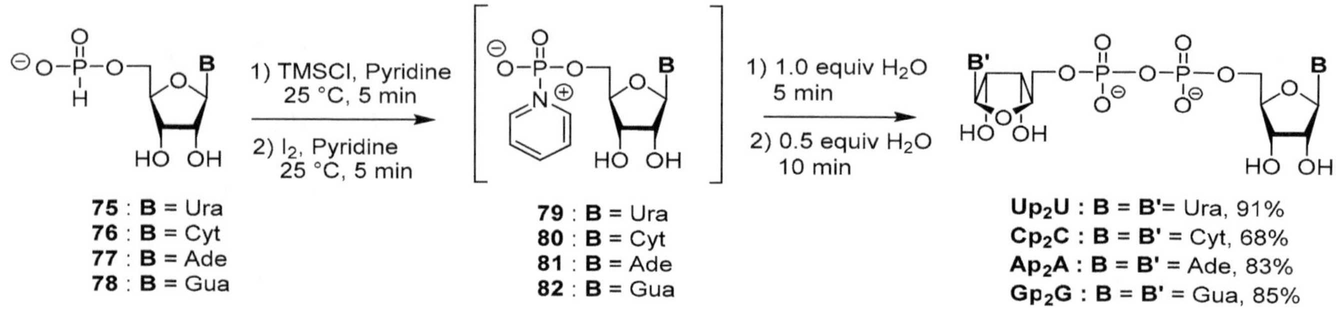

Scheme 24. Synthesis of symmetrical dinucleotides from nucleoside 5′-*H*-phosphonates [79].

4.5. Synthesis via a Phosphoramidite Intermediate

In 2015, the Jessen research group developed a very efficient strategy based on an iterative phosphoramidite approach to access nucleoside 5′-polyphosphates [81,82] and some derivatives [83]. The phosphoanhydride bond is formed by coupling a phosphoramidite (the donor) to a phosphate (the acceptor). In contrast to the other P(III) methods, all steps were performed under ambient conditions, i.e., in open flasks and using non-dried solvents. Regarding of the synthesis of dinucleotides, 2′,3′-*O*-diacetyl uridine **83** was treated with phosphorodiamidite **84** in the presence of 1*H*-tetrazole to afford phosphoramidite **85** (Scheme 25). Addition of the tetra-*n*-butylammonium salt of UMP and DCI, followed by oxidation with *m*CPBA, treatment with piperidine, and cleavage of the acetyl groups under basic conditions, gave rise to Up_2U as a mixed piperidinium/tetra-*n*-butylammonium salt. The mixed dinucleotide Ap_4U was also obtained by using the tetra-*n*-butylammonium salt of ATP and 5-ethylthio-1*H*-tetrazole (ETT) as an activator in the second step of the reaction (Scheme 25). It should be noted that these experiments were performed on a small scale (10 mg).

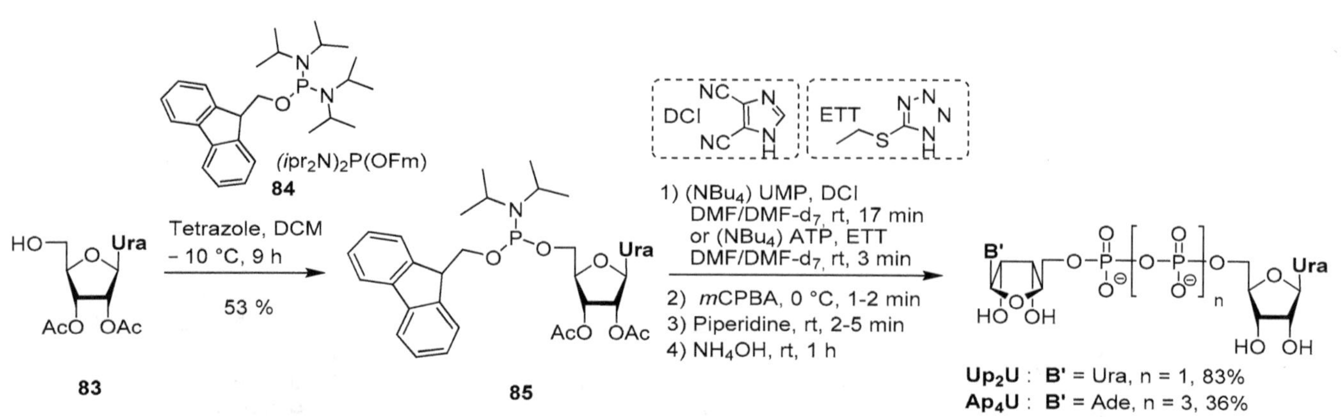

Scheme 25. Synthesis of dinucleotides via a phosphoramidite intermediate [83].

Symmetrical Np_nNs ($n = 3, 5, 7$) could also be accessed via chemoselective homologative dimerization of two phosphate monoesters with phosphorodiamidites [84]. The general strategy for the synthesis of dinucleoside triphosphates is shown in Figure 7. First, phosphordiamidites **86** and **87** (donors) were coupled with a NMP (the acceptor), affording terminal mixed phosphoric anhydride phosphoramidites **88–89** as unstable intermediates. Reaction with another phosphate monoester resulted in the formation of **90** containing a $P^V–P^{III}–P^V$ bridge, which was then oxidized to yield **91**. The corresponding Np_3N was obtained after a final deprotection step.

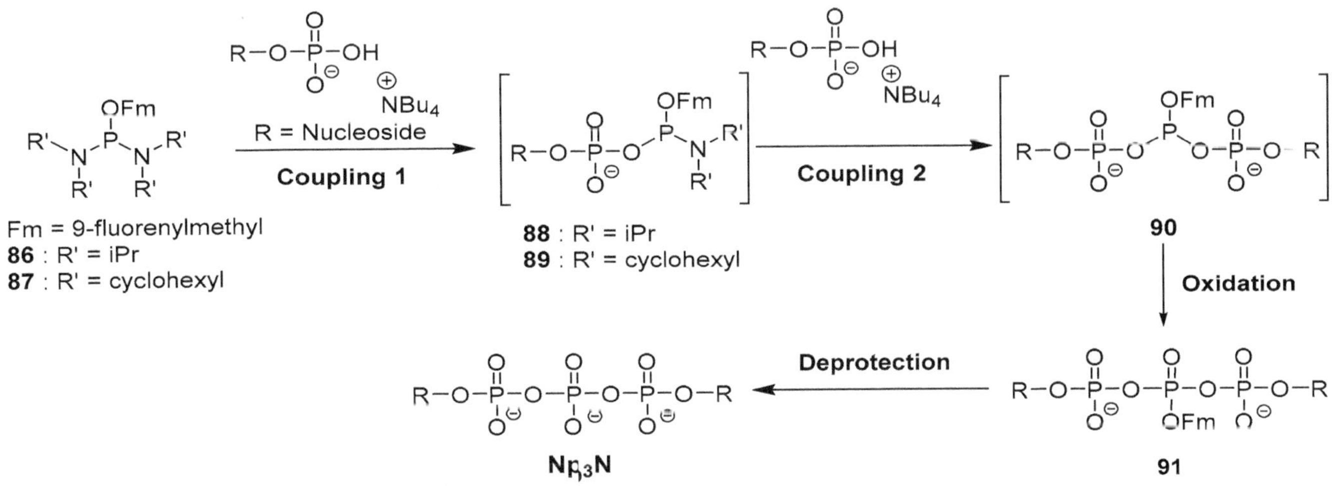

Figure 7. General approach for the preparation of symmetrical Np_3Ns based on dimerization [84].

The tetra-*n*-butylammonium salts of the canonical nucleoside 5′-monophosphates were treated in DMF with 9-fluorenylmethyl-*N*,*N*,*N*′,*N*′-tetraalkyl phosphorodiamidites **86–87** in the presence of ETT, followed by oxidation with *m*CPBA (Scheme 26). Progress of the reaction was monitored by ^{31}P NMR. The Fm-protected intermediates were isolated by precipitation with Et_2O/hexane, treated with piperidine, and purified by anion exchange chromatography to isolate the symmetrical dinucleoside triphosphates as ammonium salts in good yields.

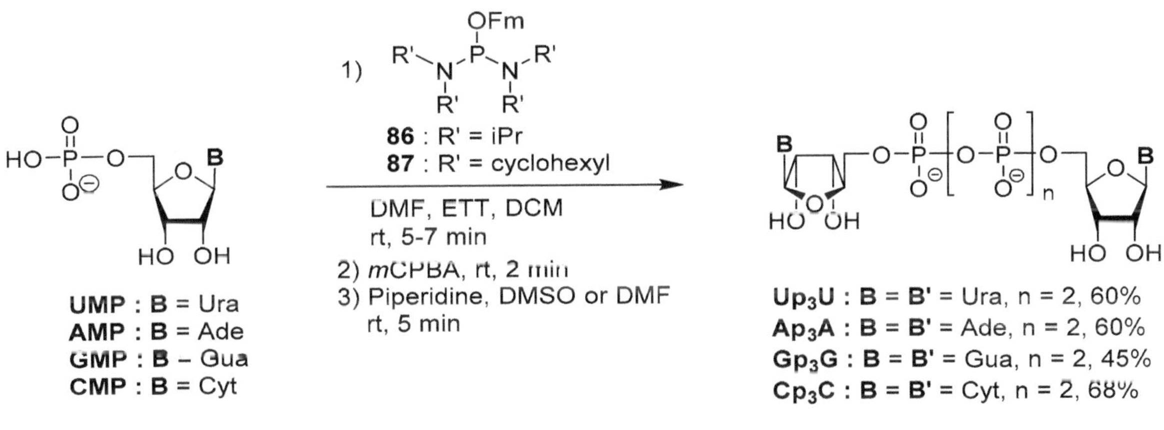

Scheme 26. Synthesis of dinucleotides via dimerization of phosphates [84].

This methodology was applied to the synthesis of symmetrical dinucleoside penta- and heptaphosphates by using as acceptor a NDP or a NTP, respectively. However, due to the water-sensivity of the intermediates, reactions had to be performed under anhydrous conditions, with a shorter reaction time and low temperature (coupling with ETT 30 s at 0 °C, oxidation 5 min at 0 °C). Accordingly, Ap_5A and Ap_7A were obtained in 50% and 18% yields, respectively. The chemoselective aspect of this method is a great advantage as it allows to prevent the use of nucleoside protecting groups. Compared to other procedures using P(III) chemistry, it offers attractive features, notably these rapid reactions can be run under ambient conditions.

5. Green Chemistry Approaches

Most of the reported methods present drawbacks, such as the use of non-volatile and toxic solvents (DMF, pyridine, *N*-methylpyrrolidone), high sensitivity to moisture, preparation of substrates or phosphorus reagents with lipophilic counterions due to solubility issues in organic media, anhydrous conditions (i.e., dry reagents and solvents), fastidious purification and possibly prior synthesis of the activated nucleotide substrates. An ideal procedure would meet the following requirements: Applicable to a wide range of substrates where protecting groups are not necessary, and giving rise to the products in short time and high yields.

In 2017, our research group reported a one-pot synthesis, in water medium, of nucleoside 5′-polyphosphates and dinucleotides starting from the corresponding NMPs [55,56]. This method uses 2-chloro-1,3-dimethylimidazolinium hexafluorophosphate (DMP) and imidazole as coupling reagents to activate the NMP into their phosphorimidazolides. Dinucleoside diphosphates were obtained by self-dimerization of the NMPs (Scheme 27).

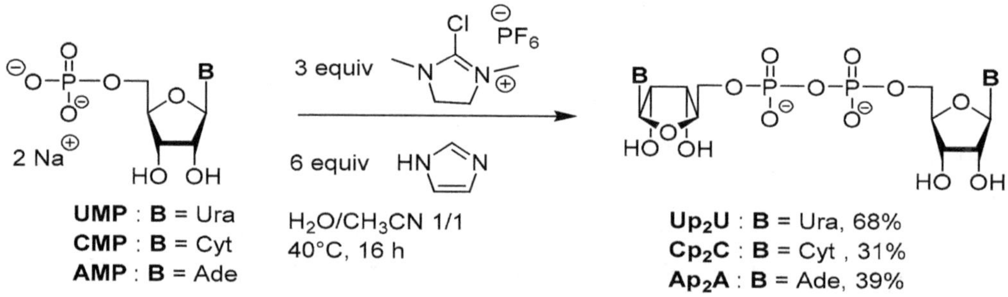

Scheme 27. Preparation of symmetrical dinucleoside diphosphates in aqueous media [56].

Our work originated from the small-scale synthesis of NDP-sugars described by Tanaka et al. [85]. The authors performed the coupling of a sugar-1-phosphate with a nucleotide in D_2O, using 2-chloro-1,3-dimethylimidazolinium chloride (DMC) and imidazole (Scheme 28). Based on ^1H and ^{31}P NMR spectroscopies, the reaction was shown to proceed via a 2-imidazolyl-1,3-dimethylimidazolinium chloride intermediate **92**, formed in situ by reaction of DMC and imidazole (Scheme 28). This key intermediate was able to activate an NMP into its corresponding phosphorimidazolide **93** (via intermediate **94**), which then reacted with a sugar-1-phosphate salt to form a sugar nucleotide.

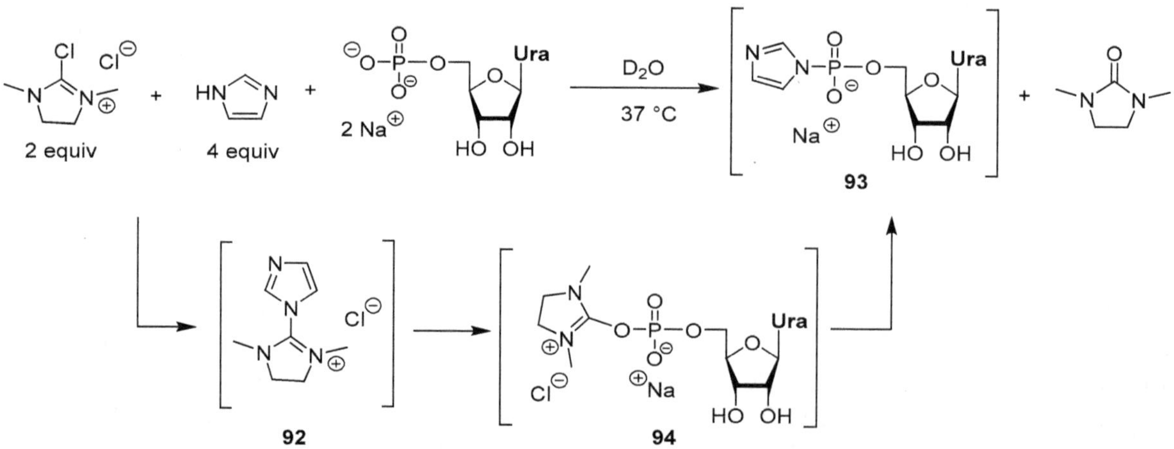

Scheme 28. Proposed activation mechanism of UMP to UMP-Im by **90** [85].

The main advantage of these reactions, which occur in D_2O [85] or in H_2O/CH_3CN [55,56] instead of the usual organic solvents, is that the NMPs are used in their commercially available and water-soluble forms, i.e., sodium and potassium salts. This greatly simplifies the handling of the reaction. Moreover, protecting groups on the nucleotide are not required and no carbonation of ribonucleotides are observed, in contrast to the use of CDI.

Alternatively, mechanochemistry has emerged in the field of nucleosides and, to a lesser extent, nucleotides [86]. This technique enables solid/solid reactions through mechanical grinding under solvent-free conditions [87–91]. In 2011, Ravalico et al. reported the mechanosynthesis of dinucleotides, namely Ap_nA (with $n = 2–4$), Ap_2dT and nicotinamide adenine dinucleotide (NAD^+), starting from adenosine 5′-phosphoromorpholidate **2**, in the presence of magnesium chloride, 1*H*-tetrazole as the acidic promoter and water (Scheme 29) [92]. Although advantageous, this methodology requires the use of costly reagents, namely tetrazole and the phosphoromorpholidate substrate, the latter may be prepared beforehand in solution using conventional solution synthesis [45].

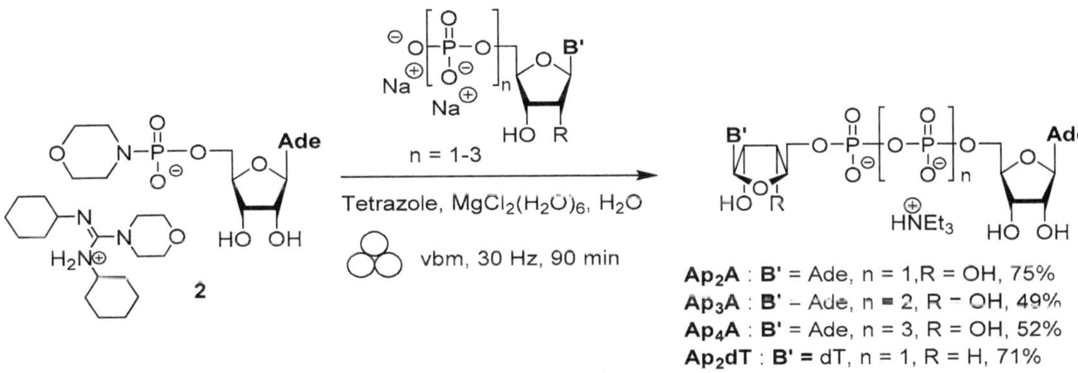

Scheme 29. Ball-milling route using commercially available nucleotide salts and morpholidate reaction partners [92].

In the course of our investigations on new strategies for phosphoanhydride bond formation, we also investigated the ball-milling technique [93]. To begin with, activation of UMP was performed according to the conditions previously reported in water medium (DMP/imidazole) [55,56], without adding any solvent in a vibratory ball-mill (vbm) at 30 Hz. A complete conversion to Up_2U was obtained within 30 min, suggesting that UMP was activated to its phosphorimidazolide and reacted promptly with the remaining UMP to form the dimer. Alternatively, we tested the use of CDI, which has recently gained interest in the field of mechanosynthesis due to the safety of its by-products (i.e., imidazole and carbon dioxide), its efficacy in *N*-acylation reactions and relatively low cost [94–97]. Activation of UMP (acidic form) to the corresponding 5′-phosphorimidazolide 2′,3-carbonate **95** was complete by grinding, for 1 h at 30 Hz, in the presence of 4 equiv. of CDI and acetonitrile (0.3 $\mu L.mg^{-1}$) as a liquid assistant (Scheme 30). The absence of reactivity observed using the disodium salt of UMP suggests that the acid form is required to protonate the imidazole of CDI and to facilitate its substitution by the phosphate monoester. This activation step was applied successfully to several ribonucleotides and a 2′-deoxyribonucleotide (dTMP).

In the second step, a nucleoside mono, di or triphosphate was added onto the activated NMP **95–96** in order to form the desired DNPs (Scheme 30). Thus, the mixed dinucleoside 5′,5′-diphosphates were successfully obtained by adding a slight excess of a NMP and a small amount of acetonitrile (0.6–0.95μL/mg) in the jar, and then ball-milling for 2 h at 30 Hz. Remarkably, the second grinding step also removed the carbonate protection. The reaction was less efficient with GMP, UDP and ATP, and, thus, required a larger excess of reagent to form the corresponding dinucleotides Gp_2U, Up_3U and Ap_4U, respectively. This user-friendly mechanochemical approach only requires a slight excess of reagents and limits the formation of side products. While isolated yields of symmetrical and mixed dinucleoside polyphosphates are similar to other approaches, the set-up of the experiments and the workup are greatly simplified, saving a considerable amount of time. In addition, compared to the previously reported mechanochemical synthesis [92], this one-pot two steps reaction starts from NMPs, which are readily available.

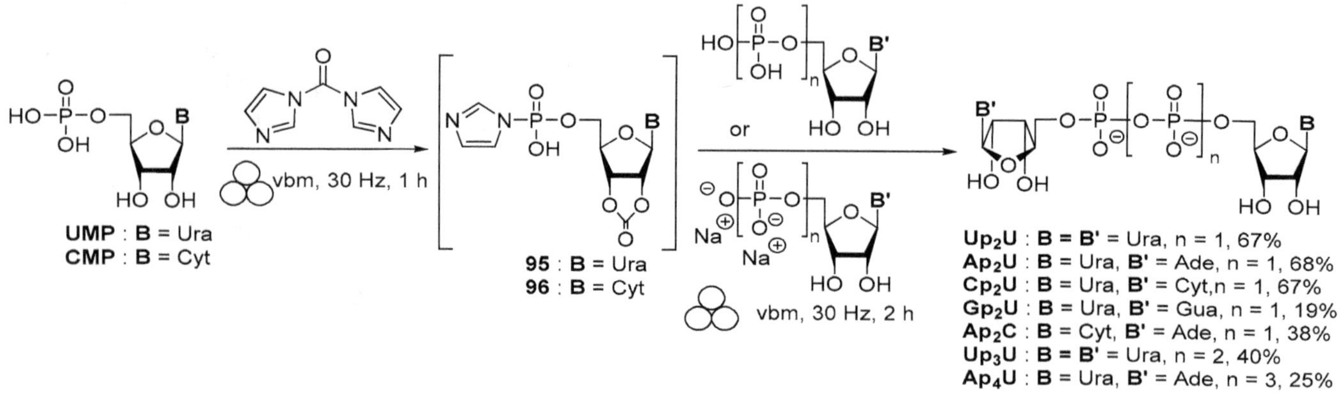

Scheme 30. Mechanochemical approach to obtain DNPs starting from NMPs [93].

In addition to their simplicity, the mechanochemical approaches alleviate issues associated with the low solubility of reagents in solution phase synthesis. Therefore, nucleotides and inorganic salts can be used in their commercially available sodium or potassium forms.

6. Purification and Characterization

6.1. Purification

In this section, we provide a brief overview of the purification procedures for dinucleotides. Due to their high polarity, they cannot be separated by normal-phase chromatography like most organic compounds. In addition, the byproducts or side products formed during synthesis share similar characteristics, resulting in a challenging separation from the desired compounds. The two main purification methods, reported in Table 1, use ion-exchange or reverse-phase (RP) chromatography.

Table 1. Main methods to isolate dinucleotides.

Types of Chromatography	Eluants	References
Anion-exchange		
DEAE-Sephadex®	0.02 M TEAB, pH 8; 10% ACN	[54]
TOYOPEARL® DEAE-650M	0.2 M TEAB, pH 8; 10% ACN	[54]
DEAE-Sephadex®	Gradient of aq. NH_4HCO_3	[67,69]
Q Sepharose® Fast Flow	Gradient of aq. NH_4HCO_3	[83,84]
DEAE Sepharose® Fast Flow	Gradient of aq. NH_4HCO_3	[83,84]
C-18 reverse-phase		
HPLC	0.01 M NH_4HCO_2, pH 4, 50% MeOH	[46]
Semi-preparative HPLC	TEAAc pH 7 or 9; 6–12% ACN	[59,60,64]
Preparative HPLC	0.1 M NH_4HCO_3; ACN	[71]
Preparative HPLC	TEAB pH 7.8; 65% ACN	[92]
RP-18 silica gel column	Water	[74]
RP-18 silica gel column	0.1 M TEAB, pH 7.5; 20% ACN	[55,56]
RP-18 silica gel column	0.1 M TEAAc, pH 7; 5–10% ACN	[93]
Gel filtration		
Sephadex® LH-20	Deionized water	[79]

The first type of ion-exchange support is positively charged, such as diethylaminoethyl (DEAE) Sephadex, and allows the chemical species to elute according to their number of charges. Elution is often performed using a gradient of triethylammonium hydrogen carbonate (TEAB) or ammonium hydrogen carbonate (NH_4HCO_3) buffers. Regarding C_{18} reverse-phase chromatography, either preparative or semi-preparative high pressure liquid chromatography (HPLC) or chromatography on RP-18 silica gel may be used. Compounds are often eluted using a gradient from an aqueous buffered solution, such as triethylammonium acetate (TEAAc) or TEAB, to acetonitrile or methanol. Alternatively, gel filtration was described for the purification of symmetrical Np_2N. Then, the samples are usually freeze-dried and converted to their sodium form by ion exchange using a Dowex-50W-Na^+ resin. The fact that purification is a time-consuming step is one of the reasons that pushed some research groups to develop solid phase strategies [76,77]. Indeed, those methods were designed in order to simplify purification of the final products, which can be collected by simple filtration. However, the recovered materials showed moderate purity and an additional purification step was required [76]. Consequently, solution-phase approaches are still predominant.

6.2. Physico-Chemical Properties

NMR, UV spectroscopy, and mass spectrometry (MS) are the main tools for the analysis of dinucleotides. Since the sugar and the phosphate moieties have no significant absorption above 230 nm, dinucleotides exhibit UV absorption profiles similar to those of their parent nucleosides with λ max values close to 260 nm [1]. MS analysis of nucleotides and derivatives such as dinucleotides has been extensively reviewed by Banoub et al. [98]. Alternatively, ^{31}P NMR is a very convenient tool to monitor reaction courses, characterize reaction intermediates, and study the conformation of dinucleotides and final compounds [1,50]. The chemical shift (δ) for most endogenous dinucleotides covers −8 to −24 ppm, and exhibits pH and counter ion dependence [1]. Table 2 summarizes the literature data for representative N_pN (n −1–7). In ^1H-decoupled ^{31}P NMR, the phosphorus atoms closest to the nucleosides are characterized by a singlet at approximately −10 ppm, while the phosphorus atoms inside the polyphosphate chain exhibit resonance signals around −22 ppm.

Table 2. ^{31}P Chemical shift ranges of dinucleoside polyphosphates in D_2O.

Np_nN	δ Ranges	References
Np_2N'	1 singlet ($P^1 + P^2$) from −9.6 to −11.5 ppm	[47,49,50,53,56,60,64,67,74,76,77,79,83,93]
Np_3N	1 doublet ($P^1 + P^3$) from −9.2 to −11.7 ppm 1 triplet (P^2) from −21.0 to −23.4 ppm	[47,49,50,54,60,64,67,69,84,93]
Np_4N	1 multiplet or 1 broad singlet or 1 doublet ($P^1 + P^4$) from −8.8 to −11.5 ppm 1 multiplet or 1 br singlet or 1 doublet ($P^2 + P^3$) from −20.6 to −23.1 ppm	[47,49,50,54,60,64,67,69,71,74,93]
Np_5N	1 multiplet or 1 br singlet or 1 doublet ($P^1 + P^5$) from −8.7 to −11.4 ppm 1 multiplet or 1 br singlet or 1 doublet ($P^2 + P^3 + P^4$) from −20.2 to −22.5 ppm	[49,50,59,69,71,84]
Np_6N	1 doublet at −10.1 ppm ($P^1 + P^6$) 1 multiplet at −21.7 ppm ($P^2 + P^3 + P^4 + P^5$)	[49]
Np_7N	1 broad singlet or 1 doublet from −10.1 to −11.6 ppm ($P^1 + P^7$) 1 multiplet from −21.7 to −23.5 ppm ($P^2 + P^3 + P^4 + P^5 + P^6$)	[49,84]

Interestingly, the study of dinucleotides using NMR techniques (^1H, ^{13}C and ^{31}P) can determine the conformation (*syn*/*anti*, sugar puckering), nucleobase properties (tautomerism, H-bonds and π-stacking interactions), as well as the anomeric α/β configuration. Such data are crucial for understanding the structure–activity relationship of either natural or synthetic dinucleotides as enzyme inhibitors and receptor ligands, and the design of potent therapeutic agents based on a dinucleotide scaffold. In this regard, Stern et al. have performed conformational studies on physiologically active Np_nN' sodium salts (N = A, G, U, C; N′ = A, G, U, C; n = 2–5) in neutral aqueous solution, using NMR and computational techniques [50,98]. While no predominant conformation was observed for the ribose moiety in any of the studied dinucleotides, natural dinucleoside polyphosphates were more frequently found in a folded (stacked) rather than an extended conformation. Purine dinucleotides showed greater stacking interactions than pyrimidine dinucleotides. In addition, dinucleotides with longer phosphate chains were found to have weaker stacking interactions. Crystal structures of Ap_4A sodium salt [99], and some dinucleotides bound to kinases or hydrolases are available [100,101]. In protein-bound dinucleotides, the polyphosphate chain adopts various conformations: extended [100–103], S-shaped [101] or folded [100]. In some cases, the protein-bound dinucleotides coordinate with a Mg^{2+} ion.

Finally, Chen and Kohler investigated the dynamics of excited electronic states formed by UV excitation of the several diadenosine polyphosphates by femtosecond transient absorption (TA) spectroscopy [104]. They found that the excimer states seen in TA experiments on nucleobase dimers are only observed in π-stacked conformations, but the lifetimes of these states are insensitive to how the stacked bases are oriented.

7. Conclusions

Naturally occurring dinucleoside 5′,5′-polyphosphates are involved in a variety of cellular processes. Among them, symmetrical purine-containing dinucleotides are the most known representatives, whereas pyrimidine-containing and mixed derivatives are still intensively studied. Therefore, the development of synthetic methods required to obtain these compounds in few steps, good yields, high purity and comfortable amounts generated interest from the chemists and remains a field of intensive research. Herein, we have focused our attention on the synthesis of dinucleoside 5′,5′-polyphosphates, either symmetrical or mixed, and including non-modified pyrophosphate bonds. These methods may involve P(III) and/or P(V) intermediates and solution-phase or supported chemistry. P(V) synthetic approaches are usually based on the activation of a 5′-nucleotide using inexpensive reagents but require the conversion of the substrates, as well as the phosphate reactants, into the corresponding tri- or tetra-alkyl salts due to solubility issues. P(III) synthetic approaches are often effective processes. However, the presence of protecting groups on the nucleoside is often mandatory, thus increasing the overall number of steps. In most cases, strictly anhydrous conditions are essential and the length of the poly-phosphate chain affects the yields of the synthesis. Recently, environmental friendly alternatives were investigated and may constitute an area of substantial and growing interest for the preparation of these derivatives in the coming years.

Author Contributions: Conceptualization: S.P. and B.R.; Writing—original draft: B.R.; Writing—review & editing: L.A., C.C., S.P. and B.R.

References

1. Roy, B.; Depaix, A.; Périgaud, C.; Peyrottes, S. Recent trends in nucleotide synthesis. *Chem. Rev.* **2016**, *116*, 7854–7897. [CrossRef] [PubMed]
2. Jankowski, V.; Van Der Giet, M.; Mischak, H.; Morgan, M.; Zidek, W.; Jankowski, J. Dinucleoside polyphosphates: Strong endogenous agonists of the purinergic system. Br. *J. Pharmacol.* **2009**, *157*, 1142–1153. [CrossRef]
3. Christie, S.M.H.; Elmore, D.T.; Kenner, G.W.; Todd, A.R.; Weymouth, F.J. Syntheses of P^1P^2-diadenosine-5′ and P^1P^2-diuridine-5′ pyrophosphates. *J. Chem. Soc.* **1953**, 2947–2953. [CrossRef]
4. Rapaport, E.; Zamecnik, P.C. Presence of diadenosine 5′,5′′′ -P^1, P^4-tetraphosphate (Ap_4A) in mamalian cells in levels varying widely with proliferative activity of the tissue: A possible positive "pleiotypic activator". *Proc. Natl. Acad. Sci. USA* **1976**, *73*, 3984–3988. [CrossRef] [PubMed]
5. Jankowski, V.; Tölle, M.; Vanholder, R.; Schönfelder, G.; Van Der Giet, M.; Henning, L.; Schlüter, H.; Paul, M.; Zidek, W.; Jankowski, J. Uridine adenosine tetraphosphate: A novel endothelium- derived vasoconstrictive factor. *Nat. Med.* **2005**, *11*, 223–227. [CrossRef]

6. Zhou, Z.; Matsumoto, T.; Jankowski, V.; Pernow, J.; Mustafa, S.J.; Duncker, D.J.; Merkus, D. Uridine adenosine tetraphosphate and purinergic signaling in cardiovascular system: An update. *Pharmacol. Res.* **2019**, *141*, 32–45. [CrossRef]
7. Delicado, E.G.; Miras-Portugal, M.T.; Carrasquero, L.M.G.; Leon, D.; Pérez-Sen, R.; Gualix, J. Dinucleoside polyphosphates and their interaction with other nucleotide signaling pathways. *Pflügers Archiv. Eur. J. Physiol.* **2006**, *452*, 563–572. [CrossRef]
8. Hampton, A.; Kappler, F.; Picker, D. Species- or isozyme-specific enzyme inhibitors. 4. Design of a two-site inhibitor of adenylate kinase with isozyme selectivity. *J. Med. Chem.* **1982**, *25*, 638–644. [CrossRef]
9. Lee, S.-Y.; Müller, C.E. Nucleotide pyrophosphatase/phosphodiesterase 1 (NPP1) and its inhibitors. *MedChemComm* **2017**, *8*, 823–840. [CrossRef]
10. Götz, K.H.; Hacker, S.M.; Mayer, D.; Dürig, J.-N.; Stenger, S.; Marx, A. Inhibitors of the Diadenosine Tetraphosphate Phosphorylase Rv2613c of Mycobacterium tuberculosis. *ACS Chem. Biol.* **2017**, *12*, 2682–2689. [CrossRef]
11. Hacker, S.M.; Mortensen, F.; Scheffner, M.; Marx, A. Selective monitoring of the enzymatic activity of the tumor suppressor Fhit. *Angew. Chem. Int. Ed.* **2014**, *53*, 10247–10250. [CrossRef] [PubMed]
12. Guranowski, A. Specific and nonspecific enzymes involved in the catabolism of mononucleoside and dinucleoside polyphosphates. *Pharmacol. Ther.* **2000**, *87*, 117–139. [CrossRef]
13. Hodgson, D.R. Physicochemical aspects of aqueous and nonaqueous approaches to the preparation of nucleosides, nucleotides and phosphate ester mimics. *Adv. Phys. Org. Chem.* **2017**, *51*, 187–219.
14. Xu, Z. A review on the chemical synthesis of pyrophosphate bonds in bioactive nucleoside diphosphate analogs. *Bioorganic Med. Chem. Lett.* **2015**, *25*, 3777–3783. [CrossRef]
15. Bezold, D.; Dürr, T.; Singh, J.; Jessen, H.J. Cyclotriphosphate: A brief history, recent developments, and perspectives in synthesis. *Chem. Eur. J.* **2019**. [CrossRef]
16. Jenal, U.; Reinders, A.; Lori, C. Cyclic di-GMP: Second messenger extraordinaire. *Nat. Rev. Microbiol.* **2017**, *15*, 271–284. [CrossRef]
17. Mlynarska-Cieslak, A.; Depaix, A.; Grudzien-Nogalska, E.; Sikorski, P.J.; Warminski, M.; Kiledjian, M.; Jemielity, J.; Kowalska, J. Nicotinamide-containing di- and trinucleotides as chemical tools for studies of NAD-capped RNAs. *Org. Lett.* **2018**, *20*, 7650–7655. [CrossRef]
18. Shanmugasundaram, M.; Senthilvelan, A.; Kore, A.R. Recent advances in synthesis and biological activity of modified cap analogs. *Curr. Org. Chem.* **2017**, *21*, 2530–2560. [CrossRef]
19. Muttach, F.; Muthmann, N.; Rentmeister, A. Synthetic mRNA capping. *Beilstein J. Org. Chem.* **2017**, *13*, 2819–2832. [CrossRef]
20. Warminski, M.; Sikorski, P.J.; Kowalska, J.; Jemielity, J. Applications of phosphate modification and labeling to study (m)RNA caps. *Top. Curr. Chem.* **2017**, *375*, 16. [CrossRef]
21. Burnstock, G. Purinergic signalling: Therapeutic developments. *Front. Pharmacol.* **2017**, *8*, 661.
22. Burnstock, G. The therapeutic potential of purinergic signalling. *Biochem. Pharmacol.* **2018**, *151*, 157–165. [CrossRef] [PubMed]
23. Rafehi, M.; Muller, C.E. Tools and drugs for uracil nucleotide-activated P2Y receptors. *Pharmacol. Ther.* **2018**, *190*, 24–80. [CrossRef] [PubMed]
24. Jacobson, K.A. Structure-Based approaches to ligands for G-protein-coupled adenosine and P2Y Receptors, from small molecules to nanoconjugates. *J. Med. Chem.* **2013**, *56*, 3749–3767. [CrossRef] [PubMed]
25. Burnstock, G. Purinergic signalling. *Br. J. Pharmacol.* **2006**, *147*, S172–S181. [CrossRef] [PubMed]
26. Burnstock, G. Pathophysiology and therapeutic potential of purinergic signaling. *Pharmacol. Rev.* **2006**, *58*, 58–86. [CrossRef] [PubMed]
27. White, N.; Burnstock, G. P2 receptors and cancer. *Trends Pharmacol. Sci.* **2006**, *27*, 211–217. [CrossRef]
28. Burnstock, G. Purinergic signalling and disorders of the central nervous system. *Nat. Rev. Drug Discov.* **2008**, *7*, 575–590. [CrossRef]
29. Erlinge, D. P2Y Receptors in health and disease. *Adv. Pharmacol.* **2011**, *61*, 417–439.
30. Xu, P.; Feng, X.; Luan, H.; Wang, J.; Ge, R.; Li, Z.; Bian, J. Current knowledge on the nucleotide agonists for the P2Y2 receptor. *Bioorg. Med. Chem.* **2018**, *26*, 366–375. [CrossRef]
31. Douglass, J.G.; Patel, R.I.; Yerxa, B.R.; Shaver, S.R.; Watson, P.S.; Bednarski, K.; Plourde, R.; Redick, C.C.; Brubaker, K.; Jones, A.C.; et al. Lipophilic modifications to dinucleoside polyphosphates and nucleotides that confer antagonist properties at the platelet $P2Y_{12}$ receptor. *J. Med. Chem.* **2008**, *51*, 1007–1025. [CrossRef] [PubMed]

32. Hoyle, C.H.; Hilderman, R.H.; Pintor, J.J.; Schlüter, H.; King, B.F. Diadenosine polyphosphates as extracellular signal molecules. *Drug Dev. Res.* **2001**, *52*, 260–273. [CrossRef]
33. Carracedo, G.; Crooke, A.; Guzman-Aranguez, A.; De Lara, M.J.P.; Martin-Gil, A.; Pintor, J. The role of dinucleoside polyphosphates on the ocular surface and other eye structures. *Prog. Retin. Eye Res.* **2016**, *55*, 182–205. [CrossRef]
34. Pintor, J.; Díaz-Hernández, M.; Gualix, J.; Gómez-Villafuertes, R.; Hernando, F.; Miras-Portugal, M.T. Diadenosine polyphosphate receptors. from rat and guinea-pig brain to human nervous system. *Pharmacol. Ther.* **2000**, *87*, 103–115. [CrossRef]
35. Lau, O.C.F.; Samarawickrama, C.; Skalicky, S.E. $P2Y_2$ receptor agonists for the treatment of dry eye disease: A review. *Clin. Ophthalmol.* **2014**, *8*, 327–334.
36. Keating, G.M. Diquafosol ophthalmic solution 3%: A review of its use in dry eye. *Drugs* **2015**, *75*, 911–922. [CrossRef]
37. Kellerman, D.; Mospan, A.R.; Engels, J.; Schaberg, A.; Gorden, J.; Smiley, L. Denufosol: A review of studies with inhaled $P2Y_2$ agonists that led to phase 3. *Pulm. Pharmacol. Ther.* **2008**, *21*, 600–607. [CrossRef]
38. Accurso, F.J.; Moss, R.B.; Wilmott, R.W.; Anbar, R.D.; Schaberg, A.E.; Durham, T.A.; Ramsey, B.W. Denufosol tetrasodium in patients with cystic fibrosis and normal to mildly impaired lung function. *Am. J. Respir. Crit. Care Med.* **2011**, *183*, 627–634. [CrossRef]
39. Yerxa, B.R.; Sabater, J.R.; Davis, C.W.; Stutts, M.J.; Lang-Furr, M.; Picher, M.; Jones, A.C.; Cowlen, M.; Dougherty, R.; Boyer, J.; et al. Pharmacology of INS37217 [P^1-(Uridine 5′)-P^4- (2′-deoxycytidine 5′)tetraphosphate, tetrasodium salt], a next-generation $P2Y_2$ receptor agonist for the treatment of cystic fibrosis. *J. Pharmacol. Exp. Ther.* **2002**, *302*, 871–880. [CrossRef]
40. Deterding, R.R.; LaVange, L.M.; Engels, J.M.; Mathews, D.W.; Coquillette, S.J.; Brody, A.S.; Millard, S.P.; Ramsey, B.W. Phase 2 randomized safety and efficacy trial of nebulized denufosol tetrasodium in cystic fibrosis. *Am. J. Respir. Crit. Care Med.* **2007**, *176*, 362–369. [CrossRef]
41. Moss, R.B. Pitfalls of drug development: Lessons learned from trials of denufosol in cystic fibrosis. *J. Pediatr.* **2013**, *162*, 676–680. [CrossRef] [PubMed]
42. Ratjen, F.; Durham, T.; Navratil, T.; Schaberg, A.; Accurso, F.J.; Wainwright, C.; Barnes, M.; Moss, R.B. Long term effects of denufosol tetrasodium in patients with cystic fibrosis. *J. Cyst. Fibros.* **2012**, *11*, 539–549. [CrossRef] [PubMed]
43. Reiss, J.R.; Moffatt, J.G. Dismutation reactions of nucleoside polyphosphates. 3. The synthesis of α,ω-Dinucleoside 5′-polyphosphates. *J. Org. Chem.* **1965**, *30*, 3381–3387. [CrossRef] [PubMed]
44. Adam, A.; Moffatt, J.G. Dismutation reactions of nucleoside polyphosphates. V. Syntheses of P^1,P^4-Di(guanosine-5′) tetraphosphate and P^1,P^3-Di(guanosine-5′) triphosphate. *J. Am. Chem. Soc.* **1966**, *88*, 838–842. [CrossRef]
45. Moffatt, J.G.; Khorana, H.G. Nucleoside polyphosphates. X. synthesis and some reactions of nucleoside-5′phosphoromorpholidates and related compounds. Improved methods for the preparation of nucleoside-5′polyphosphates. *J. Am. Chem. Soc.* **1961**, *83*, 649–658. [CrossRef]
46. Millo, E.; Zocchi, E.; Galatini, A.; Benatti, U.; Damonte, G. Simple Synthesis of P^1P^2-Diadenosine 5′-Pyrophosphate. *Synth. Commun.* **2008**, *38*, 3260–3269. [CrossRef]
47. Shaver, S.R.; Rideout, J.L.; Pendergast, W.; Douglass, J.G.; Brown, E.G.; Boyer, J.L.; Patel, R.I.; Redick, C.C.; Jones, A.C.; Picher, M.; et al. Structure–Activity relationships of dinucleotides: Potent and selective agonists of P2Y receptors. *Purinergic Signal.* **2005**, *1*, 183–191. [CrossRef]
48. Ko, H.; Carter, R.L.; Cosyn, L.; Petrelli, R.; De Castro, S.; Besada, P.; Zhou, Y.; Cappellacci, L.; Franchetti, P.; Grifantini, M.; et al. Synthesis and potency of novel uracil nucleotides and derivatives as $P2Y_2$ and $P2Y_6$ receptor agonists. *Bioorganic Med. Chem.* **2008**, *16*, 6319–6332. [CrossRef]
49. Pendergast, W.; Yerxa, B.R.; Douglass, J.G.; Shaver, S.R.; Dougherty, R.W.; Redick, C.C.; Sims, I.F.; Rideout, J.L. Synthesis and P2Y receptor activity of a series of uridine dinucleoside 5′-polyphosphates. *Bioorganic Med. Chem. Lett.* **2001**, *11*, 157–160. [CrossRef]
50. Stern, N.; Major, D.T.; Gottlieb, H.E.; Weizman, D.; Fischer, B. What is the conformation of physiologically-active dinucleoside polyphosphates in solution? Conformational analysis of free dinucleoside polyphosphates by NMR and molecular dynamics simulations. *Org. Biomol. Chem.* **2010**, *8*, 4637–4652. [CrossRef]

51. Liu, K.K.-C.; Sakya, S.M.; O'Donnell, C.J.; Flick, A.C.; Ding, H.X. Synthetic approaches to the 2010 new drugs. *Bioorganic Med. Chem.* **2012**, *20*, 1155–1174. [CrossRef] [PubMed]
52. Zatorski, A.; Goldstein, B.M.; Colby, T.D.; Jones, J.P.; Pankiewicz, K.W. Potent inhibitors of human inosine monophosphate dehydrogenase type II. Fluorine-substituted analogues of thiazole-4-carboxamide adenine dinucleotide. *J. Med. Chem* **1995**, *38*, 1098–1105. [CrossRef] [PubMed]
53. Chen, L.; Rejman, D.; Bonnac, L.; Pankiewicz, K.W.; Patterson, S.E. Nucleoside-5′-phosphoimidazolides: Reagents for facile synthesis of dinucleoside pyrophosphates. *Curr. Protoc. Nucleic Acid Chem.* **2006**, *23*, 13.4.1–13.4.10. [CrossRef] [PubMed]
54. Yanachkov, I.B.; Dix, E.J.; Yanachkova, M.I.; Wright, G.E. P^1, P^2-diimidazolyl derivatives of pyrophosphate and bis-phosphonates - synthesis, properties, and use in preparation of dinucleoside tetraphosphates and analogs. *Org. Biomol. Chem.* **2011**, *9*, 730–738. [CrossRef] [PubMed]
55. Depaix, A.; Peyrottes, S.; Roy, B. Water Medium synthesis of nucleoside 5′-polyphosphates. *Curr. Protoc. Nucleic Acid Chem.* **2017**, *69*, 13.16.1–13.16.11.
56. Depaix, A.; Peyrottes, S.; Roy, B. One-Pot synthesis of nucleotides and conjugates in aqueous medium. *Eur. J. Org. Chem.* **2017**, *2*, 241–245. [CrossRef]
57. Smith, M.; Khorana, H.G. Nucleoside Polyphosphates. VI. 1 An improved and general method for the synthesis of Ribo-and Deoxyribonucleoside 5′-Triphosphates. *J. Am. Chem. Soc.* **1958**, *80*, 1141–1145. [CrossRef]
58. Ng, K.-M.E.; Orgel, L.E. The action of a water-soluble carbodiimide on adenosine-5′-polyphosphates. *Nucleic Acids Res.* **1987**, *15*, 3573–3580. [CrossRef]
59. Mohamady, S.; Taylor, S.D. Synthesis of nucleoside tetraphosphates and dinucleoside pentaphosphates via activation of cyclic trimetaphosphate. *Org. Lett.* **2013**, *15*, 2612–2615. [CrossRef]
60. Mohamady, S.; Desoky, A.; Taylor, S.D. Sulfonyl imidazolium salts as reagents for the rapid and efficient synthesis of nucleoside polyphosphates and their conjugates. *Org. Lett.* **2012**, *14*, 402–405. [CrossRef]
61. Mohsen, M.G.; Ji, D.; Kool, E.T. Polymerase synthesis of four-base DNA from two stable dimeric nucleotides. *Nucleic Acids Res.* **2019**, *47*, 9495–9501. [CrossRef] [PubMed]
62. Kanavarioti, A.; Lu, J.; Rosenbach, M.T.; Hurley, T.B. Unexpectedly Facile Synthesis of Symmetrical P^1,P^2-Dinucleoside-5′pyrophosphates. *Tetrahedron Lett.* **1991**, *32*, 6065–6068. [CrossRef]
63. Bogachev, V.S. Synthesis of deoxynucleoside 5′-triphosphates using trifluoroacetic anhydride as an activating reagent. *Bioorg. Khim* **1996**, *22*, 699–705.
64. Mohamady, S.; Taylor, S.D. General Procedure for the Synthesis of Dinucleoside Polyphosphates. *J. Org. Chem.* **2011**, *76*, 6344–6349. [CrossRef]
65. Mohamady, S.; Taylor, S.D. Rapid and efficient synthesis of nucleoside polyphosphates and their conjugates using sulfonyl imidazolium salts. *Curr. Protoc. Nucleic Acid Chem.* **2012**, *51*, 13.11.1–13.11.24. [CrossRef]
66. Sun, Q.; Gong, S.S.; Sun, J.; Wang, C.J.; Liu, S.; Liu, G.D.; Ma, C. Efficient synthesis of nucleoside 5′-triphosphates and their β,γ-bridging oxygen-modified analogs from nucleoside 5′-phosphates. *Tetrahedron Lett.* **2014**, *55*, 2114–2118. [CrossRef]
67. Sun, Q.; Gong, S.-S.; Liu, S.; Sun, J.; Liu, G.-D.; Ma, C. 4,5 Dicyanoimidazole-promoted synthesis of dinucleoside polyphosphates and their analogs. *Tetrahedron* **2014**, *70*, 4500–4506. [CrossRef]
68. Sun, Q.; Sun, J.; Gong, S.-S.; Wang, C.-J.; Wang, X.-C. Synthesis of nucleoside tetraphosphates and dinucleoside pentaphosphates from nucleoside phosphoropiperidates via the activation of P(V)-N bond. *Chin. Chem. Lett.* **2015**, *26*, 89–92. [CrossRef]
69. Sun, Q.; Sun, J.; Gong, S.-S.; Wang, C.-J.; Wang, X.-C. One-pot synthesis of symmetrical dinucleoside polyphosphates and analogs via 4,5-dicyanoimidazole-promoted tandem P–O coupling reactions. *Tetrahedron Lett.* **2014**, *55*, 5785–5788. [CrossRef]
70. Ludwig, J.; Eckstein, F. Rapid and efficient synthesis of nucleoside 5′-*O*-(1-thiotriphosphates), 5′-triphosphates, and 2′,3′-cyclophosphorothioates using 2-chloro-4*H*-1,3,2-benzodioxaphosphorin-4-one. *J. Org. Chem.* **1989**, *54*, 631–635. [CrossRef]
71. Han, Q.; Gaffney, B.L.; Jones, R.A. One-Flask synthesis of dinucleoside tetra- and pentaphosphates. *Org. Lett.* **2006**, *8*, 2075–2077. [CrossRef] [PubMed]
72. Meier, C. cycloSal-pronucleotides design of chemical trojan horses. *Mini-Rev. Med. Chem.* **2002**, *2*, 219–234. [CrossRef] [PubMed]

73. Meier, C. cycloSal Phosphates as chemical trojan horses for intracellular nucleotide and glycosylmonophosphate delivery — Chemistry meets biology. *Eur. J. Org. Chem.* **2006**, *2006*, 1081–1102. [CrossRef]
74. Warnecke, S.; Meier, C. Synthesis of Nucleoside Di- and Triphosphates and Dinucleoside Polyphosphates with cycloSal-Nucleotides. *J. Org. Chem.* **2009**, *74*, 3024–3030. [CrossRef]
75. Wendicke, S.; Warnecke, S.; Meier, C. Efficient Synthesis of Nucleoside Diphosphate Glycopyranoses. *Angew. Chem. Int. Ed.* **2008**, *47*, 1500–1502. [CrossRef]
76. Tonn, V.C.; Meier, C. Solid-Phase synthesis of (poly)phosphorylated nucleosides and conjugates. *Chem. Eur. J.* **2011**, *17*, 9832–9842. [CrossRef]
77. Ahmadibeni, Y.; Parang, K. Solid-Phase Synthesis of Symmetrical 5′,5′-Dinucleoside Mono-, Di-, Tri-, and Tetraphosphodiesters. *Org. Lett.* **2007**, *9*, 4483–4486. [CrossRef]
78. Kistemaker, H.A.V.; Meeuwenoord, N.J.; Overkleeft, H.S.; Van Der Marel, G.A.; Filippov, D.V. On the synthesis of oligonucleotides interconnected through pyrophosphate linkages. *Eur. J. Org. Chem.* **2015**, *2015*, 6084–6091. [CrossRef]
79. Sun, Q.; Liu, S.; Sun, J.; Gong, S.; Xiao, Q.; Shen, L. One-pot synthesis of symmetrical P^1,P^2-dinucleoside-5′-diphosphates from nucleoside-5′-*H*-phosphonates: Mechanistic insights into reaction path. *Tetrahedron Lett.* **2013**, *54*, 3842–3845. [CrossRef]
80. Sun, Q.; Edathil, J.P.; Wu, R.; Smidansky, E.D.; Cameron, C.E.; Peterson, B.R. One-Pot Synthesis of Nucleoside 5′-Triphosphates from Nucleoside 5′-*H*-Phosphonates. *Org. Lett.* **2008**, *10*, 1703–1706. [CrossRef]
81. Jessen, H.J.; Ahmed, N.; Hofer, A. Phosphate esters and anhydride—recent strategies targeting nature's favoured modifications. *Org. Biomol. Chem.* **2014**, *12*, 3526–3530. [CrossRef] [PubMed]
82. Cremosnik, G.S.; Hofer, A.; Jessen, H.J. Iterative synthesis of nucleoside oligophosphates with phosphoramidites. *Angew. Chem. Int. Ed.* **2014**, *53*, 286–289. [CrossRef] [PubMed]
83. Hofer, A.; Cremosnik, G.S.; Müller, A.C.; Giambruno, R.; Trefzer, C.; Superti-Furga, G.; Jessen, H.J. Modular synthesis of modified phosphoanhydrides. *Chem. Eur. J.* **2015**, *21*, 10116–10122. [CrossRef] [PubMed]
84. Hofer, A.; Marques, E.; Kieliger, N.; Gatter, S.-K.N.; Jordi, S.; Ferrari, E.; Hofmann, M.; Fitzpatrick, T.B.; Hottiger, M.O.; Jessen, H.J. Chemoselective dimerization of phosphates. *Org. Lett.* **2016**, *18*, 3222–3225. [CrossRef]
85. Tanaka, H.; Yoshimura, Y.; Jørgensen, M.R.; Cuesta-Seijo, J.A.; Hindsgaul, O. A simple synthesis of sugar nucleoside diphosphates by chemical coupling in water. *Angew. Chem. Int. Ed.* **2012**, *51*, 11531–11534. [CrossRef]
86. Eguaogie, O.; Vyle, J.S.; Conlon, P.F.; Gîlea, M.A.; Liang, Y. Mechanochemistry of nucleosides, nucleotides and related materials. *Beilstein J. Org. Chem.* **2018**, *14*, 955–970. [CrossRef]
87. James, S.L.; Adams, C.J.; Bolm, C.; Braga, D.; Collier, P.; Friščić, T.; Grepioni, F.; Harris, K.D.M.; Hyett, G.; Jones, W.; et al. Mechanochemistry: Opportunities for new and cleaner synthesis. *Chem. Soc. Rev.* **2012**, *41*, 413–447. [CrossRef]
88. Bowmaker, G.A. Solvent-assisted mechanochemistry. *Chem. Commun.* **2013**, *49*, 334–348. [CrossRef]
89. Takacs, L. The historical development of mechanochemistry. *Chem. Soc. Rev.* **2013**, *42*, 7649–7659. [CrossRef]
90. Wang, G.W. Mechanochemical organic synthesis. *Chem. Soc. Rev.* **2013**, *42*, 7668–7700. [CrossRef]
91. Friščić, T.; Mottillo, C.; Titi, H.M. Mechanochemistry for synthesis. *Angew. Chem. Int. Ed. Engl.* **2019**, in press.
92. Ravalico, F.; Messina, I.; Berberian, M.V.; James, S.L.; Migaud, M.E.; Vyle, J.S. Rapid synthesis of nucleotide pyrophosphate linkages in a ball mill. *Org. Biomol. Chem.* **2011**, *9*, 6496–6497. [CrossRef] [PubMed]
93. Appy, L.; Depaix, A.; Bantreil, X.; Lamaty, F.; Peyrottes, S.; Roy, B. Straightforward ball-milling access to dinucleoside 5′,5′-polyphosphates via phosphorimidazolide intermediates. *Chem. Eur. J.* **2019**, *25*, 2477–2481. [PubMed]
94. Verma, S.K.; Ghorpade, R.; Pratap, A.; Kaushik, M. Solvent free, *N,N′*-carbonyldiimidazole (CDI) mediated amidation. *Tetrahedron Lett.* **2012**, *53*, 2373–2376. [CrossRef]
95. Metro, T.-X.; Bonnamour, J.; Reidon, T.; Sarpoulet, J.; Martinez, J.; Lamaty, F. Mechanosynthesis of amides in the total absence of organic solvent from reaction to product recovery. *Chem. Commun.* **2012**, *48*, 11781–11793. [CrossRef]

96. Métro, T.-X.; Bonnamour, J.; Reidon, T.; Duprez, A.; Sarpoulet, J.; Martinez, J.; Lamaty, F. Comprehensive study of the organic-solvent-free CDI-Mediated acylation of various nucleophiles by mechanochemistry. *Chem. Eur. J.* **2015**, *21*, 12787–12796. [CrossRef]
97. Métro, T.-X.; Martinez, J.; Lamaty, F. 1,1′-Carbonyldiimidazole and mechanochemistry: A shining green combination. *ACS Sustain. Chem. Eng.* **2017**, *5*, 9599–9602. [CrossRef]
98. Banoub, J.H.; Newton, R.P.; Esmans, E.; Ewing, D.F.; Mackenzie, G. Recent developments in mass spectrometry for the characterization of nucleosides, nucleotides, oligonucleotides, and nucleic acids. *Chem. Rev.* **2005**, *105*, 1869–1915. [CrossRef]
99. Watanabe, D.; Ishikawa, M.; Yamasaki, M.; Ozaki, M.; Katayama, T.; Nakajima, H. Tetrasodium P^1,P^4-Bis(5′-adenosyl)tetraphosphate Dodecahydrate. *Acta Crystallogr. Sect. C Cryst. Struct. Commun.* **1996**, *52*, 338–340. [CrossRef]
100. Simanshu, D.K.; Savithri, H.S.; Murthy, A.R.N. Crystal structures of Salmonella typhimurium propionate kinase and its complex with Ap4A: Evidence for a novel Ap_4A synthetic activity. *Proteins* **2008**, *70*, 1379–1388. [CrossRef]
101. Baker, M.D.; Holloway, D.E.; Swaminathan, G.J.; Acharya, K.R. Crystal structures of eosinophil-derived neurotoxin (EDN) in complex with the inhibitors 5′-ATP, Ap_3A, Ap_4A, and Ap_5A. *Biochemistry* **2006**, *45*, 416–426. [CrossRef] [PubMed]
102. Scheffzek, K.; Kliche, W.; Wiesmuller, L.; Reinstein, J. Crystal structure of the complex of UMP/CMP kinase from Dictyostelium discoideum and the bisubstrate inhibitor P^1-(5′-adenosyl) P^5-(5′-uridyl) pentaphosphate (UP_5A) and Mg^{2+} at 2.2 A: Implications for water-mediated specificity. *Biochemistry* **1996**, *35*, 9716–9727. [CrossRef] [PubMed]
103. Berry, M.B.; Phillips, G.N., Jr. Crystal structures of Bacillus stearothermophilus adenylate kinase with bound Ap_5A, Mg^{2+} Ap_5A, and Mn^{2+} Ap_5A reveal an intermediate lid position and six coordinate octahedral geometry for bound Mg^{2+} and Mn^{2+}. *Proteins* **1998**, *32*, 276–288. [CrossRef]
104. Chen, J.; Kohler, B. Base stacking in adenosine dimers revealed by femtosecond transient absorption spectroscopy. *J. Am. Chem. Soc.* **2014**, *136*, 6362–6372. [CrossRef] [PubMed]

8

Synthetic Procedures Leading towards Aminobisphosphonates

Ewa Chmielewska * and Paweł Kafarski

Department of Bioorganic Chemistry, Faculty of Chemistry, Wrocław University of Science and Technology, Wrocław 50-370, Poland; pawel.kafarski@pwr.edu.pl
* Correspondence: ewa.chmielewska@pwr.edu.pl

Academic Editor: György Keglevich

Abstract: Growing interest in the biological activity of aminobisphosphonates has stimulated the development of methods for their synthesis. Although several general procedures were previously elaborated to reach this goal, aminobisphosphonate chemistry is still developing quite substantially. Thus, innovative modifications of the existing commonly used reactions, as well as development of new procedures, are presented in this review, concentrating on recent achievements. Additionally, selected examples of aminobisphosphonate derivatization illustrate their usefulness for obtaining new diagnostic and therapeutic agents.

Keywords: bisphosphonates; bisphosphonylation procedures; Vilsmayer-Haack-like reactions; functionalization of aminoalkylbisphosphonates

---

1. Introduction

Bisphosphonates are a class of compounds that are currently receiving significant attention. Over 50 new papers are seen each week when searching the literature with the keyword "bisphosphonate" via Web of Science. More than 17,000 various bisphosphonate structures have been synthesized and described in the literature [1]. Most papers concern their important subclass, namely aminobisphosphonates. This strong interest results from these compounds acting as strong inhibitors of bone resorption, with several representatives of this class already commercialized as drugs of choice for the treatment of osteoporosis, skeletal complications of malignancy, Paget's disease, multiple myeloma, hypercalcemia and fibrous dysplasia [2–7] Consequently, most of the papers are devoted to various clinical aspects of the anti-resorptive effects that bisphosphonates exert towards bone tissues; however, there is also a growing interest in their applications as anticancer and antibacterial agents [5–7]. Additionally, aminobisphosphonic acids have found important industrial applications, largely as inhibitors of scale formation and as corrosion inhibitors, actions which result from their ability to complex metal ions [8–10].

Thus, simple and effective procedures for their synthesis are becoming increasingly important. However, only a few general reactions leading to these compounds have been described to date and are only partially reviewed in the literature [11,12]. Novel reports are mostly concentrated on modifications and improvement of these procedures, and there are only a few papers aiming at new reactions, which results from the commonly applied procedures being simple, economical and effective.

In this paper, we comprehensively review the recent studies (supplemented by older papers if necessary) on reactions applied to synthesize the most important class of bisphosphonates—aminobisphosphonates—and discuss their scope and limitations.

2. Overview of Synthetic Procedures

There are contradictory reports of the origin of bisphosphonates [13]. Most likely, the first one was obtained in the 19th century by Nikolay Menschutkin and/or Theodor Saltzer as an impurity in reactions designed to obtain different compounds. It was further identified by Hans von Baeyer and Wilhelm Heideprim as 1-hydroxyethanebisphosphonic acid.

There are only a few general methods for the synthesis of aminobisphosphonates. However, there are many individual procedures described for their preparation [11,12]. In this review, the following general reactions are presented: (i) starting from carboxylic acids and (ii) their amides; (iii) reactions using nitriles and (iv) isonitriles as substrates; (v) syntheses based on addition of phosphites to oxophosphonates; (vi) three-component condensation of amines, trialkyl orthoformates and dialkyl phosphites; (vii) addition of amines to vinylidenebisphosphonates; and (viii) functionalization of simple bisphosphonates treated as building blocks for the preparation of more complex structures. Additionally, some specific and non-conventional procedures that have been elucidated will be presented in this review.

2.1. Synthesis from Carboxylic Acids

1-Hydroxyethylidene-1,1-bisphosphonic acids are perhaps the oldest group of bisphosphonates. They are standardly prepared by a large-scale, one-step reaction of carboxylic acids with phosphorus trichloride and phosphorous or phosphoric acids, followed by hydrolysis with water; the procedure was optimized by Kieczykowski et al. [14]. The reaction is carried out in selected solvents (phenylsulphonic acid, various phenols, chlorobenzene, diphenyl ether or ionic liquids) with sulfone and methanesulphonic acid being preferred choices [11,15–20]. Despite many theories [19,21–23], the exact mechanism of this reaction is not fully understood; however, the formation of acid chloride as a first intermediate has been undoubtedly demonstrated (Scheme 1). This intermediate may react with methanesulphonic acid (when used as a solvent), and the formed mixed anhydride is also considered a potential intermediate for the next step [24], which is an Arbuzov-like reaction of phosphorus acid or one of its several derivatives (including anhydrides of variable structure [17]) formed during the reaction. The formed derivative of ketophosphonate is a substrate for the addition reaction of trivalent P-OH species, and bisphosphonate is obtained (Scheme 1). It has also been documented that the use of phosphorous acid could be omitted if water was added to the reaction medium, and thus this compound was formed in situ.

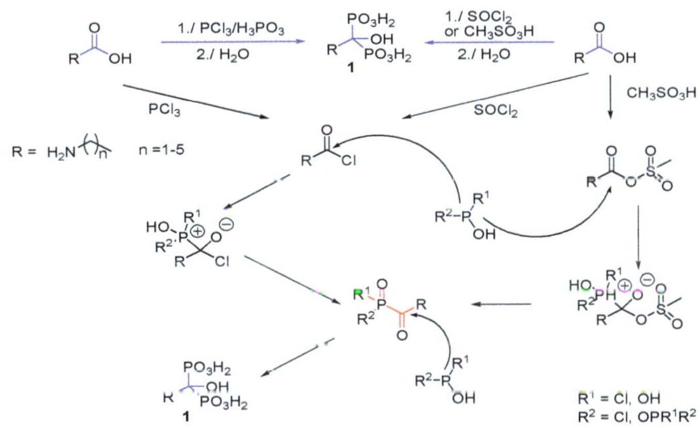

Scheme 1. Presumable mechanism of the synthesis of 1-hydroxy-1,1-bisphosphonic acids.

This reaction was commonly used for the preparation of a wide variety of anti-osteoporotic bisphosphonic acids containing free amino groups (so called dronic acids, compound **1**). In this case, free unblocked amino acids are used as substrates, and the final neutralization of reaction mixtures to a pH of approximately 4 causes precipitation of the desired products, which are formed in satisfactory yields and are of good purity [16,23–26]. Because of its technological importance, this reaction is still quite intensively studied and optimized; however, most of the studies are done in industrial laboratories, and their results are mostly disseminated as patents. Consequently, it is difficult, if not impossible, to determine which conditions are optimal for the synthesis of individual drugs. It is important because, as in the case of any multicomponent reaction, a synthetic course is strongly dependent on applied conditions and molar ratios of reagents. For example, one of the patents reports that the omission of phosphorus chloride in the reaction of 4-aminobutyric acid as the substrate in methanesulphonic acid provides mixtures of anhydrides as final products (compounds **2** and **3**, Scheme 2) [11]. This corresponds well to the known tendency of phosphonic acids to cyclize.

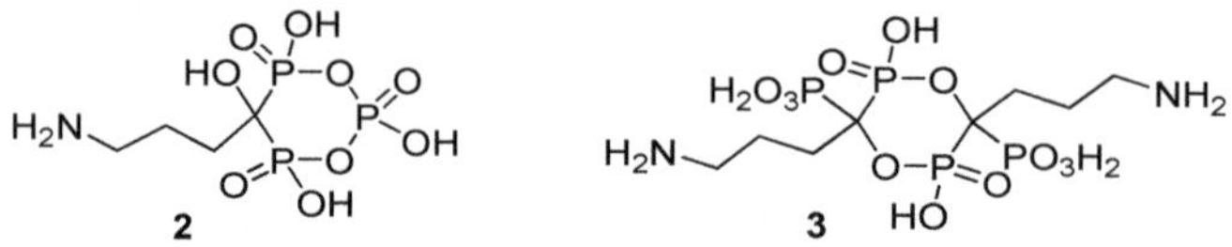

2 **3**

Scheme 2. Cyclic products of the reaction between carboxylic acids and phosphorous acid in methanesulphonic acid.

Among modifications of this procedure are (i) replacement of phosphorus trichloride with thionyl chloride to generate acid chloride in the first reaction step [11]; (ii) replacement of acid by its chloride or anhydride [11] or (iii) t-butyl ester [27]; (iv) blocking the amino moiety in the case of α-amino acids [28]; and (v) application of microwave-assisted procedures [29].

2.2. Synthesis from Amides

Although studies on the biological activity of 1-amino-1,1-bisphosphonic acids are scarce, a significant number of procedures for their preparation have been described. They have recently been reviewed by Romanenko and Kukhar [12]. Amides, being highly stable and easily available compounds, are substrates of choice, and the most commonly applied procedures are simple modifications of those elaborated for carboxylic acids. The most straightforward are reactions of *N*-acylamines and *N*-formylamines with phosphorus trichloride and phosphorous acid [30–32], phosphorus tribromide [33], or triphosgene [34], which provide a wide structural variety of 1-amino-1,1-bisphosphonic acids (compound **4**, Scheme 3). They are based on modification of the first procedure elaborated by Plöger et al. with formamide as a substrate [35].

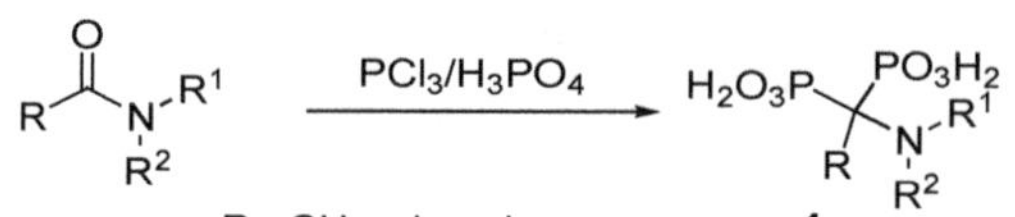

4

R= CH_3, phenyl
R^1= H, alkyl
R^2= alkyl, aryl, heteroaryl, cycloalkyl

Scheme 3. Bisphosphonylation of amides.

The exact reaction mechanism in this case is also not fully known; however, a Vilsmayer-Haack-like route via iminium ion is postulated here (Scheme 4).

Another possibility is using trialkyl or trimethylsilyl phosphites in reactions prompted by phosphoryl chloride [36–39], trifluoromethanesulphonic anhydride (Tf_2O) [40], zinc chloride [41] or trimethylsilyl trifluoromethanesulfonate [42,43]. Unfortunately, primary amides are not suitable substrates for this reaction [38]. The use of triethyl phosphite and phosphoryl chloride appeared to be especially useful when lactams were used as substrates, resulting in high yields of aminomethylene-*gem*-bisphosphonates (Scheme 5) [36–39]. They were readily hydrolyzed, yielding corresponding bisphosphonic acids (representative example compound **5**). On the other hand, this reaction is not suitable for the conversion of benzoannulated lactams, and mixtures of the desired bisphosphonates and monophosphonates of variable structures were obtained (compounds **6**, **7**, **8** and **9**, Scheme 5). In fact, monophosphonates are usually major products, and the reaction course is dependent on the size of the substrate aliphatic ring [44]. Additionally, these bisphosphonates and monophosphonates appeared to be unstable upon acid hydrolysis and upon storage, and undergo degradation with cleavage of the carbon-to-phosphorus bond (for example compound **10**, Scheme 5). Thus, corresponding acids have not yet been obtained.

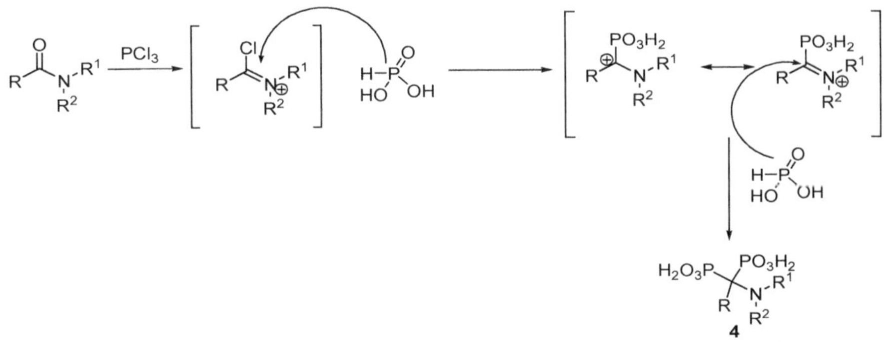

Scheme 4. Bisphosphonylation of amides and the presumable mechanism of this reaction.

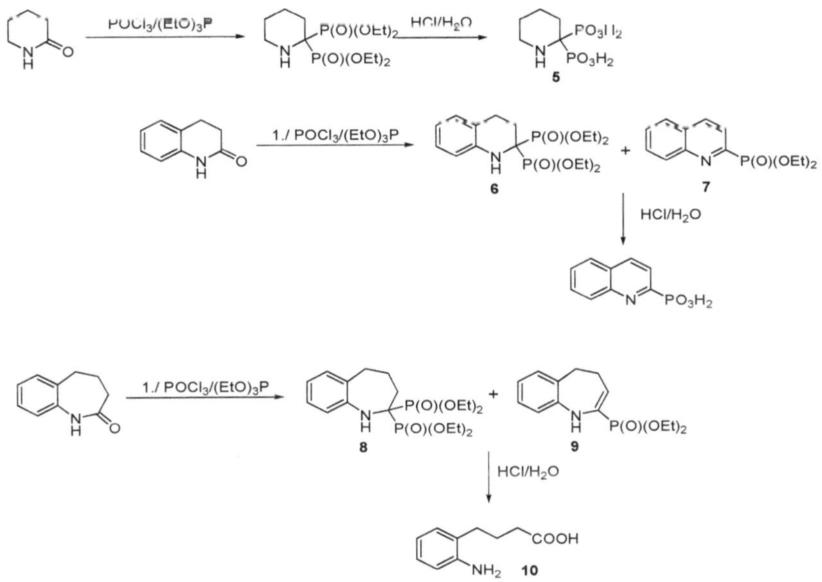

Scheme 5. Representative reactions of lactams with triethyl phosphite prompted by phosphoryl chloride.

The formation of a Vilsmayer-Haack-like intermediate is also proposed in this case with the further addition of a nucleophilic phosphorus reagent. This assumption is additionally supported by studies on the use of structurally variable imine salts as substrates for synthesis of compounds **11** (Scheme 6) [41–45].

The reaction of dibutoxyphosphine or bis(trimethylsiloxy)phosphine with dimethylformamides in the presence of trimethylsilyl trifluoromethanesulfonate affords the corresponding aminomethylenebisphosphonites **12** in high yields (Scheme 7) [41–43]. Similar products were obtained by reacting diethyl pivaloylphosphonite with dialkylformamides in the presence of excess ethanol and catalytic amounts of zinc chloride (Scheme 7). Pivaloylphosphonite decomposes in these reaction conditions, yielding diethoxyphosphine, which is the real substrate of the reaction [41].

N-Octylpyrrolidinone treated with LDA, followed by the addition of diethyl phosphorochloridite and oxidation of the reaction mixture with 30% hydrogen peroxide resulted in bisphosphonylated lactam **13** with excellent yield (Scheme 8). This reaction was then applied for the synthesis of *N*-geranylated lactams **14** and **15** and linear amides, potential inhibitors of farnesyl:protein transferase [46]. In the case of cyclic imides, in which two equivalent positions for enolate formation are present on the imide rings, and thus two carbon atoms may be phosphonylated, the structure of the product was dependent on the mode of reaction. If phosphonylation was carried out in one step, two carbon atoms were phosphonylated and vicinal bisphosphonate was obtained. Step-by-step reaction resulted in the predomination of the phosphonylation of one carbon atom, and the desired gem-bisphophonate was the major product.

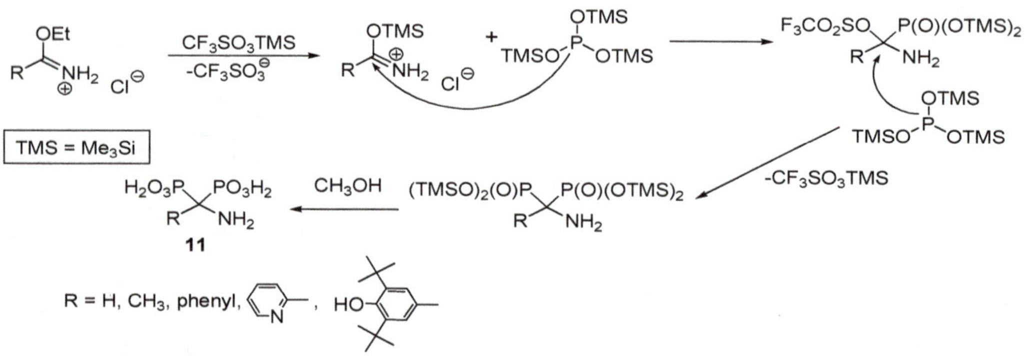

Scheme 6. Imine salts as substrates for preparation of bisphosphonic acids.

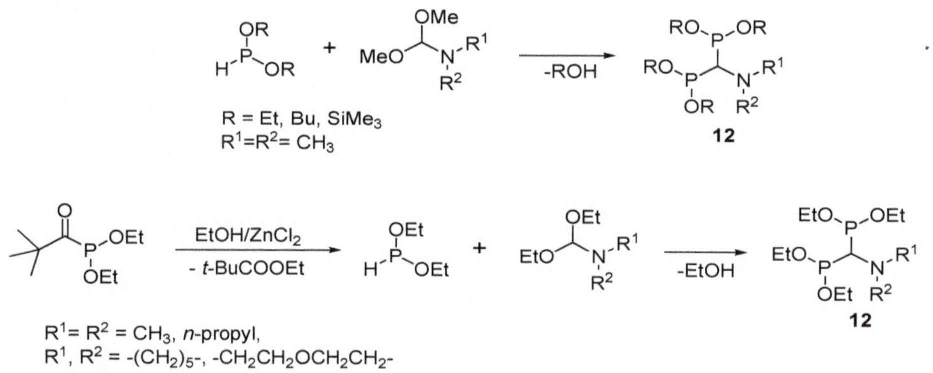

Scheme 7. Reaction of diethoxyphosphine with the acetal of dimethylformamide.

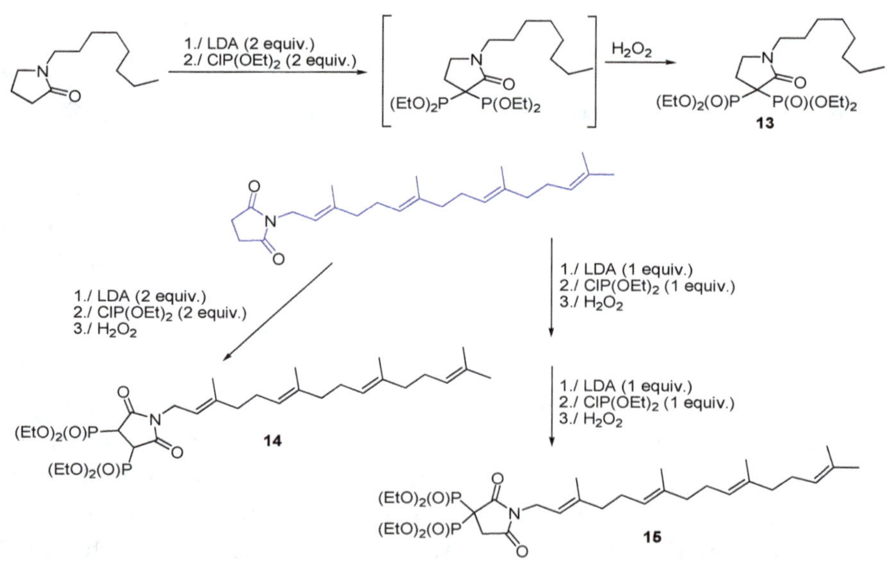

Scheme 8. Synthesis of bisphosphonate lactams.

2.3. *Synthesis from Nitriles*

1-Aminoalkylidene-1,1-bisphosphonic acids could also be prepared from nitriles by applying the same procedures as in syntheses starting from amides. Despite being readily available substrates, there is limited literature on their use for that purpose. The most popular route is the reaction of nitriles with phosphorous acid, in some cases prompted by phosphorus trichloride and phenylsulphonic or methylsulphonic acids [12,31,47,48]. The conditions of this reaction are sufficiently delicate to use peptidyl nitriles as substrates, which was demonstrated using cyanoethyl derivatives of dipeptides (for example, peptide **16**, Scheme 9) [49].

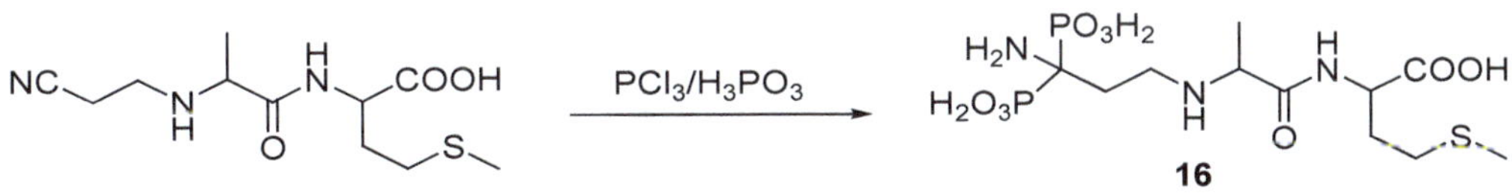

Scheme 9. Representative example of the synthesis of peptidylbisphosphonates from nitriles.

Recently, elegant, mild, and atom economical double phosphonylation of nitriles in the presence of titanocene has been described. This procedure used induced phosphorus-centered radicals mediated by titanocene dichloride (Cp_2TiCl_2) (Scheme 10) and provided a wide structural variety of aminobisphosphonates **17** [50]. This is a double radical transfer reaction initiated by the reaction of titanocene with zinc dust and the transfer of the obtained radical to epoxypropane by cleavage of its oxirane ring, followed by transfer of this radical to diethyl phosphite, which in turn reacts with nitrile (Scheme 10). The reaction carried out without epoxide, under microwave stimulation, also results in the desired bisphosphonates, although it is accompanied by formation of many side products.

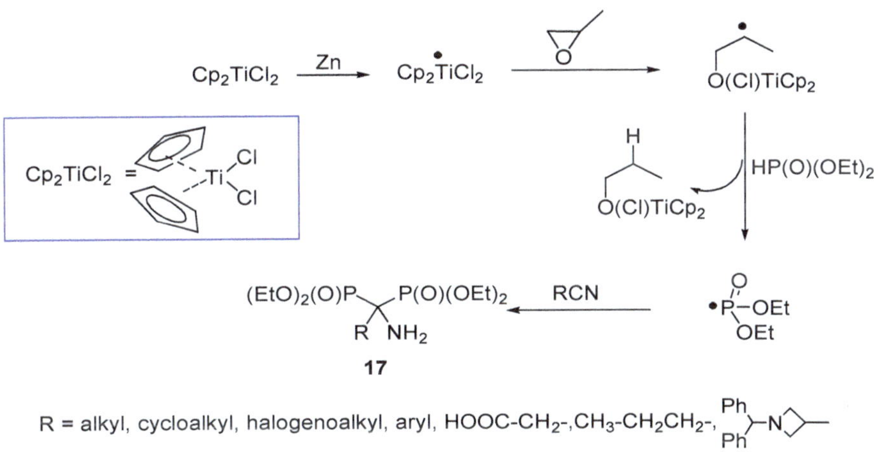

Scheme 10. Radical procedure for the synthesis of bisphosphonates from nitriles.

2.4. *Synthesis from Isonitriles*

Isonitriles, which are quite common substrates, especially in multicomponent reactions [51], have been seldom used for the preparation of aminobisphosphonates **18**. Their reactions with *H*-phosphine oxides carried out in the presence of typical palladium catalysts (Pd_2dba_3) afforded the product **19** of monophosphonylation, whereas the use of various rhodium catalysts afforded products of diphosphonylation (Scheme 11) [52].

Far simpler is the addition of diethyl phosphite to isonitriles. This reaction is carried out with an excess of hydrogen chloride in aprotic solvents. Under these conditions, nitrile is converted to an onium salt, and after addition of the first phosphite molecule, iminium salt is formed, a substrate for the addition of a second phosphite molecule (Scheme 12) [53]. Using this procedure, two distinct libraries of bisphosphonates **20** and **21** have been prepared [54–56].

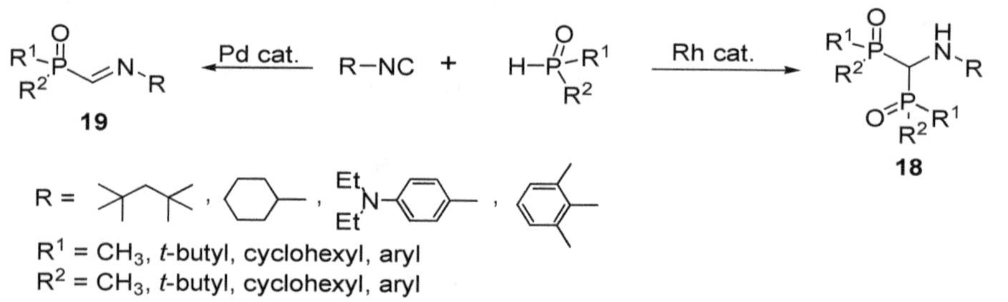

Scheme 11. Addition of *H*-phosphine oxides to isonitriles catalyzed by metallocatalysts.

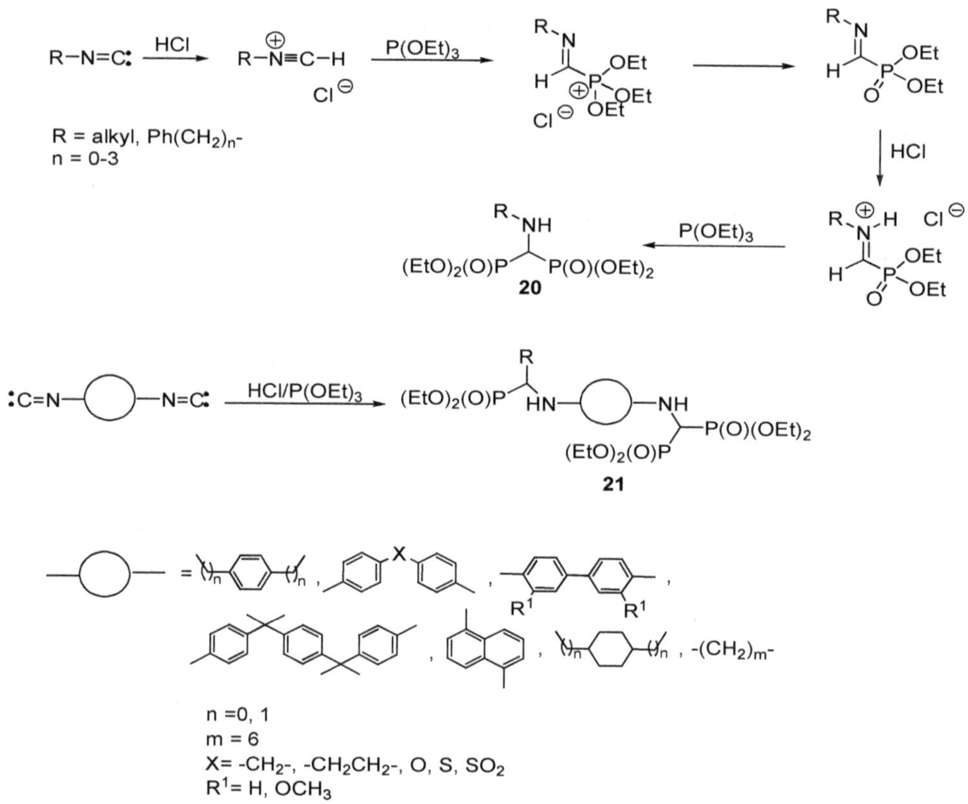

Scheme 12. Addition of phosphites to isonitriles.

2.5. Synthesis via Ketophosphonates

Presumably, the addition of phosphites to ketophosphonates was the first procedure for preparation of dialkyl 1-hydroxy-1,1-bisphosphonates [16,19,57]. Starting ketophosphonates are obtained by the Arbuzov reaction of acyl chlorides with trialkyl phosphites and used as substrates in the next step, namely, the addition of dialkyl phosphite to carbonyl double bonds (a representative example for preparation of tetramethyl pamidronate **22** is shown in Scheme 13). Since ketophosphonates are unstable species [58], the desired bisphosphonates are usually obtained via one-pot procedures applying mixtures of trialkyl- and dialkylphosphonates at elevated temperature [59]. Additionally, in situ generation of dialkyl- from trialkyl phosphites is possible by the addition of a protic solvent to the reaction mixture. A useful modification of this procedure is the application of tris(trimethylsilyl) phosphite, followed by the easy removal of ester groups by methanolysis [23,60,61]. This was used to prepare a wide variety of 1-hydroxy-1,1-bisphosphonates containing amino groups [19,23,61–63]. Additionally, the use of bis((trimethylsilyl)oxy)phosphine generated from ammonium hypophosphite was applied to prepare hydroxybisphosphinic acids (also presented in Scheme 13) [64]. Another modification was the use of acyl phosphonamidates readily prepared from phosphoramidite type reagents and a range of acid chlorides, followed by reaction with trimethyl phosphite in the presence of pyridinium perchlorate [65].

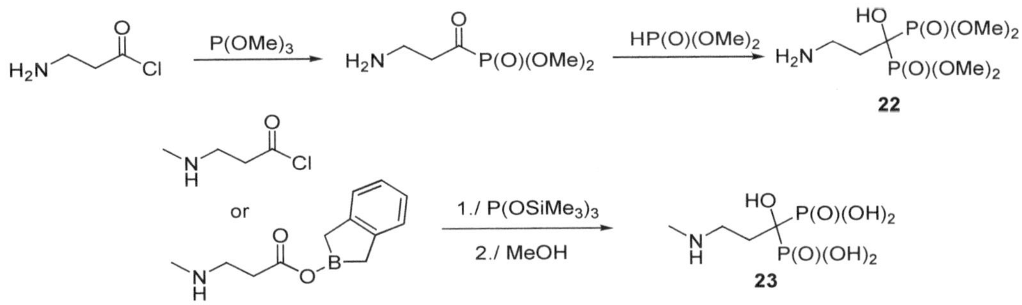

Scheme 13. Representative syntheses via ketophosphonates.

Recently, a one-pot synthesis was described reacting carboxylic acids with catecholborane, followed by the treatment of the formed acyloxy-benzodioxaborolane with tris(trimethylsilyl)phosphite (Scheme 13) [66]. The efficiency of that simple methodology was proved by the syntheses of alendronate and *N*-methyl pamidronate (compound **23**) without additional steps for the protection/deprotection of their amine functions.

2.6. Three-Component Condensation of Amines with Triethyl Orthoformate and Diethylphosphite

Simple three-component condensation of stoichiometric ratios of amines, diethyl phosphite and triethyl orthoformate, first reported in patent literature by Suzuki [67] and further extended by Maier [68], is perhaps the most common procedure for the preparation of a wide variety of aminomethylenebisphosphonic acids (Scheme 14). Since this reaction usually gives a complex mixture of products [56] that are difficult to separate, the resulting esters are not isolated but rather, the crude reaction mixtures are hydrolyzed, yielding bisphosphonic acids that are isolated after the hydrolytic step. Some modifications of this classic procedure have also been reported. They include the use of a solvent-free, microwave-assisted reaction [69–71] and reactions catalyzed by titanium dioxide [72] and by crown ethers (increasing the selectivity of the process by 10%–20%) [73]. Additionally, a reaction carried out in a micellar environment was described [74].

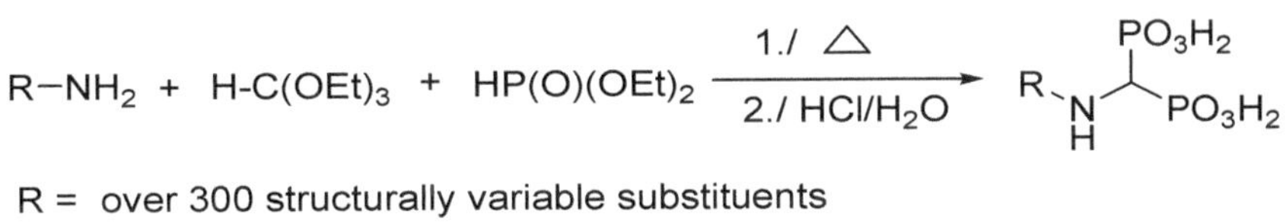

Scheme 14. Condensation of amines with triethyl orthoformate and diethyl phosphite.

The three-component reaction was applied to synthesize a large series of physiologically active compounds, in some cases of very complex chemical structures, including bone antiresorptive drug candidates [75–82]; bone imaging [83,84], antiprotozoal [85–88], antibacterial [72,89–92], anti-HIV [93] and anti-inflammatory [94] agents; herbicides [95–97], and complex ones for various metals [10,98].

In some cases, this reaction appears to be quite capricious and affords unexpected side-products along with the expected aminomethylenebisphosphonates (see representative examples in Scheme 15), the compositions of which are dependent on the applied conditions (molar ratio of substrates, temperature and reaction time). Most often, alkylation (for example, compounds **24** and **25**) or formylation (compound **26**) of amine moieties is observed [71,99], while in selected cases the formation of aminomethylenebisphosphonic acid (compound **27**) is observed. Because this compound was previously prepared by acid hydrolysis of *N*-benzhydrylaminomethylenebisphosphonic acid [83,100], it may be that it is formed upon hydrolysis of *N*-aryl derivatives.

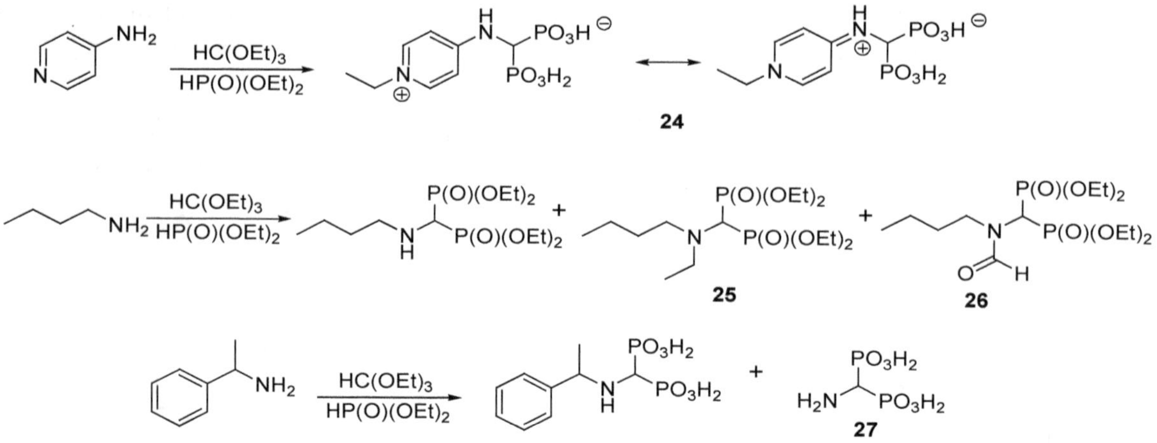

Scheme 15. Representative side products of the three-component procedure for the synthesis of aminomethylenebisphosphonic acids.

The mechanism of this useful reaction had been thoroughly studied by using ^{31}P-NMR and by isolation of all intermediates [71,72,99] and identifying interrelations between them (Scheme 16) [99]. The mechanism appeared to be quite complex because the intermediates exist in thermodynamic equilibrium. Thus, its course is strongly dependent on the properties of the used amine and applied reaction conditions.

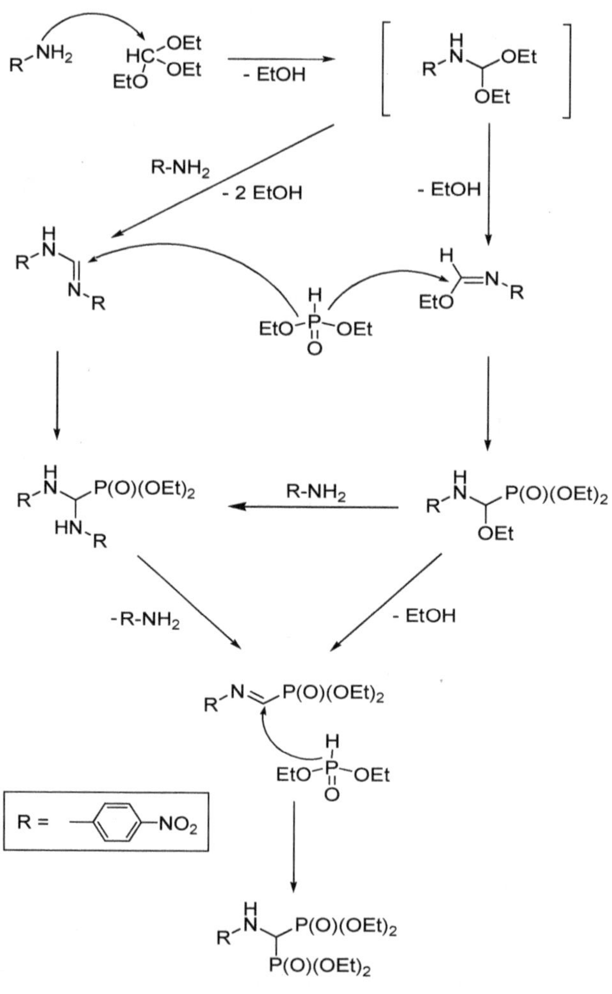

Scheme 16. Mechanism of the three-component reaction of amines with orthoformates and phosphites.

Modification of this procedure and the use of ethyl diethoxymethyl-*H*-phosphinate instead of diethyl phosphite allowed obtaining structurally variable bisphosphinates **28** (Scheme 17) [101]. Better results were obtained if the reaction was carried out under nitrogen (air oxidizes substrates and products). Notably, aromatic amines provided the desired bisphosphinates in high yields (78%–92%), whereas the reaction yields when using cyclic and aliphatic derivatives were lower. Interestingly, of many methods for the synthesis of corresponding acids **29**, acidolysis of the obtained esters **28** appeared to be optimal. The use of other conditions, including mild transesterification with trimethylsilyl bromide followed by methanolysis, resulted in the formation of side products **30** and **31**.

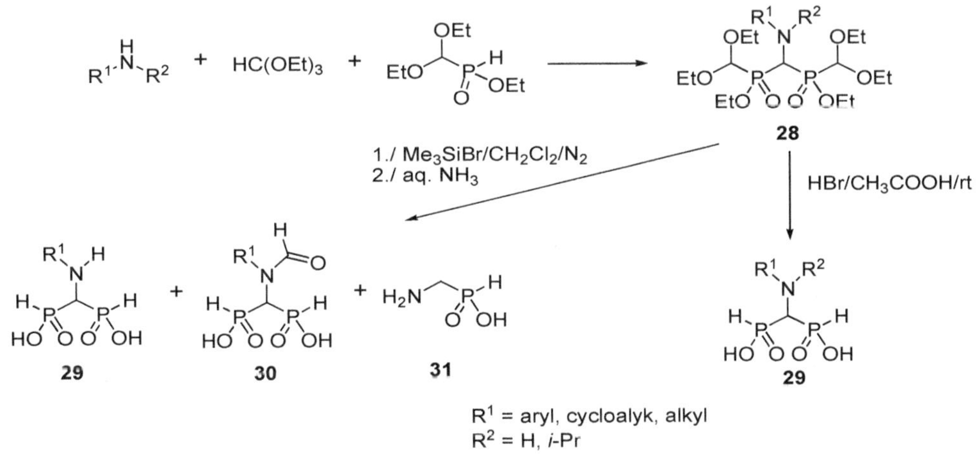

Scheme 17. Three-component synthesis of aminobisphosphonates.

2.7. Addition of Amines to Vinylidenebisphosphonates

Tetraethyl vinylidenebisphosphonate **32** is a versatile synthon useful for preparation of a wide variety of bisphosphonates. Its usefulness is a subject of recent comprehensive review [102]. As an electron-deficient alkene, it can undergo conjugate addition of strong and mild nucleophiles, with amines belonging to the latter class. Primary amines undergo smooth Michael addition (Scheme 18), however, the obtained compounds **33** must be quickly purified and hydrolyzed since they tend to undergo a retro-Michael reaction [103,104]. Fortunately, free acids **34** are substantially more stable.

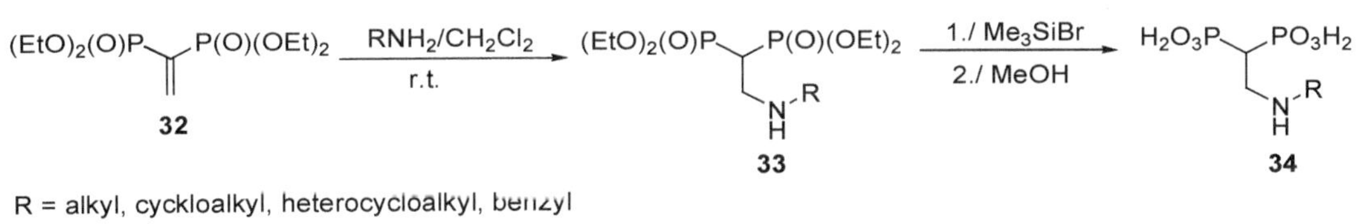

Scheme 18. Michael addition of amines to vinylidenebisphosphonate.

The reactivity of vinylidenebisphosphonates has been used for the synthesis of derivatives of various analogues of fluoroquinolone antibacterial agents [105,106], heteroaromatics with potential pharmaceutical applications [107,108], simple antiplasmodial agents [103,104], HIV reverse transcriptase inhibitors [59,93], potential anti-osteoporotic agents [109,110], and *N*-alkylated antitumor pyridines [111]. Some representatives of these compounds (**35**–**39**) are shown in Scheme 19.

Another example, albeit far less developed, is the addition of organometallic amines to vinylidenebisphosphonate **22** (Scheme 20) [112]. Huisgen copper-catalyzed 1,3-dipolar cycloaddition of this bisphosphonate to azides was used to obtain substrate **40** for a "click reaction". This reaction yielded compounds that may be considered aza-analogues **41** of zoledronate (Scheme 20) [113]. Similar substrates could also be obtained by addition of propargylamine to the double bond of vinylidenebisphosphonate [114].

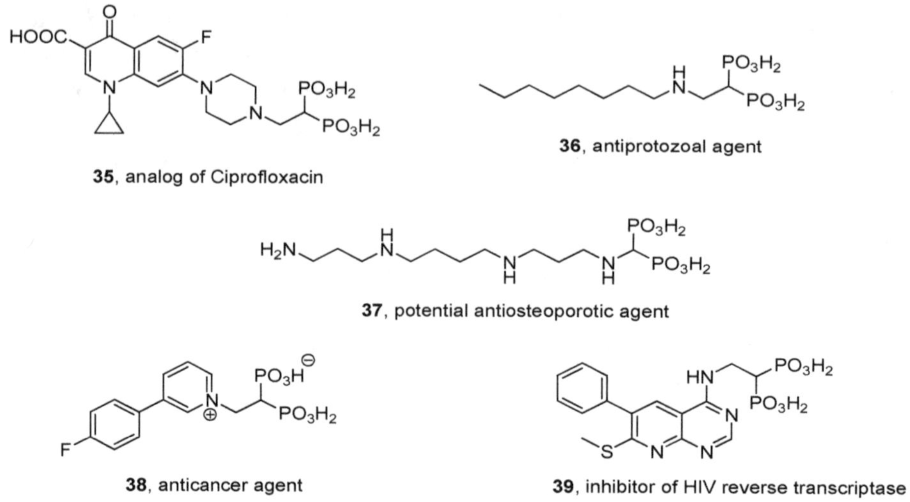

Scheme 19. Representatives of useful bisphosphonates obtained via addition of amines to vinylidenebisphosphonate.

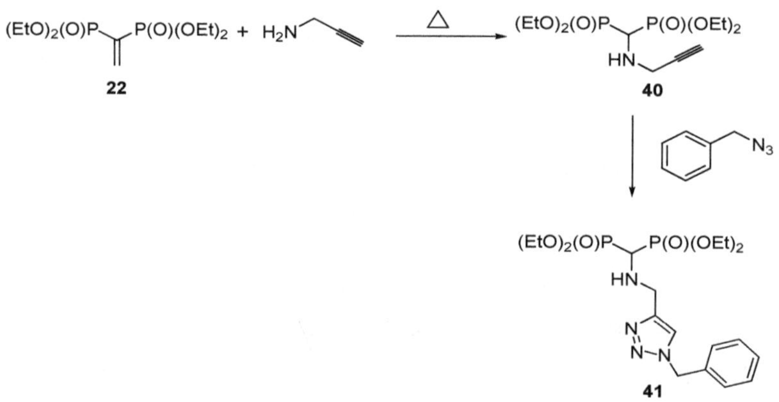

Scheme 20. An example of the synthesis of novel bisphosphonates via "click chemistry".

2.8. Miscellaneous Procedures

The desire to obtain new aminobisphosphonate scaffolds for biological studies has stimulated numerous studies on general methods of their synthesis. The methods discussed here are mostly designed to prepare specific scaffolds or are applicable only to specific, if not unusual, substrates. Only some of them may be considered as novel, general procedures.

Radical addition of sodium hypophosphite to terminal alkynes in the presence of triethylborane, which produced 1-alkyl-1,1-bis-*H*-phosphinates in moderate yields, gives access to a wide structural variety of bisphosphonates [115]. When using propargylamino acids, the corresponding bisphosphinates were obtained in satisfactory yields (a representative example is given in Scheme 21 for analogue **42** of pamidronate). Bisphosphinates are easily converted to bisphosphonates by ozonolysis.

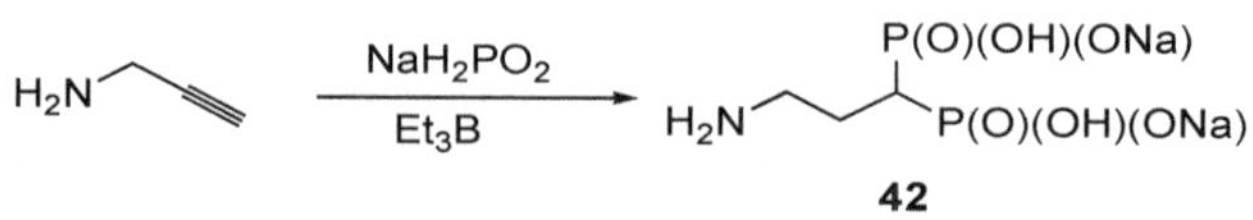

Scheme 21. Procedure for the synthesis of bisphosphinates and their conversion into bisphosphonates.

1-(*N*-acylamino)alkylphosphonates, easily accessible from *N*-acyl-α-amino acids using a two-step transformation, underwent electrophilic activation at the α-carbon by electrochemical α-methoxylation in methanol in a process mediated by NaCl. Attempts to carry out a Michaelis-Arbuzov-like reaction of the obtained diethyl 1-(*N*-acetylamino)-1-methoxy-alkylphosphonates with triethyl phosphite failed; however, they readily reacted with triphenylphosphine (Scheme 22) [116]. The resulting diethyl 1-(*N*-acetylamino)-1-triphenylphosphoniumalkylphosphonate tetrafluoroborates **43** reacted smoothly with trialkyl phosphites, dialkyl phosphonites or alkyl phosphinites in the presence of Hünig's base and methyltriphenylphosphonium iodide as catalysts. This gave bisphosphonates **44**, 1-phosphinylalkylphosphonates or 1-phosphinoylalkylphosphonates in good yields. This reaction has potential to become a general procedure.

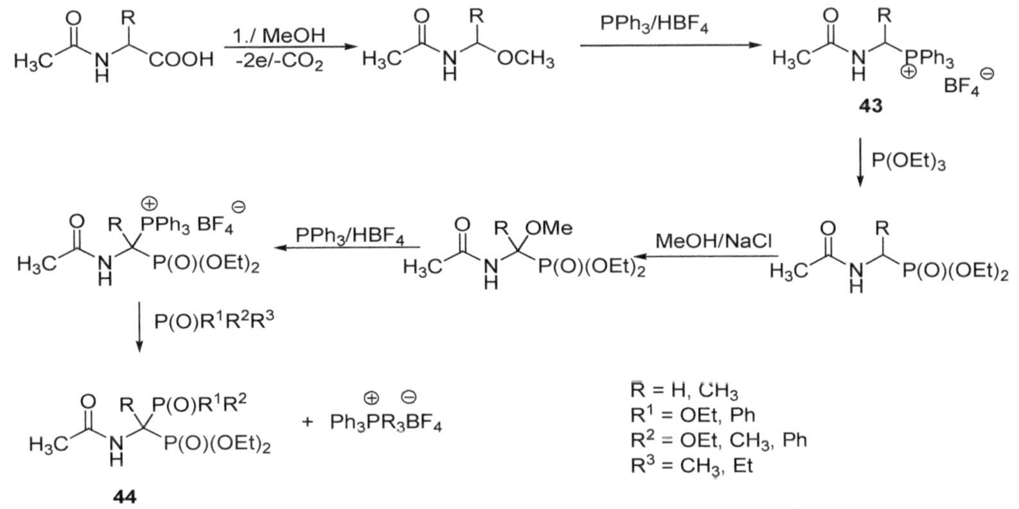

Scheme 22. Transformation of diethyl 1-(*N*-acetylamino)-1-alkylphosphonates into bisphosphoric acid esters via the corresponding phosphonium salts.

Alkylation of tetraalkyl methylenebisphosphonate **45**, a classical method for synthesizing variable bisphosphonic acids, has been rarely used for the synthesis of amino derivatives. Thus, its direct amination with the hydroxylamine ester of diphenylphosphinic acid, followed by its bromoacetylation (Scheme 23), provided substrate **46** for functionalization into glycopeptide antibiotics, in the hope that they will find an application as medication for osteomyelitis [117]. Another classical example is the monoalkylation of tetraisopropyl methylenebisphosphonate with 1,6-dibromohexane, followed by fluorination with *Selectofluor* and conversion of the remaining bromide into amine, which resulted in compound **47** (Scheme 23) [118].

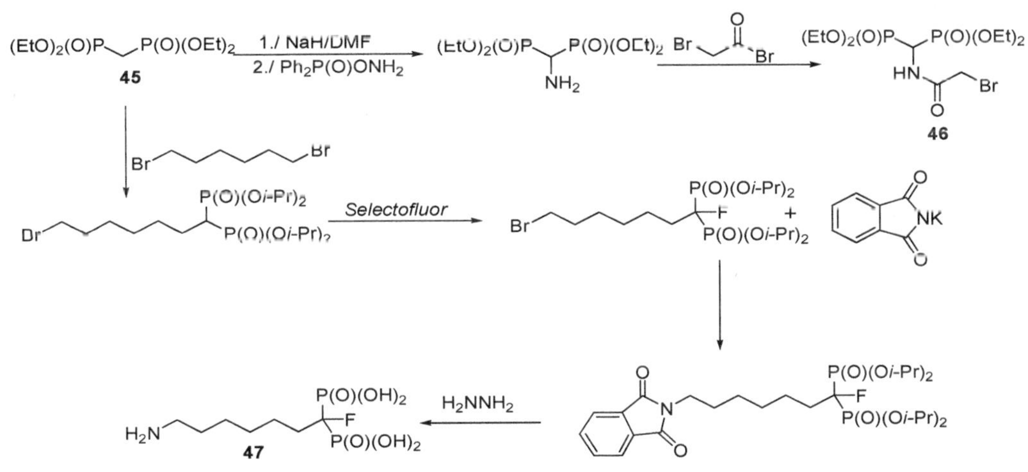

Scheme 23. Synthesis of aminobisphosphonates via alkylation of tetraalkyl methylenebisphosphonates.

An unusual procedure is metallacarbenoid insertion of aromatic amines into tetraethyl diphosphonodiazomethane **48**, which yields corresponding aminomethylenebisphosphonates **49** (Scheme 24) [119]. Among the tested catalysts, $Rh_2(NHCOCF_3)_4$ was found to be the best.

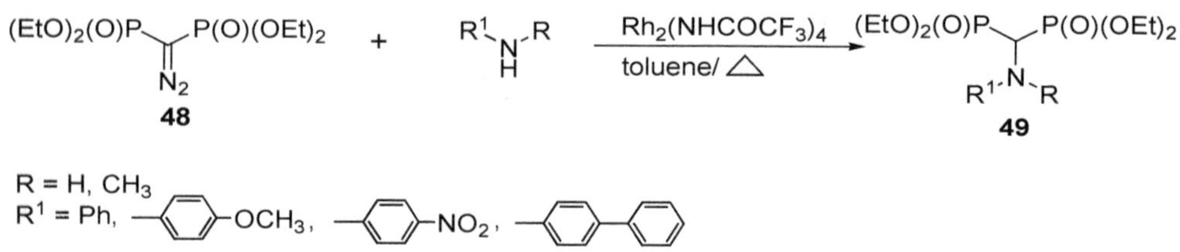

Scheme 24. N-H insertion of the reaction of aromatic amines.

A specific and unexpected reaction was the addition of silylated dialkyl phosphites to 4-phosphono-1-aza-1,3-dienes **50**, which resulted in γ-phosphono-α–aminobisphosphonates **51** (Scheme 25) [120]. This reaction is interesting in that, depending on the steric demand of the substituent present on nitrogen, double 1,2-addition or tandem 1,4-1,2-addition with formation of bisphosphonate **52** occurred (Scheme 25).

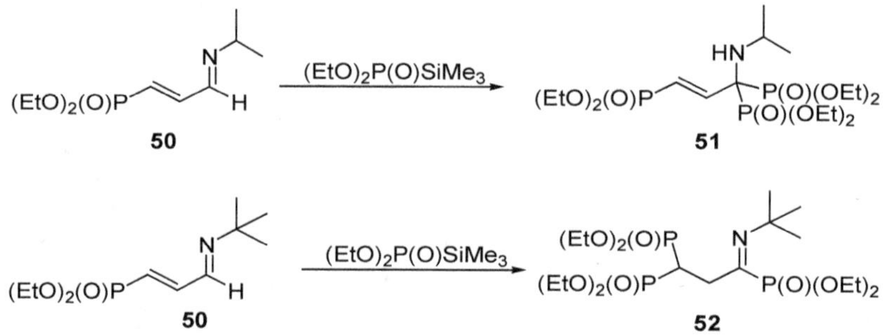

Scheme 25. Addition of phosphites to 4-phosphono-1-aza-1,3-dienes.

Additionally, the addition of phosphites to iminophosphonate esters has been studied. This reaction is limited to specific substrates, and usually the obtained bisphosphonates **53** are unstable, and phosphoryl C-N transfer to compounds **54** is observed (Scheme 26), which may be considered an example of an aza-Perkov reaction [121–123].

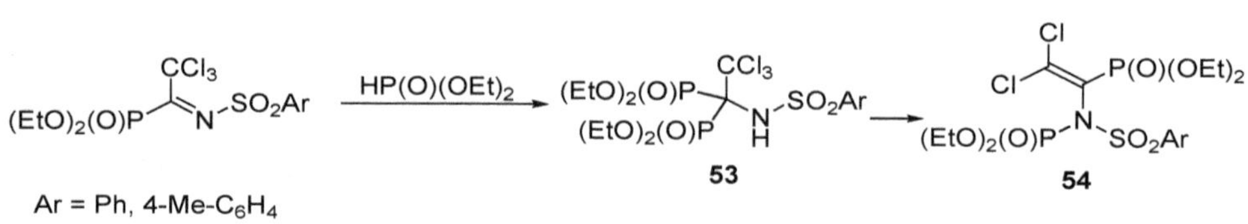

Scheme 26. Addition of phosphites to iminophosphonates.

Finally, syntheses of rare aminomethyltrisphosphonates are presented in a recent review by Romanenko and Kukhar devoted to applications of methylidynetrisphosphonates [124]. Two procedures for the synthesis of bis- and trisphosphonates **55** and **56** taken from this review and described in patent literature are depicted in Scheme 27.

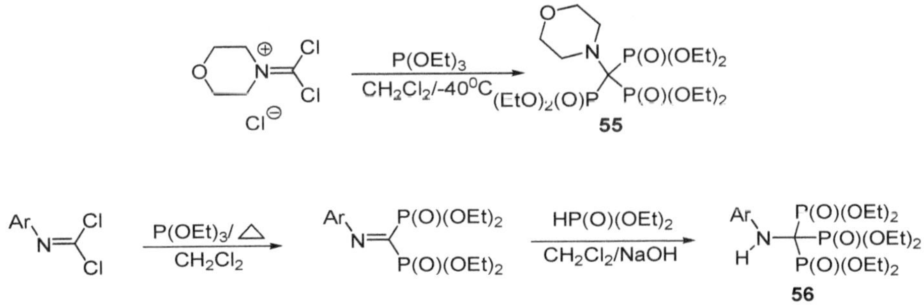

Scheme 27. Syntheses of aminomethyltrisphosphonates.

3. Functionalization of Aminobisphosphonates

Bisphosphonates are known for their affinity to bone tissue, and thus their conjugation to various drugs has been quite intensively studied; they are expected to serve as system-targeting drugs [6]. Because dronic acids are produced at industrial scale and are thus readily available and inexpensive, they are most frequently used for that purpose. However, their extreme hydrophilic character means that they are practically insoluble in most organic solvents, which limits the use of aqueous media or requires their conversion into phosphonate esters prior to functionalization. Unfortunately, the methods for direct esterification of phosphonate moieties are scarce and usually give unsatisfactory results; these esters must generally be synthesized independently.

Methods for functionalizing aminobisphosphonic acids with structurally variable molecules were recently reviewed [12]; therefore, in this review, the methods arbitrarily chosen as the most representative are reported.

3.1. Direct Acylation of Aminobisphosphonic Acids

Direct acylation of aminobisphosphonic acids is difficult because this reaction is accompanied by the possible competitive acylation of phosphonic groups [125,126] and hydroxylic groups when amino-1-hydroxy-1,1-bisphosphonic acids are substrates [126,127]. A representative example of this reaction is given in Scheme 28 for alendronic acid **57**.

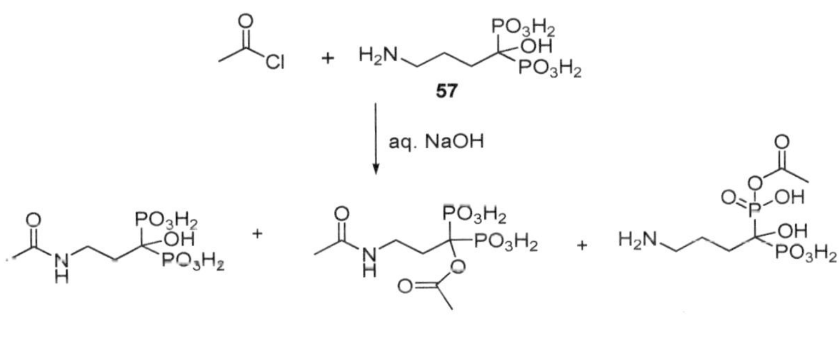

Scheme 28. Direct acylation of alendronic acid with indication of possible side-products.

Acylation of dronic acid salts with acyl chlorides in sodium hydroxide in water or in water/propanol solutions is the simple Schotten-Baumann variant [127,128]. Acidification of the solution results in the precipitation of the desired acids or their monosodium salts. To conjugate aminobisphosphonic acids with molecules bearing carboxylic groups, classical coupling protocols used in peptide synthesis have also been used. These include the activation of carboxylic groups with *N,N'*-dicyclohexylcarbodiimide (DCC) [129,130]; the use of previously prepared succinimidate esters to obtain 21 extremely complex fluorescent imaging probes (representative examples of a green fluorescent dye **58** and a red fluorescent dye **59** are shown in Scheme 29) [131]; and *N,N'*-dicarbonylimidazole, which was used to conjugate pamidronate to pullan (polysaccharide composed of maltotriose units) [132] to obtain a system for bone regeneration.

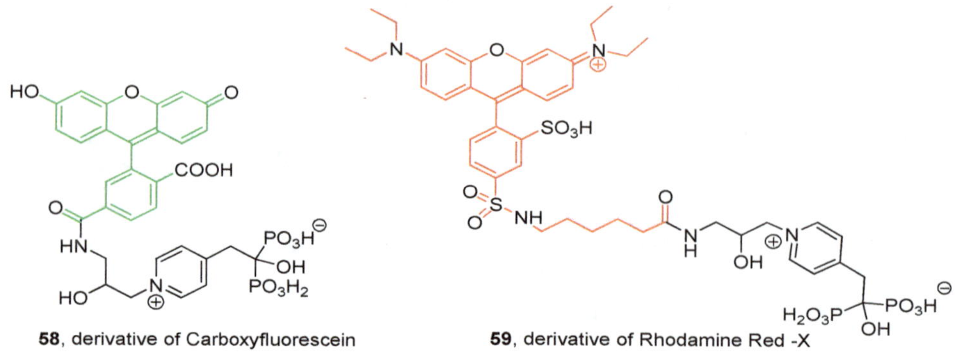

Scheme 29. Representative fluorescent imaging probes obtained via acylation of bisphosphonic acids (green—carboxyfluoresceine fragment, red—rhodamine Red-X fragment).

Acylation of aminobisphosphonic acids with small acids, such as acrylic acid [133], chloroacetic acid [134] or succinic acid [135], provided useful substrates for further functionalization of larger molecules, such as macrocycles devised for lanthanide ion complexation, chitosan, and hyaluronan.

3.2. Direct Acylation of Tetraethyl Aminobisphosphonates

Esters are far more suitable substrates for acylation than free phosphonic acids, with tetraethyl aminomethylenebisphosphonate being the most popular substrate. It has been used to obtain structurally variable conjugates with estradiol (compounds **60**, **61** and **62** in Scheme 30) using classical peptide synthesis coupling agents such as DCC and DPPA (diphenylphosphoryl azide) [136]. These conjugates were synthesized as bone-specific estrogens in the hope that they will protect elderly women from bone loss resulting from osteoporosis.

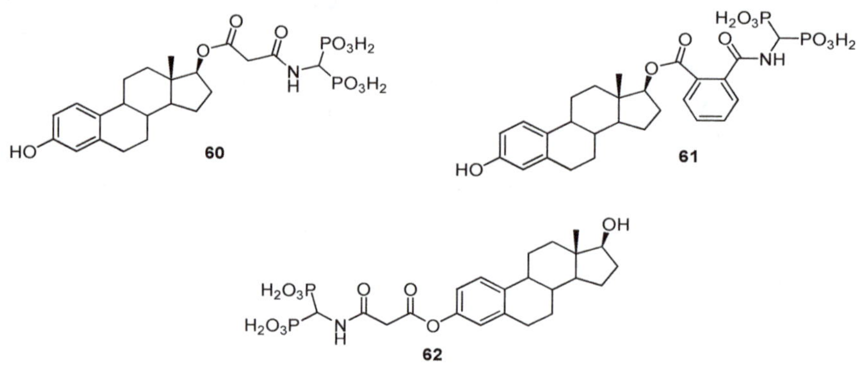

Scheme 30. Representative structures of bone-specific estrogens.

A derivative of raloxifene, a selective estrogen receptor or modulator, was obtained using a suitable acid chloride as substrate [137], whereas radioligands, which are selectively bound to bone tissue, have been synthesized by using DCC/HOBt (hydroxybenzotriazol) activation [138] and antibacterial bisphosphonated benzoxazinorifamycin prodrugs using EDCl (1-ethyl-3-(3-dimethylaminopropyl)carbodiimide) [139].

In this case, acylation of aminobisphosphonic acids with small acids was also applied as a starting step to produce antibacterial agents against osteomyelitis [88] and bone imaging macrocycles [81], using acid chlorides as acylating agents.

Using DCC as a coupling agent, diethyl 1-(2-aminoethylamino)-1,1-ethylbisphosphonate was acylated by structurally variable natural acids, such as folic acid [140,141], ursulonic and betulinic acids [142], and trolox [141,143].

Acylation of pamidronic acid ester **22** by using HBTU (2-(1*H*-benzotriazol-1-yl)-1,1,3,3-tetramethyluronium hexafluorophosphate) was used to prepare a catecholic derivative, which was then bound to the surface of magnetic iron oxide nanoparticles (Scheme 31) [144]. The particle **63** has been designed to remove uranyl ions from blood.

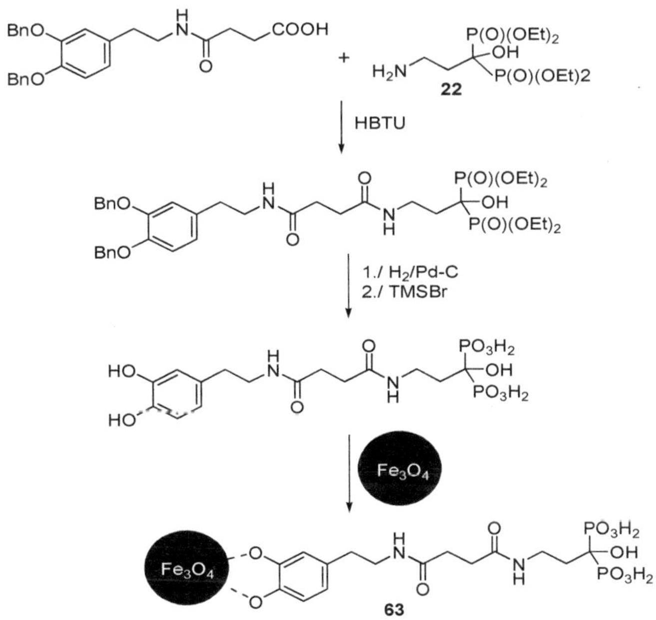

Scheme 31. Synthesis of bisphosphonate functionalized magnetic nanoparticles.

3.3. *Synthesis of Building Blocks for Polymer Chemistry*

Self-etching adhesives are polymeric materials containing phosphonate groups that have become popular in restorative dentistry because they allow strong bonds between dental hard tissues (enamel and dentin). One possibility for their preparation is to obtain monomers containing phosphonic groups [145]. Such methacrylamide monomers **64** and **65** were synthesized by acylation with suitable acryloylchlorides (Scheme 32) [146]. Studies of their photopolymerization indicated that they may be suitable for potential use in dentistry.

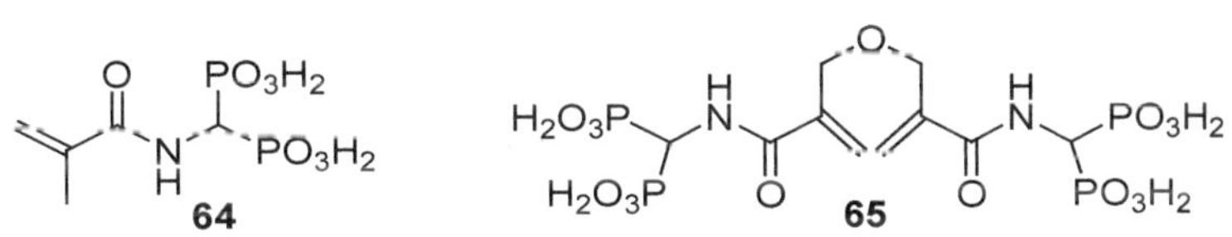

Scheme 32. Bisphosphonylated methacrylamide monomers.

A similar monomer, acryloylated pamidronate, has been used to obtain a hyaluronic acid derivative that is dually polymerized with cross-linkable hydrazide groups and bisphosphonate ligands. By mixing bisphosphonate polymer with calcium ions and aldehyde-derivatized hyaluronic acid, a hybrid hydrogel was obtained, which quickly mineralizes [147]. Such a system is of interest as a mediator for fast bone regeneration.

By dispersion copolymerization of three monomers (methacrylate bisphosphonate **66**, *N*-(3-aminopropyl) methacrylamide **67**, and tetra(ethylene glycol) diacrylate **68**) (Scheme 33), polymeric nanoparticles **69** were obtained [148]. Their size distribution was controlled by changing various polymerization parameters. By covalent attachment of a drug and/or a dye to amino groups (such as a near IR fluorescent dye), a theranostic system may be obtained.

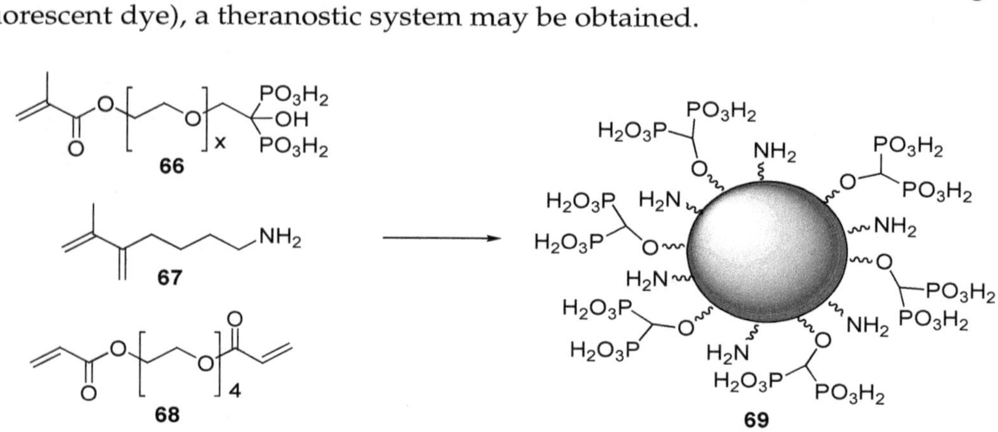

Scheme 33. Synthesis of polymeric nanoparticles as potential dye and drug carriers.

Macromolecular co-conjugates of bisphosphonate and ferrocene were synthesized by means of Michael addition copolymerization of methylenebisacrylamide (MBA) with primary amines-6-amino-1-hydroxyhexylidene-1,1-bisphosphonate and 4-ferrocenyl-butamidopropylamine [149]. The mass percentage incorporation of ferrocene analogues was found to be between 4%–5%, and 10%–12% for bisphosphonate. Such polymers could be selectively bound to bone tissue and slowly release anticancer ferrocene derivatives at this target site.

3.4. Miscellaneous

Other means to functionalize amino moieties are quite scarce and dispersed. Classical ones include the formation of Schiff bases, which are then alkylated [150] or reduced [126], and synthesis of thioureido derivatives as intermediates in the preparation of heterocyclic compounds [151–153].

The functionalization of polysaccharides is the gateway of aminobisphosphonates into nanoscience. For example, phosphonated cellulose was utilized to obtain nanocellulose with good thermal stability and potential intumescent properties. It was synthesized from birch pulp via sequential periodate oxidation and reductive amination using alendronate **57** as a phosphonating reagent (Scheme 34) [154]. After high-pressure homogenization, bisphosphonate cellulose nanofibers or nanocrystals of the general formula **70** were obtained, depending on the initial oxidation degree.

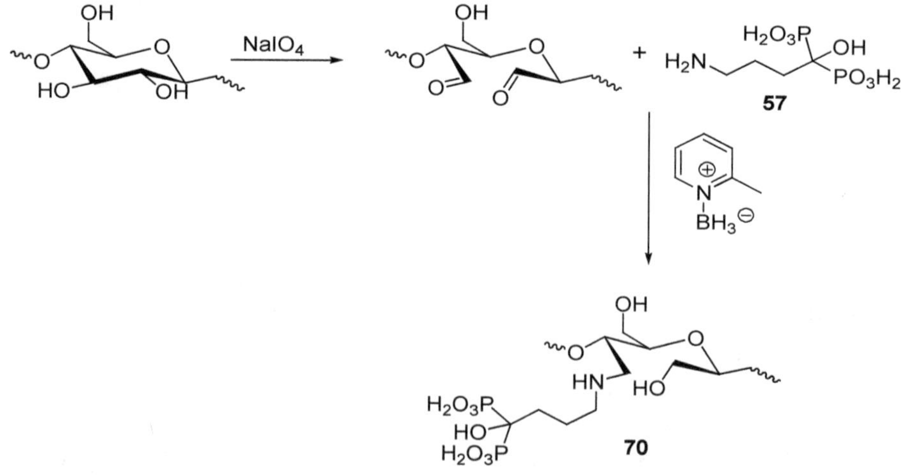

Scheme 34. Periodate oxidation of cellulose followed by reductive amination with sodium alendronate.

Alendronate was also bound to conjugates of pullulan and paclitaxel, which were bound to the polysaccharide by a cathepsin K-sensitive tetrapeptide spacer. This should ensure release of paclitaxel in bones. Then, bisphosphonate was covalently conjugated to the sugar chain via a polyethyleneglycol chain, using a technique similar to that described above (identical to those shown in Scheme 34) [155]. This system exhibited strong antiproliferative action against several cancer cell lines.

4. Conclusions

Aminobisphosphonic acids are gaining significant interest as a class of compounds with promising physiologic activity, with some representatives already commercialized as bone resorption inhibitors, and therefore useful drugs against osteoporosis and related bone disorders. This leads to both the modification of existing and the elaboration of novel procedures for their preparation. There are several commonly used reactions for this purpose; however, they have also been modified in the last decade to be tailored to specific biological needs. A few novel procedures have also been developed. Functionalization of simple aminobisphosphonic acids is a difficult task because of their strongly polar character. However, successful examples of functionalization of the amino moieties of these compounds have provided promising diagnostics and novel therapeutic agents.

Acknowledgments: This work was supported by a statuary activity subsidy from the Polish Ministry of Science and Higher Education.

References

1. Turhanen, P.A. Synthesis of triple bond containing 1-hydroxy-1,1-bisphosphonic acid derivatives to be used as precursors in "click" chemistry: Two examples. *J. Org. Chem.* **2014**, *79*, 6330–6335. [CrossRef] [PubMed]
2. Russell, R.G. Bisphosphonates: The first 40 years. *Bone* **2011**, *49*, 2–19. [CrossRef] [PubMed]
3. Ebetino, F.H.; Hogan, A.M.; Sun, S.; Tsuompra, M.K.; Duan, X.; Triffitt, J.T.; Kwaasi, A.A.; Dunford, J.E.; Barnett, B.L.; Oppermann, U.; et al. The relationship between the chemistry and biological activity of the bisphosphonates. *Bone* **2011**, *49*, 20–33. [CrossRef] [PubMed]
4. Maraka, S.; Kennel, K.A. Bisphosphonates for the prevention and treatment of osteoporosis. *Br. Med. J.* **2015**, *351*, h3783. [CrossRef] [PubMed]
5. Shi, C.G.; Zhang, Y.; Yuan, W. Efficacy of bisphosphonates on bone mineral density and fracture Rate in patients with osteogenesis imperfecta: A systematic review and meta-analysis. *Am. J. Ther.* **2016**, *3*, e894–e904. [CrossRef] [PubMed]
6. Chmielewska, E.; Kafarski, P. Physiologic Activity of Bisphosphonates—Recent Advances. *Open Pharm. Sci. J.* **2016**, *3*, 56–78. [CrossRef]
7. Demkowicz, S.; Rachoń, J.; Daśko, M.; Kozak, W. Selected organophosphorus compounds with biological activity. Applications in medicine. *RSC Adv.* **2016**, *6*, 7101–7112. [CrossRef]
8. Studnik, H.; Liebsch, S.; Forlani, G.; Wieczorek, D.; Kafarski, P.; Lipok, J. Amino polyphosphonates—Chemical features and practical uses, environmental durability and biodegradation. *New Biotechnol.* **2015**, *32*, 1–6. [CrossRef] [PubMed]
9. Turhanen, P.A.; Vepsäläinen, J.J.; Peräniemi, S. Advanced material and approach for metal ions removal from aqueous solutions. *Sci. Rep.* **2015**, *5*. [CrossRef] [PubMed]
10. Gałęzowska, J.; Gumienna-Kontecka, E. Phosphonates, their complexes and bio-applications: A spectrum of surprising diversity. *Coord. Chem. Rev.* **2012**, *256*, 105–124. [CrossRef]
11. Abdou, W.M.; Shaddy, A.A. The development of bisphosphonates for therapeutic uses, and bisphosphonate structure-activity consideration. *Arch. Org. Chem.* **2008**, *2009*, 143–182.
12. Romaneneko, V.D.; Kukhar, V.A. 1-Amino-1,1-bisphosphonates. Fundamental syntheses and new developments. *Arch. Org. Chem.* **2012**, *2012*, 127–166.
13. Petroianu, G.A. Pharmacist Theodor Salzer (1833–1900) and the discovery of bisphosphonates. *Pharmazie* **2011**, *66*, 804–807. [PubMed]

14. Kieczykowski, G.R.; Jobson, R.B.; Melillo, D.G.; Reinhold, D.F.; Grenda, V.J.; Shinkai, I. Preparation of (4-amino-1-hydroxybutylidene)bisphosphonic acid sodium salt, MK-217 (alendronate sodium). An improved procedure for the preparation of 1-hydroxy-1,1-bisphosphonic acids. *J. Org. Chem.* **1995**, *60*, 8310–8312. [CrossRef]
15. Teixeira, F.C.; Antunes, I.F.; Curto, M.J.M.; Neves, M.; Teixeira, A.P.S. Novel 1-hydroxy-1,1-bisphosphonates derived from indazole: Synthesis and characterization. *Arch. Org. Chem.* **2009**, *2009*, 69–84.
16. Yanvarev, D.V.; Koovina, A.N.; Usanov, N.N.; Kochetkov, S.N. Non-hydrolysable analogues of inorganic pyrophosphate as inhibitors of Hepatitis C virus RNA-dependent RNA polymerase. *Russ. J. Bioorg. Chem.* **2012**, *38*, 224–229. [CrossRef]
17. Srinivasa Rao, D.V.N.; Dandala, R.; Narayan, G.K.A.S.S.; Lenin, R.; Sivakumaran, N.; Naidu, A. Novel procedure for the synthesis of 1-hydroxy-1,1-bisphosphonic acids using phenols as medium. *Synth. Commun.* **2007**, *37*, 4359–4365. [CrossRef]
18. Grün, A.; Nagy, D.I.; Németh, O.; Mucsi, Z.; Garadnay, S.; Greiner, I.; Keglevich, G. The Synthesis of 3-Phenylpropidronate Applying Phosphorus Trichloride and Phosphorous Acid in Methanesulfonic Acid. *Curr. Org. Chem.* **2016**, *20*, 1745–1752. [CrossRef]
19. Lecouvey, M.; Leroux, I. Synthesis of 1-hydroxy-1,1-bisphosphonates. *Heteroat. Chem.* **2000**, *11*, 556–561. [CrossRef]
20. Keglevich, G.; Grün, A.; Kovács, R.; Garadnay, S.; Greiner, I. Green chemical synthesis of bisphosphonic/dronic derivatives. *Phosphorus Sulfur Silicon Relat. Elem.* **2015**, *190*, 664–667. [CrossRef]
21. Troev, K.; Todorov, P.; Naydenova, E.; Mitova, V.; Vassiliev, N. A study of the reaction of phosphorus trichloride with paraformaldehyde in the presence of carboxylic acids. *Phosphorus Sulfur Silicon Relat. Elem.* **2013**, *188*, 1147–1155. [CrossRef]
22. Nagy, D.I.; Grün, A.; Garadnay, S.; Greiner, I.; Keglevich, G. Synthesis of hydroxymethylenebisphosphonic acid derivatives in different solvents. *Molecules* **2016**, *21*, 1046. [CrossRef] [PubMed]
23. Kachbi Khelfallah, S.; Monteil, M.; Deschamp, J.; Gager, O.; Migianu-Griffoni, E.; Lecouvey, M. Synthesis of novel polymerisable molecules bearing bisphosphonate. *Org. Biomol. Chem.* **2015**, *13*, 11382–11392. [CrossRef] [PubMed]
24. Grün, A.; Kovács, R.; Nagy, D.L.; Garadnay, S.; Greiner, I.; Keglevich, G. Efficient synthesis of benzidronate applying of phosphorus trichloride and phosphorous acid. *Lett. Drug Des. Discov.* **2015**, *12*, 78–84. [CrossRef]
25. Keglevich, G.; Grün, A.; Aradi, K.; Garadnay, S.; Greiner, I. Optimized synthesis of *N*-heterocyclic dronic acids; closing a black-box era. *Tetrahedron Lett.* **2011**, *21*, 2744–2746. [CrossRef]
26. Keglevich, G.; Grün, A.; Garadnay, S.; Greiner, I. Rational synthesis of dronic acid derivatives. *Phosphorus Sulfur Silicon Relat. Elem.* **2015**, *190*, 2116–2124. [CrossRef]
27. Ratrout, S.S.; Al Sarabi, A.L.; Sweidan, K.A. A one-pot and efficient synthesis of zolendronic acid starting from tert-butyl imodazol-1-yl acetate. *Pharm. Chem. J.* **2015**, *48*, 837–841. [CrossRef]
28. Mizrahi, D.A.; Waner, T.; Segall, Y. α-Amino acid derived bisphosphonates: Synthesis and anti-resorptive activity. *Phosphorus Sulfur Silicon Relat. Elem.* **2001**, *173*, 1–25. [CrossRef]
29. Lenin, R.; Raju, M.R.; Srinisava Rao, D.V.N.; Ray, U.K. Microvawe-assistant efficient synthesis of bisphosphonate libraries: A useful procedure for the preparation of bisphosphonates containing nitrogen and sulfur. *Med. Chem. Res.* **2013**, *22*, 1624–1629. [CrossRef]
30. Roth, A.G.; Drescher, A.; Yang, Y.; Redmer, S.; Uhling, S.; Arenz, C. Potent and selective inhibition of acid sphingomyelinase by bisphosphonates. *Angew. Chem. Int. Ed.* **2009**, *48*, 7560–7563. [CrossRef] [PubMed]
31. Szajnman, S.H.; Ravaschino, E.L.; Docampo, R.; Rodrigues, J.B. Synthesis and biological evaluation of 1-amino-1,1-bisphosphonates derived from fatty acids against *Trypanosoma cruzi* targeting farnesyl pyrophosphate synthase. *Bioorg. Med. Chem. Lett.* **2005**, *15*, 4685–4690. [CrossRef] [PubMed]
32. Wu, M.; Chen, P.; Huang, Y. Convenient synthesis of analogs of aminomethylene gem-diphosphonic acid from amines without catalyst. *Synth. Commun.* **2004**, *34*, 1393–1398. [CrossRef]
33. Fukuda, M.; Okamoto, Y.; Sakurai, H. Synthesis of dialkylaminomethylenediphosphonic acids. *Bull. Chem. Soc. Jpn.* **1975**, *48*, 1030–1031. [CrossRef]
34. Yu, C.M.; Wang, B.; Chen, Z.W. A Novel synthesis of alkylamino substituted methylenediphosphonates using bis(trichloromethyl) carbonate and $RCONR^1R^2$. *Chin. J. Chem.* **2008**, *26*, 1899–1901. [CrossRef]
35. Plöger, W.; Schindler, N.; Wollmann, K.; Worms, K.H. Herstellung von 1-Aminoalkan-1,1-diphosphonsäuren. *Z. Anorg. Allg. Chem.* **1972**, *389*, 119–128. [CrossRef]

36. Olive, G.; le Moigne, F.; Mercier, A.; Rockenbauer, A.; Tordo, P. Synthesis of tetraalkyl (pyrrolidine-2,2-diyl) bisphosphonates and 2,2-bis(diethoxyphosphoryl)-3,4-dihydro-2*H*-pyrrole 1-oxide; ESR study of derived nitroxides. *J. Org. Chem.* **1998**, *63*, 9095–9099. [CrossRef]
37. Olive, G.; Jacques, A. Tetraethyl(pyrrolidine-2,2-diyl)bisphosphonate. *Molecules* **2001**, *6*, M275. [CrossRef]
38. Olive, G.; Jacques, A. Optimization, continuation and lack of the one-step diphosphorylation reaction assay of modification of the tetraethyl(pyrrolidine-2,2-diyl)bisphosphonate. *Phosphorus Sulfur Silicon Relat. Elem.* **2003**, *178*, 33–46. [CrossRef]
39. Olive, G.; Rockenbauer, A.; Rozanska, X.; Jacques, A.; Peeters, D.; German, A. Synthesis of new tetraethyl(*N*-alkyl-1-aminoethan-1,1-diyl)bisphosphonates and ESR analysis of chemical exchange of derived nitroxides of acyclic aminobisphosphonates. *Phosphorus Sulfur Silicon Relat. Elem.* **2008**, *182*, 2359–2369. [CrossRef]
40. Wang, A.-E.; Chang, Z.; Sun, W.-T.; Huang, P.-Q. General and chemoselective bisphosphonylation of secondary and tertiary amides. *Org. Lett.* **2015**, *17*, 732–735. [CrossRef] [PubMed]
41. Prishchenko, A.A.; Livantsov, N.V.; Novikova, O.P.; Livantsova, L.I.; Petrosyan, V.S. Synthesis of the new types of *N*-substituted aminomethylenebisorganophosphorus acids and their derivatives. *Heteroatom Chem.* **2009**, *20*, 319–324. [CrossRef]
42. Prishchenko, A.A.; Livantsov, N.V.; Novikova, O.P.; Livantsova, L.I.; Ershov, I.R.; Petrosyan, V.S. Synthesis and reactivity of the new trimethylsilyl esters of aminomethylenebisorganophosphorus acids. *Heteroatom Chem.* **2013**, *24*, 355–360. [CrossRef]
43. Prishchenko, A.A.; Livantsov, N.V.; Novikova, O.P.; Livantsova, L.I.; Petrosyan, V.S. Synthesis of aminomethylenediphosphonates and their derivatives containing $PCNH_2$ fragments. *J. Gen. Chem.* **2014**, *83*, 608–610. [CrossRef]
44. Chmielewska, E.; Miszczyk, P.; Kozłowska, J.; Prokopowicz, M.; Młynarz, P.; Kafarski, P. Reaction of benzolactams with triethyl phosphite prompted by phosphoryl chloride affords benzoannulated monophosphonates instead of expected bisphosphonates. *J. Organomet. Chem.* **2015**, *785*, 84–91. [CrossRef]
45. Prishchenko, A.A.; Livantsov, N.V.; Novikova, O.P.; Livantsova, L.I.; Ershov, I.R.; Petrosyan, V.S. Synthesis of new types of aminomethylenediphosphorus-containing acids and their derivatives. *Russ. J. Gen. Chem.* **2015**, *85*, 370–379. [CrossRef]
46. Du, Y.; Jung, K.Y.; Wimer, D.F. A one-flask synthesis of α,α-bisphosphonates via enolate chemistry. *Tetrahedron Lett.* **2002**, *43*, 8665–8668. [CrossRef]
47. Srinivasa Rao, D.V.N.; Dandala, R.; Lenin, R.; Sivakumaran, N.; Shivashankar, S.; Naidu, A. A facile one pot synthesis of bisphosphonic acids and their sodium salts from nitriles. *Arch. Org. Chem.* **2007**, *2007*, 34–38.
48. Kaabak, L.V.; Kuz'mina, N.E.; Khudneko, A.V.; Tomilov, A.P. Improved synthesis of 1-aminoethylidenediphosphonic acid. *Russ. J. Gen. Chem.* **2006**, *76*, 1673–1674. [CrossRef]
49. Bandurina, T.A.; Konyukhov, V.N.; Panomareva, O.A.; Barybin, O.S.; Pushkareva, Z.N. Synthesis and antitumor activity of aminophosphonic acids. *Pharm. Chem. J.* **1978**, *12*, 1428–1431. [CrossRef]
50. Midier, C.; Lantsoght, M.; Volle, J.-N.; Pirat, J.L.; Virieux, D.; Stevens, C.V. Hydrophosphonylation of alkenes or nitriles by double radical transfer mediated by titanocene/propylene oxide. *Tetrahedron Lett.* **2011**, *52*, 6693–6696. [CrossRef]
51. Váradi, A.; Palmer, T.C.; Notis Dardashti, R.; Majumdar, S. Isocyanide-based multicomponent reactions for the synthesis of heterocycles. *Molecules* **2016**, *21*, 19. [CrossRef] [PubMed]
52. Hirai, T.; Han, L.-B. Palladium-catalyzed insertion of isocyanides into P(O)–H bonds: Selective formation of phosphinoyl imines and bisphosphinoylaminomethanes. *J. Am. Chem. Soc.* **2006**, *128*, 7422–7423. [CrossRef] [PubMed]
53. Goldemen, W.; Kluczyński, A.; Soroka, M. The preparation of *N*-substituted aminomethylidenebisphosphonates and their tetraalkyl esters via reaction of isonitriles with trialkyl phosphites and hydrogen chloride. Part 1. *Tetrahedron Lett.* **2012**, *53*, 5290–5292.
54. Goldeman, W.; Nasulewicz-Goldeman, A. Synthesis and antiproliferative activity of aromatic and aliphatic bis(aminomethylidene(bisphosphonic)) acids. *Bioorg. Med. Chem. Lett.* **2014**, *24*, 3475–3479. [CrossRef] [PubMed]
55. Kurzak, B.; Goldeman, W.; Szpak, M.; Matczak-Jon, E.; Kamecka, E. Synthesis of *N*-methyl alkylaminomethane-1,1-diphosphonic acids and evaluation of their complex-formation abilities towards copper(II). *Polyhedron* **2015**, *85*, 675–684. [CrossRef]

56. Goldeman, W.; Nasulewicz-Goldeman, A. Synthesis and biological evaluation of aminomethylidenebisphosphonic derivatives of β-arylethylamines. *Tetrahedron* **2015**, *71*, 3282–3289. [CrossRef]
57. Bhushan, K.R.; Tanaka, E.; Frangioni, J.V. Synthesis of conjugatable bisphosphonates for molecular imaging of large animals. *Angew. Chem. Int. Ed.* **2007**, *46*, 7969–7971. [CrossRef] [PubMed]
58. Ziora, Z.; Maly, A.; Lejczak, B.; Kafarski, P.; Holband, J.; Wójcik, G. Reactions of *N*-phthalylamino acid chlorides with trialkyl phosphites. *Heteroatom Chem.* **2000**, *11*, 232–239. [CrossRef]
59. Yanvarev, D.V.; Korovina, A.N.; Usanov, N.N.; Khomich, O.A.; Vepsäläinen, J.; Puljula, E.; Kukhanova, M.K.; Kochetkov, S.N. Data on synthesis of methylene bisphosphonates and screening of their inhibitory activity towards HIV reverse transcriptase. *Data Brief* **2016**, *8*, 1157–1167. [CrossRef] [PubMed]
60. Guenin, E.; Degache, E.; Liquier, J.; Lecouvey, M. Synthesis of 1-hydroxymethylene-1,1-bis(phosphonic acids) from acid anhydrides: Preparation of a new cyclic 1-acyloxymethylene-1,1-bis(phosphonic acid). *Eur. J. Org. Chem.* **2004**, 2983–2987. [CrossRef]
61. Kachbi-Khelfallah, S.; Monteil, M.; Cortes-Clerget, M.; Migianu-Griffoni, E.; Pirat, J.-L.; Gager, O.; Deschamp, J.; Lecouvey, M. Towards potential nanoparticle contrast agents: Synthesis of new functionalized PEG bisphosphonates. *Beilstein J. Org. Chem.* **2016**, *12*, 1366–1371. [CrossRef] [PubMed]
62. Teixeira, F.C.; Rangel, C.M.; Teixeira, A.P.S. Synthesis of new azole phosphonate precursors for fuel cells proton exchange membranes. *Heteroatom Chem.* **2015**, *26*, 238–248. [CrossRef]
63. Teixeira, F.C.; Lucas, C.; Curto, M.J.M.; Neves, M.; Duarte, M.T.; André, V.; Teixeira, A.P.S. New 1-Hydroxy-1,1-bisphosphonates derived from 1*H*-pyrazolo[3,4-*b*]pyridine: Synthesis and characterization. *J. Braz. Chem. Soc.* **2013**, *24*, 1295–1306. [CrossRef]
64. Kaboudin, B.; Ezzati, A.; Faghihi, M.R.; Barati, A.; Kazemi, F.; Abdollahi, H.; Yokomatsu, T. Hydroxy-bisphosphinic acids: Synthesis and complexation properties with transition metals and lanthanide ions in aqueous solution. *J. Iran. Chem. Soc.* **2016**, *13*, 747–752. [CrossRef]
65. Crossey, K.; Migaud, M.E. Solventless synthesis of acyl phosphonamidates, precursors to masked bisphosphonates. *Chem. Commun.* **2015**, *51*, 11088–11091. [CrossRef] [PubMed]
66. Egorov, M.; Aoun, S.; Padrines, M.; Redini, F.; Heymann, D.; Lebreton, J.; Mathé-Allainmat, M.A. One-pot synthesis of 1-hydroxy-1,1-bis(phosphonic acid)s starting from the corresponding carboxylic acids. *Eur. J. Org. Chem.* **2011**, 7148–7154. [CrossRef]
67. Suzuki, F.; Fujikawa, Y.; Yamamoto, S.; Mizutani, H.; Funabashi, C.; Ohya, T.; Ikai, T.; Oguchi, T. Pharmaceutical Compositions Containing Geminal Diphosphonates. US 5583122 A, 1 February 1979.
68. Maier, L. Organishe phosphorverbindungen 75: Herstellung und Eigenschaften von Aminomethylendiphosphinaten und -diphosphonaten, $RR^1NCH[P(O)R^2(OR^3)]_2$ und Derivaten. *Phosphorus Sulfur Silicon Relat. Elem.* **1981**, *11*, 311–332. [CrossRef]
69. Kaboudin, B.; Kalipour, S. A microwave-assisted solvent- and catalyst-free synthesis of aminomethylene bisphosphonates. *Tetrahedron Lett.* **2009**, *50*, 4243–4245. [CrossRef]
70. Minaeva, L.; Patrikeeva, L.S.; Kabachnik, M.M.; Beletskaya, I.P.; Orlinson, B.S.; Novakov, I.A. Synthesis of novel aminomethylenebisphosphonates and bisphosphonic acids, containing adamantyl fragment. *Heteroatom Chem.* **2011**, *22*, 55–58. [CrossRef]
71. Bálint, E.; Tajti, A.; Dziełak, A.; Hägele, G.; Keglevich, G. Microwave-assisted synthesis of (aminomethylene) bisphosphine oxides and (aminomethylene)bisphosphonates by a three-component condensation. *Beilstein J. Org. Chem.* **2016**, *12*, 1493–1502. [CrossRef] [PubMed]
72. Reddy, N.B.; Sundar, C.S.; Krishna, B.S.; Reddy, K.M.K.; Reddy, C.S. Synthesis and in-vitro antimicrobial activity of aminomethylene bisphosphonates. *Iran. J. Org. Chem.* **2014**, *6*, 1227–1234.
73. Krutikov, V.I.; Erkin, A.V.; Pautov, P.A.; Zolotukhina, M.M. Heteryl- and arylaminomethylenebisphosphonates: Synthesis and biologic activity. *Russ. J. Gen. Chem.* **2003**, *73*, 187–191. [CrossRef]
74. Siva Prasad, S.; Jayaprakash, S.H.; Syamasundar, Ch.; Sreelakshmi, P.; Bhuvaneswar, C.; Vijaya Bhaskar, B.; Rajendra, W.; Nayak, S.K.; Suresh Reddy, C. Tween 20-/H_2O promoted green synthesis, computational and antibacterial activity of amino acid substituted methylene bisphosphonates. *Phosphorus Sulfur Silicon Relat. Elem.* **2015**, *190*, 2040–2050. [CrossRef]
75. Mimura, M.; Hayashida, M.; Nomiyama, K.; Ikegami, S.; Iida, Y.; Tamura, M.; Hiyama, Y.; Ohishi, Y. Synthesis and evaluation of (piperidinomethylene)bis(phosphonic acid) derivatives as anti-osteoporosis agents. *Chem. Pharm. Bull.* **1993**, *41*, 1973–1986. [CrossRef]

76. De Schutter, J.W.; Shaw, J.; Lin, Y.-S.; Tsantrizos, Y.S. Design of potent bisphosphonate inhibitors of the human farnesyl pyrophosphate synthase via targeted interactions with the active site "capping" phenyls. *Bioorg. Med. Chem.* **2012**, *20*, 5583–5591. [CrossRef] [PubMed]
77. Leung, C.Y.; Langille, A.M.; Mancuso, J.; Tsantrizos, Y.S. Discovery of thienopyrimidine-based inhibitors of the human farnesyl pyrophosphate synthase—Parallel synthesis of analogs via a trimethylsilyl ylidene intermediate. *Bioorg. Med. Chem.* **2013**, *21*, 2229–2240. [CrossRef] [PubMed]
78. Leung, C.Y.; Park, J.; de Schutter, J.W.; Sebag, M.; Berghuis, A.M.; Tsantrizos, Y.S. Thienopyrimidine bisphosphonate (ThPBP) inhibitors of the human farnesyl pyrophosphate synthase: Optimization and characterization of the mode of inhibition. *J. Med. Chem.* **2013**, *56*, 7939–7950. [CrossRef] [PubMed]
79. Rubino, M.T.; Agamennone, M.; Campestre, C.; Campiglia, P.; Cremasco, V.; Faccio, R.; Laghezza, A.; Loiodice, F.; Maggi, D.; Panza, E.; et al. Biphenyl sulfonylamino methyl bisphosphonic acids as inhibitors of matrix metalloproteinases and bone resorption. *ChemMedChem* **2011**, *6*, 1258–1268. [CrossRef] [PubMed]
80. Tauro, M.; Lagezza, A.; Loiodice, F.; Agamennone, M.; Campestre, C.; Tortorella, P. Arylamino methylene bisphosphonate derivatives as bone seeking matrix metalloproteinase inhibitors. *Bioorg. Med. Chem.* **2013**, *21*, 6456–6465. [CrossRef] [PubMed]
81. Chmielewska, E.; Mazur, Z.; Kempińska, K.; Wietrzyk, J.; Piątek, A.; Kuryszko, J.J.; Kiełbowicz, Z.; Kafarski, P. *N*-Arylaminomethylenebisphosphonates bearing fluorine atoms: Synthesis and antiosteoporotic activity. *Phosphorus Sulfur Silicon Relat. Elem.* **2016**, *190*. [CrossRef]
82. Chmielewska, E.; Kempińska, K.; Wietrzyk, J.; Piątek, A.; Kuryszko, J.J.; Kiełbowicz, Z.; Kafarski, P. Novel Bisphosphonates and Their Use. WO2015159153 A1, 22 October 2015.
83. Kubíček, V.; Rudovský, J.; Kotek, J.; Hermann, P.; Elst, L.V.; Muller, R.N.; Kolar, Z.I.; Wolterbeek, H.Th.; Peters, J.A.; Lukeš, I. A Bisphosphonate monoamide analogue of DOTA: A potential agent for bone targeting. *J. Am. Chem. Soc.* **2005**, *127*, 16477–16485. [CrossRef] [PubMed]
84. Årstad, E.; Hoff, P.; Skattebøl, L.; Skretting, A.; Breistøl, K. Studies on the synthesis and biological properties of non-carrier-added [^{125}I and ^{131}I]-labeled arylalkylidenebisphosphonates: Potent bone-seekers for diagnosis and therapy of malignant osseous lesions. *J. Med. Chem.* **2003**, *46*, 3021–3032. [CrossRef] [PubMed]
85. Martin, M.B.; Sanders, J.M.; Kendrick, H.; de Luca-Fradley, K.; Lewis, J.C.; Grimley, J.S.; van Brussel, E.M.; Olsen, J.R.; Meints, G.A.; Burzyńska, A.; et al. Activity of Bisphosphonates against *Trypanosoma brucei rhodesiense*. *J. Med. Chem.* **2002**, *45*, 2904–2914. [CrossRef] [PubMed]
86. Sanders, J.M.; Gomez, A.O.; Mao, J.; Meints, G.A.; van Brussel, E.M.; Burzyńska, A.; Kafarski, P.; Gonzalez-Pacanowska, D.; Oldfield, E. 3-D QSAR investigations of the inhibition of *Leishmania major* farnesyl pyrophosphate synthase by bisphosphonates. *J. Med. Chem.* **2003**, *46*, 5171–5183. [CrossRef] [PubMed]
87. Ghosh, S.; Chan, J.M.; Lea, C.R.; Meints, G.A.; Lewis, J.C.; Tovian, Z.S.; Flessner, R.M.; Loftus, T.C.; Bruchhaus, I.; Kendrick, H.; et al. Effects of bisphosphonates on the growth of *Entamoeba histolytica* and Plasmodium species in vitro and in vivo. *J. Med. Chem.* **2004**, *47*, 175–187. [CrossRef] [PubMed]
88. Ling, Y.; Sahota, D.; Odeh, S.; Chan, J.M.; Araujo, F.G.; Moreno, S.N.; Oldfield, E. Bisphosphonate inhibitors of Toxoplasma gondi growth: In vitro, QSAR, and in vivo investigations. *J. Med. Chem.* **2005**, *48*, 3130–3140. [CrossRef] [PubMed]
89. Tanaka, K.S.E.; Houghton, T.J.; Kang, T.; Dietrich, E.; Delorme, D.; Ferreira, S.S.; Caron, L.; Viens, F.; Arhin, S.S.; Sarmiento, I.; et al. Bisphosphonated fluoroquinolone esters as osteotropic prodrugs for the prevention of osteomyelitis. *Bioorg. Med. Chem.* **2008**, *16*, 9217–9229. [CrossRef] [PubMed]
90. Forlani, G.; Petrollino, D.; Fusetti, M.; Romanini, L.; Nocek, B.; Joachimiak, A.; Berlicki, L.; Kafarski, P. δ^1-Pyrroline-5-carboxylate reductase as a new target for therapeutics: Inhibition of the enzyme from *Streptococcus pyogenes* and effects in vivo. *Amino Acids* **2012**, *42*, 2283–2291. [CrossRef] [PubMed]
91. Brel, V.K. Synthesis of *gem*-bisphosphonates with (3-aryl-4,5-dihydroxisoxazol-5-yl)methylamino moiety. *Mendeleev Commun.* **2015**, *25*, 234–235. [CrossRef]
92. Kosikowska, P.; Bochno, M.; Macegoniuk, K.; Forlani, G.; Kafarski, P.; Berlicki, Ł. Bisphosphonic acids as effective inhibitors of *Mycobacterium tuberculosis* glutamine synthetase. *J. Enzyme Inhib. Med. Chem.* **2016**, *31*, 931–938. [CrossRef] [PubMed]
93. Lacbay, C.M.; Mancuso, J.; Lin, Y.-S.; Bennett, N.; Götte, M.; Tsantrizos, Y.S. Modular assembly of purine-like bisphosphonates as inhibitors of HIV-1 reverse transcriptase. *J. Med. Chem.* **2014**, *57*, 7435–7449. [CrossRef] [PubMed]

94. Shaddy, A.A.; Kamel, A.A.; Abdou, W.M. Synthesis, quantitative structure-activity relationship, and anti-inflammatory profiles of substituted 5- and 6-*N*-heterocycle bisphosphonate esters. *Chem. Commun.* **2013**, *43*, 236–252. [CrossRef]
95. Kafarski, P.; Lejczak, B.; Forlani, G. Herbicidally active aminomethylenebisphosphonic acids. *Heteroat. Chem.* **2000**, *11*, 449–453. [CrossRef]
96. Forlani, G.; Berlicki, Ł.; Duò, M.; Dzięździoła, G.; Giberti, S.; Bertazzini, M.; Kafarski, P. Synthesis and Evaluation of Effective Inhibitors of Plant δ^1-Pyrroline-5-carboxylate Reductase. *J. Agric. Food Chem.* **2013**, *61*, 6792–6798. [CrossRef] [PubMed]
97. Giberti, S.; Bertazzini, M.; Liboni, M.; Berlicki, Ł.; Kafarski, P.; Forlani, G. Phytotoxicity of aminobisphosphonates targeting both δ1-pyrroline-5-carboxylate reductase and glutamine synthetase. *Pest Manag. Sci.* **2016**, in press. [CrossRef] [PubMed]
98. Dobosz, A.; Spychała, J.; Ptak, T.; Chmielewska, E.; Maciejewska, G.; Kafarski, P.; Młynarz, P. Electrochemical and spectroscopic investigations of selected *N*-heteroalkylaminomethylene-bisphosphonic acids with Pb(II) ions. *Coord. Chem. Rev.* **2016**, in press. [CrossRef]
99. Dąbrowska, E.; Burzyńska, A.; Mucha, A.; Matczak-Jon, E.; Sawka-Dobrowolska, W.; Berlicki, Ł.; Kafarski, P. Insight into the mechanism of three component condensation leading to aminomethylenebisphosphonates. *J. Organomet. Chem.* **2009**, *6943*, 3806–3813. [CrossRef]
100. Lejczak, B.; Boduszek, B.; Kafarski, P.; Forlani, G.; Wojtasek, H.; Wieczorek, P. Mode of action of herbicidal derivatives of aminomethylenebisphosphonic acid. I. Physiologic activity and inhibition of anthocyanin biosynthesis. *J. Plant Growth Regul.* **1996**, *15*, 109–113. [CrossRef]
101. Bochno, M.; Berlicki, Ł. A three-component synthesis of aminomethylene bis-*H*-phosphinates. *Tetrahedron Lett.* **2014**, *55*, 219–223. [CrossRef]
102. Rodriguez, J.B. Tetraethyl vinylidenebisphosphonate: A versatile synthon for the preparation of bisphosphonates. *Synthesis* **2014**, *46*, 1129–1142. [CrossRef]
103. Rosso, V.S.; Szajman, S.H.; Malayil, L.; Galizzi, M.; Moreno, S.N.J.; Docampo, R.; Rodriguez, J.B. Synthesis and biological evaluation of new 2-alkylaminoethyl-1,1-bisphosphonic acids against *Trypanosoma cruzi* and *Toxoplasma gondii* targeting farnesyl diphosphate synthase. *Bioorg. Med. Chem.* **2011**, *19*, 2211–2217. [CrossRef] [PubMed]
104. Szajnman, S.H.; Liñares, G.E.G.; Li, Z.-H.; Jiang, C.; Galizzi, M.; Bontempi, E.; Ferella, M.; Moreno, S.N.J.; Docampo, R.; Rodriguez, J.B. Synthesis and biological evaluation of 2-alkylaminoethyl-1,1-bisphosphonic acids against *Trypanosoma cruzi* and *Toxoplasma gondii* targeting farnesyl diphosphate synthase. *Bioorg. Med. Chem.* **2008**, *15*, 3283–3290. [CrossRef] [PubMed]
105. Herczegh, P.; Buxton, T.B.; McPherson, J.C., III; Kovács-Kulyassa, A.; Brewer, P.D.; Sztaricskai, F.; Stroebel, G.C.; Plowman, K.M.; Farcasiu, D.; Hartmann, J.F. Osteoadsorptive bisphosphonate derivatives of fluoroquinolone antibacterials. *J. Med. Chem.* **2002**, *45*, 2338–2341. [CrossRef] [PubMed]
106. Houghton, T.J.; Tanaka, K.S.E.; Kang, T.; Dietrich, E.; Lafontaine, Y.; Delorme, D.; Ferreira, S.S.; Viens, F.; Arhin, F.F.; Sarmiento, I.; et al. Linking bisphosphonates to the free amino groups in fluoroquinolones: Preparation of osteotropic prodrugs for the prevention of osteomyelitis. *J. Med. Chem.* **2008**, *51*, 6955–6969. [CrossRef] [PubMed]
107. Dzięgielewski, M.; Hejmanowska, J.; Albrecht, Ł. A convenient approach to a novel group of quaternary amino acids containing a geminal bisphosphonate moiety. *Synthesis* **2014**, *46*, 3233–3239. [CrossRef]
108. Simoni, D.; Gebbia, N.; Invidiata, M.P.; Eleopra, M.; Marchetti, P.; Rondanin, R.; Baruchello, R.; Provera, S.; Marchioro, C.; Tolomeo, M.; et al. Design, synthesis, and biological evaluation of novel aminobisphosphonates possessing an in vivo antitumor activity through a γδ-T lymphocytes-mediated activation mechanism. *J. Med. Chem.* **2008**, *51*, 6800–6807. [CrossRef] [PubMed]
109. Sankala, E.; Weisell, J.M.; Huhtala, T.; Närvänen, A.T.O.; Vepsäläinen, J.J. Synthesis of novel bisphosphonate polyamine conjugates and their affinity to hydroxyapatite. *Arch. Org. Chem.* **2012**, *2012*, 233–241.
110. Chen, K.M.C.; Hudock, M.P.; Zhang, Y.; Guo, R.T.; Cao, R.; No, J.H.; Liang, P.H.; Ko, T.-P.; Chang, T.H.; Chang, S.; et al. Inhibition of geranylgeranyl diphosphate synthase by bisphosphonates: A crystallographic and computational investigation. *J. Med. Chem.* **2008**, *51*, 5594–5607.
111. Zhang, Y.; Leon, A.; Song, Y.; Studer, D.; Haase, C.; Kościelski, Ł.A.; Oldfield, E. Activity of nitrogen-containing and non-nitrogen-containing bisphosphonates on tumor cell lines. *J. Med. Chem.* **2006**, *49*, 5804–5814. [CrossRef] [PubMed]

112. Makarov, M.V.; Rybalkina, E.Y.; Röschenthaler, G.-V. New 3,5-bis(arylidene)-4-piperidones with bisphosphonate moiety: Synthesis and antitumor activity. *Russ. Chem. Bull. Int. Ed.* **2014**, *63*, 1181–1186. [CrossRef]
113. Skarpos, H.; Osipov, S.N.; Vororb'eva, D.V.; Odinets, I.L.; Lork, E.; Röschenthaler, G.-V. Synthesis of functionalized bisphosphonates via click chemistry. *Org. Biomol. Chem.* **2007**, *5*, 2361–2367. [CrossRef] [PubMed]
114. Liu, S.; Bi, W.; Li, X.; Chen, X.; Qu, N.; Zhao, Y. A Practical method to synthesize 1,2,3-triazole-amino-bisphosphonate derivatives. *Phosphorus Sulfur Relat. Elem.* **2015**, *190*, 1735–1742. [CrossRef]
115. Gouault-Bironneau, S.; Deprèle, S.; Sutor, A.; Montchamp, J.-L. Radical reaction of sodium hypophosphite with terminal alkynes: Synthesis of 1,1,-bis-*H*-phosphinates. *Org. Lett.* **2005**, *7*, 5909–5912. [CrossRef] [PubMed]
116. Kuźnik, A.; Mazurkiewicz, R.; Grymel, M.; Zielińska, K.; Adamek, J.; Chmielewska, E.; Bochno, M.; Kubica, S. A new method for the synthesis of α-aminoalkylidenebisphosphonates and their asymmetric phosphonyl-phosphinyl and phosphonyl-phosphinoyl analogues. *Beilstein J. Org. Chem.* **2015**, *11*, 1418–1424. [CrossRef] [PubMed]
117. Tanaka, K.S.; Dietrich, E.; Ciblat, S.; Métayer, C.; Arhin, F.F.; Sarmiento, I.; Moeck, G.; Parr, T.R., Jr.; Far, A.R. Synthesis and in vitro evaluation of bisphosphonate glycopeptide prodrugs for the treatment of osteomyelitis. *Bioorg. Med. Chem. Lett.* **2010**, *20*, 1355–1359. [CrossRef] [PubMed]
118. Bala, J.L.F.; KAshemirov, B.A.; McKenna, C.E. Synthesis of novel bisphosphonic acid alkene monomer. *Synth. Commun.* **2010**, *40*, 3577–3584. [CrossRef]
119. Lecerclé, D.; Gabillet, S.; Gomis, J.-M.; Taran, F. A facile synthesis of aminomethylene bisphosphonates through rhodium carbenoid mediated N-H insertion reaction. Application to the preparation of powerful uranyl ligands. *Tetrahedron Lett.* **2008**, *49*, 2083–2087. [CrossRef]
120. Masschelein, K.G.R.; Stevens, C.V. Double nucleophilic 1,2-addition of silylated dialkyl phosphites to 4-phosphono-1-aza-1,3-dienes: Synthesis of γ-phosphono-α–aminobisphosphonates. *J. Org. Chem.* **2007**, *72*, 9248–9252. [CrossRef] [PubMed]
121. Rassukana, Y.V.; Davydova, K.O.; Onys'ko, P.P.; Sinitsa, A.D. Synthesis and rearrangements of *N*-trichloroacetylfluoroacetimidolyl chloride and its phosphorylation products. *J. Fluor. Chem.* **2002**, *117*, 107–113. [CrossRef]
122. Rassukana, Y.V.; Onys'ko, P.P.; Davydova, K.O.; Sinitsa, A.D. A new reaction of phosphorylated *N*-sulfonylimines with hydrophosphoryl agents involving C-N transfer of phosphoryl groups. *Tetrahedron Lett.* **2004**, *45*, 3899–3902. [CrossRef]
123. Kolotilo, N.V.; Sinitsa, A.D.; Rassukanaya, Y.V.; Onys'ko, P.P. *N*-Sulfonyl- and *N*-phosphorylbenzimidoylphosphonates. *Russ. J. Gen. Chem.* **2006**, *76*, 1210–1218. [CrossRef]
124. Romanenko, V.D.; Kukhar, V.P. Methylidynetrisphosphonates: Promising C1 building block for the design of phosphate mimetics. *Beilstein J. Org. Chem.* **2013**, *9*, 991–1001. [CrossRef] [PubMed]
125. Kafarski, P.; Lejczak, B. A facile conversion of aminoalkanephosphonic acids into their diethyl esters. The use of unblocked aminophosphonic acids in phosphono peptide synthesis. *Synthesis* **1988**, *4*, 307–310. [CrossRef]
126. Turhanen, P.A.; Weisell, J.; Vepsäläinen, J.J. Preparation of mixed trialkyl alkylcarbonate derivatives of etidronic acid via an unusual route. *Beilstein J. Org. Chem.* **2012**, *8*, 2019–2024. [CrossRef] [PubMed]
127. Turhanen, P.A.; Vepsäläinen, J.J. Synthesis of novel (1-alkanoyloxy-4-alkanoylaminobutylidene)-1,1-bisphosphonic acid derivatives. *Beilstein J. Org. Chem.* **2006**, *2*, 2. [CrossRef] [PubMed]
128. Xu, G.; Yang, C.; Bo, L.; Wu, X.; Xie, Y. Synthesis of new potential chelating agents: Catechol-bisphosphonate conjugates for metal intoxication therapy. *Heteroatom Chem.* **2004**, *15*, 251–257. [CrossRef]
129. Agyin, J.K.; Santhamma, B.; Roy, S.S. Design, synthesis, and biological evaluation of bone targeted proteasome inhibitors for multiple myeloma. *Bioorg. Med. Chem.* **2013**, *23*, 5455–6458. [CrossRef] [PubMed]
130. Ezra, A.; Hoffman, A.; Breuer, E.; Alferiev, A.S.; Mönkkönen, J.; El Hanany-Rozen, N.; Weiss, G.; Stepensky, D.; Gati, I.; Cohen, H.; et al. A Peptide prodrug approach for improving bisphosphonate oral absorption. *J. Med. Chem.* **2000**, *43*, 3641–3652. [CrossRef] [PubMed]
131. Sun, S.; Błażewska, K.; Kadina, A.P.; Kashemirov, B.A.; Duan, X.; Triffitt, J.T.; Dunford, J.E.; Russell, R.G.R.; Ebetino, F.H.; Roelofs, A.J.; et al. Fluorescent bisphosphonate and carboxyphosphonate probes: A versatile imaging toolkit for applications in bone biology and biomedicine. *Bioconj. Chem.* **2016**, *27*, 329–340. [CrossRef] [PubMed]

132. Liu, J.; Jo, J.; Kawai, Y.; Aoki, I.; Tanaka, C.; Yamamoto, M.; Tabata, Y. Preparation of polymer-based multimodal imaging agent to visualize the process of bone regeneration. *J. Control. Release* **2012**, *157*, 398–405. [CrossRef] [PubMed]
133. Varghese, O.P.; Sun, W.; Hilborn, J.; Ossipov, D.A. In situ cross-linkable high molecular weight hyaluronan-bisphosphonate conjugate for localized delivery and cell-specific targeting: A hydrogel linked prodrug approach. *J. Am. Chem. Soc.* **2009**, *131*, 8781–8783. [CrossRef] [PubMed]
134. Vitha, T.; Kubiček, V.; Hermann, P.; Elst, L.V.; Muller, R.N.; Kollar, Z.I.; Wolterbeek, H.T.; Breeman, W.A.P.; Lukeš, I.; Peters, J.A. Lanthanide(III) Complexes of Bis(phosphonate) Monoamide Analogues of DOTA: Bone-Seeking Agents for Imaging and Therapy. *J. Med. Chem.* **2008**, *51*, 677–683. [CrossRef] [PubMed]
135. Shao, X.; Xu, Y.; Jiao, Y.; Zhou, C. Synthesis and characterization of an alendronate-chitosan conjugate. *Appl. Mech. Mater.* **2012**, *140*, 53–57. [CrossRef]
136. Morioka, M.; Kamizono, A.; Takikawa, H.; Mori, A.; Ueno, H.; Kadowaki, S.; Nakano, Y.; Kato, K.; Umezawa, K. Design, synthesis, and biological evaluation of novel estradiol-bisphosphonate conjugates as bone-specific estrogens. *Bioorg. Med. Chem.* **2010**, *18*, 1143–1148. [CrossRef] [PubMed]
137. Dadiboyena, S. Recent advances in the synthesis of raloxifene: A selective estrogen receptor modulator. *Eur. J. Med. Chem.* **2012**, *51*, 17–34. [CrossRef] [PubMed]
138. Palma, E.; Oliveira, B.M.; Correira, J.D.G.; Gano, L.; Maria, L.; Santos, I.C.; Santos, I. A new bisphosphonate-containing ^{99m}Tc(I) tricarbonyl complex potentially useful as bone-seeking agent: Synthesis and biological evaluation. *J. Biol. Inorg. Chem.* **2007**, *12*, 667–679. [CrossRef] [PubMed]
139. Reddy, R.; Dietrich, E.; Lafontaine, Y.; Houghton, T.J.; Belanger, O.; Dubois, A.; Arhin, F.F.; Sarmiento, I.; Fadhil, I.; Laquerre, K.; et al. Bisphosphonated benzoxazinorifamycin prodrugs for the prevention and treatment of osteomyelitis. *ChemMedChem* **2008**, *3*, 1863–1868. [CrossRef] [PubMed]
140. Bekker, K.S.; Chukanov, N.V.; Grigor'ev, I.A. Synthesis of a bisphosphonate derivative of folic acid. *Chem. Nat. Prod.* **2013**, *49*, 495. [CrossRef]
141. Bekker, K.S.; Chukanov, N.V.; Popov, S.A.; Grigor'ev, I.A. New Amino-Bisphosphonate Building Blocks in the Synthesis of Bisphosphonic Derivatives Based on Lead Compounds. *Phosphorus Sulfur Silicon Relat. Elem.* **2015**, *190*, 1209–1212. [CrossRef]
142. Bekker, K.S.; Chukanov, N.V.; Popov, S.A.; Grigor'ev, I.A. Conjugates of bisphosphonates with derivatives of ursolic, betulonic, and betulinic acids. *Chem. Nat. Prod.* **2013**, *49*, 581–582. [CrossRef]
143. Wang, L.; Yang, Z.; Gao, J.; Xu, K.; Gu, H.; Zhang, B.; Zhang, X.; Xu, B. A biocompatible method of decorporation: Bisphosphonate-modified magnetite nanoparticles to remove uranyl ions from blood. *J. Am. Chem. Soc.* **2006**, *131*, 13358–13359. [CrossRef] [PubMed]
144. Bekker, K.S.; Chukanov, N.V.; Grigor'ev, I.A. Synthesis of a bis-phosphonic acid derivative of trolox, a new potential antioxidant. *Chem. Nat. Prod.* **2013**, *49*, 785–786. [CrossRef]
145. Altin, A.; Akgun, B.; Bilgici, Z.S.; Turker, S.B.; Avci, D. Synthesis, photopolymerization, and adhesive properties of hydrolytically stable phosphonic acid-containing (meth)acrylamides. *J. Polym. Sci. A Polym. Chem.* **2014**, *52*, 512–522. [CrossRef]
146. Akgun, B.; Avci, D. Synthesis and evaluations of bisphosphonate-containing monomers for dental materials. *J. Polym. Sci. A Polym. Chem.* **2012**, *50*, 4854–4863. [CrossRef]
147. Yang, X.; Akhtar, X.; Rubino, S.; Leifer, K.; Hilborn, J.; Ossipov, D. Direct "click" synthesis of hybrid bisphosphonate—Hyaluronic acid hydrogel in aqueous solution for biomineralization. *Chem. Mater.* **2012**, *24*, 1690–1697. [CrossRef]
148. Gluz, E.; Grinberg, I.; Corem-Salkmon, E.; Mizrahi, D.; Margel, S. Engineering of new crosslinked near-infrared fluorescent polyethylene glycol bisphosphonate nanoparticles for bone targeting. *J. Polym. Sci. A Polym. Chem.* **2013**, *51*, 4282–4291. [CrossRef]
149. Mukaya, H.E.; Mbianda, X.Y. Macromolecular co-conjugate of ferrocene and bisphosphonate: Synthesis, characterization and kinetic drug release study. *J. Inorg. Organomet. Polym.* **2015**, *25*, 411–418. [CrossRef]
150. Beck, J.; Gharbi, S.; Herteg-Fernea, A.; Vercheval, L.; Berbone, C.; Lassaux, P.; Zervosen, A.; Marchant-Bryanert, J. Aminophosphonic acids and aminobis(phosphonic acids) as potential inhibitors of penicillin-binding proteins. *Eur. J. Org. Chem.* **2009**, *2009*, 85–97. [CrossRef]
151. Chuiko, A.L.; Filonenko, L.P.; Borisevich, A.M.; Lozinskii, M.O. pH-Dependent isomerism of (iso)thioureidomethylenebisphosphonates. *Russ. J. Gen. Chem.* **2009**, *79*, 72–77. [CrossRef]

152. Chuiko, A.L.; Filonenko, L.P.; Lozinskii, M.O. Reaction of thioureidomethylenebisphosphonic acids with α-haloketones: II. Effect of pH on the reaction pathway. *Russ. J. Gen. Chem.* **2009**, *79*, 57–66. [CrossRef]
153. Chuiko, A.L.; Filonenko, L.P.; Lozinskii, M.O. Synthesis of ((thiazol-2-ylamino)-methylene)bisphosphonic acids. *Chem. Heterocycl. Compd.* **2011**, *47*, 1137–1140. [CrossRef]
154. Sirviö, J.A.; Hasa, T.; Ahola, J.; Liimatainen, H.; Niinimäki, J.; Hormi, O. Phosphonated nanocelluloses from sequential oxidative reductive treatment—Physicochemical characteristics and thermal properties. *Carbohydr. Polym.* **2015**, *133*, 524–532. [CrossRef] [PubMed]
155. Bonzi, G.; Salmaso, S.; Scomparin, A.; Edar-Boock, A.; Satchi-Fainaro, R.; Caliceti, P. Novel pullulan bioconjugate for selective breast cancer bone metastases treatment. *Bioconj. Chem.* **2015**, *26*, 489–501. [CrossRef] [PubMed]

9

Synthesis and Reactions of α-Hydroxyphosphonates

Zita Rádai and György Keglevich *

Department of Organic Chemistry and Technology, Budapest University of Technology and Economics, 1521 Budapest, Hungary; radai.zita@mail.bme.hu
* Correspondence: gkeglevich@mail.bme.hu

Abstract: This review summarizes the main synthetic routes towards α-hydroxyphosphonates that are known as enzyme inhibitors, herbicides and antioxidants, moreover, a number of representatives express antibacterial or antifungal effect. Special attention is devoted to green chemical aspects. α-Hydroxyphosphonates are also versatile intermediates for other valuable derivatives. *O*-Alkylation and *O*-acylation are typical reactions to afford α-alkoxy-, or α-acyloxyphosphonates, respectively. The oxidation of hydroxyphosphonates leads to ketophosphonates. The hydroxy function at the α carbon atom of hydroxyphosphonates may be replaced by a halogen atom. α-Aminophosphonates formed in the nucleophilic substitution reaction of α-hydroxyphosphonates with primary or secondary amines are also potentially bioactive compounds. Another typical reaction is the base-catalyzed rearrangement of α-hydroxy-phosphonates to phosphates. Hydrolysis of the ester function of hydroxyphosphonates leads to the corresponding phosphonic acids.

Keywords: α-hydroxyphosphonate; Pudovik reaction; *O*-derivatization; oxidation; substitution; rearrangement; hydrolysis

---

1. Synthetic Routes towards α-Hydroxyphosphonates

Due to the bioactivity of α-hydroxyphosphonates **1**, the synthesis of these derivatives is an evergreen topic in organophosphorus chemistry [1–8]. The main approaches to obtain α-hydroxyphosphonates **1** are shown in Scheme 1. The most commonly studied method is the addition of a dialkyl phosphite to an oxo compound (method "A") [9]. The good atom economy makes this method the most appealing way to synthesize α-hydroxyphosphonates **1**. In the majority of cases, the addition is carried out in the presence of a base catalyst, but a few acid-catalyzed variations are also known. An alternative route is the condensation of an oxo compound and a trialkyl phosphite (method "B") [10]. In contrast to method "A", this reaction is usually catalyzed by different acids. In the literature, the denomination of the reactions is not consistent, as both methods "A" and "B" are referred to as the Pudovik and Abramov reaction, and also as the phospha-aldol reaction. The third main approach towards α-hydroxyphosphonates **1** involves the reaction of α-ketophosphonates **2** with Grignard reagents [11] or other organometallic compounds [11,12], followed by hydrolysis (method "C"). α-Alkyl phosphonates **3** may also be converted to the corresponding α-hydroxy derivatives by oxidation (method "D") [13,14]. In this review, methods "A" and "B" are discussed in detail. Although the asymmetric synthesis of α-hydroxyphosphonates **1** is of special importance [4–8], in this review the discussion is limited to racemic derivatives.

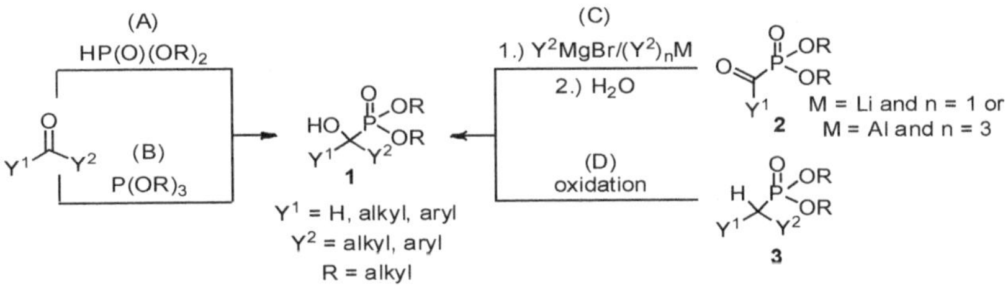

Scheme 1. Synthetic pathways towards α-hydroxyphosphonates **1**.

1.1. *Synthesis of α-Hydroxyphosphonates by the Reaction of Aldehydes/Ketones and Dialkyl Phosphites*

The reaction of oxo compounds and dialkyl phosphites catalyzed by alkali alcoholates (Scheme 1, method "A") was first reported by Pudovik [15]. Since then, a number of variants involving different catalysts and conditions were elaborated to obtain α-hydroxyphosphonates by the reaction under discussion (Table 1). Recent methods have targeted the use of inexpensive and simple catalysts, and mild reaction conditions in the spirit of green chemistry. In the great majority of the cases, the addition was carried out without using any solvent. It is worth mentioning that starting from ketones, the accomplishment of the reaction is more challenging than in the cases applying aldehydes. The synthesis of α-alkyl-α-hydroxyphosphonates often requires the use of "exotic" catalysts [16–20], or an excess of a base. Procedures also applicable for the conversion of ketones are marked by an asterisk in Table 1.

A base-catalyzed variation of the Pudovik reaction was carried out in the presence of 5 mol % potassium phosphate as the catalyst (Table 1/Entry 1) [21]. This method has the advantage of using an inexpensive catalyst, requiring short reaction times and an easy work-up procedure that comprises simple extraction, and, in most cases, provides the products in high yields. Barium hydroxide was also an efficient catalyst in the addition [22,23]. One method employed 10 mol % $Ba(OH)_2$ (Table 1/Entry 2) [22], while according to another protocol, $Ba(OH)_2 \cdot 8H_2O$ was used (Table 1/Entry 3) [23]. The first $Ba(OH)_2$-catalyzed variation suffers from the drawback of a tedious work-up procedure involving extraction, and washing, followed by crystallization [22], while in the latter case, the use of THF as the solvent is a disadvantage [23]. One equivalent of magnesium oxide efficiently catalyzed the addition of diethyl phosphite to substituted benzaldehydes (Table 1/Entry 4) [24]. The method could be extended to the conversion of γ-cyanoketones by applying the base in a twofold quantity (Table 1/Entry 5) [25]. The reaction of dimethyl phosphite with oxo compounds including ketones afforded the corresponding α-hydroxyphosphonates **1** in the presence of three equivalents of aluminum oxide after stirring for 72 h (Table 1/Entry 6) [26]. The reaction time could be shortened dramatically by using potassium fluoride with only one equivalent of Al_2O_3. However, this latter method was efficient only in case of aldehydes as the starting compounds (Table 1/Entry 7) [27]. The addition was also performed in the presence of one equivalent of triethylamine as the catalyst (Table 1/Entry 8) [28]. A simple crystallization afforded the desired products in good to excellent yields. Another method employed three equivalents of triethylamine together with one equivalent of magnesium chloride (Table 1/Entry 9) [29]. In the latter case, an extraction was inserted before the crystallization step in the work-up procedure [29]. In agreement with the requirements of environmentally friendly approaches, green activation methods, such as microwave (MW) irradiation [30,31] and grinding [32,33] were also employed in the synthesis of α-hydroxyphosphonates **1**. According to a MW-assisted procedure, the reaction of benzaldehydes and diethyl phosphite was carried out without the use of any catalyst or solvent (Table 1/Entry 10) [30], while another method combined MW irradiation with sodium carbonate as the base (Table 1/Entry 11) [31]. Na_2CO_3 was also used efficiently in another protocol, when the contact among the reaction components was promoted by grinding (Table 1/Entry 12) [32]. The drawback of the method is the complicated work-up procedure comprising washing, extraction, and finally recrystallization. A similar protocol utilizes the piperazine-catalyzed reaction of aldehydes and diethyl phosphite in a mill (Table 1/Entry 13) [33]. In the latter case, the work-up involved washing, extraction and column chromatography. The synthesis of α-hydroxyphosphonates **1** was also carried out in the presence of 10 mol % choline hydroxide as an IL catalyst (Table 1/Entry 14) [34]. According to a plausible mechanism suggested by the authors, the IL promotes the reaction by activation through *H*-bridges. An interesting protocol is a special fluoroapatite-catalyzed Pudovik reaction (Table 1/Entry 15) [35]. A mixture of the starting dialkyl phosphite and oxo compound was stirred with a spatula, and then left standing on the catalyst over 1–1.5 min. The work-up involved extraction with dichloromethane, and recrystallization of the crude product. Acid-catalyzed variations of the Pudovik reaction occur rarely in the literature [36,37]. Potassium hydrogensulfate proved to be an efficient catalyst in an amount of 20 mol % in addition of diethyl phosphite to substituted aldehydes (Table 1/Entry 16) [36]. According to another method, silica-supported tungstic acid was applied as the catalyst (Table 1/Entry 17) [37]. Organometallic compounds may also play the role of the catalyst during the synthesis of α-hydroxyphosphonates **1** [38–40]. Butyllithium was applied in the amount of 0.1 mol % in hexane as the solvent (Table 1/Entry 18) [38]. The method not only bears the disadvantages of using an exotic catalyst in an inert atmosphere, but also the need for a complicated work-up involving quenching, washing, and finally column chromatography. As for the advantages, mild reaction conditions and wide substrate scope including ketones can be mentioned. Titanium tetraisopropylate was also applied to catalyze the reaction of ketones and dimethyl phosphite (Table 1/Entry 19) [39]. Column chromatography of the crude product afforded α-alkyl α-hydroxyphosphonates **1** in good to excellent yields. Molybdenum dioxide dichloride (MoO_2Cl_2) was also tested as the catalyst in the Pudovik reaction (Table 1/Entry 20) [40]. Using it in an amount of 5 mol % under solvent-free conditions at 80 °C, the hydroxyphosphonates **1** were obtained good yields.

1.2. *Synthesis of α-Hydroxyphosphonates by the Reaction of Aldehydes/Ketones and Trialkyl Phosphites*

Besides the reaction of oxo compounds with dialkyl phosphites discussed in Section 1.1 (Scheme 1, method "A"), the other widespread method for the synthesis of α-hydroxyphosphonates is the condensation of aldehydes or ketones with trialkyl phosphites (Scheme 1, method "B") [10]. This way is of somewhat less importance due to the reduced atom economy, as compared to the major route outlined. While the reaction involving dialkyl phosphites is usually catalyzed by different bases, the typical catalysts for the condensation of oxo compounds with trialkyl phosphites exhibit acidic character (Table 2).

A method of choice is a solvent- and catalyst-free accomplishment under ultrasonic irradiation (Table 2/Entry 1) [41]. The reaction of aldehydes and trialkyl phosphites at 25 °C was complete after 10–35 min and provided the α-hydroxyphosphonates **1** in excellent yields after crystallization (Table 2/Entry 1). The method was successfully applied to ketones as starting materials as well. Another ultrasound-assisted variation is the potassium dihydrogen phosphate-catalyzed reaction (Table 2/Entry 2) [42]. According to a plausible mechanism, the role of the acid catalyst is the protonation of the carbonyl function of the aldehyde making the C=O bond more electrophilic, and hence facilitating the nucleophilic attack of the $(RO)_3P$ reagent. A similar procedure employed sulphamic acid as the catalyst (Table 2/Entry 3) [43]. Again, the catalytic effect of the sulphamic acid was attributed to its capability to protonate the oxo compound. The camphorsulfonic acid-catalyzed syntheses of α-hydroxyphosphonates **1** from aldehydes and triethyl phosphite was carried out under ultrasonic irradiation on stirring (Table 2/Entry 4) [44]. Applying the ultrasound technique, the reaction times could be shortened from 30–75 min to 8–15 min after comparing the results with those obtained without ultrasonication. At the same time, the yields were similar (85–93%) for the two procedures [44]. Oxalic acid also proved to be an efficient catalyst in the condensation under discussion at 80 °C (Table 2/Entry 5) [45]. A series of organic solvents were tested as the reaction medium, but the desired products were obtained in low yields (<30%). Under neat conditions, the α-hydroxyphosphonates **1** were formed efficiently as suggested by the yields of 83–98% obtained after quenching, extraction and column chromatography. In the hope of synthesizing a new class of antibiotics, α-hydroxy-phosphonates **1** containing a β-lactam scaffold were prepared in the presence of tartaric acid as the catalyst (Table 2/Entry 6) [46].

Table 1. Selected examples for the synthesis of α-hydroxyphosphonates (**1**) from aldehydes or ketones and dialkyl phosphites (method "A").

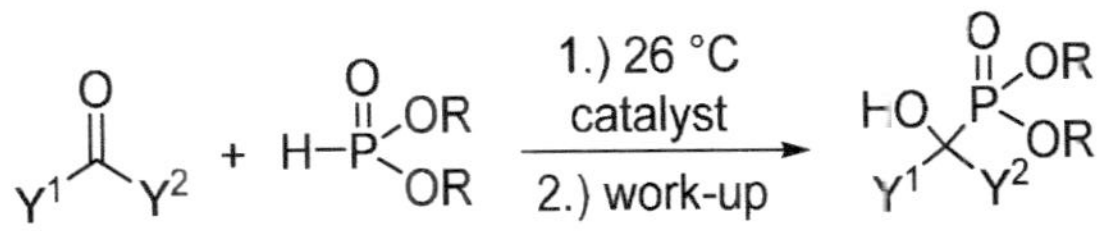

1

Entry	Y^1	Y^2	R	Catalyst	Amount of Catalyst	Conditions	Yield (%)	Ref.
1	^{i}Pr, ^{c}Hex, C_6H_5, 4-MeC_6H_4, 4-$^i PrC_6H_4$, 4-$MeOC_6H_4$, 3,4-di$MeOC_6H_3$, 4-ClC_6H_4, 2-$NO_2C_6H_4$, 4-CNC_6H_4, C_6H_5CH=CH, 3,4-$OCH_2OC_6H_3$, 4-CH_2=$CHCH_2OC_6H_4$, 4-$PhCH_2OC_6H_4$, 2-furyl, 2-thienyl	H	Me, Et, ^{i}Pr	K_3PO_4	5 mol %	25 °C, 4–8 min for aromatic aldehydes, 24 h for aliphatic aldehydes	20–98	[21]
2	C_6H_5, 4-$MeOC_6H_4$, 3,4-di$MeOC_6H_3$, 2-ClC_6H_4, 4-ClC_6H_4, 2,3-diClC_6H_3, 2,4-diClC_6H_3, 2,6-diClC_6H_3, 2-BrC_6H_4, 4-BrC_6H_4, 3-FC_6H_4, 4-FC_6H_4, 2-$NO_2C_6H_4$, 3-$NO_2C_6H_4$, 4-$NO_2C_6H_4$, 4-CNC_6H_4, C_6H_5CH=CH, 2-thienyl, 4-pyridyl	H	Et	$Ba(OH)_2$	10 mol %	25 °C, 4–10 min	70–98	[22]
3	^{i}Bu, C_6H_5, 2-MeC_6H_4, 3-MeC_6H_4, 4-MeC_6H_4, 4-EtC_6H_4, 4-$MeOC_6H_4$, 2,3,4-tri$MeOC_6H_2$, 2-ClC_6H_4, 4-ClC_6H_4, 2,6-diClC_6H_3, 3-BrC_6H_4, 4-BrC_6H_4, 3-FC_6H_4, 4-FC_6H_4, 2-$NO_2C_6H_4$, 3-$NO_2C_6H_4$, 4-$NO_2C_6H_4$, 3-CNC_6H_4, 4-CNC_6H_4, 2-furyl, 2-thienyl, 2-pyridyl, 3-pyridyl, 4-pyridyl, 1-naphthyl, 2-naphthyl, 9-anthryl, C_6H_5CH=CH	H	Et, Bu, Bn	$Ba(OH)_2 \cdot 3H_2O$	2–7 mol %	25 °C, 15 min, THF	72–99	[23]
4	C_6H_5, 4-MeC_6H_4, 3-$MeOC_6H_4$, 4-$MeOC_6H_4$, 3-HOC_6H_4, 2-ClC_6H_4, 4-ClC_6H_4, 2,6-diClC_6H_3, 2-$NO_2C_6H_4$, 3-$NO_2C_6H_4$, 4-$NO_2C_6H_4$, 4-$Me_2NC_6H_4$, C_6H_5CH=CH, 2-furyl, 1-naphthyl, 2-naphthyl	H	Et	MgO	1 equiv.	25 °C, 2 min–4 h	80–100	[24]
5 *	C_6H_5, 4-MeC_6H_4, 4-$MeOC_6H_4$	$(CH_2)_2CN$, $MeCHCH_2CN$	Me, Et	MgO	2 equiv.	25 °C, 1–6 h	70–82	[25]
6 *	Pr, Bu, C_6H_5, 4-MeC_6H_4, 4-$MeOC_6H_4$, 4-$^i PrC_6H_4$, 2,4,6-triMeC_6H_2, 2-OHC_6H_4, 4-ClC_6H_4, 4-BrC_6H_4, 3-FC_6H_4, 4-FC_6H_4, 4-Br-2-OHC_6H_3, 3,4-di$MeOC_6H_3$, 3,4-$OCH_2OC_6H_3$, 2-$NO_2C_6H_4$, 4-$NO_2C_6H_4$, 2-furyl, 2-thienyl, 3-thienyl, 1-naphthyl, 2-naphthyl	H, Me	Me	Al_2O_3	3 equiv.	25 °C, 72 h	52–98	[26]
7	C_6H_5, 4-MeC_6H_4, 4-$MeOC_6H_4$, 4-ClC_6H_4, 4-FC_6H_4, 4-$NO_2C_6H_4$	H	Me, Et	Al_2O_3 + KF	1 + 2 equiv.	25 °C, 30 min	53–88	[27]
8 *	Me, $ClCH_2$, C_6H_5, 4-MeC_6H_4, 4-$MeOC_6H_4$, 2-ClC_6H_4, 3-ClC_6H_4, 4-ClC_6H_4, 3-BrC_6H_4, 4-BrC_6H_4, 2-FC_6H_4, 4-FC_6H_4, 2,4-diClC_6H_3, 2-furyl, 2-thienyl	Me, Pr, Ph	Me	Et_3N	1 equiv.	40 °C, 2 h	63–99	[28]

Table 1. *Cont.*

Entry	Y^1	Y^2	R	Catalyst	Amount of Catalyst	Conditions	Yield (%)	Ref.
9 *	Pr, C_6H_5, 4-MeC_6H_4, 4-$MeOC_6H_4$, 2-ClC_6H_4, 4-ClC_6H_4, 3-$NO_2C_6H_4$, 4-$NO_2C_6H_4$, 4-CNC_6H_4, $(CH_2)_2C_6H_5$, C_6H_5CH=CH, 2-furyl, 2-naphthyl, 9-anthryl	H, Me	Me	$Et_3N + MgCl_2$	3 + 1 equiv.	50 °C, 1–2 h	85–98	[29]
10	C_6H_5, 4-MeC_6H_4, 4-$MeOC_6H_4$, 4-OHC_6H_4, 2-ClC_6H_4, 4-ClC_6H_4, 4-BrC_6H_4, 2-$NO_2C_6H_4$, 3-$NO_2C_6H_4$, 4-$NO_2C_6H_4$	H	Et	–	–	MW, 90–100 °C, 10–20 min	79–95	[30]
11	C_6H_5, 4-MeC_6H_4, 4-$MeOC_6H_4$, 4-ClC_6H_4, 4-$NO_2C_6H_4$	H	Me, Et	Na_2CO_3	0.75 equiv.	MW, 110 °C, 20 min	62–88	[31]
12	C_6H_5, 4-MeC_6H_4, 4-$MeOC_6H_4$, 4-ClC_6H_4, 3-$NO_2C_6H_4$, 4-$NO_2C_6H_4$	H	Et	Na_2CO_3	1 equiv.	Grinding, 25 °C, 10 min	75–83	[32]
13	C_6H_5, 4-MeC_6H_4, 4-$^i PrC_6H_4$, 4-$MeOC_6H_4$, 3,4-$diMeOC_6H_3$, 2-ClC_6H_4, 4-ClC_6H_4, 2-$NO_2C_6H_4$, 3-$NO_2C_6H_4$, 4-$NO_2C_6H_4$, 4-$Me_2NC_6H_4$, 4-$PhCH_2OC_6H_4$, 4-$C_5H_4NC_6H_4$, 9-anthryl	H	Et	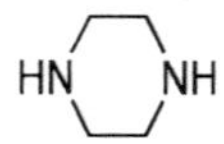	1 equiv.	Grinding, 25 °C, 2–10 min	78–96	[33]
14	C_6H_5, 2-MeC_6H_4, 4-MeC_6H_4, 2-$^i PrC_6H_4$, 3,5-$diMeC_6H_3$, 2-$MeOC_6H_4$, 4-$MeOC_6H_4$, 3,5-$diMeOC_6H_3$, 3,4,5-$triMeOC_6H_2$, 4-$EtOC_6H_4$, 2-ClC_6H_4, 2-BrC_6H_4, 3-BrC_6H_4, 4-BrC_6H_4, 2-Cl-6-$furylC_6H_3$, 2-F-4-BrC_6H_3, 2-$NO_2C_6H_4$, 4-$NO_2C_6H_4$, 9-anthryl, 1-pyrenyl, 2-furyl, 1-C_6H_5-4-pyrazolyl	H	Et	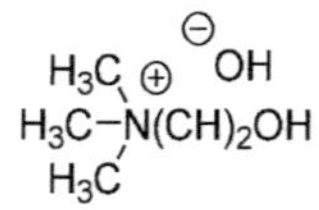	10 mol %	25 °C, 5–10 min	90–98	[34]
15 *	C_6H_5, 3-$MeOC_6H_4$, 4-$MeOC_6H_4$, 4-ClC_6H_4, 3-$NO_2C_6H_4$, 4-$NO_2C_6H_4$, C_6H_5CH=CH	H, Me	Me, Et	Na-modified fluoroapatite	1 g/1.25 mol acetophenone	20–25 °C, 1–1.5 min	75–98	[35]
16	4-$^i PrC_6H_4$, 4-$MeOC_6H_4$, 2,6-$diMeOC_6H_3$, 3,5-$diMeOC_6H_3$, 4-OHC_6H_4, 3-$NO_2C_6H_4$, 4-$Me_2NC_6H_4$, 2-$PhCH_2OC_6H_4$, 2-imidazyl, 3-indolyl	H	Et	$KHSO_4$	20 mol %	25 °C, 2–4 h	82–91	[36]
17	3-FC_6H_4, 4-$NO_2C_6H_4$, 3,4-$OCH_2OC_6H_3$, 4-$MeSC_6H_4$, $C_5H_{10}N$, 3-MeO-4-OHC_6H_3, 4-$C_4H_8NC_6H_4$, 2-furyl, 2-thienyl, 4-imidazyl, 2-pyrrolyl, 4-pyridyl	H	Me	silica-supported tungstic acid	20 mol %	25 °C, 30 min	85–96	[37]
18 *	Me, Pr, ^{i}Bu, Pent, C_6H_5, 2-MeC_6H_4, 4-MeC_6H_4, 2-$MeOC_6H_4$, 3-$MeOC_6H_4$, 4-$MeOC_6H_4$, 2-ClC_6H_4, 3-ClC_6H_4, 4-ClC_6H_4, 2-BrC_6H_4, 4-BrC_6H_4, 4-FC_6H_4, 3-$NO_2C_6H_4$, 4-$NO_2C_6H_4$, 1-naphthyl, 2-naphthyl, 2-furyl, 2-thienyl, 3-pyridyl	H, Me, Ph, CF_3, $(CH_2)_{10}CH_3$, C(O)Ph, $CH_2C(O)Ph$	Et, ^{i}Pr, Ph	BuLi	0.1 mol %	10–25 °C, 5 min, hexane	35–99	[38]
19 *	Et, ^{c}Hex, C_6H_5, 3-MeC_6H_4, 4-MeC_6H_4, 3-$MeOC_6H_4$, 4-ClC_6H_4, 3,4-$diClC_6H_3$, 2,3,4-$triClC_6H_2$, 2-FC_6H_4, 4-FC_6H_4, 4-$NO_2C_6H_4$, 3-$CF_3C_6H_4CH2$=CH, 3-$CF_3C_6H_4CH(OH)CH_2$, 2-thienyl, 2-naphthyl, EtOC(O)CHBn	Me, Et, Ph, $CH(OEt)_2$, COOMe, $(CH_2)_2Cl$	Me	$Ti(O^iPr)_4$	5 mol %	30 °C, 15 min	74–98	[39]
20	C_6H_5, 4-$MeOC_6H_4$, 4-FC_6H_4, 4-$NO_2C_6H_4$, 4-$CF_3C_6H_4$, 4-CNC_6H_4, 4-$MeOC(O)C_6H_4$, $(CH_2)_2C_6H_5$, C_6H_5CH=CH	H	Et	MoO_2Cl_2	5 mol %	80 °C, 1–24 h	70–96	[40]

* The procedure is also suitable starting from ketones.

Table 2. Selected procedures to synthesize α-hydroxyphosphonates (**1**) from aldehydes or ketones and trialkyl phosphites (method "B").

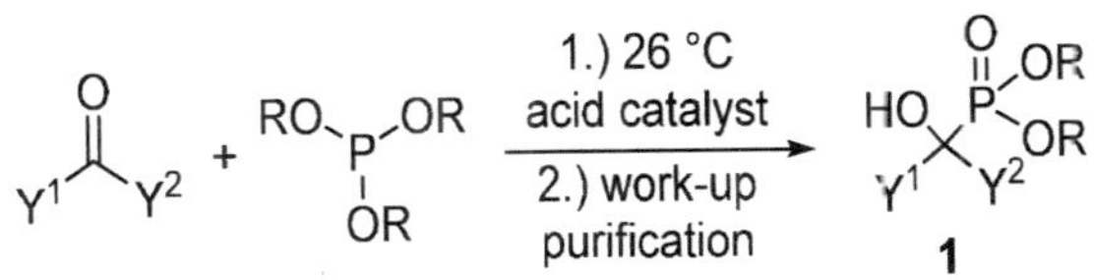

Entry	Y^1	Y^2	R	Catalyst	Amount of Catalyst	Conditions	Yield (%)	Ref.
1 *	C_6H_5, 4-MeC_6H_4, 4-$^iPrC_6H_4$, 4-$MeOC_5H_4$, 2-ClC_6H_4, 4-ClC_6H_4, 4-BrC_6H_4, 4-FC_6H_4, 4-$NO_2C_6H_4$, 4-$CF_3C_6H_4$, 4-$Me_2NC_6H_4$, $CH_2C_6H_5$, $(CH_2)_2C_6H_5$	H, Me, Et, Ph	Me, Et	–	–	Sonication, 25 °C, 10–35 min	84–94	[41]
2	Pr, ^iBu, C_6H_5, 4-MeC_6H_4, 4-$MeOC_6H_4$, 3-OHC_6H_4, 2-ClC_6H_4, 4-ClC_6H_4, $C_6H_5CH{=}CH$, 3-chromonyl, 6-Cl-7-Me-3-chromonyl, 2-Cl-3-quinolinyl	H	Et	KH_2PO_4	5 mol %	Sonication, 25 °C, 5–45 min	48–92	[42]
3	C_6H_5, 4-MeC_6H_4 4-$MeOC_6H_4$, 4-ClC_6H_4, 3-pyridyl, 3-chromonyl, 6,8-diMe-3-chromonyl, 2-Cl-3-quinolinyl, 6-Cl-3-chromonyl, 6,7-diCl-3-chromonyl, 6,8-diCl-3-chromonyl, 6-Br-3-chromonyl, 6-Cl-7-Me-3-chromonyl, 2-Cl-6-Me-3-quinolinyl, 2-Cl-7-Me-3-quinolinyl, 2-Cl-8-Me-3-quinolinyl, 2-Cl-6-MeO-3-quinolinyl, 2-Cl-6-EtO-3-quinolinyl, 2-Cl-8-Et-3-quinolinyl	H	Me, Et	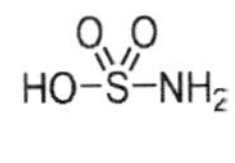	25 mol %	Sonication, 25 °C, 1–60 min	78–98	[43]
4	C_6H_5, 4-MeC_6H_4, 4-$MeOC_6H_4$, 4-OHC_6H_4, 4-ClC_6H_4, 4-$NO_2C_6H_4$, 3,4-$OCH_2OC_6H_3$, $C_6H_5CH{=}CH$, 2-furyl, 2-thienyl, 2-Cl-3-quinolinyl, 4-tetrazolo[1,5-*a*]quinolinyl	H	Et	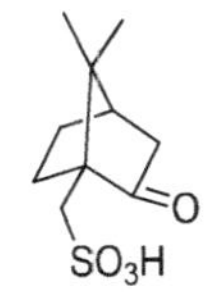	10 mol %	Sonication, 25 °C, 8–20 min	85–93	[44]
5	Pr, ^iPr, Pent, ^cHex, C_6H_5, 4-MeC_6H_4, 4-$MeOC_6H_4$, 4-OHC_6H_4, 2-ClC_6H_4, 4-ClC_6H_4, 4-CNC_6H_4, $C_6H_5CH{=}CH$, 2-furyl	H	Me	$(COOH)_2$	10 mol %	80 °C, 3 h	83–98	[45]
6	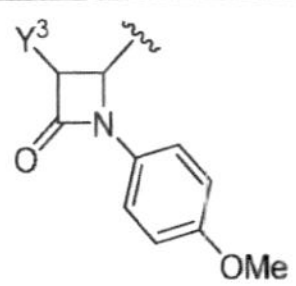 Y^3 = Me, Bu, Ph, OPh	H	Me, Et	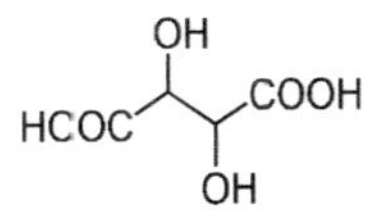	10 mol %	Δ, 30 min, acetonitrile	41–69	[46]
7 *	Pr, C_6H_5, 4-MeC_6H_4, 4-$MeOC_6H_4$, 2-ClC_6H_4, 4-ClC_6H_4, 3-$NO_2C_6H_4$, 4-$NO_2C_6H_4$, 4-CNC_6H_4, $(CH_2)_2C_6H_5$, $C_6H_5CH{=}CH$, 2-furyl	H, Me	Me	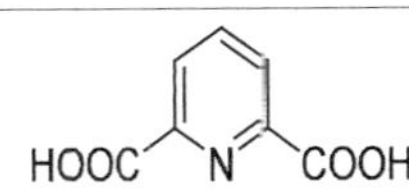	10 mol %	50 °C, 1.3–3 h, H_2O	60–95	[47]

Table 2. *Cont.*

Entry	Y^1	Y^2	R	Catalyst	Amount of Catalyst	Conditions	Yield (%)	Ref.
8	Pr, ^{i}Pr, Bu, ^{t}Bu, Hex, ^{c}Hex, C_6H_5, 4-ClC_6H_4, 2-furyl, 2-pyridyl	H	Me	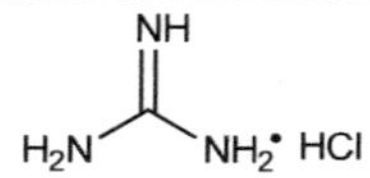	0.5 mol %	50 °C, 2 h, H_2O	60–95	[48]
9	Me, Et, ^{i}Pr, C_6H_5, 4-MeC_6H_4, 4-MeOC_6H_4, 2-ClC_6H_4, 4-ClC_6H_4, 2,4-diClC_6H_3, 2-$NO_2C_6H_4$, 3-$NO_2C_6H_4$, 4-$NO_2C_6H_4$, C_6H_5CH=CH, 2-furyl, 2-thienyl	H	Et	I_2	10 mol %	80 °C, 15–120 min, H_2O	83–97	[49]
10	C_6H_5, 4-MeC_6H_4, 4-MeOC_6H_4, 2-OHC_6H_4, 4-ClC_6H_4, 4-BrC_6H_4, 4-FC_6H_4, 3-$NO_2C_6H_4$, 4-$NO_2C_6H_4$, 2-pyridyl, 2-naphthyl, 2-furyl, 2-thienyl	H	Et	β-cyclodextrin	1 equiv.	60–70 °C, 8–12 h, H_2O	80–93	[50]
11	C_6H_5, 4-MeC_6H_4, 4-MeOC_6H_4, 3-OHC_6H_4, 2-ClC_6H_4, 4-ClC_6H_4, 4-$NO_2C_6H_4$, C_6H_5CH=CH, 3-chromonyl, 6-Cl-3-chromonyl, 6,7-diCl-3-chromonyl, 6,8-diCl-3-chromonyl, 6-Cl-7-Me-3-chromonyl, 2-Cl-3-quinolinyl, 2-Cl-6-Me-3-quinolinyl, 2-Cl-7-MeO-3-quinolinyl, 2-Cl-8-Et-3-quinolinyl, 2-Cl-6-EtO-3-quinolinyl	H	Et	NH_4VO_3	10 mol %	25 °C, 5–40 min	80–94	[51]
12	4-ClC_6H_4, 2,4-diClC_6H_3, 4-BrC_6H_4, 4-MeOC(O)C_6H_4, 4-$CF_3C_6H_4$, 2-NO_2-3,6-diMeOC_6H_2, 1-naphthyl	H	Et	$ZnBr_2$	10 mol %	25 °C, 10–30 min	68–91	[52]
13	C_6H_5, 4-MeC_6H_4, 4-MeOC_6H_4, 4-HOC_6H_4, 2-ClC_6H_4, 4-ClC_6H_4, 4-$NO_2C_6H_4$, 4-$Me_2NC_6H_4$, C_6H_5CH=CH, 2-Cl-3-quinolinyl, 2-Cl-6-Me-3-quinolinyl	H	Et	$Bi(NO_3)_3{\cdot}5H_2O$	10 mol %	MW, 70 °C, 10–15 min	88–95	[53]
14 *	Et, Pr, ^{i}Pr, ^{t}Bu, ^{c}Hex, CH_3CH=CH, C_6H_5, 4-MeOC_6H_4, 4-ClC_6H_4, $(CH_2)_2C_6H_5$, C_6H_5CH=CH	H, Me	Me, Et	$NbCl_5$, TMSCl	0.05 mol %	25 °C, 20–90 min	44–96	[54]

* The procedure is also suitable starting from ketones.

Among a series of acids (e.g., acetic acid, trifluoroacetic acid, *p*-toluenesulfonic acid, lactic acid, fumaric acid and tartaric acid) screened as potential catalysts of the reaction, fumaric acid and tartaric acid were found to be the most efficient. It was concluded that weak acids (with a pKa value around 4.5) are the best catalysts of the reaction [46]. According to another procedure, pyridine-2,6-dicarboxylic acid was applied as the catalyst (Table 2/Entry 7) [47]. The reaction of oxo compounds including ketones with trimethyl phosphite was carried out in water at 50 °C to furnish hydroxyphosphonates. In another approach, the pyridine-based catalyst was replaced by guanidine hydrochloride (Table 2/Entry 8) [48]. The drawback of this protocol is the rather complicated work-up comprising quenching, extraction, and then chromatography. An interesting finding is that iodine may also be a catalyst in the reaction of aldehydes and triethyl phosphite (Table 2/Entry 9) [49]. The catalytic effect of I_2 was explained similarly as that of the acid catalysts: iodine is able to interact with the oxygen atom of the carbonyl group, making the C=O bond more electrophilic. β-Cyclodextrin was found to be a promising and recoverable catalyst of the condensation (Table 2/Entry 10) [50]. However, it was necessary to use this catalyst in a one equivalent quantity, and completion of the reaction required longer times (8–12 h) as compared with other protocols. Ammonium metavanadate was tested as the catalyst in the reaction of benzaldehydes with triethyl phosphite, as well as with diethyl phosphite [51]. In the first case, the reaction was complete after 12–25 min (Table 2/Entry 11). However, in the second case, the desired products were only formed in traces. Metal salts were also applied in the reaction of oxo compounds and trialkyl phosphites with success (Table 2/Entries 12–14) [52–54]. The reaction of slightly electron-rich benzaldehydes and triethyl phosphite in the presence of $ZnBr_2$ afforded the corresponding α-hydroxyphosphonates (Table 2/Entry 12) [52]. Bismuth(III) nitrate pentahydrate was applied efficiently as the catalyst in the reaction of aldehydes and triethyl phosphite under MW irradiation at 70 °C (Table 2/Entry 13) [53], while the niobium(V)chloride-trimethylsilyl chloride system was used in an amount of only 0.05 mol % at 25 °C to prepare hydroxyphosphonates (Table 2/Entry 14) [54].

1.3. The "Greenest" Protocol for the Synthesis of α-Hydroxyphosphonates

As it was presented in Sections 1.1 and 1.2, a great number of procedures have been elaborated for the synthesis of α-hydroxyphosphonates **1** from oxo compounds by reaction with dialkyl or trialkyl phosphites. Although, several methods involved the use of inexpensive and easily available catalysts together with mild and solvent-free reaction conditions, the work-up procedures still suffered from drawbacks. In most cases, the crude product was extracted with organic solvents, and then purified by column chromatography or recrystallization. The combination of the purification steps mentioned is also a frequent option [22,32,33,35,38,48].

We aimed at the elaboration of a new, environmentally-friendly, procedure for the synthesis of α-hydroxyphosphonates **1**. As opposed to the procedures already published, we wished to reduce the amount of volatile organic solvents used not only during the reaction, but also in the course of the work-up procedure. According to our method, an equimolar mixture of substituted benzaldehyde and dialkyl phosphite was stirred at reflux in the presence of 10 mol % of triethylamine as the catalyst, in a minimal amount of acetone (1.0 mL/11.0 mmol of the reagents). Then, *n*-pentane was added to the mixture, which was then cooled to 5 °C. The desired product crystallized from the reaction mixture, and the α-hydroxyphosphonates **1** could be obtained easily by a simple filtration in good to excellent yields (78–99%) in a pure form. The main novelty of our protocol is the absence of further purification steps, a consequence of the "one-pot" synthesis and precipitation from the reaction mixture (Scheme 2) [55].

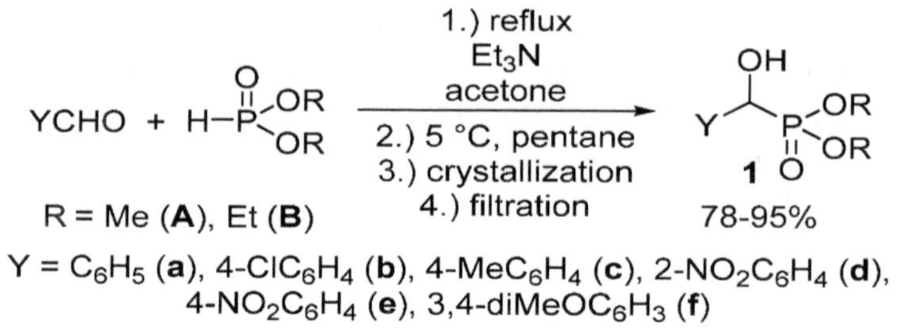

Scheme 2. A green, solvent-economical synthetic method towards α-hydroxyphosphonates (**1**).

To reveal the role of triethylamine in the addition, the reaction of benzaldehyde with diethyl phosphite was investigated by quantum chemical calculations [56]. It was found that in the absence of a catalyst, the enthalpy of activation is around 85.9 kJ/mol, and the addition is thermoneutral. The calculations revealed that triethylamine is able to promote the proton transfer from the >P(O)H reagent to the oxo compound. The amine-catalyzed reaction goes with a decreased enthalpy of activation of 68.8 kJ/mol, and follows an exothermic route ($\Delta H = -17.2$ kJ/mol). The suggested mechanism involving transition state **4** and intermediate **5** is shown in Scheme 3.

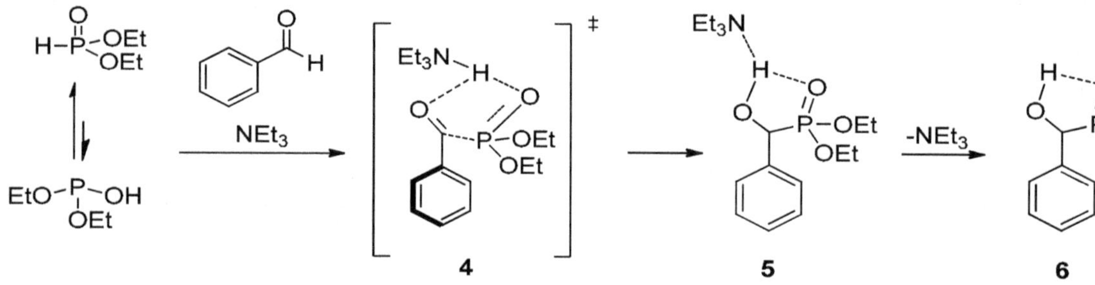

Scheme 3. Calculated mechanism of the triethylamine-catalyzed Pudovik reaction.

In conclusion, we were successful in the elaboration of another green protocol for the synthesis of α-hydroxyphosphonates **1** from substituted benzaldehydes and a series of dialkyl phosphites in the presence of triethylamine as the catalyst [55]. The new method involves the preparation in boiling acetone followed by crystallization of the product in high purity. There was no need for additional purification. The role of triethylamine in the reaction was justified by quantum chemical calculations [56].

2. Reactions of α-Hydroxyphosphonates

α-Hydroxyphosphonates **1** deserve interest not only as potential enzyme inhibitors [57], herbicides [58], antibiotics [59] and antibacterial or antifungal [60] agents, but also as starting materials for other valuable derivatives [61]. Section 2.1 summarizes the alkylation reactions of α-hydroxyphosphonates (Scheme 4, Route "A"). Acylation of the hydroxy function is one of the most extensively studied transformations within this family of compounds (Scheme 4, Route "B", Section 2.2). A series of α-acyloxyphosphonates (**8**) was synthesized applying different acylating agents, like carboxylic and sulfonyl chlorides, as well as carboxylic anhydrides and acids. The oxidation of α-hydroxyphosphonates leads to α-ketophosphonates **9** (Scheme 4, Route "C", Section 2.3). Another thoroughly studied field is nucleophilic substitution at the α carbon atom of α-hydroxyphosphonates to afford α-halo-, and α-amino or α-alkylphosphonates **10** (Scheme 4, Route "D", Section 2.4). The rearrangement of hydroxyphosphonates to benzyl phosphates **11** is an interesting field (Scheme 4, Route "E", Section 2.5). The curiosity of this transformation is the high substrate-dependence. Last but not least, α-hydroxyphosphonic acids **12** may be obtained by the hydrolysis of the ester function of α-hydroxyphosphonates (Scheme 4, Route "F", Section 2.6).

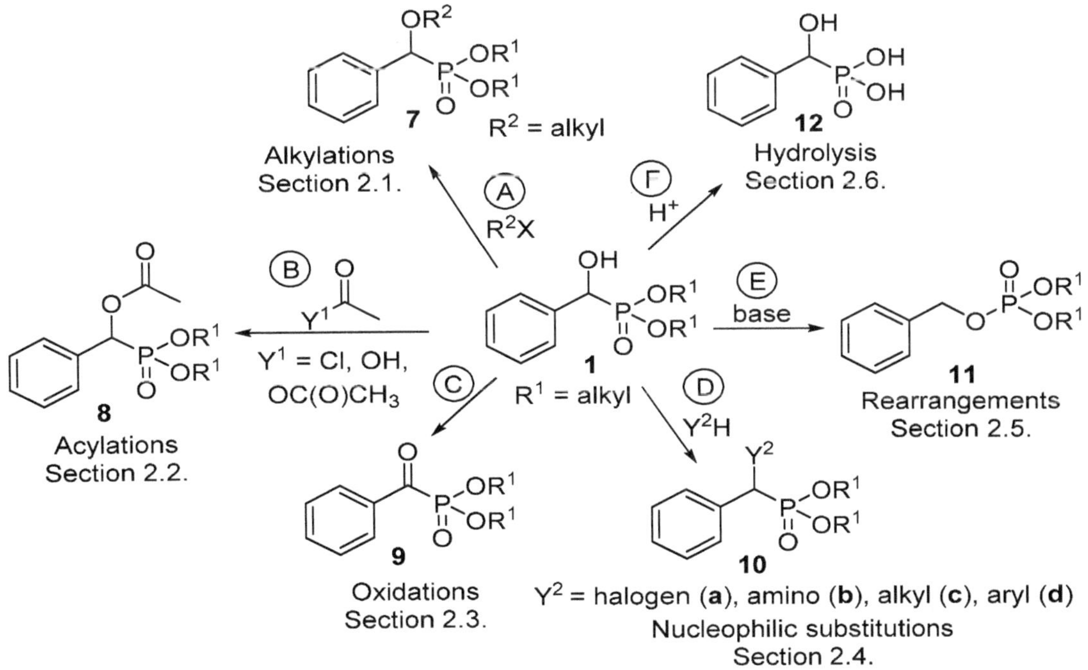

Scheme 4. Family tree of compounds **7**–**12** derived from α-hydroxyphosphonates (**1**).

The derivatives **7**–**12** synthesized from α-hydroxyphosphonates were summarized in the family tree shown in Scheme 4. Classes **7**–**12** represent potentially bioactive compounds.

2.1. *Alkylations*

Alkylation of the hydroxy function of α-hydroxyphosphonates **1** does not belong to their extensively studied reactions, as only a few articles reported the synthesis of α-alkoxyphosphonates **7**. A related article targeted antimalarial drugs, and the synthesis involved the *O*-alkylation of hydroxyphosphonates **1B** with benzyl bromoacetate in the presence of silver oxide (Scheme 5) [62].

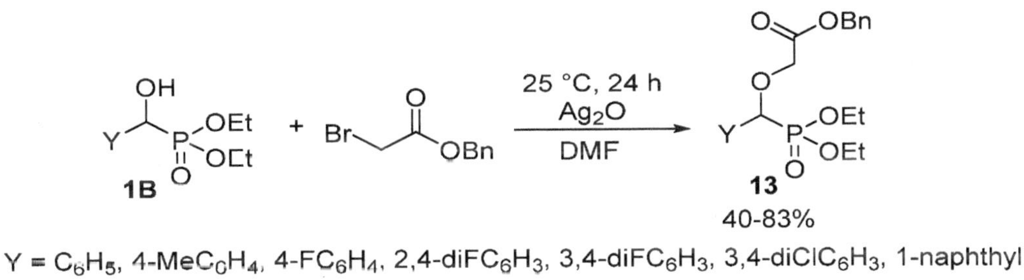

Scheme 5. *O*-alkylation of hydroxyphosphonates **1B** with benzyl bromoacetate.

The protection of the α-hydroxy function may also occur in the literature as another utilization of *O*-alkylation. A tetrahydropyranyl (THP) protecting group was introduced to an α-hydroxyphosphonate **1B** by treatment with 3,4-dihydro-2*H*-pyran applying *p*-toluenesulfonic acid as the catalyst [63]. The silylation of α-hydroxyphosphonates **1** is also a useful tool to protect the hydroxy function [64–67]. The reaction of hydroxyphosphonates **1B** with hexamethyldisilazane (HMDS) was studied using a great variety of catalysts including iodine (Scheme 6, Route "a") [65], copper triflate (Scheme 6, Route "b") [66] or iron(III) trifluoroacetate (Scheme 6, Route "c") [67].

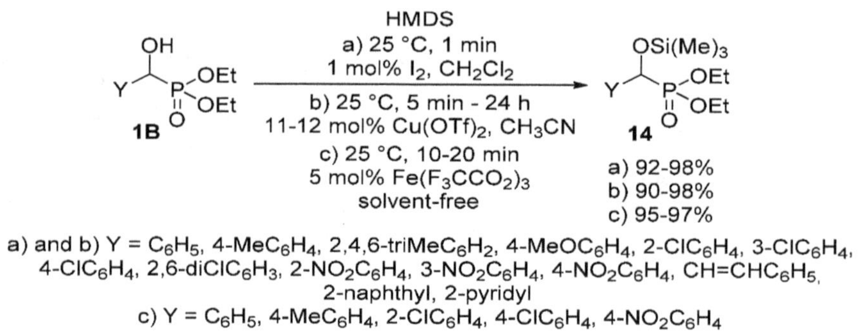

Scheme 6. Possible ways of introducing a silyl protecting group to α-hydroxyphosphonates **1B**.

A unique protocol has been reported for the cyclization of α-hydroxyphosphonates bearing an alkynyl function in the *ortho*-position of the aromatic ring (**15**) [68]. Applying Pd(II) acetate as the catalyst, the nucleophilic attack of the α-*O* atom occurred on the more distant *C*-atom of the triple bond of the alkynyl substituent to form the six-membered product **16** selectively. Surprisingly, changing the $Pd(OAc)_2$ catalyst to DBU used in 2 equivalents led to a five-membered cyclic product **17**. In the latter case, the attack of the hydroxy group occurred on the nearer *C*-atom of the triple bond, followed by isomerization (Scheme 7).

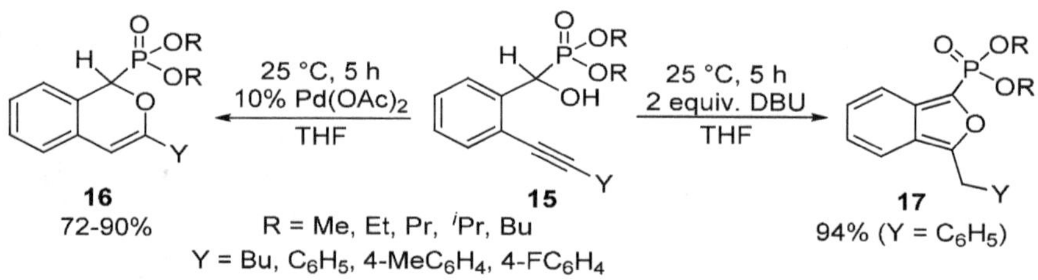

Scheme 7. Selective cyclization of *ortho*-alkynyl α-hydroxyphosphonates **15**.

2.2. Acylations

The synthesis of α-acyloxyphosphonates **8** is one of the most thoroughly studied transformations of α-hydroxyphosphonates **1**. α-Acyloxyphosphonates **8** attracted attention as potential herbicides [69–71], fungicides [69], and insecticides [72]. A number of articles reported the acylation of α-hydroxyphosphonates **1** with carboxylic acid chlorides in the presence of triethylamine [73–75] or pyridine [76,77] as the hydrochloric acid scavenger. Among the –C(O)Cl reagents, phenoxyacetyl chloride derivatives **18** are frequently applied acylating agents [70,71,76–84].

In general, the reaction of α-hydroxyphosphonates **1** with phenoxyacetyl chlorides **18** required a reaction time of 3–7 h, and the corresponding α-acyloxyphosphonates **19** were isolated in yields of 53–89% (Scheme 8) [78]. The compounds **19** thus-obtained are popular target molecules, as a series of related derivatives exhibited herbicidal activity [70,76–80,83,84].

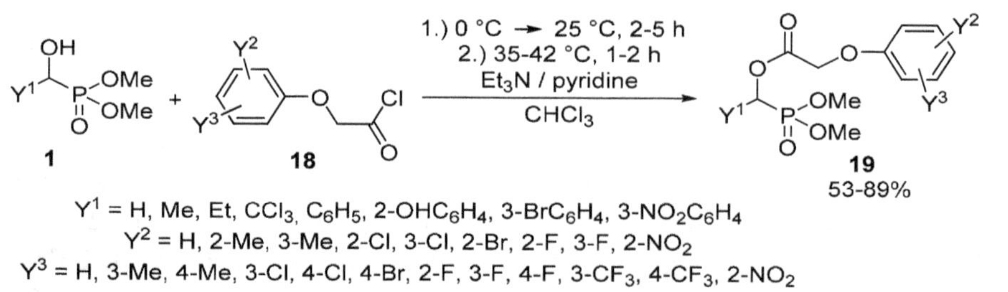

Scheme 8. Synthesis of α-acyloxyphosphonates (**19**) from α-hydroxyphosphonates (**1**).

The reaction of α-hydroxy-phosphonates **1A** with 2,6-pyridinedicarboxylic acid chloride (**20**) led to bis[dimethyl phosphonomethyl pyridine-2,6-dicarboxylates] **21** containing two α-acyloxyphosphonate moieties (Scheme 9). The reaction took place under mild conditions, and the products **21** were isolated with variable yields (38–80%). The derivatives so-obtained **21** also showed herbicidal activity [85].

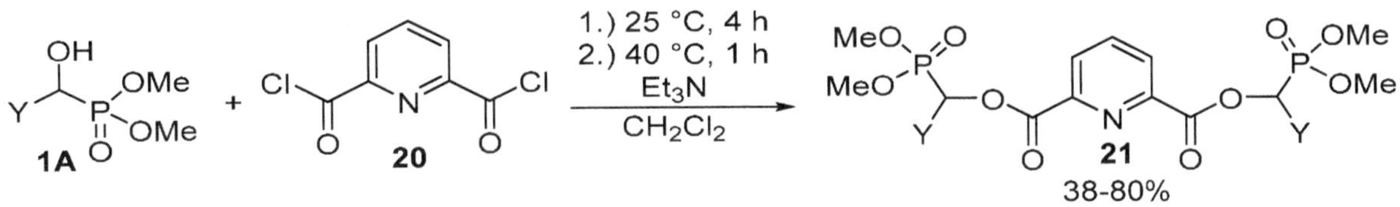

Y = H, Me, Et, Pr, ^{i}Pr, CCl_3, C_6H_5, 4-MeC_6H_4, 4-$MeOC_6H_4$, 2-ClC_6H_4, 4-ClC_6H_4, 2,4-diClC_6H_3, 3-$NO_2C_6H_4$, 3,4-$OCH_2OC_6H_3$, 2-furyl, 2-thienyl

Scheme 9. Acylation of α-hydroxyphosphonates (**1A**) with 2,6-pyridinedicarboxylic acid chloride (**20**).

The acylation of α-hydroxyphosphonates **1B** has been also performed with acetic anhydride [86–88]. This reaction was carried out in the presence of a catalytic amount (1.3–8 mol %) of $Cu(OTf)_2$ at 25 °C to afford the desired α-acetyloxyphosphonates **22** almost quantitatively (Scheme 10, Route "a") [86]. In another instance, 1 mol % of $TiCl_3(OTf)$ was employed as the catalyst (Scheme 10, Route "b"). In this case, completion of the reaction required a somewhat higher temperature (70 °C) [87]. A more benign synthetic route to α-acetyloxyphosphonates **22** is the reaction of α-hydroxyphosphonates (**1B**) and acetic anhydride under MW irradiation (Scheme 10, Route "c") [88]. The acylation was complete after a short reaction time of 5 min, and the products were obtained in excellent yields (90–98%). However, a lack of control of the reaction temperatures is a shortcoming that prevents reproduction.

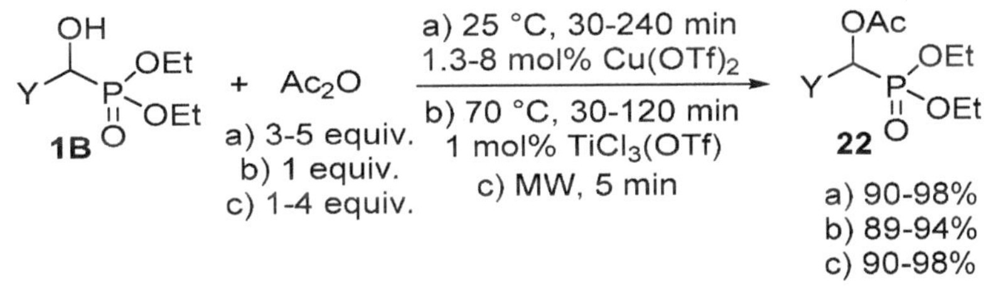

a) and c) Y = C_6H_5, 4-MeC_6H_4, 2,4,6-triMeC_6H_2, 4-$MeOC_6H_4$, 2-ClC_6H_4, 3-ClC_6H_4, 4-ClC_6H_4, 2,6-diClC_6H_3, 2-$NO_2C_6H_4$, 3-$NO_2C_6H_4$, 4-$NO_2C_6H_4$, 2-naphthyl, 2-pyridyl, $CH_3CH{=}CH$, $C_6H_5CH{=}CH$
b) Y = C_6H_5, 4-MeC_6H_4, 4-ClC_6H_4

Scheme 10. Possible ways for the synthesis of α-acetyloxyphosphonates **22**.

The synthesis of α-acyloxyphosphonates **8** was also carried out starting from substituted benzaldehydes, dimethyl phosphite and acetic anhydride in the presence of magnetic nanoparticle-supported guanidine [89]. In this one-pot procedure, the formation of the α-hydroxyphosphonate **1** in the Pudovik reaction was followed by acylation of the hydroxy function.

Carboxylic acids may also be efficient acylating agents of α-hydroxyphosphonates **1** in the presence of *N,N′*-dicyclohexylcarbodiimide (DCC) as the coupling reagent, and 4-dimethyl- aminopyridine (DMAP) as the base [27,90,91]. This is exemplified by the reaction of 2-(4,6-dimethoxypyrimidin- 2-yloxy)benzoic acid (**23**) at 25 °C to give α-acyloxyphosphonates **24** possessing herbicidal activity (Scheme 11) [27].

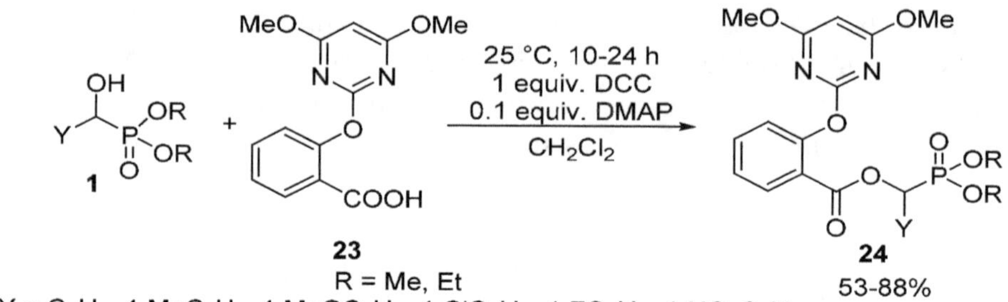

R = Me, Et

Y = C_6H_5, 4-MeC_6H_4, 4-$MeOC_6H_4$, 4-ClC_6H_4, 4-FC_6H_4, 4-$NO_2C_6H_4$

Scheme 11. Coupling reaction of α-hydroxyphosphonates (**1**) with carboxylic acid **23**.

Another method of choice for the acylation of α-hydroxyphosphonates **1B** with carboxylic acids, is the Mitsunobu reaction applying 4,4′-azopyridine (**25**) and triphenylphosphine in acetonitrile as the solvent (Scheme 12) [92]. Completion of the reaction required a reaction time of 9–24 h at 80 °C, and the products **26** were obtained in yields of 65–88%.

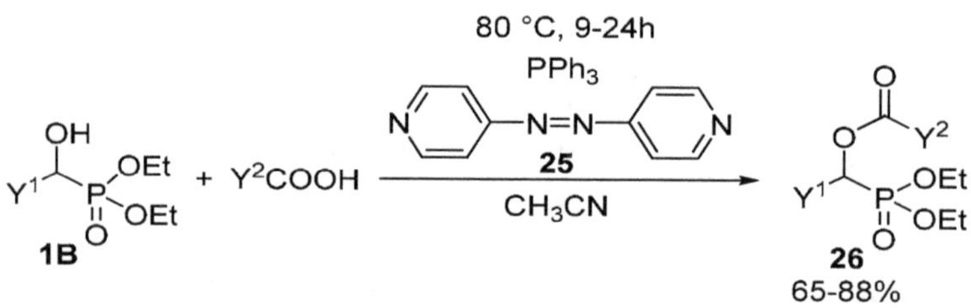

Y^1 = C_6H_5, 4-MeC_6H_4, 4-ClC_6H_4, 2,6-diClC_6H_3

Y^2 = Et, C_6H_5, 4-MeC_6H_4, 2-ClC_6H_4, 3-ClC_6H_4, 4-ClC_6H_4, 4-$NO_2C_6H_4$, $CH_3CH{=}CH$

Scheme 12. Synthesis of α-acyloxyphosphonates (**26**) via Mitsunobu reaction.

The reaction of α-hydroxyphosphonates **1** with isocyanates **27** or isothiocyanates **29** resulted in the formation of α-carbamoyloxy-**28** or α-thiocarbamoyloxyphosphonates **30**, respectively [93,94]. One study targeted herbicides by the acylation of α-hydroxyphosphonates (**1**) with pyrimidine-based isocyanates **27** (Scheme 13, Route "A") [93].

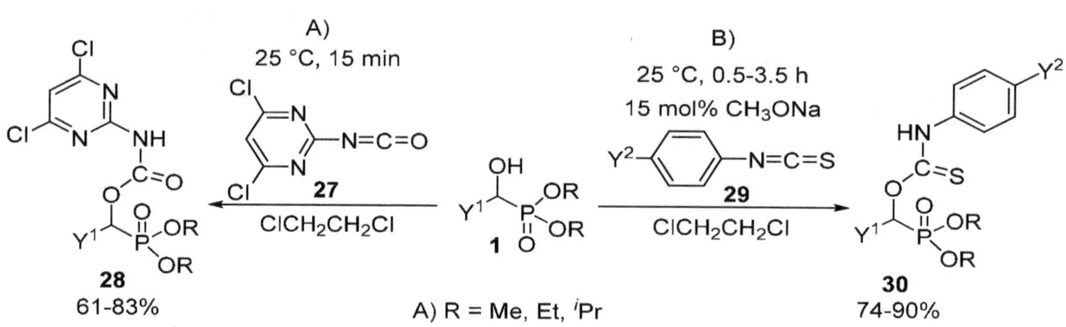

A) R = Me, Et, ^{i}Pr

Y^1 = H, Me, Et, Bu, C_6H_5, 2-ClC_6H_4, 4-ClC_6H_4, 4-$NO_2C_6H_4$, 2-furyl

B) R = Me, Et, ^{i}Pr; Y^2 = EtO, Cl, Br

Y^1 = C_6H_5, 4-MeC_6H_4, 2-ClC_6H_4, 4-ClC_6H_4

Scheme 13. The reaction of α-hydroxyphosphonates (**1**) with isocyanates **27** and isothiocyanates **29**.

The reaction took place under benign conditions, and the products **28** so-obtained revealed a weak herbicidal activity. On reacting hydroxyphosphonates (**1**) and isothiocyanates **29** in the absence of a catalyst, the corresponding α-thiocarbamoyl-oxyphosphonates **30** were obtained in modest yields of 20–30%. However, the application of 15 mol % of sodium methoxide promoted the acylation, and the products **30** could be isolated in yields of 74–90% (Scheme 13, Route "B") [94]. The authors of the article emphasize the need for mild reaction conditions (25 °C, 0.5–3.5 h), as heating of the reaction mixture resulted in the formation of a by-product through the trimerization of the starting isothiocyanate. The α-thiocarbamoyl-oxyphosphonates **30** obtained in this way expressed plant growth regulating activity.

As shown above, the acylation of α-hydroxyphosphonates (**1B**) was carried out with a wide range of carboxylic acid derivatives. Besides carboxylic chlorides, anhydrides and acids, sulfonic acid derivatives may also function as the acylating agents. By analogy with the reaction of α-hydroxyphosphonates (**1B**) with carboxylic chlorides, the acylation was also attempted with sulfonyl chlorides [95–97]. Methanesulfonyl chloride (Scheme 14, Route "a") [96] and *p*-toluenesulfonyl chloride (Scheme 14, Route "b") [97] worked well as sulfonylating agents in the presence of triethylamine.

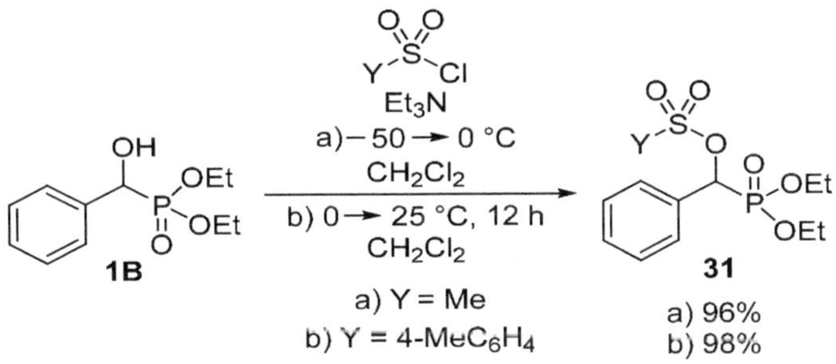

Scheme 14. Sulfonylation of α-hydroxyphosphonates **1B**.

In contrast to the wide range of α-acyloxy and α-sulfonyloxy phosphonates, α-phosphoryloxy analogues appear rarely in the literature. A study from the early 1960's reported the synthesis of a >P(O)OCHP(O)< type compound **33** by a thermo-induced rearrangement (Scheme 15) [98]. However, the phosphorylation of α-hydroxyphosphonates (**1**) with *P*-chlorides presented a challenge.

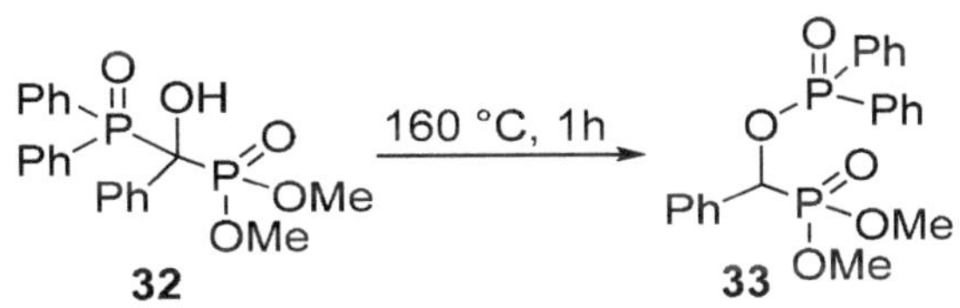

Scheme 15. Synthesis of *O*-phosphoryloxy phosphonate **33** by rearrangement.

Recognizing the lack of *O*-phosphorylated α-hydroxyphosphonates **34** in the literature, we aimed at the synthesis of new classes of derivatives by the acylation of α-hydroxyphosphonates **1** with phosphoryl and phosphinic chlorides. It was observed that by stirring the reaction components at ambient temperature, the *O*-phosphorylated phosphonates **34** were formed with conversions of 55–90% (Scheme 16) [99].

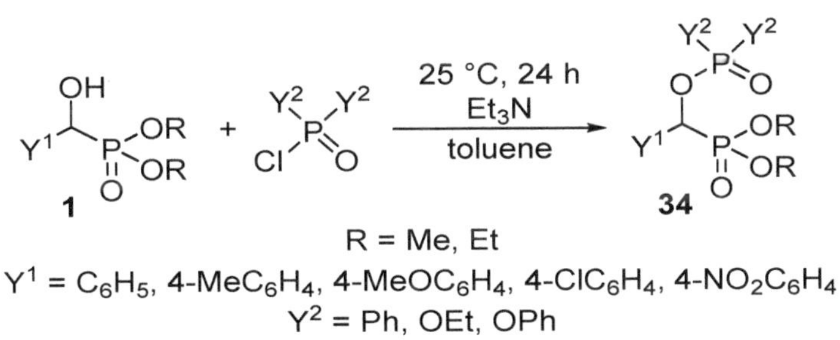

Scheme 16. Phosphorylation of α-hydroxyphosphonates **1** with phosphoryl and phosphinic chlorides.

2.3. Oxidations

The oxidation of α-hydroxyphosphonates **1B** affords the corresponding α-ketophosphonates (**35**) that are versatile precursors of a series of organophosphorus compounds [100–102]. The application of metal compounds with variable valency as the oxidizing agent is predominating in the literature. Alumina-supported chromium(III) oxide (CrO_3/Al_2O_3) (Scheme 17, Route "a") [103], and chromium salts, like zinc dichromate trihydrate ($ZnCr_2O_7 \cdot 3H_2O$) (Scheme 17, Route "b") [104] were found to be efficient reagents in the oxidation of α-hydroxyphosphonates **1B**. In both cases, the reaction took place at ambient temperature under solvent-free conditions. The procedures provided the desired products **35** in good to excellent yields (up to 96%).

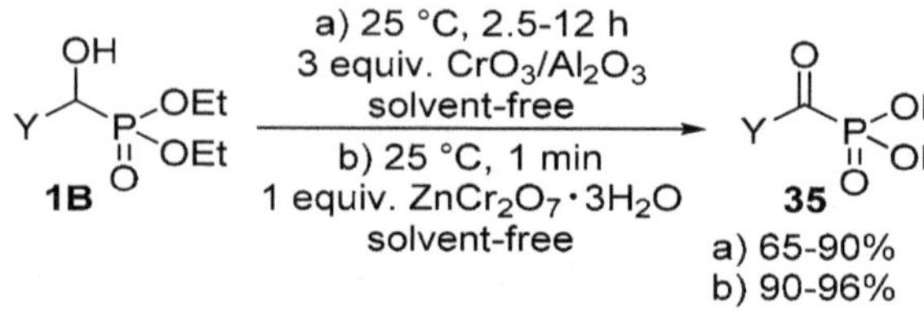

a) Y = Bu, Pent, C_6H_5, 4-MeC_6H_4, 4-$MeOC_6H_4$, 2-ClC_6H_4, 4-ClC_6H_4, 2,4-diClC_6H_3, 2-$NO_2C_6H_4$, 3-$NO_2C_6H_4$, 4-$NO_2C_6H_4$, 2-furyl, 1-naphthyl, 2-naphthyl, $CH_3CH{=}CH$, $C_6H_5CH{=}CH$, $C_6H_5CHCH_3$, $C_6H_5(CH_2)_2$
b) Y = ^{i}Pr, C_6H_5, 4-MeC_6H_4, 4-$MeOC_6H_4$, 2,4,6-triMeC_6H_2, 2-ClC_6H_4, 3-ClC_6H_4, 4-ClC_6H_4, 2,6-diClC_6H_3, 2-$NO_2C_6H_4$, 3-$NO_2C_6H_4$, 4-$NO_2C_6H_4$, 2-naphthyl

Scheme 17. Oxidation of α-hydroxyphosphonates **1B** using metal compounds with variable valency.

In other variations, pyridinium dichromate [105] or pyridinium chlorochromate [106] were applied successfully to give α-ketophosphonates **35**. The reaction under discussion was also reported using $KMnO_4$ as the oxidant (Scheme 18). A shortcoming of this latter protocol is the need for dry benzene as the reaction medium [107]. Another problem of the chemical oxidating agents is the low atom efficiency and the lack of "greenness".

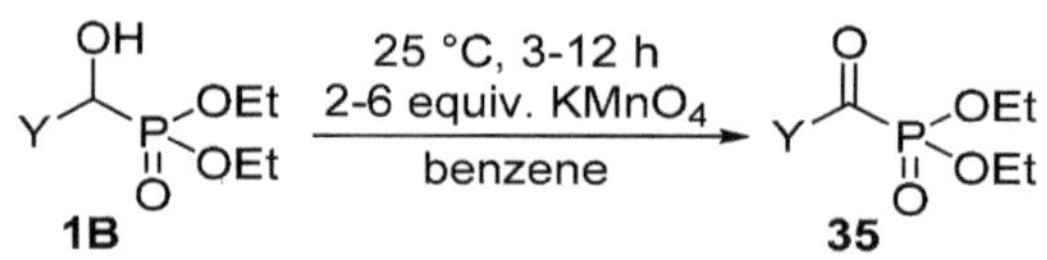

85-98%

Y = C_6H_5, 4-MeC_6H_4, 4-$MeOC_6H_4$, 2,4,6-triMeC_6H_2, 2-ClC_6H_4, 3-ClC_6H_4, 4-ClC_6H_4, 2,6-diClC_6H_3, 2-$NO_2C_6H_4$, 3-$NO_2C_6H_4$, 4-$NO_2C_6H_4$, $C_6H_5CHCH_3$, 2-naphthyl, 3-pyridyl

Scheme 18. Synthesis of α-ketophosphonates **35** from α-hydroxyphosphonates **1B** with $KMnO_4$ oxidant.

An elegant achievement is the oxidation of dibenzyl α-hydroxyphosphonates **1C** with oxygen, employing a chiral vanadyl(V) complex (**37**) as the catalyst (Scheme 19) [108]. After a reaction time of 6–150 h, the corresponding α-ketophosphonates **36** were obtained in conversions of 49–51%, while due to the presence of the optically active catalyst, the (*S*) enantiomer of the α-hydroxyphosphonate (*S*)-**1C** remained untouched. This protocol offered an efficient way for the resolution of α-hydroxyphosphonates **1C** with high enantiomeric excesses up to 99% (with two exceptions), in parallel with the formation of α-ketophosphonates **36**.

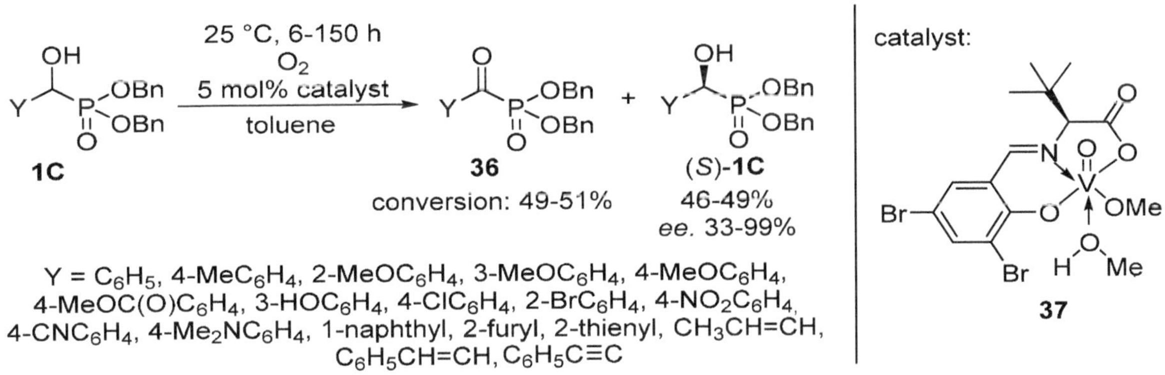

Scheme 19. Tandem resolution and oxidation of α-hydroxyphosphonates **1C**.

2.4. Nucleophilic Substitutions

The nucleophilic substitution of the hydroxy group of α-hydroxyphosphonates **1** represents one of the most widely studied reaction types of this class of compounds. A typical transformation is the change of the hydroxy function to a halogen atom. α-Halobenzylphosphonates **38** deserve interest as potential inhibitors of protein tyrosine phosphatases [109,110]. A chloro or bromo substituent may be introduced into an α-hydroxyphosphonate **1** molecule by treatment with thionyl chloride [111] or thionyl bromide (Scheme 20, Route "a") [110], respectively. Another method of choice to obtain α-halobenzylphosphonates **38** is the joint application of triphenylphosphine and the corresponding carbon tetrahalide [112–114]. One procedure applied this synthetic route to obtain α-chlorobenzylphosphonates (Scheme 20, Route "b") [114], while another paper reported the introduction of a chloro atom to both aliphatic and aromatic hydroxyphosphonates under similar conditions (Scheme 20, Route "c") [113].

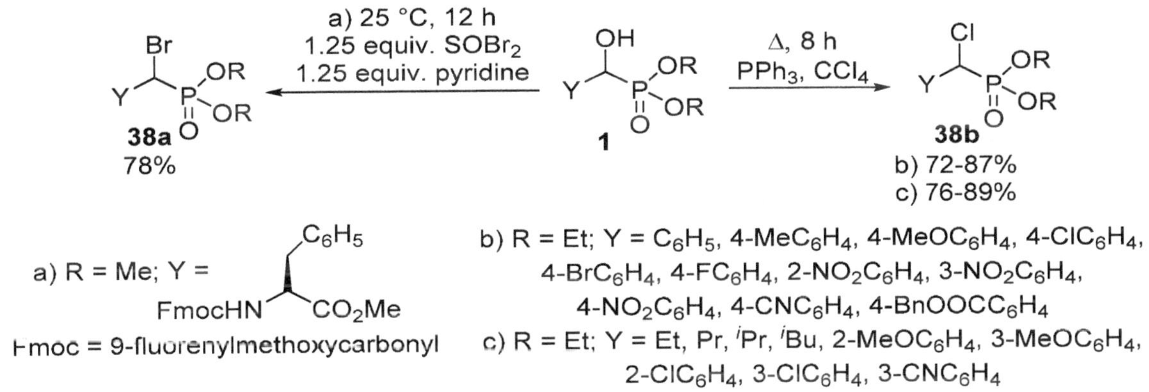

Scheme 20. Synthetic routes to obtain α-halophosphonates **38**.

A further variation involves the use of tetrabutylammonium bromide as the nucleophile in the presence of triphenylphosphine—DDQ (2,3-dichloro-5,6-dicyanobenzoquinone) system, affording α-bromophosphonates [115,116].

In another protocol, hydroxyphosphonates **1** were converted to their corresponding α-halo analogues **38** by reaction with molecular halogens (Br_2 or I_2) applying 4-aminophenyl-diphenylphosphinite **39** as the acid scavenger (Scheme 21) [117]. The amino function of the phosphinite **39** made possible the removal of the forming hydrogen halogenide as an ammonium salt after a simple filtration.

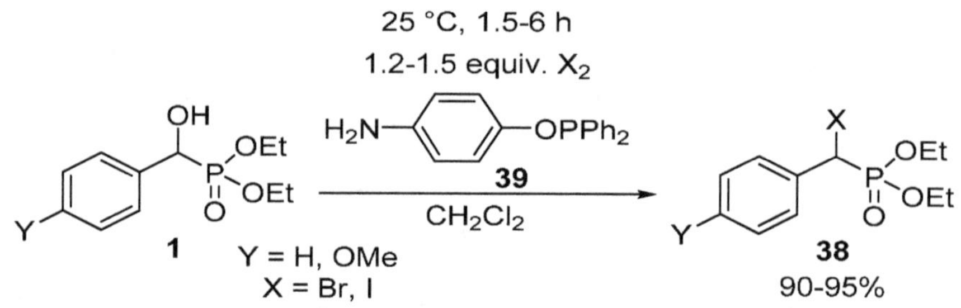

Scheme 21. Synthesis of α-halophosphonates **38** from α-hydroxyphosphonates **1** with molecular halogens.

To synthesize α-fluorophosphonates, the most commonly used fluorinating agent is diethylaminosulfur trifluoride (Et_2NSF_3) [111,114,118–121]. As a variation of this method, a protocol applying morpholinosulfur trifluoride was elaborated for the OH → F change in α-hydroxy-bisphosphonates [122]. *N*-fluorobisbenzenesulfonimide was used efficiently to introduce a fluoro function to the α-carbon atom of α-monohalophosphonates **38** in the presence of sodium hexamethyldisilazane [113,114,116].

Halogens are not the sole nucleophiles that the α-hydroxy function of benzylphosphonates **1B** may be replaced with. One method operated with the above mentioned triphenylphosphine—DDQ system to catalyze the reaction of α-hydroxyphosphonates **1B** with ammonium thiocyanate to obtain α-thiocyanatophosphonates **40** (Scheme 22, Route "A") [115]. Another study reported the Mitsunobu reaction of α-hydroxyphosphonates **1B** with hydrazoic acid as the nucleophile. The treatment of the so-formed α-azidophosphonates **41** with triphenylphosphine and carbon disulfide resulted in the formation of α-isothiocyanatophosphonates **42** (Scheme 22, Route "B") [123].

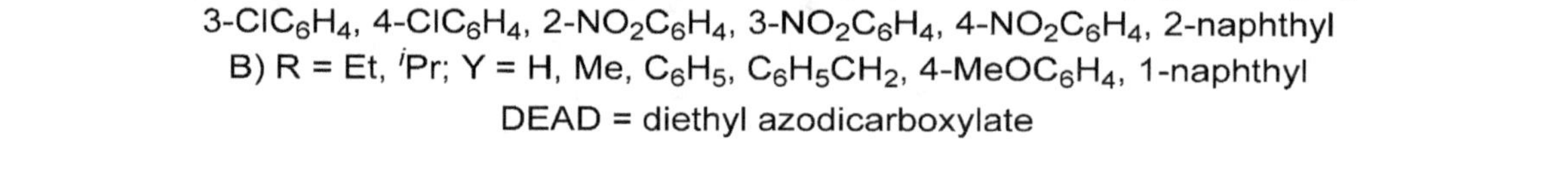

Scheme 22. Nucleophilic substitution of α-hydroxyphosphonates **1B** with NH_4SCN and hydrazoic acid.

α-Aminophosphonates **43** are usually obtained by the three-component condensation (Kabachnik-Fields reaction) of oxo compounds, amines and dialkyl phosphites or secondary phosphine oxides [124–127]. However, these compounds may also be synthesized via the nucleophilic substitution of α-hydroxyphosphonates **1B** with primary or secondary amines. According to a simple protocol, this latter reaction was performed under MW irradiation, in the presence of Al_2O_3 [128]. After 3–7 min, a complete conversion was attained (Scheme 23). However, reaction temperatures are missing from the articles, as the experiments were carried out in a kitchen MW oven.

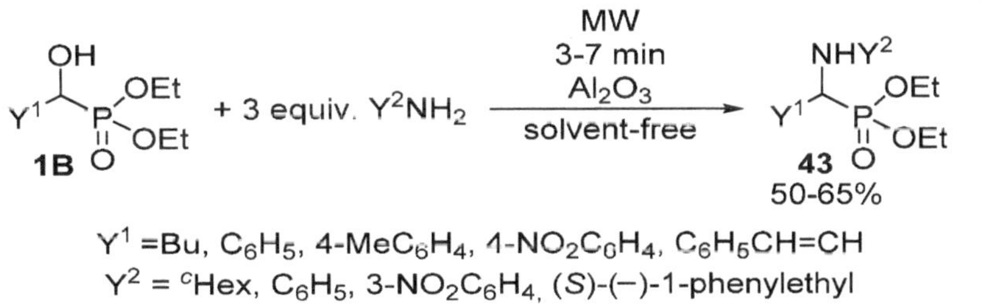

Scheme 23. Synthesis of α-aminophosphonates **43** by nucleophilic substitution of **1B** with amines.

We found that α-hydroxyphosphonates **1B** may be converted to the corresponding α-aminophosphonates **44–46** under MW conditions without the use of any catalyst or solvent (Scheme 24) [56,129]. The starting hydroxyphosphonates **1B** were reacted with three equivalents of a primary amine (propyl-, butyl- or cyclohexyl-amine) applying MW irradiation at 100 °C. Following this method, the substitution took place with surprising ease. Starting from substituted hydroxyphosphonates **1Bb**, **1Bc** and **1Bg**, a short reaction time of 15–30 min was sufficient for complete conversion (Table 3/Entries 4–12). The unsubstituted analogue **1Ba** was even more reactive, and in this case the reaction took place after 10–15 min (Table 3/Entries 1–3).

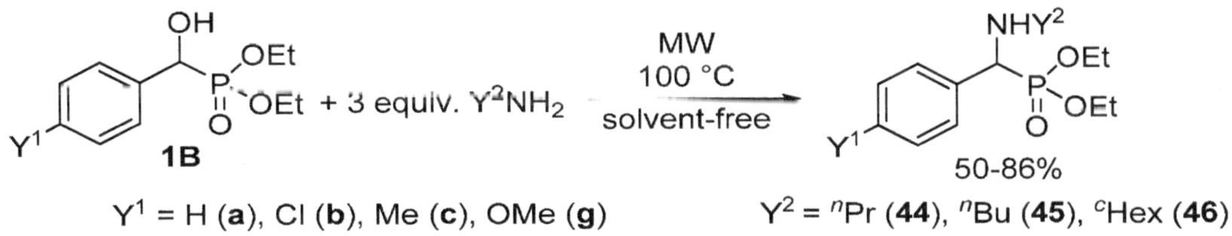

Scheme 24. MW-Assisted synthesis of α-aminophosphonates **44–46** through nucleophilic substitution.

Table 3. Details of the synthesis of α-aminophosphonates **44–46** from α-hydroxyphosphonates **1B**.

Entry	Y^1	Y^2	Reaction Time (min)	Yield (%)	Product	Ref.
1	H	^{n}Pr	10	78	**44a**	[129]
2	H	^{n}Bu	15	86	**45a**	[129]
3	H	^{c}Hex	10	84	**46a**	[129]
4	Cl	^{n}Pr	15	60	**44b**	[50]
5	Cl	^{n}Bu	20	50	**45b**	[56]
6	Cl	^{c}Hex	30	54	**46b**	[56]
7	Me	^{n}Pr	15	58	**44c**	[56]
8	Me	^{n}Bu	15	79	**45c**	[56]
9	Me	^{c}Hex	30	73	**46c**	[56]
10	OMe	^{n}Pr	15	72	**44g**	[56]
11	OMe	^{n}Bu	30	66	**45g**	[56]
12	OMe	^{c}Hex	30	70	**46g**	[56]

Encouraged by the recognition that the reaction of α-hydroxyphosphonates **1B** with primary amines takes place rather easily, we wished to study the mechanism of the substitution by quantum chemical (DFT) calculations [129]. The nucleophilic attack of methylamine on diethyl α-hydroxybenzylphosphonate **1Ba** was chosen as the model reaction. It was found that the reaction follows an S_N2 mechanism. It was also revealed that a favorable neighboring group effect facilitates the transfer of the oxygen atom of the hydroxy group from the α-carbon atom to the phosphorus (transition state **47**) [129]. The proposed mechanism is shown on Scheme 25.

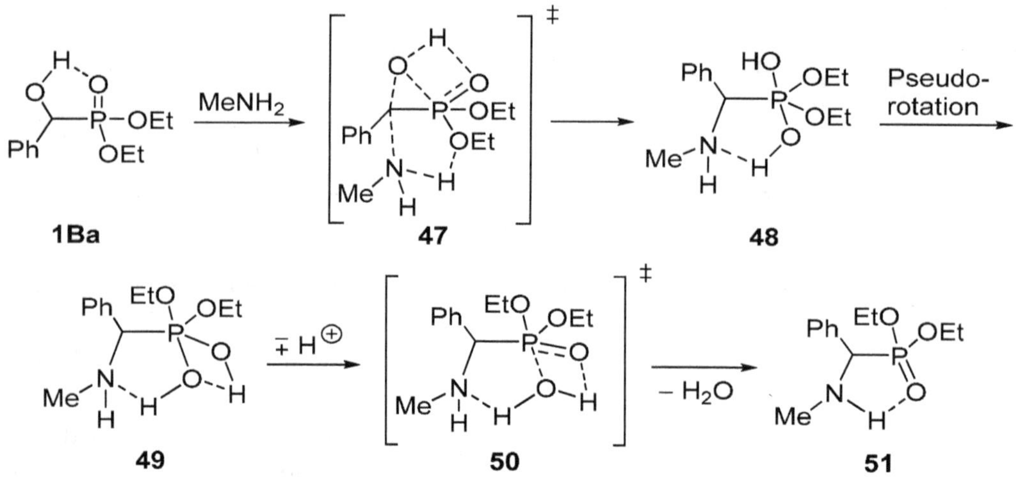

Scheme 25. Calculated mechanism of the substitution of hydroxyphosphonate **1Ba** with methylamine.

α-Sulfonamidophosphonates were obtained through the reaction of α-hydroxyphosphonates and sulfonamides in the presence of HOTf in dioxane. The reaction took place at ambient temperature within 5 h, and the desired products were obtained in yields of 70–94% [130]. The products so-obtained were found to be as promising corrosion inhibitors for mild steel [131].

The Friedel-Crafts alkylation using α-hydroxyphosphonates **1** in the presence of an acid catalyst (e.g., $FeCl_3$ or HOTf) led to α,α-diarylphosphonates **53** [132–134]. Benzene and naphthalene served as the aromatic substrate. According to the proposed mechanism, the first step is the formation of benzylphosphonate carbocation **52** that is followed by its electrophilic attack on the aromatic substrate (Scheme 26) [132].

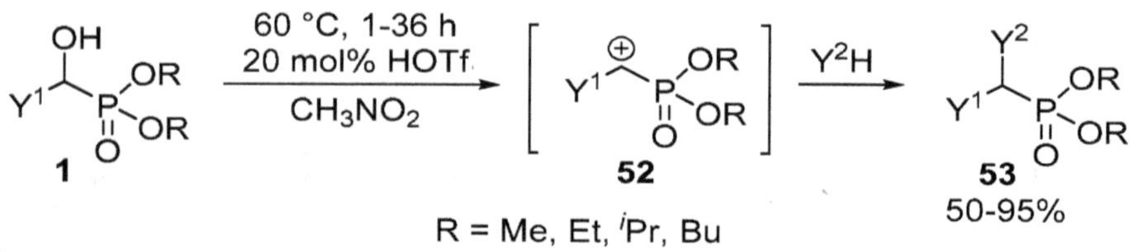

R = Me, Et, ^{i}Pr, Bu

Y^1 = 4-MeC_6H_4, 2,4-diMeC_6H_3, 3,5-diMe-4-OHC_6H_2, 2-$MeOC_6H_4$, 4-$MeOC_6H_4$, 2,4-di$MeOC_6H_3$, 3,4,5-tri$MeOC_6H_2$, 4-$EtOC_6H_4$, 4-$MeSC_6H_4$, 4-$Me_2NC_6H_4$

Y^2 = 2-MeC_6H_4, 4-MeC_6H_4, 4-$^tBuC_6H_4$, 2,4,6-triMeC_6H_2, 3,5-diMe-4-OHC_6H_2, 4-$MeOC_6H_5$, 2,4-di$MeOC_6H_3$, 2,5-di$MeOC_6H_3$, 2,4,6-tri$MeOC_6H_2$, 2-OH-4-$^tBuC_6H_3$, 4-$MeSC_6H_4$, 1-naphthyl, 4-Me-1-naphthyl, 1-Me-3-indolyl, 1-Me-2-pyrrolyl, 2-Me-5-thienyl, 3-benzofuryl

Scheme 26. Friedel-Crafts alkylation of α-hydroxyphosphonates **1** with arene nucleophiles.

An interesting reaction is when α-hydroxyphosphonates **1B** function as an alkylating agent of 1,3-diketones to afford γ-ketophosphonates **54** (Scheme 27) [135]. The reaction of the starting components in organic solvents in the presence of $FeCl_3$ or $Cu(OTf)_2$ Lewis acids resulted in the formation of compound **54** (Scheme 27, Route "A"). The optimization of the alkylations was a real challenge for the authors, as each hydroxyphosphonate—diketone combination required different reaction conditions regarding the Lewis acid, temperature, time and solvent. However, applying $FeCl_3 \cdot 6H_2O$ as the catalyst, a C–C bond cleavage of the desired product **54** also occurred due to the presence of the crystal hydrate of the iron salt. In this case, a mixture of compounds **54** and **55** was attained (Scheme 27, Route "B").

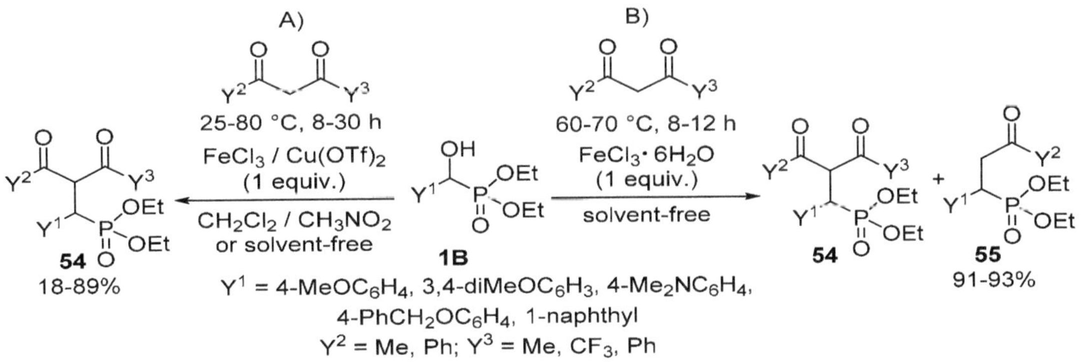

Scheme 27. Alkylation of 1,3-diketones with α-hydroxyphosphonates **1B**.

2.5. Rearrangements

The rearrangement of α-hydroxyphosphonates **1** to benzyl phosphates was first discovered through the example of the widely known insecticide prodrug, trichlorofon (*O,O*-dimethyl (2,2,2-trichloro-1-hydroxyethyl)phosphonate) that is rearranged to 2,2-dichlorovinyl dimethyl phosphate (DDVP) acting as an acetylcholinesterase inhibitor [136]. Phosphonate-phosphate rearrangements were investigated in the presence of strong bases, such as sodium hydroxide [136], sodium ethoxide [137] and sodium hydride [138]. A number of protocols operated with triethylamine as the catalyst [139–141]. However, the Et_3N-catalyzed rearrangement is not a versatile method, as it was reported to apply to only six-membered cyclic phosphonates [139–141].

A new approach of this transformation is when the Pudovik reaction of benzaldehydes and dialkyl phosphites and the subsequent rearrangement of the so-formed α-hydroxyphosphonate **1** take place in "one-pot", under the same conditions. The synthesis of benzyl phosphates **56** directly from an oxo compound and a *P*-reagent was efficiently catalyzed by butyllithium (Scheme 28, Route "a") [142] as well as by DBU (Scheme 28, Route "b") [143]. The tandem Pudovik reaction and rearrangement was also reported in the presence of superbases **57** (Scheme 28, Route "c") [144]. The authors of this latter article emphasized that both the reaction media and the substituent in the aromatic ring had a great impact on the reaction. The rearrangement took place faster in alcohols (EtOH or ^{i}PrOH) than in acetonitrile. It was also shown that the presence of electron-withdrawing substituents facilitated the formation of the benzyl phosphate **56** [144].

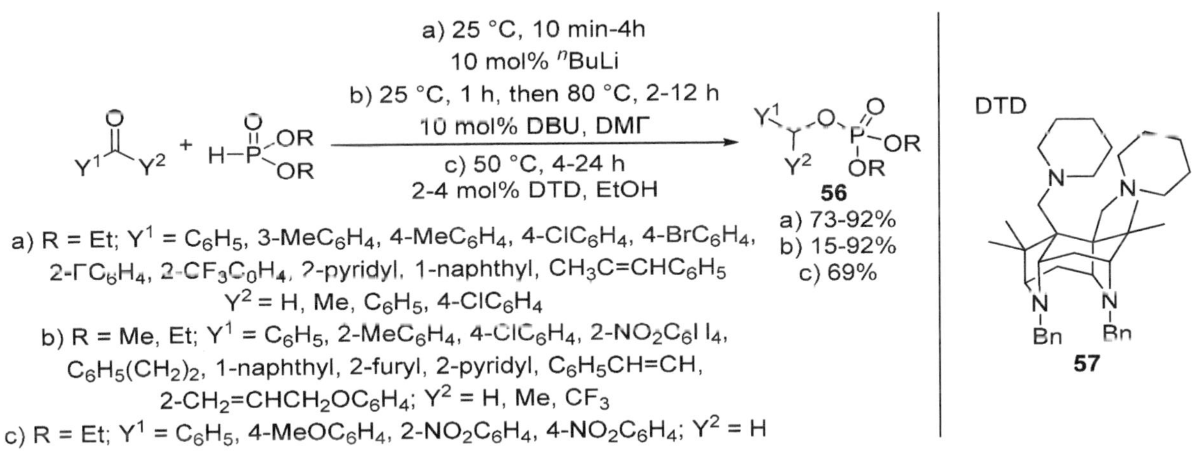

Scheme 28. Methods of the tandem Pudovik reaction and rearrangement affording benzyl phosphates **56**.

In this field, our research targets the accomplishment of the rearrangement of α-hydroxyphosphonates **1** using bases under phase transfer catalytic conditions (Scheme 29) [145].

Scheme 29. Rearrangement of α-hydroxyphosphonates **1** under phase-transfer catalytic conditions.

Our preliminary results highlighted that the substituent in the aromatic ring has a significant impact on the reaction. In the presence of electron-withdrawing substituents (Cl and F), the reaction took place at ambient temperature, whereas the rearrangement of α-hydroxyphosphonates bearing electron-releasing substituents required 2–3 h of heating [145].

2.6. Hydrolysis

The hydrolysis of α-hydroxyphosphonates **1** is the most common way for the synthesis of α-hydroxyphosphonic acids **12** that may be inhibitors of a wide variety of enzymes including protein tyrosine phosphatases [110,146,147], mandelate racemase [148], phosphoglycerate kinase [149], plant P5C reductase [150] and plant glutamine synthetase [151]. A series of α-hydroxyphosphonic acid (**12**) and α-hydroxybisphosphonic acid was recognized as antibiotics against Gram-positive bacteria [152]. α-Hydroxyphosphonic acids **12** bearing an amide function in the *para*-position of the aromatic ring were found to be in vivo inhibitors of autotaxin [153].

As for the synthesis of α-hydroxyphosphonic acids **12**, most of the related articles reported the acidic hydrolysis of the corresponding methyl or ethyl ester [149,150,152,154]. A number of protocols applied hydrochloric acid in an excess at reflux, and the hydrolysis usually required a prolonged reaction time of 6–24 h [149,150,152]. According to another method, beside the excess of 6 M hydrochloric acid, dioxane was also added as a co-solvent [154].

In the hope of selective hydrolysis of one of the ester functions, the hydrolytic reaction of dimethyl 1-hydroxy-1-phenylmethylphosphonate **1Aa** was carried out in aqueous sodium hydroxide at reflux for 5 h (Scheme 30, Route "A") [155]. After pH adjustment, the desired product (**57**) was obtained in a yield of 85%. This method was suitable only for the synthesis of the racemic target compound **57**. In order to maintain the C chirality center of the starting α-hydroxyphosphonate (*R*)-**1Aa**, a milder protocol had to be elaborated. On stirring the starting optically active α-hydroxyphosphonate (*R*)-**1Aa** in the presence of sodium iodide in acetone at reflux for 22 h, racemization of the hydroxyphosphonate (*R*)-**1Aa** could be avoided (Scheme 30, Route "B") [155].

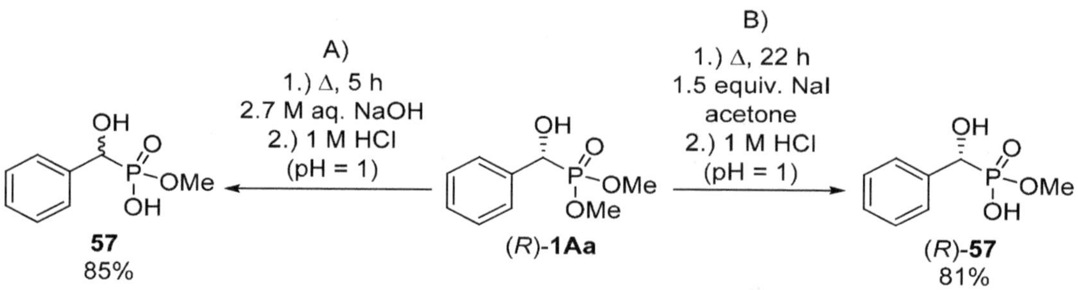

Scheme 30. Selective hydrolysis of one ester function of hydroxyphosphonate (*R*)-**1Aa**.

Another way of synthesizing α-hydroxyphosphonic acids (**12**) from α-hydroxyphosphonates **1** involves the cleavage of the C–O bond of the ester function by trimethylsilyl chloride [77,138] or trimethylsilyl bromide [110]. This reaction is usually carried out at ambient temperature within 2–18 h, generally in acetonitrile as the solvent.

3. Conclusions

This review summarizes the synthesis and reactions of α-hydroxyphosphonates that are of importance due to their bioactivity. The first half of the review presents the most common synthetic routes towards α-hydroxyphosphonates, with special stress on the addition of dialkyl phosphite to an oxo compound, and the condensation of trialkyl phosphites with aldehydes or ketones. The overview is followed by the discussion of the green synthetic protocols for α-hydroxyphosphonates. Then, the reactions of α-hydroxyphosphonates involving *O*-alkylation, *O*-acylation and oxidation of the hydroxy function, substitutions of the hydroxy group by a chloro or amino function, rearrangements to phosphate derivatives and hydrolysis of the ester function are discussed. Beside the literature examples, our own synthetic results are also included.

Acknowledgments: The authors are indebted to David W. Allen for his advice. The above project was supported by the National Research Development and Innovation Fund (K119202). Z. Rádai is grateful for the fellowship provided by the Chinoin-Sanofi Pharmaceuticals and the József Varga Foundation.

References

1. Olszewski, T.K. Environmentally benign syntheses of α-substituted phosphonates: Preparation of α-amino- and α-hydroxyphosphonates in water, in ionic liquids, and under solvent-free conditions. *Synthesis* **2014**, *46*, 403–429. [CrossRef]
2. Failla, S.; Finocchiaro, P.; Consiglio, G.A. Syntheses, characterization, stereochemistry and complexing properties of acyclic and macrocyclic compounds possessing α-amino- or α-hydroxyphosphonate units: A review article. *Heteroatom. Chem.* **2000**, *11*, 493–504. [CrossRef]
3. Rádai, Z.; Kiss, N.Z.; Keglevich, G. *Synthesis of α-Hydroxyphosphonates, an Important Class of Bioactive Compounds*; György, K., Ed.; Organophosphorus Chemistry: Novel Developments; Walter de Gruyter: Berlin, Germany; Boston, MA, USA, 2018; pp. 91–107. 315p.
4. Gröger, H.; Hammer, B. Catalytic concepts for the enantioselective synthesis of α-amino and α-hydroxy phosphonates. *Chem. Eur. J.* **2000**, *6*, 943–948. [CrossRef]
5. Kolodiazhnyi, O.I. Chiral hydroxyl phosphonates: Synthesis, configuration and biological properties. *Russ. Chem. Rev.* **2006**, *75*, 227–253. [CrossRef]
6. Merino, P.; Marqués-López, E.; Herrera, R.P. Catalytic enantioselective hydrophosphonylation of aldehydes and imines. *Adv. Synth. Catal.* **2008**, *350*, 1195–1208. [CrossRef]
7. Spilling, C.D.; Malla, R.K. Synthesis of non-racemic α-hydroxyphosphonates via asymmetric phosphor-aldol reaction. *Top. Curr. Chem.* **2015**, *361*, 83–136. [PubMed]
8. Phillips, A.M.F. Organocatalytic asymmetric synthesis of chiral phosphonates. *Mini Rev. Org. Chem.* **2014**, *11*, 164–185. [CrossRef]
9. Pudovik, A.N.; Konovalova, I.V. Addition reactions of esters of phosphorus(III) acids with unsaturated systems. *Synthesis* **1979**, *2*, 81–96. [CrossRef]
10. Abramov, V.S.; Kiryukhina, L.I.; Kudryavtseva, N. Reaction of dialkyl esters of phosphorus acids with aldehydes and ketones. XXV. Esters of α-hydroxy-β-(2,2,3-trimethyl3-cyclopentenyl)ethylphosphonic and α-hydroxynitrobenzylphosphonic acids. *Zh. Obshch. Khim.* **1961**, *31*, 2235–2238.
11. Maeda, H.; Takahashi, K.; Ohmori, H. Reactions of acyl tributylphosphonium chlorides and dialkyl acylphosphonates with Grignard and organolithium reagents. *Tetrahedron* **1998**, *54*, 12233–12242. [CrossRef]
12. Seven, O.; Polat-Cakir, S.; Hossain, M.S.; Emrullahoglu, M.; Demir, A.S. Reactions of acyl phosphonates with organoaluminum reagents: A new method for the synthesis of secondary and tertiary α-hydroxy phosphonates. *Tetrahedron* **2011**, *67*, 3464–3469. [CrossRef]
13. Gu, L.; Jin, C.; Zhang, H. The catalytic aerobic synthesis of quaternary α-hydroxy phosphonates via direct hydroxylation of phosphonate compounds. *New J. Chem.* **2015**, *39*, 1579–1582. [CrossRef]
14. Li, X.; Jin, C.; Gu, L. C–H hydroxylation of phosphonates with oxygen in [bmIm]OH to produce quaternary α-hydroxy phosphonates. *J. Org. Chem.* **2015**, *80*, 2443–2447. [CrossRef] [PubMed]

15. Pudovik, A.N.; Zametaeva, G.A. New synthesis of esters of phosphonic and thiophosphonic acids. XIII. Addition of diethyl thiophosphite to ketones and aldehydes. *Izv. Akad. Nauk SSSR Ser. Khim.* **1952**, *1952*, 932–939.
16. Zhou, S.; Wu, Z.; Rong, J.; Wang, S.; Yang, G.; Zhu, X.; Zhang, L. Highly efficient hydrophosphonylation of aldehydes and unactivated ketones catalyzed by methylene-linked pyrrolyl rare earth metal amido complexes. *Chem. Eur. J.* **2012**, *18*, 2653–2659. [CrossRef] [PubMed]
17. Zhao, L.; Ding, H.; Zhao, B.; Lu, C.; Yao, Y. Synthesis and characterization of amidate rare-earth metal amides and their catalytic activities toward hydrophosphonylation of aldehydes and unactivated ketones. *Polyhedron* **2014**, *83*, 50–59. [CrossRef]
18. Yang, S.; Zhu, X.; Zhou, S.; Wang, S.; Feng, Z.; Wei, Y.; Miao, H.; Guo, L.; Wang, F.; Zhang, G.; et al. Synthesis, structure, and catalytic activity of novel trinuclear rare-earth metal amido complexes incorporating μ–η^5:η^1 bonding indolyl and μ_3-oxo groups. *Dalton Trans.* **2014**, *43*, 2521–2533. [CrossRef] [PubMed]
19. Miao, H.; Zhou, S.L.; Wang, S.W.; Zhang, L.J.; Wei, Y.; Yang, S.; Wang, F.H.; Chen, Z.; Chen, Y.; Yuan, Q.B. Rare-earth metal amido complexes supported by bridged bis(β-diketiminato) ligand as efficient catalysts for hydrophosphonylation of aldehydes and ketones. *Sci. China Chem.* **2013**, *56*, 329–336. [CrossRef]
20. Liu, B.; Carpentier, J.F.; Sarazin, Y. Highly effective alkaline earth catalysts for the sterically governed hydrophosphonylation of aldehydes and nonactivated ketones. *Chem. Eur. J.* **2012**, *18*, 13259–13264. [CrossRef] [PubMed]
21. Kulkarni, M.A.; Lad, U.P.; Desai, U.V.; Mitragotri, S.D.; Wadgaonkar, P.P. Mechanistic approach for expeditious and solvent-free synthesis of α-hydroxy phosphonates using potassium phosphate as catalyst. *C. R. Chim.* **2013**, *16*, 148–152. [CrossRef]
22. Nandre, K.P.; Nandre, J.P.; Patil, V.S.; Bhosale, S.V. Barium hydroxide catalyzed greener protocol for the highly efficient and rapid synthesis of α-hydroxyphosphonates under solvent free conditions. *Chem. Biol. Interface* **2012**, *2*, 314–321.
23. Pandi, M.; Chanani, P.K.; Govindasamy, S. An efficient synthesis of α-hydroxy phosphonates and 2-nitroalkanols using $Ba(OH)_2$ as catalyst. *Appl. Catal. A* **2012**, *441–442*, 119–123. [CrossRef]
24. Sardarian, A.R.; Kaboudin, B. Surface-mediated solid phase reactions: Preparation of diethyl 1-hydroxyarylmethylphosphonates on the surface of magnesia. *Synth. Commun.* **1997**, *27*, 543–551. [CrossRef]
25. Aouani, I.; Lahbib, K.; Touil, S. Green synthesis and antioxidant activity of novel γ-cyano-α-hydroxyphosphonate derivatives. *Med. Chem.* **2015**, *11*, 206–213. [CrossRef] [PubMed]
26. Hudson, H.R.; Yusuf, R.O.; Matthews, R.W. The preparation of dimethyl α-hydroxyphosphonates and the chemical shift non-equivalence of their diastereotopic methyl ester groups. *Phosphorus Sulfur Silicon Relat. Elem.* **2008**, *183*, 1527–1540. [CrossRef]
27. Jin, C.; He, H. Synthesis and herbicidal activity of novel dialkoxyphosphoryl aryl methyl 2-(4,6-dimethoxypyrimidin-2-yloxy) benzoate derivatives. *Phosphorus Sulfur Silicon Relat. Elem.* **2011**, *186*, 1397–1403. [CrossRef]
28. Wang, C.; Zhou, J.; Lv, X.; Wen, J.; He, H. Solvent-free synthesis of tertiary α-hydroxyphosphates by the triethylamine-catalyzed hydrophosphonylation of ketones. *Phosphorus Sulfur Silicon Relat. Elem.* **2013**, *188*, 1334–1339. [CrossRef]
29. Tajbakhsh, M.; Samad, K.; Zahra, T.; Ahmadreza, B. $MgCl_2/Et_3N$ base system as a new catalyst for the synthesis of α-hydroxyphosphonate. *Chin. J. Chem.* **2012**, *30*, 827–829. [CrossRef]
30. Kumari, S.; Shekhar, A.; Pathak, D.D. A new catalyst and solvent-free green synthesis of α-hydroxy phosphonates and α-aminophosphonates. *Chem. Sci. Trans.* **2014**, *3*, 45–54.
31. Keglevich, G.; Tóth, V.R.; Drahos, L. Microwave-assisted synthesis of α-hydroxy-benzylphosphonates and –benzylphosphine oxides. *Heteroatom Chem.* **2011**, *22*, 15–17. [CrossRef]
32. Kong, D.L.; Liu, R.D.; Li, G.Z.; Zhang, P.W.; Wu, M.S. A rapid, convenient, solventless green approach for the synthesis of α-hydroxyphosphonates by grinding. *Asian J. Chem.* **2014**, *26*, 1246–1248.
33. Kumar, K.S.; Reddy, C.B.; Reddy, M.V.N.; Rani, C.R.; Reddy, C.S. Green chemical synthesis of α-hydroxyphosphonates. *Org. Commun.* **2012**, *5*, 50–57.
34. Kalla, R.M.N.; Zhang, Y.; Kim, I. Highly efficient green synthesis of α-hydroxyphosphonates using a recyclable choline hydroxide catalyst. *New J. Chem.* **2017**, *41*, 5373–5379. [CrossRef]

35. Ramananarivo, H.R.; Solhy, A.; Sebti, J.; Smahi, A.; Zahouily, M.; Clark, J.; Sebti, S. An eco-friendly paradigm for the synthesis of α-hydroxyphosphonates using sodium-modified fluorapatite under solventless conditions. *ACS Sustain. Chem. Eng.* **2013**, *1*, 403–409. [CrossRef]
36. Rao, K.U.M.; Sundar, C.S.; Prasad, S.S.; Rani, C.R.; Reddy, C.S. Neat synthesis and anti-oxidant activity of α-hydroxyphosphonates. *Bull. Korean Chem. Soc.* **2011**, *32*, 3343–3347. [CrossRef]
37. Santhisudha, S.; Sreelakshimi, P.; Jayaprakash, S.H.; Kumar, B.V.; Reddy, C.S. Silica-supported tungstic acid catalyzed synthesis and antioxidant activity of α-hydroxyphosphonates. *Phosphorus Sulfur Silicon Relat. Elem.* **2015**, *190*, 1479–1488. [CrossRef]
38. Liu, C.; Zhang, Y.; Qian, Q.; Yuan, D.; Yao, Y. n-BuLi as a highly efficient precatalyst for hydrophosphonylation of aldehydes and unactivated ketones. *Org. Lett.* **2014**, *16*, 6172–6175. [CrossRef] [PubMed]
39. Zhou, X.; Liu, Y.; Chang, L.; Zhao, J.; Shang, D.; Liu, X.; Lin, L.; Feng, X. Highly efficient synthesis of quaternary α-hydroxy phosphonates via Lewis acid-catalyzed hydrophosphonylation of ketones. *Adv. Synth. Catal.* **2009**, *351*, 2567–2572. [CrossRef]
40. de Noronha, R.G.; Costa, P.J.; Romão, C.C.; Calhorda, M.J.; Fernandes, A.C. MoO_2Cl_2 as a novel catalyst for C–P bond formation and for hydrophosphonylation of aldehydes. *Organometallics* **2009**, *28*, 6206–6212. [CrossRef]
41. Bouzina, A.; Aouf, N.E.; Berredjem, M. Ultrasound assisted green synthesis of α-hydroxyphosphonates under solvent-free conditions. *Res. Chem. Intermed.* **2016**, *42*, 5993–6002. [CrossRef]
42. Mandhane, P.G.; Joshi, R.S.; Nagargoje, D.R.; Gill, C.H. Ultrasound-promoted greener approach to synthesize α-hydroxy phosphonates catalyzed by potassium dihydrogen phosphate under solvent-free condition. *Tetrahedron Lett.* **2010**, *51*, 1490–1492. [CrossRef]
43. Sadaphal, A.S.; Sonar, S.S.; Pokalwar, R.U.; Shitole, N.V.; Shingare, M.S. Sulphamic acid: An efficient catalyst for the synthesis of α-hydroxy phosphonates using ultrasound irradiation. *J. Korean Chem. Soc.* **2009**, *53*, 536–541.
44. Shinde, P.V.; Kategaonkar, A.H.; Shingate, B.B.; Shingare, M.S. An organocatalyzed facile and rapid access to α-hydroxy and α-amino phosphonates under conventional/ultrasound technique. *Tetrahedron Lett.* **2011**, *52*, 2889–2892. [CrossRef]
45. Vahdat, S.M.; Baharfar, R.; Tajbakhsh, M.; Heydari, A.; Baghbanian, S.M.; Khaksar, S. Organocatalytic synthesis of α-hydroxy and α-aminophosphonates. *Tetrahedron Lett.* **2008**, *49*, 6501–6504. [CrossRef]
46. Vangala, V.B.; Pati, H.N. Efficient synthesis of β-lactam containing α-hydroxy phosphonates using tartaric acid and fumaric acid as mild catalysts. *Synth. Commun.* **2016**, *46*, 374–378. [CrossRef]
47. Jahani, F.; Zamenian, B.; Khaksar, S.; Tajbakhsh, M. Pyridine 2,6-dicarboxylic acid as a bifunctional organocatalyst for hydrophosphonylation of aldehydes and ketones in water. *Synthesis* **2010**, *19*, 3315–3318. [CrossRef]
48. Heydari, A.; Arefi, A.; Khaksar, S.; Tajbakhsh, M. Hydrophosphonylation of aldehydes catalyzed by guanidine hydrochloride in water. *Catal. Commun.* **2006**, *7*, 982–984. [CrossRef]
49. Wang, H.S.; Zeng, J.E. Iodine-catalyzed, efficient synthesis of α-hydroxy phosphonates in water. *Phosphorus Sulfur Silicon Relat. Elem* **2010**, *185*, 1425–1428. [CrossRef]
50. Ramesh, K.; Madhav, B.; Murthy, S.N.; Nageswar, Y.V.D. Aqueous-phase synthesis of α-hydroxyphosphonates catalyzed by β-cyclodextrin. *Synth. Commun.* **2012**, *42*, 258–265. [CrossRef]
51. Sonar, S.S.; Kategaonkar, A.H.; Ware, M.N.; Gill, C.H.; Shingate, B.B.; Shingare, M.S. Ammonium metavanadate: An effective catalyst for synthesis of α-hydroxyphosphonates. *Arkivoc* **2009**, *2*, 138–148.
52. Raju, P.; Rajeshwaran, G.G.; Nandakumar, M.; Mohanakrishnan, A.K. Unusual reactivity of aryl aldehydes with triethyl phosphite and zinc bromide: A facile preparation of epoxides, benzisoxazoles, and α-hydroxy phosphonate esters. *Eur. J. Org. Chem.* **2015**, *2015*, 3513–3523. [CrossRef]
53. Kumar, A.; Jamwal, S.; Khan, S.; Singh, N.; Rai, V.K. $Bi(NO_3)_3{\cdot}5H_2O$ catalyzed phosphonylation of aldehydes: An efficient route to α-hydroxyphosphonates. *Phosphorus Sulfur Silicon Relat. Elem.* **2016**, *192*, 381–385. [CrossRef]
54. Thottempudi, V.; Chung, K.H. Niobium(V) chloride catalyzed Abramov reaction: An efficient protocol for the preparation of α-hydroxy phosphonates. *Bull. Korean Chem. Soc.* **2008**, *29*, 1781–1783. [CrossRef]
55. Keglevich, G.; Rádai, Z.; Kiss, N.Z. To date the greenest method for the preparation of α-hydroxyphosphonates from substituted benzaldehydes and dialkyl phosphites. *Green Process Synth.* **2017**, *6*, 197–201. [CrossRef]

56. Kiss, N.Z.; Rádai, Z.; Mucsi, Z.; Keglevich, G. Synthesis of α-aminophosphonates from α-hydroxyphosphonates; a theoretical study. *Heteroat. Chem.* **2016**, *27*, 260–268. [CrossRef]
57. Lorenz, W.; Henglein, A.; Schrader, G. The new insecticide *O,O*-dimethyl 2,2,2-trichloro-1-hydroxyethylphosphonate. *J. Am. Chem. Soc.* **1955**, *77*, 2554–2556. [CrossRef]
58. Song, H.; Mao, H.; Shi, D. Synthesis and herbicidal activity of α-hydroxy phosphonate derivatives containing pyrimidine moiety. *Chin. J. Chem.* **2010**, *28*, 2020–2024. [CrossRef]
59. Pokalwar, R.U.; Hangarge, R.V.; Maske, P.V.; Shingare, M.S. Synthesis and antibacterial activities of α-hydroxyphosphonates and α-acetyloxyphosphonates derived from 2-chloroquinoline-3-carbaldehyde. *Arkivoc* **2006**, *11*, 196–204.
60. Kategaonkar, A.H.; Pokalwar, R.U.; Sonar, S.S.; Gawali, V.U.; Shingate, B.B.; Shingare, M.S. Synthesis, in vitro antibacterial and antifungal evaluations of new α-hydroxyphosphonate and new α-acetoxyphosphonate derivatives of tetrazolo [1, 5-a] quinolone. *Eur. J. Med. Chem.* **2010**, *45*, 1128–1132. [CrossRef] [PubMed]
61. Sobhani, S.; Tashrifi, Z. Synthesis of α-functionalized phosphonates from α-hydroxyphosphonates. *Tetrahedron* **2010**, *66*, 1429–1439. [CrossRef]
62. Brücher, K.; Illarionov, B.; Held, J.; Tschan, S.; Kunfermann, A.; Pein, M.K.; Bacher, A.; Gräwert, T.; Maes, L.; Mordmüller, B.; et al. α-Substituted β-oxa isosteres of fosmidomycin: Synthesis and biological evaluation. *J. Med. Chem.* **2012**, *55*, 6566–6575. [CrossRef] [PubMed]
63. Guan, Q.; Yang, F.; Guo, D.; Xu, J.; Jiang, M.; Liu, C.; Bao, K.; Wu, Y.; Zhang, W. Synthesis and biological evaluation of novel 3,4-diaryl-1,2,5-selenadiazol analogues of combretastatin A-4. *Eur. J. Med. Chem.* **2014**, *87*, 1–9. [CrossRef] [PubMed]
64. Sekine, M.; Nakajima, M.; Kume, A.; Hashizume, A.; Hata, T. Silyl phosphites. XVIII. Versatile utility of α-(trimethylsilyloxy)-alkylphosphonates as key intermediates for transformation of aldehydes into several carbonyl derivatives. *Bull. Chem. Soc. Jpn.* **1982**, *55*, 224–238. [CrossRef]
65. Firouzabadi, H.; Iranpoor, N.; Sobhani, S. A high yielding preparation of α-trimethylsilyloxyphosphonates by silylation of α-hydroxyphosphonates with HMDS catalyzed by iodine. *Tetrahedron Lett.* **2002**, *43*, 3653–3655. [CrossRef]
66. Firouzabadi, H.; Iranpoor, N.; Sobhani, S.; Ghassamipour, S.; Amoozgar, Z. Copper triflate [$Cu(OTf)_2$] is an efficient and mild catalyst for the silylation of α-hydroxyphosphonates to α-trimethylsilyloxyphosphonates with HMDS at room temperature. *Tetrahedron Lett.* **2003**, *44*, 891–893. [CrossRef]
67. Firouzabadi, H.; Iranpoor, N.; Jafari, A.A.; Jafari, M.R. Iron(III) trifluoroacetate [$Fe(F_3CCO_2)_3$] as an easily available, non-hygroscopic, non-corrosive, highly stable and a reusable Lewis acis catalyst: Efficient *O*-silylation of α-hydroxyphosphonates, alcohols and phenols by hexamethyldisilazane (HMDS) under solvent-free conditions. *J. Organomet. Chem.* **2008**, *693*, 2711–2714.
68. Wang, F.; Miao, Z.; Chen, R. Efficient syntheses of phosphonylated isochromenes by regioselective 6-*endo-dig* addition to carbon-carbon triple bond catalyzed by $Pd(OAc)_2$. *Org. Biomol. Chem.* **2009**, *7*, 2848–2850. [CrossRef] [PubMed]
69. Chen, X.B.; Shi, D.Q. Synthesis and biological activity of novel phosphonate derivatives containing of pyridyl and 1,2,3-triazole rings. *Phosphorus Sulfur Silicon Relat. Elem.* **2008**, *183*, 1134–1144. [CrossRef]
70. Chen, T.; Shen, P.; Li, Y.; He, H. Synthesis and herbicidal activity of *O,O*-dialkyl phenoxyacetoxyalkylphosphonates containing fluorine. *J. Fluorine Chem.* **2006**, *127*, 291–295. [CrossRef]
71. Peng, H.; Wang, T.; Xie, P.; Chen, T.; He, H.W.; Wan, J. Molecular docking and three-dimensional quantitative structure-activity relationship studied on the binding modes of herbicidal 1-(substituted phenoxyacetoxy)alkylphosphonates to the E1 component of pyruvate dehydrogenase. *J. Agric. Food Chem.* **2007**, *55*, 1871–1880. [CrossRef] [PubMed]
72. Wang, W.; Wang, L.P.; Ning, B.K.; Mao, M.Z.; Xue, C.; Wang, H.Y. Synthesis and insecticidal activities of *O,O*-dialkyl-2-[3-bromo-1-(3-chloropyridin-2-yl)-1H-pyrazole-5-carbonyloxy] (aryl) methylphosphonates. *Phosphorus Sulfur Silicon Relat. Elem.* **2016**, *191*, 1362–1367. [CrossRef]
73. Fang, H.; Wang, D.; Chen, W.; Zhao, Y.; Fang, M. Synthesis, crystal structure and fragmentation pathway of arylcarboxy ester of α-hydroxyphosphonate. *Chin. J. Org. Chem.* **2010**, *30*, 1377–1382.
74. Hudson, H.R.; Jászay, Z.M.; Pianka, M. The preparation and properties of some α-acyloxy- and α-carbamoyloxy-phosphonothionates. *Phosphorus Sulfur Silicon Relat. Elem.* **2003**, *178*, 1571–1582. [CrossRef]
75. Shi, D.Q.; Li, X.J.; Wei, J. 5-Fluorouracil derivatives containing α-hydroxy phosphonates. *Phosphorus Sulfur Silicon Relat. Elem.* **2007**, *182*, 405–412. [CrossRef]

76. Peng, H.; Long, Q.; Deng, X.; He, H. Synthesis and herbicidal activities of lithium or potassium hydrogen 1-(substituted phenoxyacetoxy)alkylphosphonates. *Phosphorus Sulfur Silicon Relat. Elem.* **2013**, *188*, 1868–1874. [CrossRef]
77. Long, Q.; Deng, X.; Gao, Y.; Xie, H.; Peng, H.; He, H. Synthesis and herbicidal activities of sodium hydrogen 1-(substituted phenoxyacetoxy)alkylphosphonates. *Phosphorus Sulfur Silicon Relat. Elem.* **2013**, *188*, 819–825. [CrossRef]
78. He, H.W.; Yuan, J.L.; Peng, H.; Chen, T.; Shen, P.; Wan, S.Q.; Li, Y.; Tan, H.L.; He, Y.H.; He, J.B.; et al. Studies of *O,O*-dimethyl α-(2,4-dichlorophenoxyacetoxy)-ethylphosphonate (HW02) as a new herbicide. 1. Synthesis and herbicidal activity of HW02 and analogues as novel inhibitors of pyruvate dehydrogenase complex. *J. Agric. Food Chem.* **2011**, *59*, 4801–4813. [CrossRef] [PubMed]
79. Meng, L.; Joshi, R.; Li, M.; He, H. Synthesis and biological activity of novel 1-(substituted phenoxyacetoxy) alkyl phosphonates and phosphinates. *J. Nepal Chem. Soc.* **2009**, *23*, 11–20.
80. Soung, M.G.; Hwang, T.Y.; Sung, N.D. Synthesis and 3D-QSARs analyses of herbicidal *O,O*-dialkyl-1-phenoxy-1-acetoxy-1-methylphosphonate analogues as a new class of potent inhibitors of pyruvate dehydrogenase. *Bull. Korean Chem. Soc.* **2010**, *31*, 1361–1367. [CrossRef]
81. Wang, T.; He, H.W. An efficient synthesis of α-(2,4-dichloro-phenoxyacetoxy)aryl methyl phosphonate monosodium salts. *Synth. Commun.* **2004**, *34*, 1415–1423. [CrossRef]
82. Wang, T.; He, H.W. Simple and improved preparation of α-oxophosphonate monolithium salts. *Phosphorus Sulfur Silicon Relat. Elem.* **2004**, *179*, 2081–2089. [CrossRef]
83. Wang, W.; He, H.W.; Zuo, N.; Zhang, X.; Lin, J.S.; Chen, W.; Peng, H. Synthesis and herbicidal activity of 2-(substituted phenoxyacetoxy)alkyl-5,5-dimethyl-1,3,2-dioxaphosphinan-2-one containing fluorine. *J. Fluorine Chem.* **2012**, *142*, 24–28. [CrossRef]
84. Wang, W.; He, H.W.; Zuo, N.; He, H.F.; Peng, H.; Tan, X.S. Synthesis and herbicidal activity of 2-(substituted phenoxyacetoxy)alkyl-5,5-dimethyl-1,3,2-dioxaphosphinan-2-one. *J. Agric. Food Chem.* **2012**, *60*, 7581–7587. [CrossRef] [PubMed]
85. Wang, T.; Lei, D.Y.; Huang, Y.; Ao, L.H. Synthesis and biological activity of *O,O*-dimethyl-2,6-pyridinyl diformyloxy alkyl phosphonates. *Phosphorus Sulfur Silicon Relat. Elem.* **2009**, *184*, 2777–2785. [CrossRef]
86. Firouzabadi, H.; Iranpoor, N.; Sobhani, S.; Amoozgar, Z. Copper triflate as a useful catalyst for the high-yielding preparation of α-acetyloxyphosphonates under solvent-free conditions. *Synthesis* **2004**, *2*, 295–297. [CrossRef]
87. Firouzabadi, H.; Iranpoor, N.; Farahi, S. Solid trichlorotitanium(IV) trifluoromethanesulfonate $TiCl_3(OTf)$ catalyzed efficient acylation of –OH and –SH: Direct esterification of alcohols with carboxylic acids and transesterification of alcohols with esters under neat conditions. *J. Mol. Catal. A: Chem.* **2008**, *289*, 61–68. [CrossRef]
88. Firouzabadi, H.; Iranpoor, N.; Sobhani, S.; Amoozgar, Z. Facile and high-yielding preparation of α-hydroxyphosphonates assisted by microwave irradiation. *Synthesis* **2004**, *11*, 1771–1774. [CrossRef]
89. Rostami, A.; Atashkar, B.; Moradi, D. Synthesis, characterization and catalytic properties of magnetic nanoparticle supported guanidine in base catalyzed synthesis of α-hydroxyphosphonates and α-acetoxyphosphonates. *Appl. Catal. A* **2013**, *467*, 7–16. [CrossRef]
90. Green, D.; Elgendy, S.; Patel, G.; Skordalakes, E.; Goodwin, C.A.; Scully, M.F.; Kakkar, V.V.; Deadman, J.J. Substrate related O,O-dialkyldipeptidyly ψ carboxybenzylphosphonates, a new type of thrombin inhibitor. *Phosphorus Sulfur Silicon Relat. Elem.* **2000**, *156*, 151–155. [CrossRef]
91. Yang, J.; Ma, J.; Che, W.; Li, M.; Li, G.; Song, B. Microwave-assisted synthesis and antitumor activity of salicyl acyloxy phosphonate derivatives. *Chin. J. Org. Chem.* **2014**, *34*, 2566–2571. [CrossRef]
92. Iranpoor, N.; Firouzabadi, H.; Khalili, D. The first Mitsunobu protocol for efficient synthesis of α-acyloxyphosphonates using 4,4′-azopyridine. *Phosphorus Sulfur Silicon Relat. Elem.* **2011**, *186*, 2166–2171. [CrossRef]
93. Xu, L.; You, G.; Peng, H.; He, H. Synthesis and biological activities of *O,O*-dialkyl 1-((4,6-dichloropyrimidin-2-yl)carbamyloxy) alkylphosphonates. *Phosphorus Sulfur Silicon Relat. Elem.* **2014**, *189*, 812–818. [CrossRef]
94. Li, J.P.; Zhu, J.G.; Liu, R.J.; Cui, F.L.; Liu, P.; Liu, G.S. Straightforward synthesis of a new series of α-(arylamino thiocarbonyloxy) hydrocarbylphosphonates. *S. Afr. J. Chem.* **2008**, *61*, 5–8.
95. Li, Z.G.; Sun, H.K.; Wang, Q.M.; Huang, R.Q. An α-hydrazinoalkylphosphonate as building block for novel *N*-phosphonoalkylheterocycles. *Heteroatom Chem.* **2003**, *14*, 384–386. [CrossRef]

96. Creary, X.; Geiger, C.C.; Hilton, K. Mesylate derivatives of α-hydroxy phosphonates. Formation of carbocations adjacent to the diethyl phosphonate group. *J. Am. Chem. Soc.* **1983**, *105*, 2851–2858. [CrossRef]
97. Kong, D.L.; Li, G.Z.; Liu, R.D. Synthesis and crystal structure of diethyl tosyloxybenzylphosphonate. *Asian J. Chem.* **2014**, *26*, 2138–2140.
98. Davidson, R.S.; Sheldon, R.A.; Trippett, S. The reaction of tetraphenyldiphosphine with aromatic carboxylic acids. *J. Chem. Soc.* **1967**, 1547–1552. [CrossRef]
99. Rádai, Z.; Hodula, V.; Kiss, N.Z.; Kegelevich, G. Unpublished results.
100. Samanta, S.; Zhao, C.G. Organocatalyzed nitroaldol reaction of α-ketophosphonates and nitromethane revisited. *Arkivoc* **2007**, *8*, 218–226.
101. Guang, J.; Zhao, C.G. Organocatalyzed asymmetric Michael reaction of β-aryl-α-ketophosphonates and nitroalkenes. *Tetrahedron Lett.* **2013**, *54*, 5703–5706. [CrossRef] [PubMed]
102. Telan, L.A.; Poon, C.D.; Evans, S.A. Diastereoselectivity in the Mukaiyama–Michael reaction employing α-acyl β,γ-unsaturated phosphonates. *J. Org. Chem.* **1996**, *61*, 7455–7462. [CrossRef] [PubMed]
103. Kaboudin, B. Surface-mediated solid-phase reactions: The preparation of acyl phosphonates by oxidation of 1-hydroxyphosphonates on the solid surface. *Tetrahedron Lett.* **2000**, *41*, 3169–3171. [CrossRef]
104. Firouzabadi, H.; Iranpoor, N.; Sobhani, S.; Sardarian, A.R. High yields preparation of α-ketophosphonates by oxidation of α-hydroxyphosphonates with zinc dichromate trihydrate ($ZnCr_2O_7 \cdot 3H_2O$) under solvent-free conditions. *Tetrahedron Lett.* **2001**, *42*, 4369–4371. [CrossRef]
105. Guliaiko, I.; Nesterov, V.; Sheiko, S.; Kolodiazhnyi, O.I.; Freytag, M.; Jones, P.G.; Schmutzler, R. Synthesis of optically active hydroxyphosphonates. *Heteroatom Chem.* **2008**, *19*, 133–139. [CrossRef]
106. Firouzabadi, H.; Iranpoor, N.; Sobhani, S. Preparation of α-ketophosphonates by oxidation of α-hydroxyphosphonates with pyridinium chlorochromate (PCC). *Phosphorus Sulfur Silicon Relat. Elem.* **2004**, *179*, 1483–1491. [CrossRef]
107. Firouzabadi, H.; Iranpoor, N.; Sobhani, S. Preparation of α-ketophosphonates by oxidation of α-hydroxyphosphonates with neutral alumina supported potassium permanganate (NASPP) under solvent-free conditions and potassium permanganate in dry benzene. *Tetrahedron Lett.* **2002**, *43*, 477–480. [CrossRef]
108. Pawar, V.D.; Bettigeri, S.; Weng, S.S.; Kao, J.Q.; Chen, C.T. Highly enantioselective aerobic oxidation of α-hydroxyphosphonates catalyzed by chiral vanadyl(V) methoxides bearing *N*-salicylidene-α-aminocarboxylates. *J. Am. Chem. Soc.* **2006**, *128*, 6308–6309. [CrossRef] [PubMed]
109. Taylor, W.P.; Zhang, Z.Y.; Widlanski, T.S. Quiescent affinity inactivators of protein tyrosine phosphatases. *Bioorg. Med. Chem.* **1996**, *4*, 1515–1520. [CrossRef]
110. Tulsi, N.S.; Downey, A.M.; Cairo, C.W. A protected L-bromophosphonomethylphenylalanine amino acid derivative (BrPmp) for synthesis of irreversible protein tyrosine phosphatase inhibitors. *Bioorg. Med. Chem.* **2010**, *18*, 8679–8686. [CrossRef] [PubMed]
111. Caplan, N.A.; Pogson, C.I.; Hayes, D.J.; Blackburn, G.M. The synthesis of novel bisphosphonates as inhibitors of phosphoglycerate kinase (3-PGK). *J. Chem. Soc. Perkin Trans. 1* **2000**, 421–437. [CrossRef]
112. Gajda, T. Preparation of diethyl 1-bromoalkylphosphonates. *Phosphorus Sulfur Silicon Relat. Elem.* **1990**, *53*, 327–331. [CrossRef]
113. Guan, Z.; Wu, D.; Fu, J.P.; He, Y.H. A facile and efficient synthesis of diethyl α,α-chlorofluoroalkanephosphonates. *Heteroatom Chem.* **2010**, *21*, 250–255. [CrossRef]
114. Wu, D.; He, Y.; Tang, R.; Guan, Z. The first synthesis of diethyl α,α-chlorofluorobenzylphosphonates. *Synlett* **2009**, *13*, 2180–2182.
115. Firouzabadi, H.; Iranpoor, N.; Sobhani, S. PPh_3/DDQ as a neutral system for the facile preparation of diethyl α-bromo, α-iodo and α-azidophosphonates from diethyl α-hydroxyphosphonates. *Tetrahedron* **2004**, *60*, 203–210. [CrossRef]
116. Fu, J.P.; He, Y.H.; Zhong, J.; Yang, Y.; Deng, X.; Guan, Z. An efficient and general route to the synthesis of diethyl α,α-bromofluorophosphonates. *J. Fluorine Chem.* **2011**, *132*, 636–640. [CrossRef]
117. Iranpoor, N.; Firouzabadi, H.; Gholinejad, M. 4-Aminophenyldiphenylphosphinite (APDPP), a new heterogeneous and acid scavenger phosphinite—Conversion of alcohols, trimethylsilyl, and tetrahydropyranyl ethers to alkyl halides with halogens or *N*-halosuccinimides. *Can. J. Chem.* **2006**, *84*, 1006–1012. [CrossRef]

118. Allmendinger, T.; Fujimoto, R.; Gasparini, F.; Schilling, W.; Satoh, Y. α-Fluoro-benzylphosphonates as reagents for the preparation of 1-fluoro-1-aryl alkenes and α-fluorostilbenes. *Chimia* **2004**, *58*, 133–137. [CrossRef]
119. Blackburn, G.M.; Kent, D.E. A novel synthesis of α- and γ-fluoroalkylphosphonates. *J. Chem. Soc. Chem. Commun.* **1981**, 511–513. [CrossRef]
120. Blackburn, G.M.; Kent, D.E. Synthesis of α- and γ-fluoroalkylphosphonates. *J. Chem. Soc. Perkin Trans.* **1986**, 913–917. [CrossRef]
121. Yokomatsu, T.; Yamagishi, T.; Matsumoto, K.; Shibuya, S. Stereocontrolled synthesis of hydroxymethylene phosphonate analogues of phosphorylated tyrosine and their conversion to monofluoromethylene phosphonate analogues. *Tetrahedron* **1996**, *52*, 11725–11738. [CrossRef]
122. Guzyr, O.I.; Zasukha, S.V.; Vlasenko, Y.G.; Chernega, A.N.; Rozhenko, A.B.; Shermolovich, Y.G. Simple route to adducts of (amino)(aryl)carbene with phosphorus pentafluoride. *Eur. J. Inorg. Chem.* **2013**, *2013*, 4154–4158. [CrossRef]
123. Psurski, M.; Błażewska, K.; Gajda, A.; Gajda, T.; Wietrzyk, J.; Oleksyszyn, J. Synthesis and antiproliferative activity of novel α- and β-dialkoxyphosphoryl isothiocyanates. *Bioorg. Med. Chem. Lett.* **2011**, *21*, 4572–4576. [CrossRef] [PubMed]
124. Keglevich, G.; Bálint, E. The Kabachnik–Fields reaction: Mechanism and synthetic use. *Molecules* **2012**, *17*, 12821–12835. [CrossRef] [PubMed]
125. Bálint, E.; Fazekas, E.; Tripolszky, A.; Tajti, Á.; Kangyal, R.; Milen, M.; Keglevich, G. Synthesis of aminophosphonate derivatives by microwave assisted Kabachnik–Fields reaction. *Organomet. Chem.* **2012**, *717*, 655–659. [CrossRef]
126. Zefirov, N.S.; Matveeva, E.D. Catalytic Kabachnik–Fields reaction: New horizons for old reaction. *Arkivoc* **2008**, *1*, 1–17.
127. Kafarski, P.; Gorny vel Gorniak, M.; Andrasiak, I. Kabachnik–Fields reaction under green conditions—A critical overview. *Curr. Green Chem.* **2015**, *2*, 218–222. [CrossRef]
128. Kaboudin, B. A convenient synthesis of 1-aminophosphonates from 1-hydroxyphosphonates. *Tetrahedron Lett.* **2003**, *44*, 1051–1053. [CrossRef]
129. Kiss, N.Z.; Kaszás, A.; Drahos, L.; Mucsi, Z.; Keglevich, G. A neighbouring group effect leading to enhanced nucleophilic substitution of amines at the hindered α-carbon atom of an α-hydroxyphosphonate. *Tetrahedron Lett.* **2012**, *53*, 207–209. [CrossRef]
130. Pallikonda, G.; Chakravarty, M. Triflic acid mediated functionalization of α-hydroxyphosphonates: Route for sulfonamide phosphates. *RSC Adv.* **2013**, *3*, 20503–20511. [CrossRef]
131. Verma, C.; Singh, A.; Pallikonda, G.; Chakravarty, M.; Quraishi, M.A.; Bahadur, I.; Ebenso, E.E. Aryl sulfonamidomethylphosphonates as new class of green corrosion inhibitors for mild steel in 1 M HCl: Electrochemical, surface and quantum chemical investigation. *J. Mol. Liq.* **2015**, *209*, 306–319. [CrossRef]
132. Chen, L.; Zou, Y.X.; Fang, X.Y.; Yin, X. Convenient synthesis of α-diarylmethylphosphonates by HOTf catalyzed Friedel–Crafts arylation of α-aryl α-hydroxyphosphonates. *Phosphorus Sulfur Silicon Relat. Elem.* **2018**, *193*, 168–177. [CrossRef]
133. Khalid, M.B.Z.; Pallikonda, G.; Tulichala, R.N.P.; Chakravarty, M. Oxy-Wittig reactions of 1-naphthyl(aryl) methylphosphonates: A new approach to naphthylarylketones. *Tetrahedron* **2016**, *72*, 2094–2101. [CrossRef]
134. Pallikonda, G.; Chakravarty, M. $FeCl_3$-mediated arylation of α-hydroxyphosphonates with unactivated arenes. Pseudo umpolung in allylic phosphonates. *Eur. J. Org. Chem.* **2013**, *2013*, 944–951. [CrossRef]
135. Pallikonda, G.; Chakravarty, M.; Sahoo, M. An easy access to α-aryl substituted γ-ketophosphonates: Lewis acid mediated reactions of 1,3-diketones with α-hydroxyphosphonates and tandem regioselective C–C bond cleavage. *Org. Biomol. Chem.* **2014**, *12*, 7140–7149. [CrossRef] [PubMed]
136. Hall, L.A.R.; Stephens, C.W.; Drysdale, J.J. A rearrangement to from diethyl 1-cyanoethyl phosphate. *J. Am. Chem. Soc.* **1957**, *79*, 1768–1769. [CrossRef]
137. Yoshino, K.; Kohno, T.; Morita, T.; Tsukamoto, G. Organic phosphorus compounds. 2. Synthesis and coronary vasodilator activity of (benzothiazolylbenzyl)phosphonate derivatives. *J. Med. Chem.* **1989**, *32*, 1528–1532. [CrossRef] [PubMed]
138. McGeary, R.P.; Vella, P.; Mak, J.Y.W.; Guddat, L.W.; Schenk, G. Inhibition of purple acid phosphatase with α-alkoxynaphthylmethylphosphonic acids. *Bioorg. Med. Chem. Lett.* **2009**, *19*, 163–166. [CrossRef] [PubMed]

139. Kumaraswamy, S.; Selvi, R.S.; Swamy, K.C.K. Synthesis of new α-hydroxy-, α-halogeno- and vinylphosphonates derived from 5,5-dimethyl-1,3,2-dioxaphosphinan-2-one. *Synthesis* **1997**, *1997*, 207–212. [CrossRef]
140. Jankowski, S.; Marczak, J.; Olczak, A.; Główka, M.L. Stereochemistry of 1-hydroxyphosphonate–phosphate rearrangement. Retention of configuration at the phosphorus atom. *Tetrahedron Lett.* **2006**, *47*, 3341–3344. [CrossRef]
141. Pallitsch, K.; Roller, A.; Hammerschmidt, F. The stereochemical course of the α-hydroxyphosphonate–phosphate rearrangement. *Chem. Eur. J.* **2015**, *21*, 10200–10206. [CrossRef] [PubMed]
142. Pallikonda, G.; Santosh, R.; Ghosal, S.; Chakravarty, M. BuLi-triggered phospha-Brook rearrangement: Efficient synthesis of organophosphates from ketones and aldehydes. *Tetrahedron Lett.* **2015**, *56*, 3796–3798. [CrossRef]
143. El Kaïm, L.; Gaultier, L.; Grimaud, L.; Santos, A.D. Formation of new phosphates from aldehydes by a DBU-catalysed phospha-brook rearrangement in a polar solvent. *Synlett* **2005**, *15*, 2335–2336. [CrossRef]
144. Galeta, J.; Potáček, M. Applications of caged-designed proton sponges in base-catalyzed transformations. *J. Mol. Catal. A: Chem.* **2014**, *395*, 87–92. [CrossRef]
145. Rádai, Z.; Szabó, R.; Kiss, N.Z.; Keglevich, G. Unpublished results.
146. Baguley, T.D.; Xu, H.C.; Chatterjee, M.; Nairn, A.C.; Lombroso, P.J.; Ellman, J.A. Substrate-based fragment identification for the development of selective, nonpeptidic inhibitors of striatal-enriched protein tyrosine phosphatase. *J. Med. Chem.* **2013**, *56*, 7636–7650. [CrossRef] [PubMed]
147. Beers, S.A.; Malloy, E.A.; Wu, W.; Wachter, M.P.; Gunnia, U.; Cavender, D.; Harris, C.; Davis, J.; Brosius, R.; Pellegrino-Gensey, J.L.; et al. Nitroarylhydroxymethylphosphonic acids as inhibitors of CD45. *Bioorg. Med. Chem.* **1997**, *5*, 2203–2211. [CrossRef]
148. Burley, R.K.M.; Bearne, S.L. Inhibition of mandelate racemase by the substrate–intermediate–product analogue 1,1-diphenyl-1-hydroxymethylphosphonate. *Bioorg. Med. Chem.* **2005**, *15*, 4342–4344. [CrossRef] [PubMed]
149. Caplan, N.A.; Pogson, C.I.; Hayes, D.J.; Blackburn, G.M. Novel bisphosphonate inhibitors of phosphoglycerate kinase. *Bioorg. Med. Chem. Lett.* **1998**, *8*, 515–520. [CrossRef]
150. Forlani, G.; Occhipinti, A.; Berlicki, Ł.; Dziedzioła, G.; Wieczorek, A.; Kafarski, P. Tailoring the structure of aminobisphosphonates to target plant P5C reductase. *J. Agric. Food Chem.* **2008**, *56*, 3193–3199. [CrossRef] [PubMed]
151. Occhipinti, A.; Berlicki, Ł.; Giberti, S.; Dziedzioła, G.; Kafarski, P.; Forlani, G. Effectiveness and mode of action of phosphonate inhibitors of plant glutamine synthetase. *Pest. Manag. Sci.* **2010**, *66*, 51–58. [CrossRef] [PubMed]
152. Desai, J.; Wang, Y.; Wang, K.; Malwal, S.R.; Oldfield, E. Isoprenoid biosynthesis inhibitors targeting bacterial cell growth. *Chem. Med. Chem.* **2016**, *11*, 2205–2215. [CrossRef] [PubMed]
153. Jiang, G.; Madan, D.; Prestwich, G.D. Aromatic phosphonates inhibit the lysophospholipase D activity of autotaxin. *Bioorg. Med. Chem.* **2011**, *21*, 5098–5101. [CrossRef] [PubMed]
154. Nesterov, V.V.; Kolodiazhnyi, O.I. Efficient method for the asymmetric reduction of α- and β-ketophosphonates. *Tetrahedron* **2007**, *63*, 6720–6731. [CrossRef]
155. Colton, I.J.; Yin, D.T.; Grochulski, P.; Kazlauskas, R.J. Molecular basis of chiral acid recognition by *candida rugosa* lipase: X-ray structure of transition state analog and modeling of the hydrolysis of methyl 2-methoxy-2-phenylacetate. *Adv. Synth. Catal.* **2011**, *353*, 2529–2544. [CrossRef]

10

Stoichiometric and Catalytic Synthesis of Alkynylphosphines

Elise Bernoud, Romain Veillard, Carole Alayrac * and Annie-Claude Gaumont *

Laboratoire de Chimie Moléculaire et Thioorganique, UMR CNRS 6507, INC3M FR3038, ENSICAEN & Université de Caen Basse-Normandie, 14050 Caen, France; e.bernoud@gmail.com (E.B.); romain.veillard@ensicaen.fr (R.V.)

* Correspondence: carole.witulski-alayrac@ensicaen.fr (C.A.); annie-claude.gaumont@ensicaen.fr (A.-C.G.)

Abstract: Alkynylphosphines or their borane complexes are available either through C–P bond forming reactions or through modification of the phosphorus or the alkynyl function of various alkynyl phosphorus derivatives. The latter strategy, and in particular the one involving phosphoryl reduction by alanes or silanes, is the method of choice for preparing primary and secondary alkynylphosphines, while the former strategy is usually employed for the synthesis of tertiary alkynylphosphines or their borane complexes. The classical C–P bond forming methods rely on the reaction between halophosphines or their borane complexes with terminal acetylenes in the presence of a stoichiometric amount of organometallic bases, which precludes the access to alkynylphosphines bearing sensitive functional groups. In less than a decade, efficient catalytic procedures, mostly involving copper complexes and either an electrophilic or a nucleophilic phosphorus reagent, have emerged. By proceeding under mild conditions, these new methods have allowed a significant broadening of the substituent scope and structure complexity.

Keywords: organophosphorus chemistry; phosphines; phosphine-boranes; homogeneous catalysis

---

1. Introduction

Alkynylphosphines are relevant versatile compounds due to the presence of two important functions, a phosphorus atom—a key-atom in metal coordination [1]—and an alkynyl moiety, which can also coordinate to metals and is known to display a rich and diverse chemistry [2].

In the past, these phosphines were mainly used in coordination chemistry for the assembly of various mono- or bi-metallic complexes and clusters [3], by taking advantage of their diverse coordination modes to transition metals, via the phosphorus atom and/or the triple bond. It is noteworthy that the range of alkynylphosphines involved in those complexes and clusters is almost exclusively covered by diphenylphosphino-phenylacetylene and bis(diphenylphosphino)acetylene.

It is only recently that the high potential of alkynylphosphines as ligands for homogeneous catalysis has emerged through a few relevant catalytic applications [4]. Notably P-stereogenic alkynylphosphines were shown to induce excellent enantioselectivity in several asymmetric transition metal-catalyzed reactions [5]. Moreover the easy transformation of alkynylphosphines into valuable functionalized alkenyl mono- or di-phosphines through addition reactions on the triple bond was recently demonstrated [6]. These new applications as ligands for homogeneous catalysis and building blocks for organophosphorus chemistry have aroused the need for new general synthetic methods allowing the access to a broad substitution pattern.

The classical method for preparing alkynylphosphines is based on the nucleophilic substitution reaction at the phosphorus atom of a halophosphine by a metal (lithium, magnesium, titanium or zinc)

acetylide at low temperature (Scheme 1, path A). This methodology was recently successfully applied to the synthesis of a few enantioenriched P-stereogenic alkynylphosphine-boranes. The reaction however suffers from severe limitations due to the weak configurational stability of halophosphines.

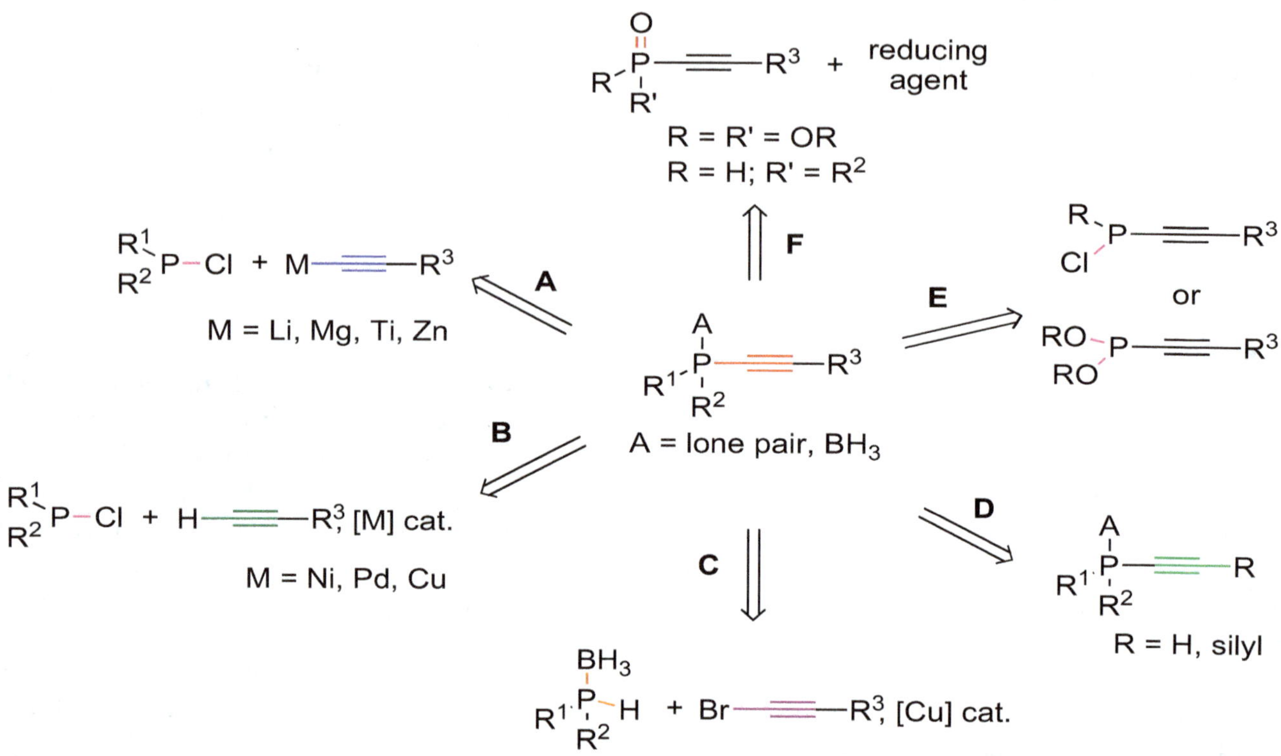

Scheme 1. Strategies A–F for the synthesis of alkynylphosphines and/or their borane complexes.

Catalytic variants of this reaction have been recently described as useful alternative approaches when sensitive functional groups are present (Scheme 1, path B). They rely on the cross-coupling reactions of terminal alkynes with chlorophosphines in the presence of a catalytic amount of nickel, palladium or copper complexes. Both stoichiometric and catalytic methods are mainly based on an electrophilic phosphorus derivative—a halophosphine—as starting material. The use of a nucleophilic phosphorus derivative has been only little explored so far. In the stoichiometric series, attempts involving treatment of alkynylphosphonium chlorides with a base failed to afford the expected tertiary alkynylphosphines because their oxidation could not be avoided under the reaction conditions [7]. On the other hand the catalytic synthesis of tertiary alkynylphosphine-boranes based on the copper(I)-catalyzed cross-coupling reaction between secondary phosphine-boranes and bromoalkynes was shown to proceed readily (Scheme 1, path C). A further strategy towards alkynylphosphines relies on the transformation of alkynylphosphorus derivatives through modification of the alkyne (Scheme 1, path D) or the phosphorus function (Scheme 1, paths E and F). Notably the reduction of the phosphoryl moiety of alkynylphosphorus derivatives (Scheme 1, path F) is highly relevant for the synthesis of primary and secondary alkynylphosphines, which are air- and heat-sensitive compounds.

The scope of this review encompasses the synthesis of alkynylphosphines and their borane complexes [8]. The stoichiometric methods for C–P bond formation will be first surveyed followed by the catalytic procedures. Finally the methods based on the transformation of alkynylphosphorus derivatives will be detailed.

2. Nucleophilic Substitution at P-atom of Halophosphines

Tertiary alkynylphosphines are classically prepared through the reaction of chlorophosphines and metal acetylides. For secondary alkynylphosphines, only those bearing sterically encumbered substituents are available through this procedure.

2.1. *Synthesis of Secondary Alkynylphosphines*

Märkl and Reitinger reported the synthesis of a set of secondary alkynylphosphines based on the reaction of the bulky secondary chlorophosphine **1** with various alkyne Grignard reagents in diethyl ether at low temperature [9] (Scheme 2).

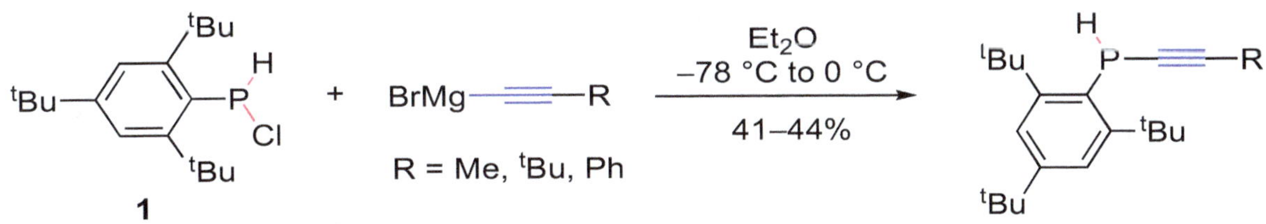

Scheme 2. Synthesis of secondary alkynylphosphines via nucleophilic substitution.

The analogous reactions of chlorophosphine **1** with lithium acetylides in THF did not afford alkynylphosphines but their 3*H*-phosphaallene tautomers, because of the stronger basicity of lithium acetylides in comparison to the related Grignard reagents. One exception was found with lithium trimethylsilylacetylide, which led to the alkynyl product under the same conditions. The reaction of secondary chlorophosphine **1** with metal acetylides proved to be not only metal- and substrate- but also solvent-dependent, probably because of the importance of the acetylide solvation. For example the reaction of **1** with lithium *tert*-butylacetylide led to the alkynyl product in diethyl ether and to the allenyl product in THF (Scheme 3).

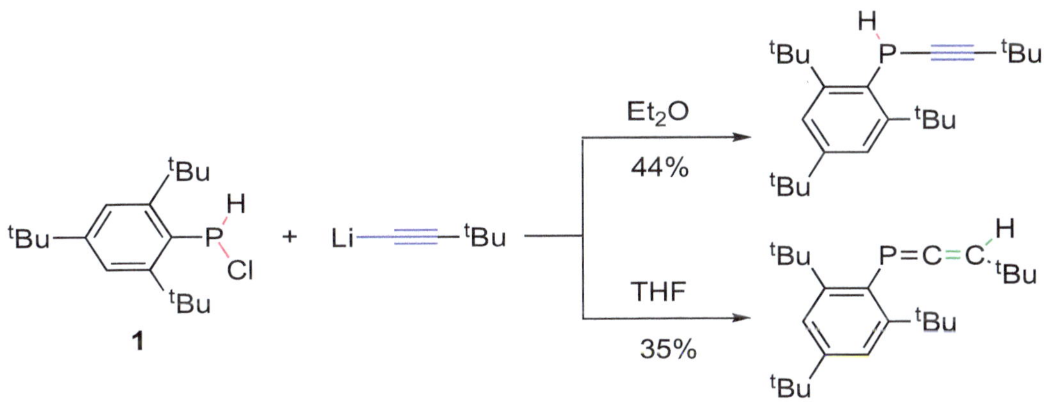

Scheme 3. Solvent-effect on the outcome of the reaction of lithium acetylides with a secondary chlorophosphine.

2.2. *Synthesis of Tertiary Alkynylphosphines*

2.2.1. Version with Achiral Reagents

The oldest methods for the preparation of tertiary alkynylphosphines are based on the nucleophilic substitution reaction at the phosphorus atom of halophosphines by lithium [10] or magnesium [11] acetylides. For example, Cadiot and Chodkiewicz described the synthesis of a set of alkynylphosphines in good to excellent yields (50%–93%) through the reaction of Grignard reagents and dialkyl- or diarylchlorophosphines in THF at −30 °C to 40 °C (Scheme 4).

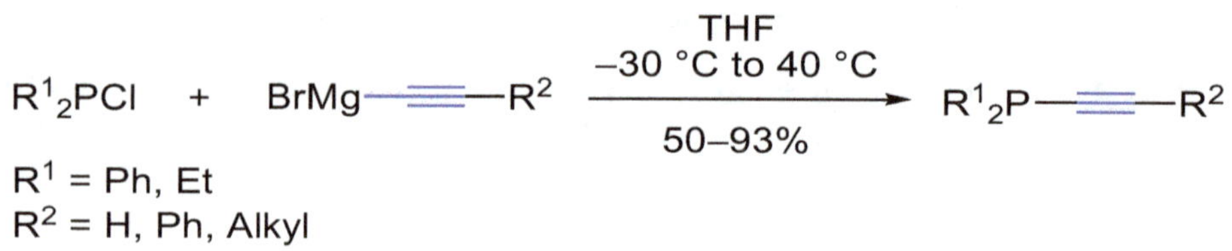

Scheme 4. Reaction of magnesium acetylides with chlorophosphines towards alkynylphosphines.

These procedures involving magnesium or lithium acetylides are still commonly used nowadays. For example Montchamp [12] described the synthesis of several alkynylphosphines using *sec*-butyllithium to generate *in situ* the lithium acetylide reagent. Subsequent treatment with $BH_3 \bullet SMe_2$ afforded the corresponding borane adducts in good yields (Scheme 5).

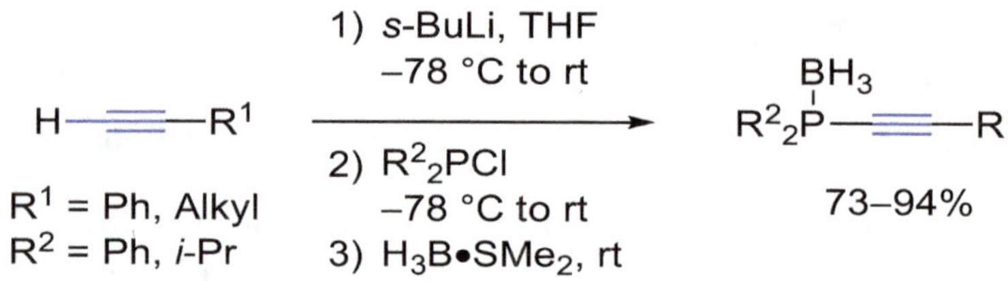

Scheme 5. Synthesis of alkynylphosphine-borane complexes from lithium acetylides.

Besides magnesium and lithium reagents, copper- [13], tin- [14], or silver- [15] acetylides have been involved in a few examples. In view of their high reactivity, magnesium and lithium acetylides are suitable for the formation of alkynylphosphines which do not bear functional groups sensitive to nucleophilic attack such as esters. To prevent the occurrence of side reactions, the use of more selective organometallic reagents like zinc [16] and titanium [17] acetylides was explored. Knochel demonstrated that zinc organometallics were effective for the synthesis of polyfunctionalized phosphines and described one example in the alkynyl series, but on a simple substrate. The alkynylzinc bromide reagent resulting from the transmetalation of 1-hexynyllithium with zinc bromide reacted with chlorodiphenylphosphine to provide the corresponding alkynylphosphine-borane in 80% yield after complexation with borane (Scheme 6).

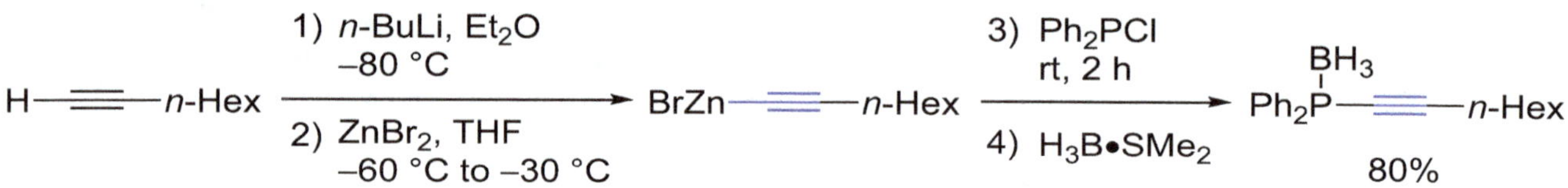

Scheme 6. Synthesis of alkynylphosphines using a zinc acetylide.

Periasamy [17] reported the synthesis of alkynylphosphines using alkynyl titanium reagents. The latter were not obtained by transmetalation of alkynyllithium reagents [18], but were generated *in situ* by direct metalation of 1-alkynes with the combination of titanium tetrachloride and triethylamine in dichloromethane (Scheme 7, Equation 1). Alkyne dimerisation may occur as a side reaction. For example, the reaction of phenyl acetylene with diphenylchlorophosphine afforded the corresponding alkynylphosphine in 54% yield, together with 2% of the diyne. Not only diphenylchlorophosphine but also triethylphosphite could serve as an electrophile. The substitution of all ethoxy groups was observed leading to tris(phenylethynyl)phosphine in 59% yield (Scheme 7, Equation 2).

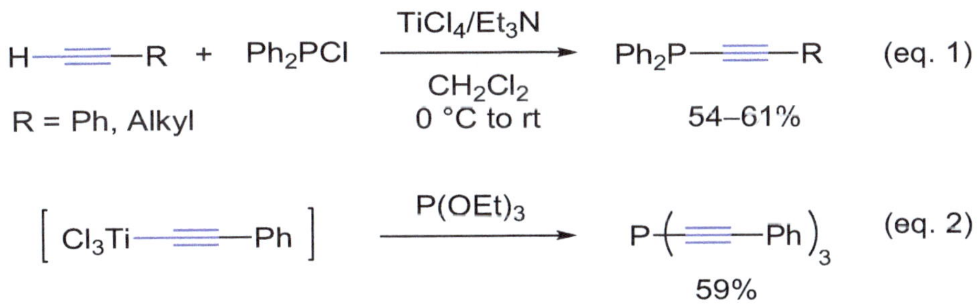

Scheme 7. Synthesis of alkynylphosphines using titanium acetylides.

2.2.2. Asymmetric Version

The single example relating the synthesis of a set of enantiopure P-stereogenic alkynylphosphines was described by Imamoto [5] (Scheme 8). The procedure relies on the *in situ* generation of a P-stereogenic bromophosphine-borane from an enantiopure secondary phosphine-borane via deprotonation by *n*-butyllithium in diethyl ether at −78 °C and subsequent halogenation of the resulting lithium phosphido-borane by treatment with dibromoethane. The bromophosphine-borane was not isolated because of its tendency to racemize at room temperature, but was treated *in situ* with a set of alkynyl lithium reagents. The nucleophilic substitution reaction was shown to occur with clean inversion of configuration at the phosphorus atom, delivering the alkynylphosphines in high yields and excellent stereospecificity.

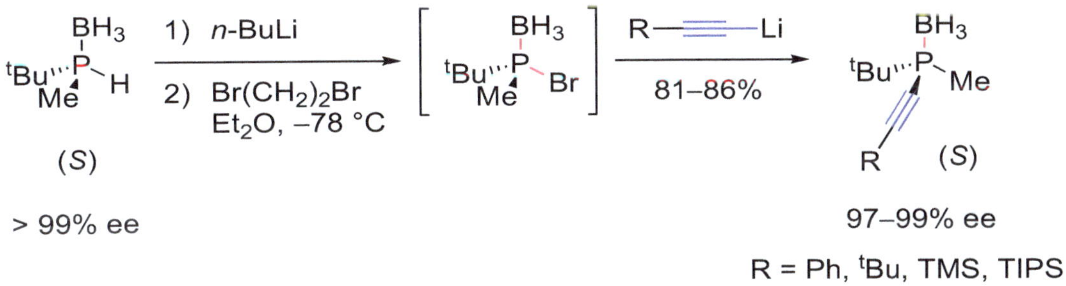

Scheme 8. Synthesis of enantioenriched P-stereogenic alkynylphosphine-boranes.

The procedure is up to now only applicable to enantiopure *tert*-butylmethylbromophosphine-borane, which displays a sufficient configurational stability at low temperature. This successful example is probably difficult to extend to a wide variety of phosphines due to the weak configurational stability of halophosphine-boranes.

3. Catalytic C-P bond Forming Reactions

Within the last ten years a few catalytic approaches towards alkynylphosphines have emerged. They rely on the use of either (i) electrophilic phosphorus derivatives, using nickel, palladium or copper complexes as catalytic system, or (ii) nucleophilic phosphorus derivatives using a copper(I) complex.

3.1. Electrophilic Phosphorus Reagent

As previously mentioned, alkynylphosphines are classically available through the reaction of chlorophosphines with terminal alkynes in the presence of stoichiometric amounts of organometallic bases. The Beletskaya group reported in 2003 [19] a catalytic variant, which was based on the use of nickel or palladium complexes in the presence of an excess of triethylamine. Nickel complexes were shown to display a higher activity than the palladium ones, the best catalysts being $Ni(cod)_2$ and $Ni(acac)_2$. The reaction with aryl or alkyl acetylenes readily proceeded in toluene at 80 °C for

10 to 30 min or at room temperature for 6 to 10 h (Scheme 9). It is noteworthy that the selective monosubstitution of one chloride atom from di- or tri-chlorosubstrates is not possible.

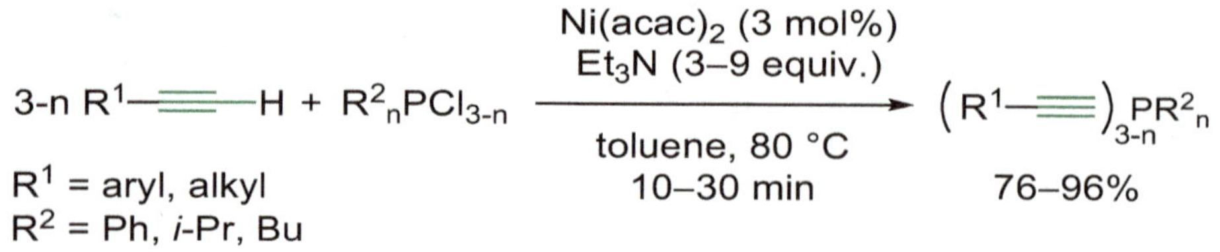

Scheme 9. Nickel-catalyzed cross-coupling of terminal alkynes with chlorophosphines.

The simplified postulated mechanism first involves the insertion of the metal into the P–Cl bond of the chlorophosphine. The resulting metal phosphido complex undergoes transmetalation with the *in situ* formed acetylide ion, followed by reductive elimination (Scheme 10).

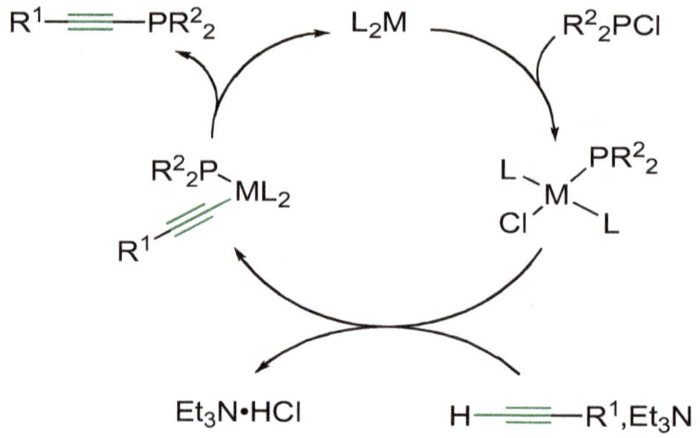

Scheme 10. Simplified postulated mechanism of nickel- or palladium-catalyzed cross-coupling between terminal alkynes and chlorophosphines.

The procedure was applied to the synthesis of a set of functionalized alkynylphosphines bearing carbonyl groups, which would have been difficult to access via the stoichiometric method (Figure 1). However in the case of aliphatic alkynes, a full conversion could not be reached and poor yields (around 30%) were obtained [20].

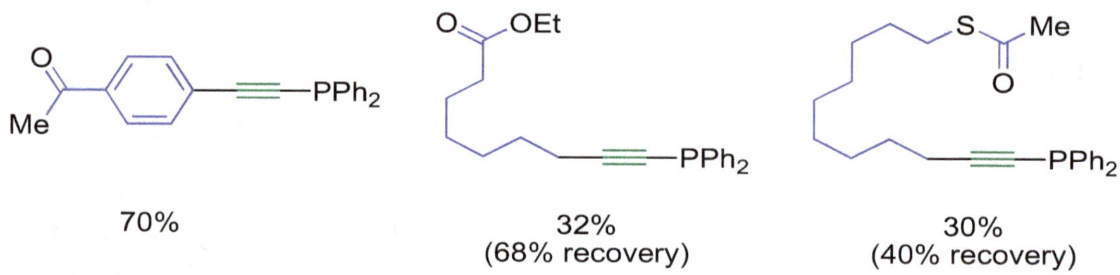

Figure 1. Functionalized alkynylphosphines synthesized via the nickel-catalyzed procedure.

Another limitation of this nickel-catalyzed procedure was its lack of efficiency with substrates bearing an alkoxy or amino group. To overcome these limitations a more versatile catalytic method was developed by the same authors. The method is based on the use of copper salts [21] and proceeds under mild conditions. Copper(I) iodide was used as the copper source in the presence of an excess of triethylamine in toluene at room temperature (Scheme 11). The reaction scope is broad and encompasses alkyl and (het)aryl acetylenes. Regarding the phosphorus coupling partner, aryl and alkyl chlorophosphines are suitable substrates.

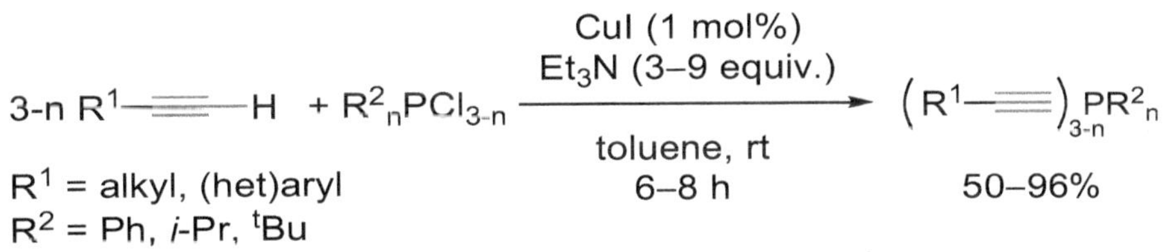

Scheme 11. Copper-catalyzed cross-coupling of terminal alkynes with chlorophosphines.

The postulated mechanism does not rely on P–Cl bond activation by the metal, as proposed in the case of Ni or Pd catalysis, but copper π-complexation of the terminal alkyne, followed by nucleophilic substitution at the phosphorus atom of the chlorophosphine in the presence of triethylamine, as outlined on Scheme 12. This hypothesis was supported by the isolation of the tetrahedral complex $(Ph_2PCl)_3CuI$, which is the first active species involved in the suggested catalytic cycle.

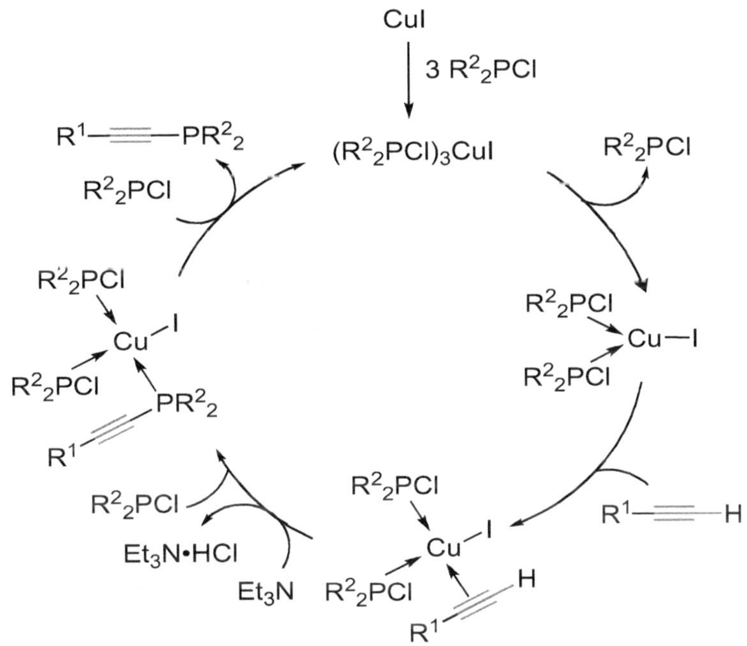

Scheme 12. Postulated mechanism of copper-catalyzed cross-coupling between terminal alkynes and chlorophosphines.

This general and mild procedure was successfully applied to the synthesis of bulky triethynylphosphine ligands [22,23] (Scheme 13).

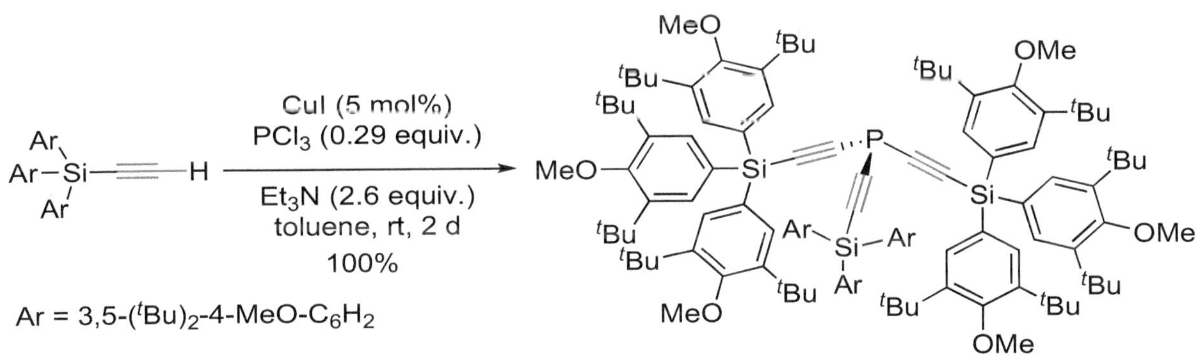

Scheme 13. Copper(I)-catalyzed synthesis of functionalized triethynylphosphines.

3.2. Nucleophilic Phosphorus Reagent

The Gaumont group reported the only example of a catalytic synthesis of alkynylphosphines, relying on nucleophilic phosphorus reagents [24]. The alkynylphosphines were obtained as their borane complexes. The methodology is based on the copper-catalyzed cross-coupling of secondary phosphine-boranes with alkyl or aryl 1-bromoalkynes using CuI/1,10-phenanthroline as catalytic source (Scheme 14). The reaction proceeds in toluene under mild conditions (20 to 60 °C, weak base).

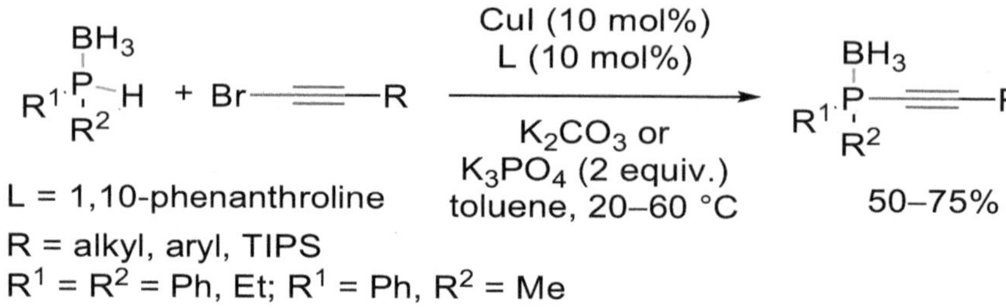

Scheme 14. Copper(I)-catalyzed reaction of secondary phosphine-boranes with 1-bromoalkynes towards alkynylphosphine-boranes.

Preliminary mechanistic studies are consistent with the involvement of a copper(I) phosphido-borane complex as the reactive species. Indeed neutral copper phosphido-boranes [$R_2PHCuPPh_2BH_3$phen], which could be isolated for the first time and fully characterized via experimental and computational methods [25], were shown to serve as pre-catalysts in the P-alkynylation of secondary phosphine-boranes with 1-bromoalkynes (Scheme 15).

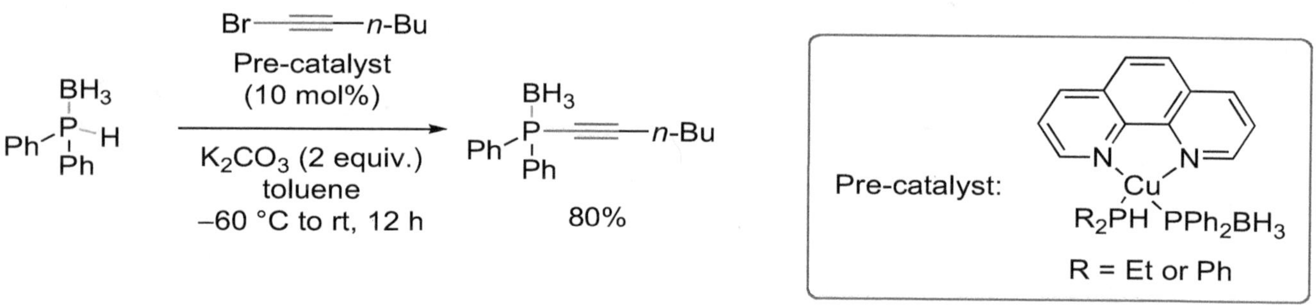

Scheme 15. P-alkynylation of secondary phosphine-boranes catalyzed by well-defined copper(I) phosphido-borane complexes.

The putative mechanism of the reaction involves oxidative addition of the copper(I) phosphido-borane complex to the 1-bromoalkyne, followed by reductive elimination, leading to the cross-coupling product and allowing the release of the copper(I) catalyst (Scheme 16).

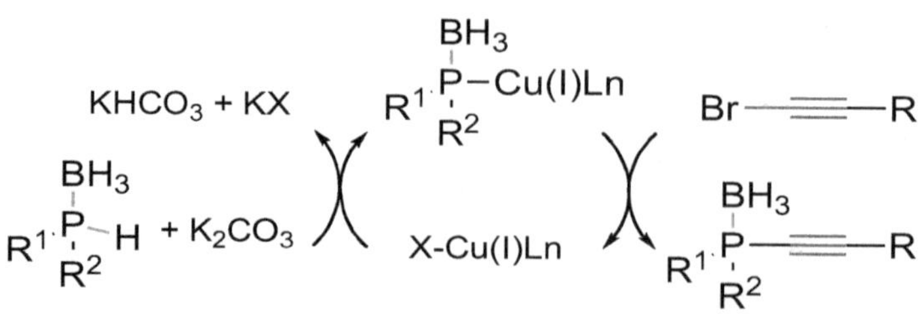

Scheme 16. Putative mechanistic cycle for the copper(I)-catalyzed P-alkynylation of secondary phosphine-boranes.

4. Transformation of Alkynylphosphorus Derivatives

Regarding the synthesis of tertiary alkynylphosphines, modification of the alkynyl or the phosphorus function of alkynyl phosphorus derivatives is a less commonly used method in comparison to C–P bond formation. On the other hand, in the case of secondary and primary alkynylphosphines, the reduction of the phosphoryl moiety of phosphine oxides and phosphonates respectively, is the method of choice for their preparation.

4.1. Functionalization of the Triple bond Moiety

Classical alkyne chemistry has been applied in the functionalization of alkynylphosphines through their metalation and subsequent reaction with an electrophile. With the aim of preparing dendrimers, Caminade and Majoral [26] generated the Grignard reagent of ethynylphosphine as building block for the synthesis of mono-, di-, tri-, and tetra-phosphines (Scheme 17). It is noteworthy that the lithium acetylide of ethynylphosphine was not suitable for these transformations, complex mixtures being formed in this case.

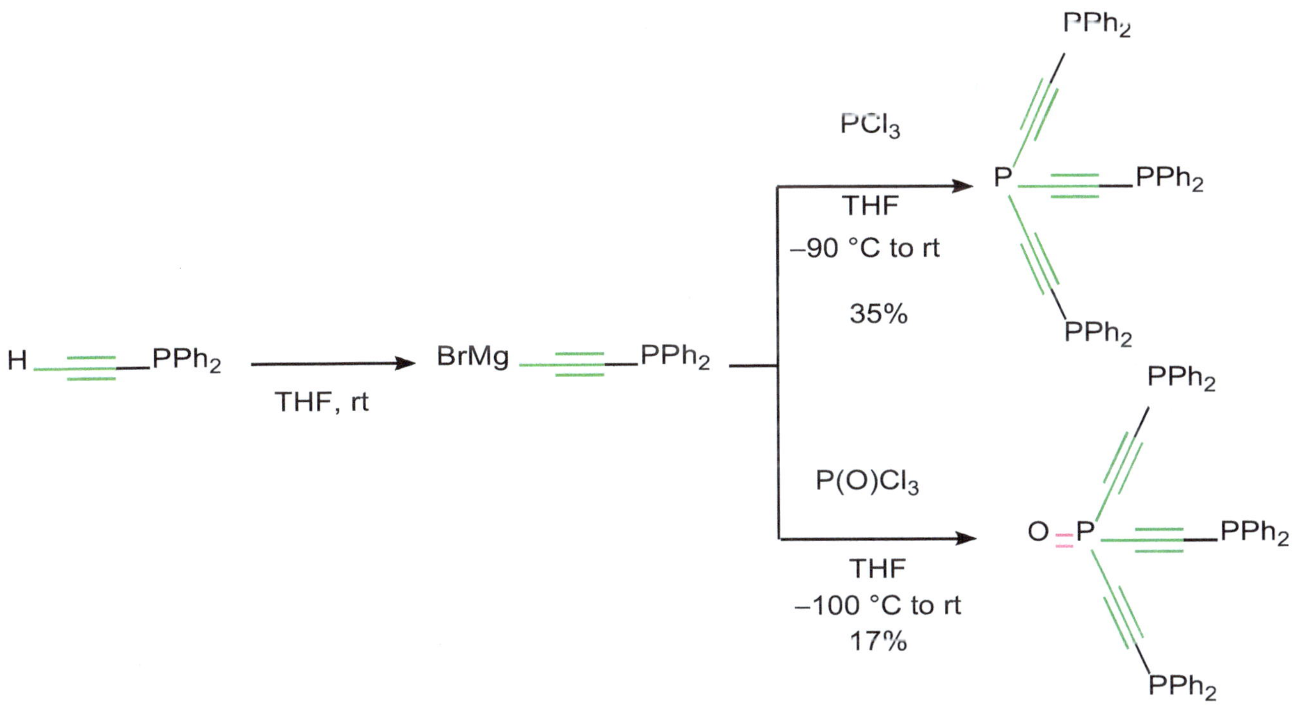

Scheme 17. Metalation and functionalization of ethynylphosphine.

The Imamoto group [5] described a 4-step reaction sequence allowing the synthesis of an alkynyl variant of BisP* ligands, analogous to DIPAMP [27], from an enantiopure silyl alkynylphosphine-borane (Scheme 18). The first step was a dimerization reaction through copper(II)-catalyzed oxidative coupling leading to the bis-alkynylphosphine-borane **2**. The next step was the silyl deprotection by treatment with TBAF. It was followed by deprotonation with *sec*-BuLi and methylation of the resulting lithium derivative with methyl iodide. Finally borane removal by treatment with DABCO (1,4-diazabicyclo[2.2.2]octane) afforded the bis-alkynylphosphine.

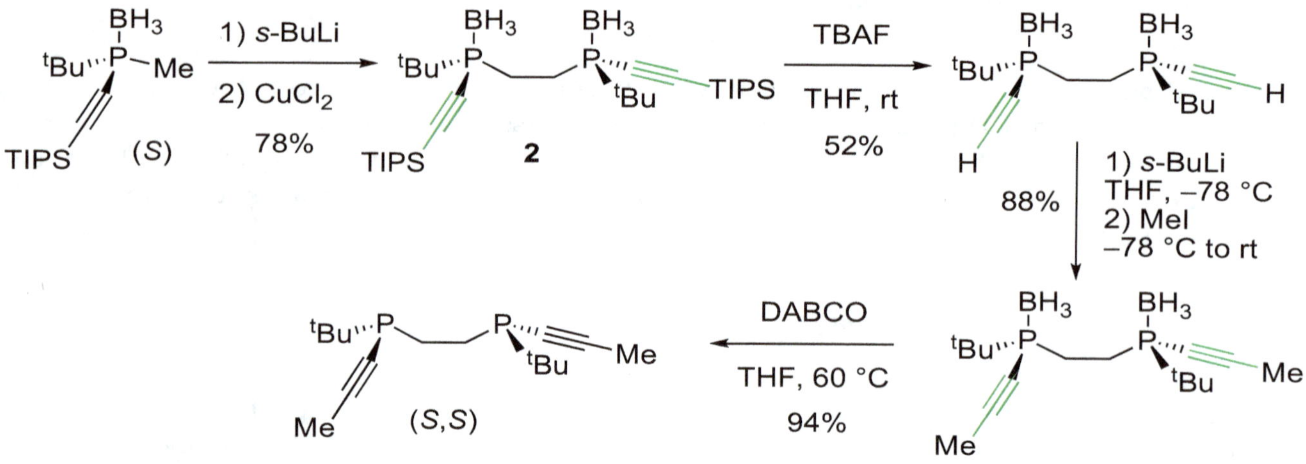

Scheme 18. Synthesis of an alkynyl analogue of DIPAMP.

4.2. Nucleophilic Substitution at Phosphorus Atom

The nucleophilic substitution at the phosphorus atom of alkynylchlorophosphines by Grignard reagents has been used for the building of complex structures. As an example, the Yoshifuji group [28] described the synthesis of the dialkynyldiphosphine **4** from alkynyldichlorophosphine **3** (Scheme 19).

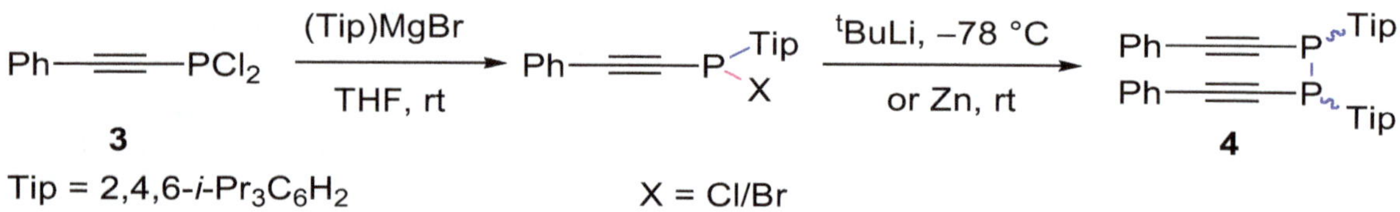

Scheme 19. Synthesis of alkynylphosphines from alkynyl-chlorophosphines.

Besides chlorophosphines, phosphonites can serve as precursors of alkynylphosphines. The Aguiar group [29] reported the synthesis of a set of alkynylphosphines through the reaction of the *in situ* generated diethyl alkynyl-1-phosphonite **5** and Grignard reagents (Scheme 20).

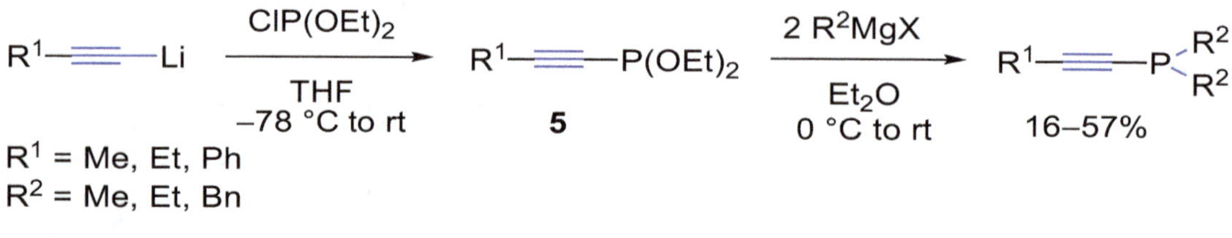

Scheme 20. Synthesis of alkynylphosphines from *in situ* generated alkynylphosphonites.

4.3. Reduction of the Phosphoryl Moiety

As described in the Section 2.1, a few secondary alkynylphosphines were prepared via the reaction of chlorophosphines and metal acetylides. However it concerned specific substrates stabilized by the presence of bulky substituents on the phosphorus atom.

A more straightforward and general access to unstabilized primary and secondary alkynylphosphines relies on the reduction of phosphoryl derivatives by alanes or silanes [30]. Guillemin and Denis [31] developed the chemoselective reduction of alkynylphosphonates using $AlHCl_2$ as reducing agent. The best yields were obtained when the reaction was performed at low temperature in diethyl ether or THF (Scheme 21, Equation 1). The same alane reducing agent was less suitable for the preparation of secondary alkynylphosphines from alkynylphosphinates. Indeed C–P bond cleavage occurred as a side-reaction [32], thus lowering the yields (Scheme 21, Equation 2).

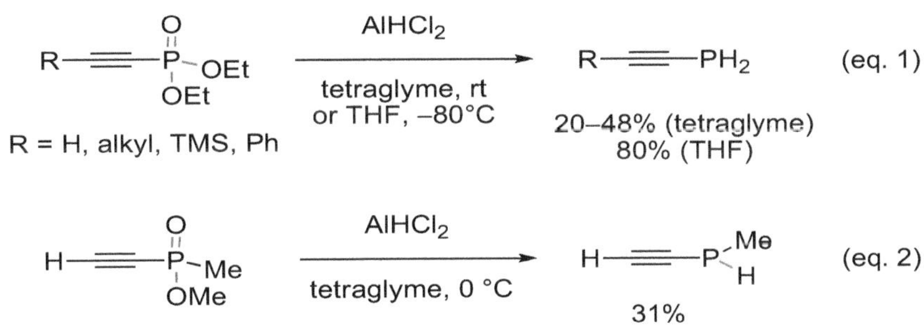

Scheme 21. Synthesis of primary and secondary alkynylphosphines via phosphoryl reduction by alanes.

A mild and practical synthesis of secondary alkynylphosphines was reported by Gaumont and Denis [33]. It is based on the reduction of secondary alkynylphosphine oxides in the presence of silane derivatives. The starting phosphine oxides were available through acidic cleavage of aminophosphine precursors. The appropriate silane reducing agent depended on the substituents at the phosphorus atom. Phenylsilane was used as reducing agent in the case of P-alkyl derivatives (Scheme 22, Equation 1) while the reduction of P-phenyl derivatives was only partial under these conditions. It was finally achieved with the combination of phenylsilane and trichlorophenylsilane in a 3:1 ratio (Scheme 22, Equation 2). The active reducing agent is believed to be the *in situ* generated $PhSiHCl_2$. The presence of chlorine atoms on silicon, by increasing the oxophilic properties of the silane reagent, would explain its higher reactivity.

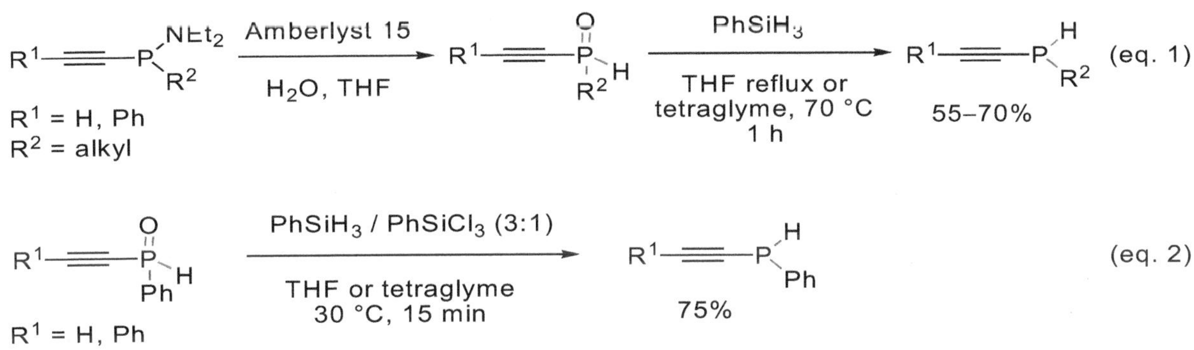

Scheme 22. Synthesis of secondary alkynylphosphines via phosphoryl reduction by silanes.

5. Conclusions

The synthetic pathways towards alkynylphosphines used to be restricted to stoichiometric procedures involving the use of organometallic bases, which limited the scope of the substitution pattern to robust functional groups. Since less than a decade, benefiting from the renaissance of copper catalysis for C–C and C–Heteroatom bond forming reactions, catalytic procedures have emerged, which involve either an electrophilic or a nucleophilic phosphorus derivative. By proceeding under mild conditions, these new catalytic methods have broadened the substituent scope and have been successfully applied to the synthesis of functionalized complex phosphine structures.

Future developments comprise catalytic procedures avoiding the use of halogen-containing derivatives, such as oxidative cross-couplings, which have proven their efficiency for the synthesis of alkynylphosphonates [34], but whose application to the synthesis of alkynylphosphines is challenging because of the high sensitivity of phosphines towards oxidation. Another challenge is the extension to the asymmetric series through the use of chiral ligands associated with the metal catalyst.

Acknowledgments: We gratefully acknowledge financial support from CNRS and Crunch (interregional organic chemistry network), the "Ministère de la Recherche et des Nouvelles Technologies", the "Région Basse-Normandie" and ERDF funding (ISCE-Chem & INTERREG IVa program). COST ACTION CM0802 "PHOSCINET" is also gratefully thanked for support.

References

1. Börner, A. *Phosphorus Ligands in Asymmetric Catalysis*; Wiley-VCH: Weinheim, Germany, 2008.
2. Diederich, F.; Stang, P.; Tykwinski, R.R. *Acetylene Chemistry*; Wiley-VCH: Weinheim, Germany, 2004.
3. Weymiens, W.; Slootweg, J.C.; Lammertsma, K. Phosphine Acetylenic Macrocycles and Cages: Synthesis and Reactivity. In *Phosphorus Compounds Advanced Tools in Catalysis and Material Science*, 1st ed.; Peruzzini, M., Gonsalvi, L., Eds.; Springer: New York, NY, USA, 2011; Volume 37, pp. 21–55.
4. Ito, H.; Ohmiya, H.; Sawamura, M. Construction of methylenecycloheptane frameworks through 7-*exo-dig* cyclisation of acetylenic silyl enol ethers catalyzed by triethynylphosphine-gold complex. *Org. Lett.* **2010**, *12*, 4380–4383. [CrossRef] [PubMed]
5. Imamoto, T.; Saitoh, Y.; Koide, A.; Ogura, T.; Yoshida, K. Synthesis and enantioselectivity of P-chiral phosphine ligands with alkynyl groups. *Angew. Chem. Int. Ed.* **2007**, *46*, 8636–8639. [CrossRef] [PubMed]
6. Kondoh, A.; Yorimitsu, H.; Oshima, K. 1-Alkynylphosphines and their derivatives as key starting materials in creating new phosphines. *Chem. Asian J.* **2010**, *5*, 398–409. [CrossRef] [PubMed]
7. D'yachkova, S.G.; Nikitin, M.V.; Beskrylaya, E.A.; Arbuzova, S.N.; Kashik, T.V.; Gusarova, N.K.; Trofimov, B.A. Reaction of alkylthio(chloro)acetylenes with bis(2-phenylethyl)phosphine. *Russ. J. Gen. Chem.* **1999**, *69*, 767–770.
8. Pietrusiewicz, K.M.; Stankevic, M. 1-Phosphorus-functionalized alk-1-ynes. In *Science of Synthesis: Houben-Weyl Methods of Molecular Transformations*; de Meijere, A., Ed.; Thieme: Stuttgart, Germany, 2005; Volume 24, pp. 1073–1085.
9. Märkl, G.; Reitinger, S. 3H-Phosphaallen—Alkinyl-1H-Phosphan—Tautomere. *Tetrahedron Lett.* **1988**, *29*, 463–466. [CrossRef]
10. Voskuil, W.; Arens, J.F. Chemistry of acetylenic ethers. LXII. Tertiary phosphines with an acetylene-phosphorus bond. *Recl. Trav. Chim. Pays-Bas* **1962**, *81*, 993–1008. [CrossRef]
11. Charrier, C.; Chodkiewicz, W.; Cadiot, P. Préparations, propriétés chimiques et spectrographiques de phosphines α-acétyléniques. *Bull. Soc. Chim. Fr.* **1966**, 1002–1011.
12. Ortial, S.; Montchamp, J.-L. Synthesis of *Z*-alkenyl phosphorus compounds through hydroalumination and carbocupration of alkynyl precursors. *Org. Lett.* **2011**, *13*, 3134–3137. [CrossRef]
13. Sladkov, A.M.; Ukhin, L.Y.; Korshak, V.V. Reaction of copper acetylides with halogen compounds. *Izv. Akad. Nauk SSR Ser. Khim.* **1963**, 2213–2215.
14. Hartmann, H. Exchange reactions of some ethynyl-, diethynyl-, and butadiynyltin compounds. *Justus Liebigs Ann. Chem.* **1968**, *714*, 1–7. [CrossRef]
15. Gol'ding, I.R.; Sladkov, A.M. Silver acetylenides in ethynylation reactions. *Izv. Akad. Nauk SSR Ser. Khim.* **1972**, 529–530.
16. Langer, F.; Püntener, K.; Stürmer, R.; Knochel, P. Preparation of polyfunctional phosphines using zinc organometallics. *Tetrahedron: Asymmetry* **1997**, *8*, 715–738. [CrossRef]
17. Bharathi, P.; Periasamy, M. Direct metalation of 1-alkynes using $TiCl_4/Et_3N$ and the reactions of the organotitanium intermediates with electrophiles. *Organometallics* **2000**, *19*, 5511–5513. [CrossRef]
18. Krause, N.; Seebach, D. Chemistry of acetylenic titanium compounds. *Chem. Ber.* **1987**, *120*, 1845–1851. [CrossRef]
19. Beletskaya, I.P.; Afanasiev, V.V.; Kazankova, M.A.; Efimova, I.V. New approach to phosphinoalkynes based on Pd- and Ni-catalyzed cross-coupling of terminal alkynes with chlorophosphanes. *Org. Lett.* **2003**, *5*, 4309–4311. [CrossRef]
20. Kondoh, A.; Yorimitsu, H.; Oshima, K. Copper-catalyzed anti-hydrophosphination reaction of 1-alkynylphosphines with diphenylphosphine providing (*Z*)-1,2-diphosphino-1-alkenes. *J. Am. Chem. Soc.* **2007**, *129*, 4099–4104. [CrossRef]
21. Afanasiev, V.V.; Beletskaya, I.P.; Kazankova, M.A.; Efimova, I.V.; Antipin, M.U. A convenient and direct route to phosphinoalkynes via copper-catalyzed cross-coupling of terminal alkynes with chlorophosphanes. *Synthesis* **2003**, 2835–2838. [CrossRef]
22. Ochida, A.; Ito, H.; Sawamura, M. Using triethynylphosphine ligands bearing bulky end caps to create a holey catalytic environment: Application to gold(I)-catalyzed alkyne cyclizations. *J. Am. Chem. Soc.* **2006**, *128*, 16486–16487. [CrossRef]

23. Ochida, A.; Sawamura, M. Phosphorus ligands with a large cavity: synthesis of triethynylphosphines with bulky end caps and application to the rhodium-catalyzed hydrosilylation of ketones. *Chem. Asian J.* **2007**, *2*, 609–618. [CrossRef]
24. Bernoud, E.; Alayrac, C.; Delacroix, O.; Gaumont, A.-C. Copper-catalyzed synthesis of alkynylphosphine derivatives: unprecedented use of nucleophilic phosphorus compounds. *Chem. Commun.* **2011**, *47*, 3239–3241. [CrossRef]
25. Abdellah, I.; Bernoud, E.; Lohier, J.-F.; Alayrac, C.; Toupet, L.; Lepetit, C.; Gaumont, A.-C. Neutral copper-phosphido-borane complexes: Synthesis, characterization, and use as precatalysts in C_{sp}-P bond formation. *Chem. Commun.* **2012**, *48*, 4088–4090. [CrossRef] [PubMed]
26. Huc, V.; Balueva, A.; Sebastian, R.-M.; Caminade, A.-M.; Majoral, J.-P. Synthesis of functionalized mono-, di-, tri- and tetraphosphines: attempted application to prepare hyperbranched polymers and dendrimers built with phosphines at each branching point. *Synthesis* **2000**, 726–730. [CrossRef]
27. Knowles, W.S.; Sabacky, M.J.; Vineyard, B.D.; Weinkauff, D.J. Asymmetric hydrogenation with a complex of rhodium and a chiral bisphosphine. *J. Am. Chem. Soc.* **1975**, *97*, 2567–2568. [CrossRef]
28. Yoshifuji, M.; Ichikawa, Y.; Yamada, N.; Toyota, K. Preparation and X-ray analysis of novel carbonyltungsten(0) complexes of diphosphinidenecyclobutenes. *Chem. Commun.* **1998**, 27–28. [CrossRef]
29. Aguiar, A.M.; Irelan, J.R.S.; Morrow, C.J.; John, J.P.; Prejean, G.W. A convenient, synthetic pathway to dialkyl-1-alkynylphosphines. *J. Org. Chem.* **1969**, *34*, 2684–2686. [CrossRef]
30. Keglevich, G.; Gaumont, A.-C.; Denis, J.-M. Selective reductions in the sphere of organophosphorus compounds. *Heteroat. Chem.* **2001**, *12*, 161–167. [CrossRef]
31. Guillemin, J.-C.; Savignac, P.; Denis, J.-M. Primary alkynylphosphines and allenylphosphines. *Inorg. Chem.* **1991**, *30*, 2170–2173. [CrossRef]
32. Guillemin, J.-C.; Janati, T.; Denis, J.-M.; Guenot, P.; Savignac, P. Unstabilized 1-phosphaallenes: synthesis and characterization. *Tetrahedron Lett.* **1994**, *35*, 245–248. [CrossRef]
33. Pilard, J.-F.; Baba, G.; Gaumont, A.-C.; Denis, J.-M. Reduction of phosphine oxides, a mild and practical synthesis of secondary P-ethynylphosphines. *Synlett* **1995**, 1168–1170. [CrossRef]
34. Gao, Y.; Wang, G.; Chen, L.; Xu, P.; Zhao, Y.; Zhou, Y.; Han, L.-B. Copper-catalyzed aerobic oxidative coupling of terminal alkynes with *H*-phosphonates leading to alkynylphosphonates. *J. Am. Chem. Soc.* **2009**, *131*, 7956–7957. [CrossRef]

11

New Rigid Polycyclic Bis(phosphane) for Asymmetric Catalysis

K. Michał Pietrusiewicz [1], Katarzyna Szwaczko [1,*], Barbara Mirosław [1,*], Izabela Dybała [2], Radomir Jasiński [3] and Oleg M. Demchuk [1,4]

1 Faculty of Chemistry, Maria Curie-Sklodowska University, 33-Gliniana St., 20-031 Lublin, Poland; Kazimierz.Pietrusiewicz@poczta.umcs.lublin.pl (K.M.P.); O.Demchuk@IFarm.eu (O.M.D.)
2 Faculty of Pharmacy, Medical University of Lublin, 4A-Chodźki St., 20-093 Lublin, Poland; izalis1@gazeta.pl
3 Department of Organic Chemistry, Cracow University of Technology, 24-Warszawska St., 31-155 Cracow, Poland; radomir@chemia.pk.edu.pl
4 Pharmaceutical Research Institute, 8-Rydygiera St., 01-793 Warsaw, Poland
* Correspondence: katarzyna.szwaczko@poczta.umcs.lublin.pl (K.S.); barbara.miroslaw@poczta.umcs.lublin.pl (B.M.)

Academic Editor: Giovanni De Feo

Abstract: A simple, highly efficient synthesis of a series of novel chiral non-racemic rigid tetracyclic phosphorus ligands, applicable in important chemical asymmetric transformations, was performed. In a tandem cross-coupling/C-H bond activation reaction, a well-recognised and readily available ligand *(R,R)*-NORPHOS was used as the starting material. The palladium complexes of new ligands were obtained and characterised on the example of a crystalline dichloropalladium complex of [(1*R*,2*R*,9*S*,10*S*,11*R*,12*R*)-4-phenyltetracyclo[8.2.1.$0^{2,9}$.$0^{3,8}$]trideca-3,5,7-triene-11,12-diyl]bis(diphenylphosphane). A notably high activity and stereoselectivity of the palladium catalysts based on the new ligands were confirmed in a model asymmetric allylic substitution reaction. Herein, we discuss the geometry of the palladium complexes formed and its impact on the efficiency of the catalysts. A comparison of their geometric features with other bis(phosphane) ligand complexes found in the Cambridge Structural Database and built density functional theory (DFT) commutated models is also presented and rationalised.

Keywords: bis(phosphane) palladium complex; metallacycle; NORPHOS; allylic alkylation; asymmetric catalysis; chiral phosphines; C-H bond activation; polycyclic compounds; stereoselective synthesis; DFT calculations

1. Introduction

One of the most challenging approaches in contemporary synthetic organic chemistry is the selective activation of C-H bonds leading to the creation of new carbon-carbon bonds in the structure of fine organic compounds of commercial value. A palladium-catalysed Heck coupling reaction is one of the major tools used for the C-C bond formation. The reaction is widely used to prepare drugs, fragrances and agrochemicals [1–3]. An extension of the regular alkenes arylation according to the Heck coupling mechanism is the palladium-catalysed Heck–Catellani reaction, reported in 1985, which involves an intramolecular C-H bond activation and functionalisation [4]. This discovery provided the foundation for the majority of recent developments of direct C-H activation/C-C coupling reactions [5–9] that are conducted to improve the efficiency and atom economy of traditional couplings.

Chiral biphosphine ligands are a powerful synthetic tool for generating complex and diverse chemical compounds. Recently, we have reported a new methodology for a highly regio- and stereoselective reaction of the NORPHOS oxide (**1**) [10,11] with substituted aryl bromides (**2**), which furnished various pentacyclic NORPHOS oxide derivatives **3** (Scheme 1) [12]. This new type of the Heck–Catellani reaction proceeded through tandem cross-coupling/cross-dehydrogenative-coupling stages, mediated by highly active palladium complexes of bulky electron-rich bidentate mono phosphines (S-Phos, *Sym*-Phos and MeO*Sym*-Phos) [12,13]. Inspired by these achievements, we have been prompted to explore some new opportunities that a reaction of the Heck–Catellani type could provide. It had been mentioned in [12] that in the reaction leading to pentacyclic NORPHOS oxide derivatives (**3**), some by-products with molecular weights that formally satisfy chemical formulae of classical mono-arylated Heck products were formed in trace amounts. Since the formation of the Heck reaction product is rather non-typical for norbornene substrates, we supposed that the observed minor by-product could have the structure of a tetracyclic compound (**4**). Interestingly, the best chiral ligands used in the transition-metal-mediated asymmetric synthesis display chirality in the backbones of the ligand and a rigid structure weakly influenced by random conformational changes [14]. This phenomenon has its roots in the fact that the catalysts with restricted freedom of conformational movements more significantly differentiate the competitive transition states leading to enantiomeric reaction products. Thus, C-chiral chelating phosphanes whose stereometry is stabilised by polycyclic systems, especially those based on five- and four-membered rings, could be considered as promising candidates for efficient ligands to be used in catalytic transition-metal-mediated asymmetric transformations. A new, efficient, highly stereo- and regioselective approach to the synthesis of tetracyclic compound (**4**), using readily available substrates and inexpensive commercial palladium catalysts, should be a welcome addition to our previously reported methods for the synthesis of valuable polycyclic bis(phosphane) ligands.

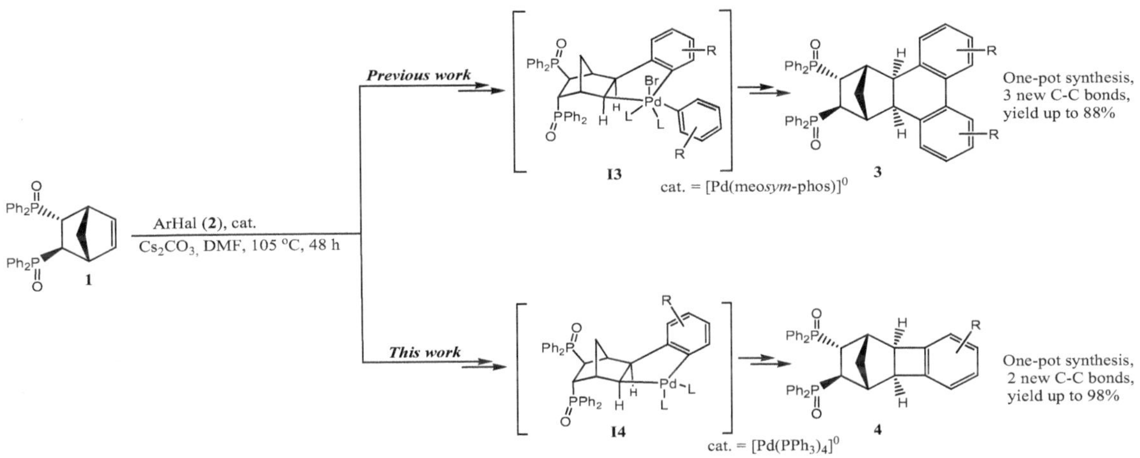

Scheme 1. Synthesis of polycyclic NORPHOS derivatives.

Herein, we disclose a transition-metal-catalysed, one-pot stereoselective synthesis of [tetracyclo$[8.2.1.0^{2,9}.0^{3,8}]$trideca-3,5,7-triene-11,12-diyl]bis(diphenylphosphane) dioxides: novel NORPHOS oxide derivatives bearing a benzocyclobutene motif. After deoxygenation, the obtained phosphanes could be used as ligands in transition-metal-mediated asymmetric transformations. We also present a short summary of the geometry in the pentametallacyclic complexes of bis(phosphones) with transition metals found in the Cambridge Structural Database, which were compared with the dichloropalladium complex of phosphane derived from **4**.

2. Results and Discussion

Utilisation of most active ligands, such as MeO*Sym*-Phos: Electron-rich and bulky, according to the basic concept [15,16], should accelerate the oxidative addition of aromatic halides to Pd(0) species and, therefore, a double addition of aromatic halides should predominate in the reaction [12]. Thus, in order to direct the reaction to an alternative path of mono-arylation, an electron-deficient ligand of a relatively small steric hindrance, which stabilises intermediate **I4** and facilitates the reductive elimination to product **4**, should be used (Scheme 1). Tetrakis(triphenylphosphine)palladium(0) was therefore selected to be used as the catalyst. The two-component reactions of racemic NORPHOS oxide (*rac*-**1**) and aryl halides (bromide or iodide) used in an equimolar amount or in a 3-fold excess were carried out in the presence of $Pd(PPh_3)_4$, bases and solvents at 105 °C for 12–48 h. We found that a proper choice of the base and solvent was crucial to achieving high yields. Among the tested bases, including K_3PO_4, KOAc, KO^{t}Bu, K_2CO_3 and others (see electronic supplementary information (ESI)), Cs_2CO_3 was selected as the most promising, which significantly increased the yield of desirable product **4**. Polar aprotic solvents are often used in the coupling reaction as they allow for the best solubility of the substrates and the catalysts. In the model reaction between **1** and *p*-bromotoluene (1.2 equiv.) run in dimethylformamide (DMF) in the presence of 5 mol% of $[Pd(PPh_3)_4]$ and caesium carbonate (3.0 equiv.) at 105 °C for 48 h, the isolated yield of product **4b** was 83%. A study of the solvent effects on the yield of the coupling reaction revealed that while the reaction proceeds well in toluene (yield 79%), in THF (yields 72%) and in acetonitrile (yield 65%), the catalytic process is greatly affected in DMF (see ESI). Compared to other combination of used bases and solvents, Cs_2CO_3 solubilised in DMF is less likely to coordinate to the palladium centre and, therefore, less likely to interfere with the catalyst activity. To obtain a satisfactory conversion in DMF, 5 mol% of $[Pd(PPh_3)_4]$ was required. The presence of 1 mol% and 0.1 mol% of the phosphine palladium complex resulted in the formation of 45% and 21% yields of the product, respectively (see ESI).

Having optimised the conditions, we prepared a series of phosphine oxides **4**. Table 1 presents the isolated yields of products **3** and **4** bearing diversified substituents in the introduced arenes.

The coupling reaction of ortho-substituted aryl halides with **1** gave cyclobutene derivatives **4** with a very good yield. A higher yield of **4** was observed when the proportion of the aryl halide was increased from 1.2 to 3.0 equiv. Thus, in the reaction of **1** with 1.2 equiv. or with 3.0 equiv. of *o*-bromotoluene (**2f**), the desired product was isolated in 78% and 85% of the yield, respectively (Table 1, entry 6). Under similar conditions, a tri-substituted aryl halide (Table 1, entry 8) allows us to achieve a similar 86% yield. Using $[Pd(PPh_3)_4]$, a generally low-activity catalyst, allows us to conduct the reaction between **1** and 2-bromochlorobenzene (**2g**) chemoselectively, and only product **4g** was isolated in a 54% yield. In the case of 2-bromobiphenyl (**2q**), the reaction furnished tetracyclic product [tetracyclo[8.2.1.0^{2,9}.0^{3,8}]trideca-3,5,7-triene-11,12-diyl]bis(diphenylphosphane) dioxides (**4q**) in an excellent 98% yield.

Surprisingly, we have found that the direction of the reaction depends not only on the catalyst used, but also on the nature of the aryl halide. The NORPHOS oxide underwent the reaction with highly active aromatic halides, e.g., possessing an electron-withdrawing groups (EWG) (**2j**, **2k**) or a trimethylsilyl (TMS) group (**2l**) and with unsubstituted phenyl bromide (**2a**) furnishing the mixture of products **3** and **4** (Table 1, entries 1, 10, and 11) with the ratio of **3** to **4** increasing in line with the increasing excess of substrates **2** in the reaction mixture. This observation supports the suggestion that the selection of the reaction path depends on the relation of the rates of conversion from intermediates **I4** to **4** and **I4** to **I3** (Scheme 2). Thus, using "fast", electron-rich and bulky catalysts and active substrates, due to rapid halo-arylation of the palladium atom in **I4**, facilitates the formation of **I4** and the next product **3**. In contrast to that, utilisation of less-active, electron-deficient and ligand-based catalysts and electron-donating group (EDG) substituted substrates resulted in the formation of product **4** in higher yields.

Our methodology can also be extended to non-benzenoid aromatic bromides (including naphthalenes and phenanthrenes). Both 1-bromonaphthalene (**2m**) and 2-bromo-6-methoxynaphthalene (**2n**) smoothly underwent the reaction with **1** to give the corresponding carbocyclisation products **4m** and **4n** in 53% and 68% yields, respectively (Table 1, entry 13, 14). A similar 54% yield of **4p** was obtained in the case of 9-bromophenanthrene (**2p**) (Table 1, entry 16). As could be expected, the halides that do not possess any hydrogen atoms in position ortho- to the aryl–halogen bond did not undergo the studied reaction (Table 1, entry 15). Moreover, our attempts to obtain heterocyclic products **3e** or **4e** failed, perhaps due to the strong coordination of the nitrogen atom of pyridine to palladium (Table 1, entry 5).

Table 1. Reaction of the NORPHOS oxide with aryl halides (**2**) in the presence of $[Pd(PPh_3)_4]$.

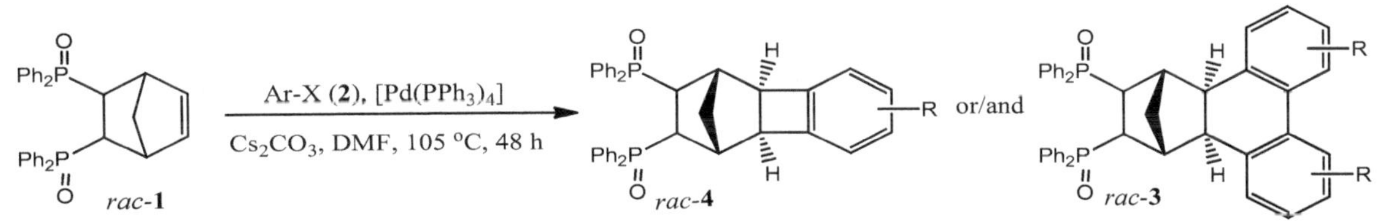

Entry	Aryl Halide	Equiv. of Aryl Halide	Product 3	Product 3 (Yield%)	Product 4	Product 4 (Yield%)
1	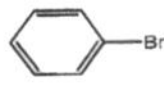 **2a**	3.0	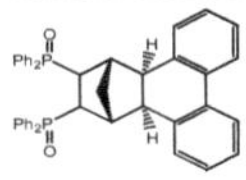	**3a** (69) [a]	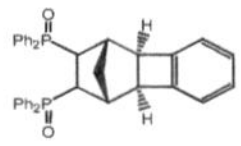	**4a** (no) [b]
2	 **2b**	1.2 3.0	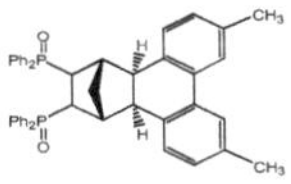	**3b** (no)	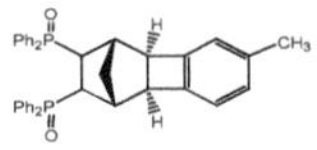	**4b** (59) **4b** (83)
3	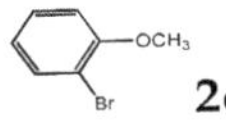 **2c**	1.2	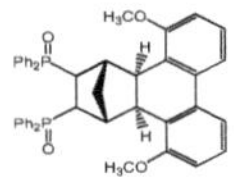	**3c** (no)	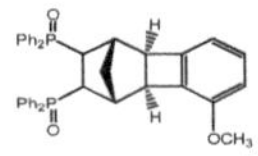	**4c** (76)
4	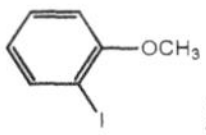 **2d**	1.2 3.0	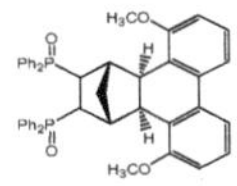	**3c** (no)	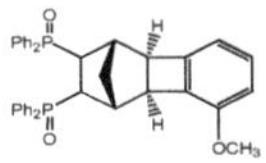	**4c** (35) **4c** (45)
5	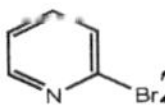 **2e**	3.0	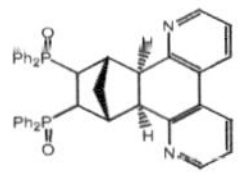	**3e** (no)	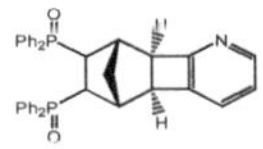	**4e** (no)
6	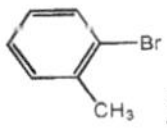 **2f**	1.2 1.2 3.0	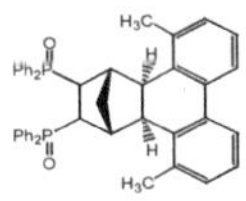	**3f** (no)	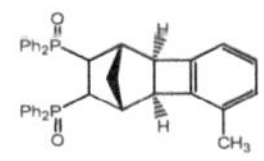	**4f** (25) [a] **4f** (78) **4f** (85)
7	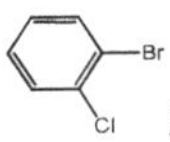 **2g**	1.2 3.0	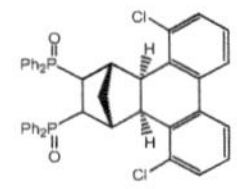	**3g** (no)	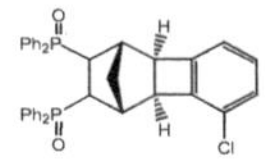	**4g** (48) **4g** (54)
8	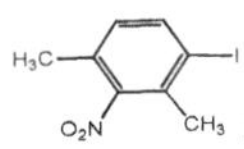 **2h**	1.2	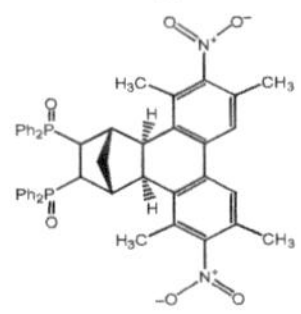	**3h** (no)	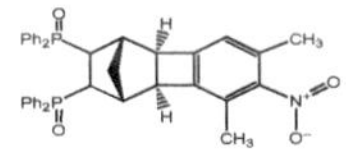	**4h** (86)
9	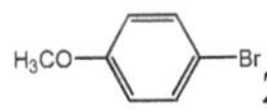 **2i**	3.0	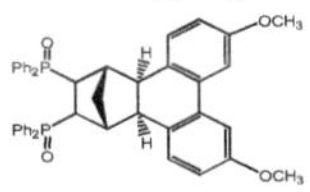	**3i** (no)	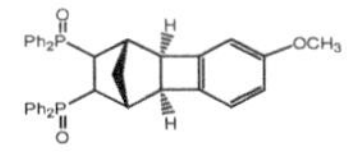	**4i** (18)
10	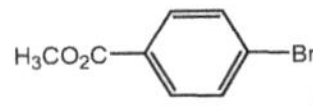 **2j**	1.2 3.0	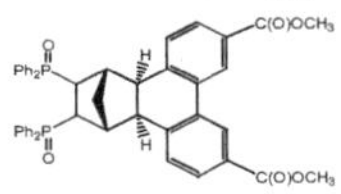	**3j** (<2) **3j** (51)	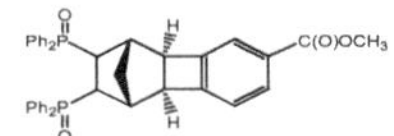	**4j** (84) **4j** (24)
11	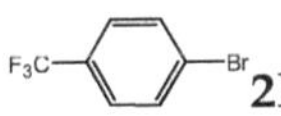 **2k**	1.2 3.0	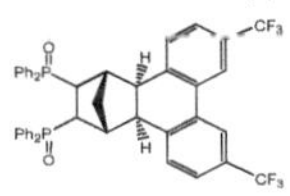	**3k** (32) **3k** (67)	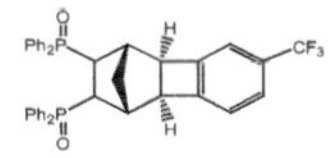	**4k** (no)
12	**2l**	3.0	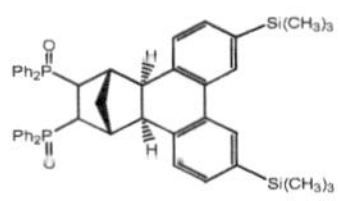	**3l** (63)	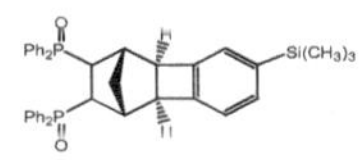	**4l** (no)
13	**2m**	1.2	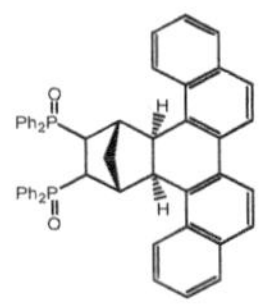	**3m** (no)	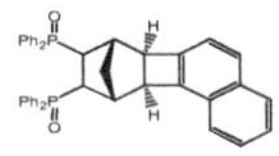	**4m** (68)
14	**2n**	1.2	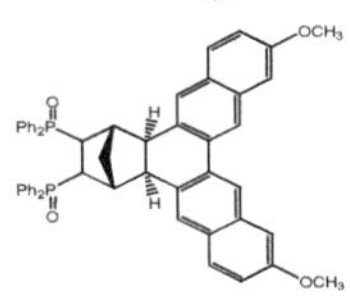	**3n** (no)	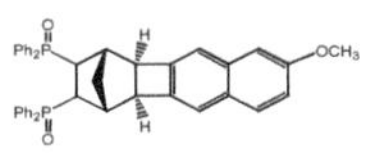	**4n** (53)

Table 1. *Cont.*

Entry	Aryl Halide	Equiv. of Aryl Halide	Product 3	Product 3 (Yield%)	Product 4	Product 4 (Yield%)
15	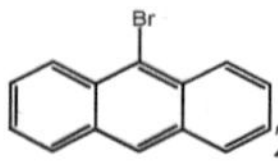 **2o**	1.2	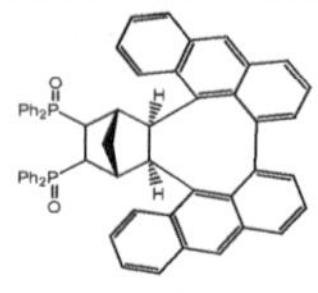	**3o** (no)	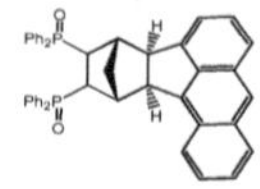	**4o** (no)
16	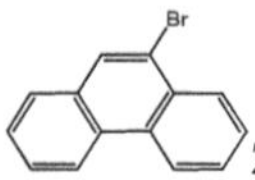 **2p**	1.2	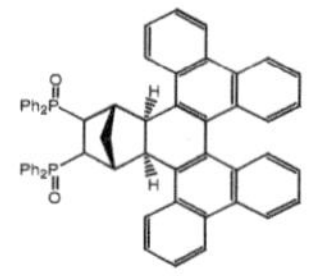	**3p** (no)	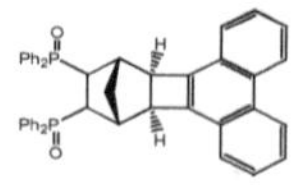	**4p** (54)
17	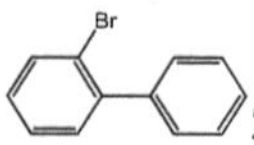 **2q**	1.2	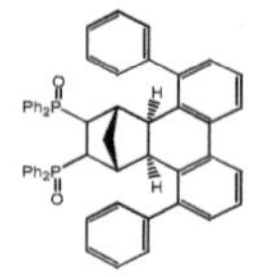	**3q** (no)	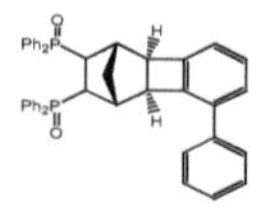	**4q** (98)

Conditions: NORPHOS oxide (0.2 mmol), aryl halide (3.0 equiv.), Cs_2CO_3 (3.0 equiv.), $[Pd(PPh_3)_4]$ (5 mol%) and DMF (4 mL) at 105 °C for 48 h. [a] The reaction time was 18 h. [b] Product was not observed.

2.1. General Mechanism and Selectivity of the Reaction

The catalytic carbocyclisation reaction between the NOPHROS oxide and aryl halides that resulted in cyclobutene derivatives **4** is highly regioselective, giving the cis-exo isomer as the exclusive product. An unexpected side reaction product **4a** (2–10%) was detected in the reaction mixtures by the ^{1}H-NMR and HPLC-HR-MS spectral analyses in almost all cases when differently substituted aromatic halides were used (Figure 1). We could assume that, as an exclusive source of phenyls, the phenylidene fragment originated from triphenylphosphane. The migration of the phenyl substituent from the phosphorus atom to palladium had already been observed in the course of other studies [17]. The molecular mechanism of this rare transformation is not clear and its elucidation requires additional theoretical as well as experimental investigations.

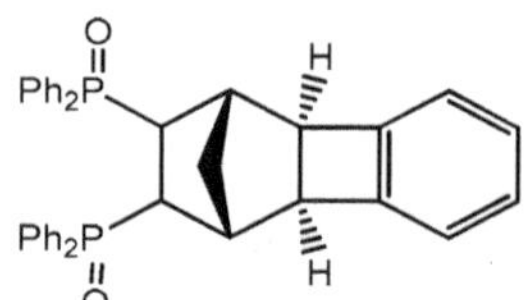

4a

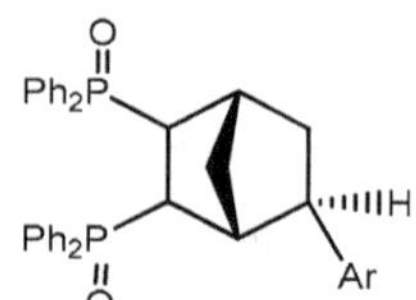

5

Figure 1. Side reaction products **4a** and **5**.

A tentative mechanism of the reaction leading to products **3–5** includes several stages (Scheme 2). The first obvious step of the reaction is an oxidative addition of **2** to the palladium catalysts; then, an insertion of the obtained arylpalladium intermediate into the double bond of the NORPHOS oxide furnished intermediate **I1**. Two possible isomeric intermediates **I1** and **I1′** could be formed. Nevertheless, only a single type of isomeric product was found in the reaction mixtures, and was identified as **3** and **4**. This could be rationalised by the steric hindrance created by $P(O)Ph_2$ groups, which forces large aryl substituents to occupy a less-hindered position. Obviously, both isomeric products of the alternative endo insertion of arylpalladium halides to **1** are significantly unfavorable and were not formed. The second reaction step is an intramolecular activation of the ortho C-H bond of the introduced arene by palladium, which resulted in the formation of intermediate **I2**. A base facilitated elimination of the hydrogen halide at the next step, which led to the reduction of the oxidation stage of palladium in intermediate **I4**. Next, when using electron-rich ligands of the *C,P*-type transition metal binding [12,13] or highly active aromatic halides, the oxidative addition of the second molecule of **2** occurs and the reaction runs via intermediate **I3** towards product **3**. Thus, in the case of less-active catalysts and non-activated substrates, intermediate **I4** undergoes reductive elimination to form product **4**.

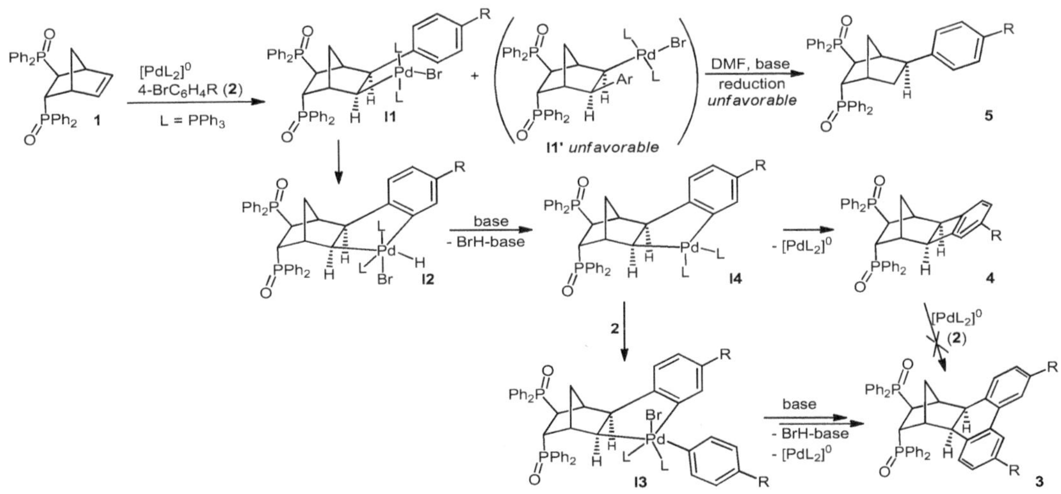

Scheme 2. A tentative suggestion for the molecular mechanism of the coupling reaction.

We have also examined the possibility of product **4**'s conversion to **3,** which theoretically may occur if an excess of the aromatic halide is present in the reaction mixture by the activation of a strained cyclobutane ring with reactive palladium species. For this purpose, in the presence of the palladium complex of the S-Phos ligand and caesium carbonate (3.0 equiv.), compound **4** was subjected to the reactions with 3.0 equiv. of aryl bromides (**2a** and **2b**). After 48 h in DMF at 105 °C, we did not observe the formation of compound **3** and did not find any evidence that such a reaction could be conducted under the studied conditions.

Since both products **3** and **4** are formed from the same intermediate **I4,** it could be anticipated that a lower rate of the oxidative addition reaction, inherent to the electron-poor, ligand-based catalyst, will facilitate the formation of product **4**.

Additionally, in the cases of Cs_2CO_3 (due to its notably strong basic character) and polycyclic aromatic substrates **2m** and **2p**, trace amounts of reductive Heck reaction products **5,** formed by reducing intermediate **I1**, presumably by a solvent [18,19], were detected in the postreaction mixtures (Figure 1).

2.2. *Preparation of Chiral Non-Racemic Ligand* **6q**

Having developed such a high-yielding procedure for the synthesis of desirable highly rigid bis(phosphane) dioxides, we aimed to conduct the synthesis of new chiral ligands to be used in asymmetric transition metal complex-catalysed reactions. We decided to use our new methodology to synthesise representative chiral non-racemic phosphane **6q**, since the corresponding bis(phosphane) dioxide **4q** was obtained in the highest yield. First of all, the scaleup of the synthesis of **4q** was performed. Notably, the synthesis conducted in a scale that increased up to one gram of *(R,R)*-**1** furnished the product in a 90% yield without the need of further optimisation of the reaction conditions. The obtained bis(phosphane) dioxide underwent a deoxygenation reaction in the presence of trichlorosilane and tributylamine, affording ligand *(R,R)*-**6q** in an 85% yield (Scheme 3). Ligand **6q** appeared to be sufficiently resistant to exposure to air and could be purified chromatographically. Using the enantiomerically pure, readily available substrate *(R,R)*-**1** in the synthesis allows us to obtain an enantiomerically pure product with no need of practical-scale enantiopurification.

2.3. *Synthesis of the Complex [Pd(6q)Cl₂]*

Since the validation of the proper geometry of the formed ligand by means of NMR spectroscopy is not a trivial matter, we have decided to perform it by analysing the single crystal X-ray diffraction data of the derived dichloropalladium complex [Pd(**6q**)Cl_2]. The complex was prepared from [Pd(^{t}-BuCN)$_2$$Cl_2$] and ligand **6q** in tetrahydrofuran (THF), and precipitated after a partial replacement of the solvent with C_6H_6. A single crystal has been grown from the hot ethanol/benzene mixture.

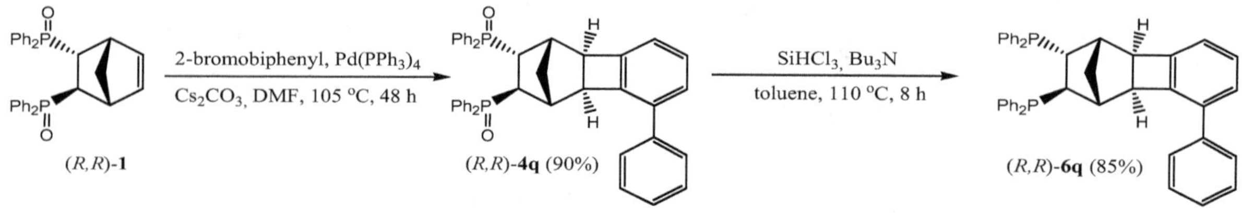

Scheme 3. Synthesis of the phosphine ligand.

2.4. Molecular and Crystal Structure of the Complex [Pd(6q)Cl_2]

The palladium coordination compound [Pd(**6q**)Cl_2] has been characterised by a single crystal X-ray diffraction analysis (CCDC No.: 1885894). [Pd(**6q**)Cl_2] crystallises in a tetragonal crystal system in the chiral $P4_32_12$ space group. The absolute structure assignment of the complex [Pd(**6q**)Cl_2] has been determined through anomalous scattering with the value of the Flack parameter refined to -0.007(3), using 2963 quotients (for more details, crystal data and structure refinements see (ESI).

The chiral centers are localised at the C1 (*R*), C2 (*R*), C9 (*S*), C10 (*S*), C11 (*R*) and C12 (*R*) atoms (Figure 2). The bond lengths and selected angles are given in Tables 2 and 3. An X-ray analysis confirmed the cis, exo configuration of the hydrocarbon skeleton at the C2–C9 joint atoms. The hydrogens attached to the C11 and C12 atoms are located in a trans configuration, which is in accordance with the synthesis route (Tables 2 and 3, Scheme 3, Figure 2). In the four-membered ring, the valence angles show strains resulting in a distortion of the sp^3- and sp^2- hybridised carbon atom centers from the ideal 109.5 and 120° to 86 and 94°, respectively, with higher values being observed at the side of the aromatic ring (Tables 2 and 3). The phenyl ring E is nearly coplanar with the ring F (torsion C2E–C1E–C4–C3 14.6(6)°). The complex [Pd(**6q**)Cl_2] is an example of metallacycles, which are known to be reactive intermediates in a catalysis. This five-membered ring is formed by one Pd, two P and two C atoms and has the shape of an envelope with the C12 atom located at the flap, which deviates by 0.835(4) Å from the mean plane of Pd1–P1–C11–P1′. The puckering amplitude (maximum out-of-plane deviation) q2 calculated for this metallacycle was 1.142(5) Å and the phase angle describing the puckering (phase shift) Φ2 was 153.8(1)° [20]. The Pd metal center has a planar square coordination with two cis-located Cl atoms. Ligand **6q** acts as a chelating bidentate agent forming two Pd–P coordination bonds of lengths equal to 2.2463(8) and 2.2722(9) Å. There are no classic hydrogen bonds in the crystal structure. The molecules interact through C–H . . . Cl contacts and T-shaped π . . . π interactions (Table 4, Figure 3).

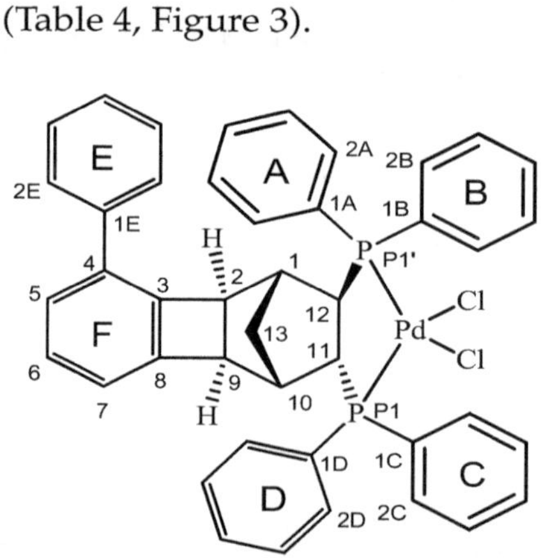

Figure 2. The molecular structure of the complex [Pd(**6q**)Cl_2].

Table 2. Bond lengths (Å) and valence angles (°) for [Pd(**6q**)Cl_2].

Bond	Length	Angle	Value
Pd1-Cl1	2.3452(12)	P1-Pd1-P1′	88.62(3)
Pd1-Cl2	2.3337(10)	P1-Pd1-Cl1	175.85(4)
Pd1-P1′	2.2463(8)	P1-Pd1-Cl2	89.30(4)
Pd1-P1	2.2722(9)	C1A-P1-Pd1′	120.58(13)
P1′-C1A	1.812(4)	C1A-P1′-C12	107.45(17)
P1′-C1B	1.802(4)	C1B-P1′-Pd1	113.58(12)
P1′-C12	1.824(3)	C1B-P1′-C1A	105.59(18)
P1-C1C	1.820(4)	C1B-P1′-C12	108.17(16)
P1-C1D	1.811(4)	C12-P1′-Pd1	100.75(11)
P1-C11	1.836(3)	C1C-P1-Pd1	118.19(13)
C1-C2	1.537(5)	C1C-P1-C11	105.81(16)
C1-C12	1.563(4)	C1D-P1-Pd1	114.56(14)
C1-C13	1.539(4)	C1D-P1-C1C	103.96(18)
C1E-C4	1.477(6)	C1D-P1-C11	110.03(16)
C2-C3	1.525(4)	C11-P1-Pd1	103.92(11)
C2-C9	1.602(5)		
C3-C4	1.388(5)	The valence angles of the four-membered ring	
C3-C8	1.386(5)	C2-C3-C8	93.9(3)
C4-C5	1.406(5)	C3-C8-C9	94.2(3)
C5-C6	1.389(6)	C8-C9-C2	85.8(3)
C6-C7	1.394(6)	C9-C2-C3	86.1(3)
C7-C8	1.374(5)		
C8-C9	1.525(5)		
C9-C10	1.540(5)		
C10-C11	1.545(4)		
C10-C13	1.550(5)		
C11-C12	1.555(5)		

Table 3. Torsion angles (°) for [Pd(**6q**)Cl_2].

Torsion	Value
Pd1-P1′-C12-C11	56.1(2)
P1′-C12-C11-P1	−61.2(2)
C12-C11-P1-Pd1	36.3(2)
C11-P1-Pd1-P1′-	−1.8(1)
P1-Pd1-P1′-C12	−26.5(1)
H11-C11-C12-H12	161
C3-C4-C1E-C2E	14.6(6)

Table 4. The geometry of intermolecular interactions in the [Pd(**6q**)Cl_2] (Å,°) crystal.

D–H ... A	D–H	D...A	H...A	<D–H...A
C2C-H2C ... Cl2	0.93	3.482(4)	2.73	139
C6-H6 ... Cl2 ^{i}	0.93	3.557(4)	2.87	132
C6E-H6E ... Cl1 ^{i}	0.93	3.623(4)	2.76	154
C7-H7 ... Cl1 ^{ii}	0.93	3.756(4)	2.99	141
C3B-H3B ... *Cg*1 ^{ii}	0.93	3.689(5)	2.77	169
C5B-H5B ... *Cg*2 ^{iii}	0.93	3.521(4)	2.71	146

Symmetry codes: ^{i} $-1 + x, y, z$; ^{ii} $-\frac{1}{2} + y, \frac{1}{2} - x, \frac{1}{4} + z$; ^{iii} $-1 + y, 1 + x, 1 - z$. *Cg*1 ring centroid C1E-C6E; *Cg*2 ring centroid C1A-C6A.

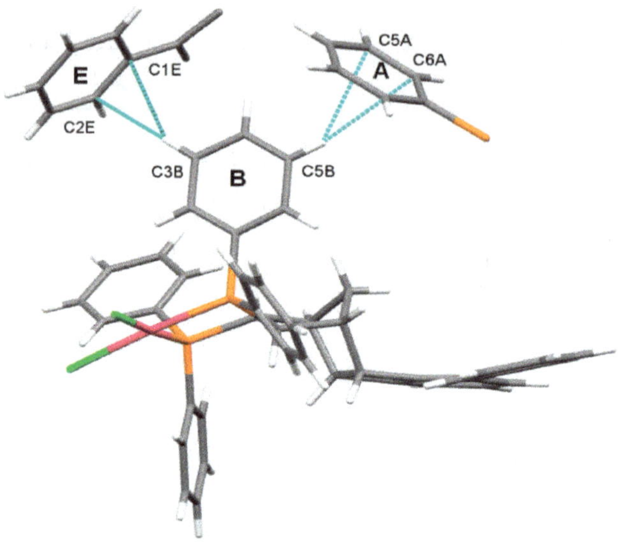

Figure 3. T-shaped $\pi \ldots \pi$ interactions in the complex [Pd(**6q**)Cl_2].

2.5. Geometry of Five-Membered Metallacycles of the Transition Metals and Bis(phosphine) Ligands Found in the Cambridge Structural Database (CSD)

In order to better understand the reactivity and stereochemistry of the bis(phosphane) metalorganic catalysts, a search of the CSD [21] for the structures of five-membered metallacycles with a transition metal and the P–C(HC)–C(HC)–P fragment of bis(phosphane) has been performed. It gave a result of 120 hits, 27 of which were the structures of complexes with Pd (CSD Refcodes: DEPXAX, GEPQUN, HOGBUA, MIDPIY, MIDPOE, NEXHOP, NIDYIK, NURKAM, OFAQET, OFAQIX, OFAQUJ, OKOHAX, PAQBIS, PAQBOY, POPZEA, QELSAB, QUBGID, QUBGOJ, RIYGUB, RONPOZ, RUDJAC, RUNVUR, TESBEZ, TIYPAS, VASXUI, WORDIQ and YOBWUH).

The endocyclic torsion angles characterising this system are summarised in Table 5, and the distribution of their absolute values is presented in Figure 4. There is a clear difference in the most frequently occurring values. The dihedral angles T4 and the corresponding one T5 (C–P–Tr–P) were most frequently synperiplanar, whereas torsions T1–T3 were usually synclinal. In most of the structures, the metallacycle had a conformation of an envelope, as in the presented complex [Pd(**6q**)Cl_2], with the exception of five entries where the ring was almost flat (CSD Refcodes: EQUDEA (Ru), HAYREE (Pt), HAYRII (Pt), VISRAR (Os) and ZOVCIY (Fe)) [22,23]. Usually, one of the carbon atoms was located at the envelope flap (the same as in [Pd(**6q**)Cl_2]), providing a favourable tetragonal flat coordination environment near the metal atom and the asymmetric structure of the metallacycle. Only in three palladium complexes was the metal atom positioned at the flap instead of the carbon atom (in the structures where the carbon atoms belonged to the four-membered hydrocarbon ring and the exocyclic H–C–C–H torsion indicated a cis configuration; CSD Refcodes: NIDYIK, QUBGID and QUBGOJ) [24,25]. However, in analogous Pd complexes with the four-membered hydrocarbon ring and the trans configuration at the same torsion, the metallacycle ring had, as usual, a carbon atom at the envelope flap (CSD Refcodes: OFAQET, OFAQIX and OFAQUJ) [26]. This asymmetry observed in the metallacycle structure is probably beneficial for transition-metal-promoted asymmetric transformations.

The analysis of the orientation of the H atoms within the metallacycle showed that cis stereoisomers were nearly half of the hits (54 of 120). In the presented complex [Pd(**6q**)Cl_2], the H–C–C–H torsion showed the trans configuration. To illustrate the differences within this group of crystal structures, they were divided into four subsets according to the value of the H–C–C–H torsion. There were 17 hits with synperiplanar torsions, 37 synclinal, 6 anticlinal and 60 antiperiplanar as in the presented complex [Pd(**6q**)Cl_2] (Tables 2–5). The results showed that the most frequent conformation for the five-membered metallacycles of transition metals and bis(phosphane) ligands was the envelope type conformation, regardless of the transition metal and phosphine substituent types.

Table 5. Absolute values of the endo- and exocyclic torsions within the five-membered metallacycles found in the Cambridge Structural Database (CSD).

Code	Torsion	Minimum	Maximum	Mean
abs(T1 or T3)	Tr–P–C–C	0.733	60.136	36.141
abs(T2)	P–C–C–P	0.353	69.677	44.108
abs(T4 or T5)	P–Tr–P–C	0.921	44.906	17.174
abs(T6)	H–C–C–H	0.164	179.96	113.52

T1–T5, endocyclic torsion angles defined in Figure 4; Tr, transition metal.

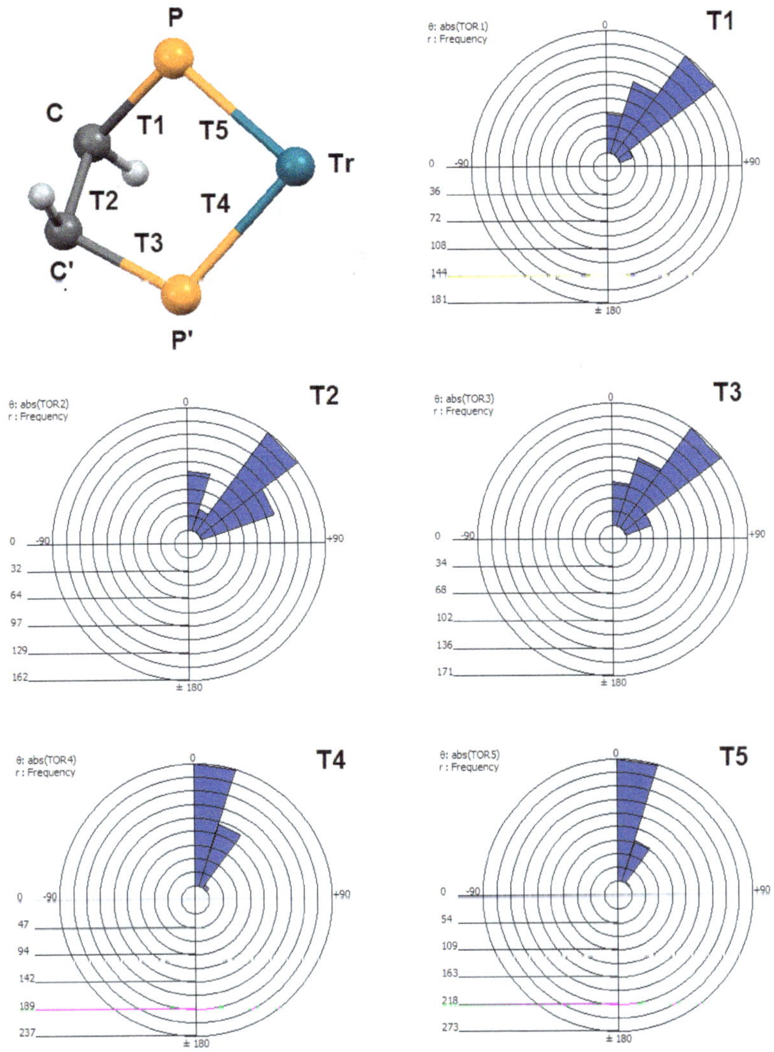

Figure 4. Distribution of the torsion values T1–T5 in the structures of five-membered metallacycles in the complexes of transition metals (Tr) with bis(phosphane) ligands found in the CSD.

2.6. *Quantum Chemical Considerations*

The quantum chemical calculation performed at the M062x/LANL2DZ theoretical level [27,28], which is specially dedicated to precise energetic considerations and has recently been applied in the simulation of energetic and structural analyses of molecules with similar structural moieties [29–33], allows us to compare the geometry of the molecules of phosphines **6**, their dioxides and complexes that were not crystallised in crystals, well-measurable by a single crystal X-ray diffraction. The geometry optimisation for the structure [Pd(**6q**)Cl_2] was therefore performed and the obtained geometry parameters of $[Pd(\textbf{6q})Cl_2]^{calc}$ were compared with those received from the X-ray diffraction experiments for $[Pd(\textbf{6q})Cl_2]^{xray}$. It was found that all key interatomic distances (Table 6) are characterised by the values that are typical for the respective bonds in similar organic moieties [13,33–35]. The comparison of the obtained geometries with the respective X-ray structures confirms that the density functional theory (DFT) calculations at the M062x/LANL2DZ theoretical level illustrate well the key structural aspects of the considered complexes.

Table 6. Selected geometrical parameters of experimental and calculated [Pd(**6**)Cl_2] and **4.**

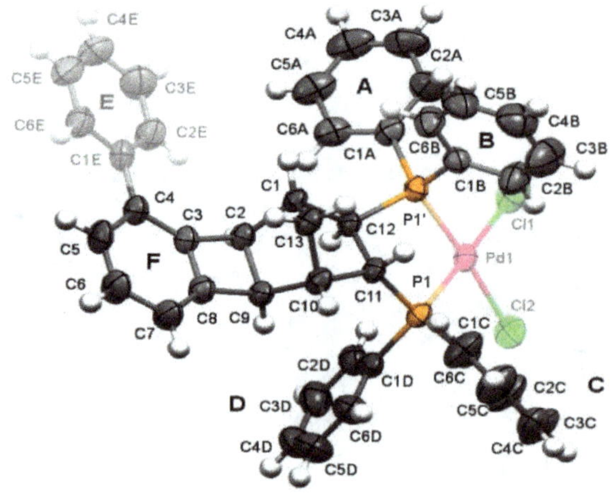

	$[Pd(\textbf{6q})Cl_2]^{xray}$	$[Pd(\textbf{6q})Cl_2]^{calc}$	$[Pd(\textbf{6a})Cl_2]^{calc}$	$\textbf{4q}^{calc}$	$\textbf{4a}^{calc}$
Angle, (°)					
A/B	73.3	81.5	81.3	68.9	70.0
C/D	74.3	67.4	67.6	66.7	65.5
B/C	39.6	39.3	43.7	4.2	8.2
A/D	39.3	45.1	39.9	45.4	44.5
Torsion angle, (°)					
C3-C4-C1E-C2E	14.6	30.4	na	29.3	na
P1-C11-C12-P2	−61.2	−67.2	−66.9	−101.5	−103.8
Bond lengths, (Å)					
C1-C2	1.537(5)	1.551	1.552	1.550	1.550
C2-C3	1.525(4)	1.529	1.530	1.532	1.532
C3-C4	1.388(5)	1.397	1.390	1.397	1.391
C4-C5	1.406(5)	1.417	1.415	1.418	1.415
C5-C6	1.389(6)	1.406	1.408	1.406	1.407
C6-C7	1.394(6)	1.414	1.413	1.414	1.414
C7-C8	1.374(5)	1.390	1.392	1.391	1.391
C8-C3	1.386(5)	1.403	1.406	1.405	1.407
C8-C9	1.525(5)	1.533	1.532	1.530	1.531
C9-C2	1.602(5)	1.606	1.608	1.600	1.602
C9-C10	1.540(5)	1.549	1.550	1.547	1.546
C10-C11	1.545(4)	1.550	1.549	1.552	1.551
C11-C12	1.555(5)	1.561	1.561	1.564	1.566
C12-C1	1.563(4)	1.560	1.563	1.555	1.555
C1-C13	1.539(5)	1.552	1.552	1.549	1.550
C13-C10	1.550(5)	1.562	1.562	1.555	1.554
C12-P1	1.823(5)	1.881	1.881	1.872	1.878
C11-P2	1.836(5)	1.883	1.884	1.859	1.850
Pd-Cl1	2.345 (1)	2.423	2.425	na	na
Pd-Cl2	2.334(1)	2.420	2.419	na	na
Pd-P1	2.246 (1)	2.381	2.379	na	na
Pd-P2	2.272(1)	2.380	2.379	na	na

The fitting of molecular structures $[Pd(\mathbf{6q})Cl_2]^{xray}$ and $[Pd(\mathbf{6q})Cl_2]^{calc}$ clearly shows in which fragments the differences are observed (Figure 5). The molecular overlay indicates that, due to its rigidity, the polycyclic skeleton is identical both in the experimental and theoretical models. Thus, some minor differentiation is observed in the position of the phenyl substituted at the phosphorus atom. This could be explained by the fact that the calculation of a single molecule in vacuum, also with a consideration of a solvent, cannot correctly reflect all crystal net forces influencing the geometry of a molecule in a solid state. The structural comparison is summarised in Table 6. The comparison of the theoretical model and the experimental structure proved that there is a high degree of similarity of their geometries in the polycyclic core.

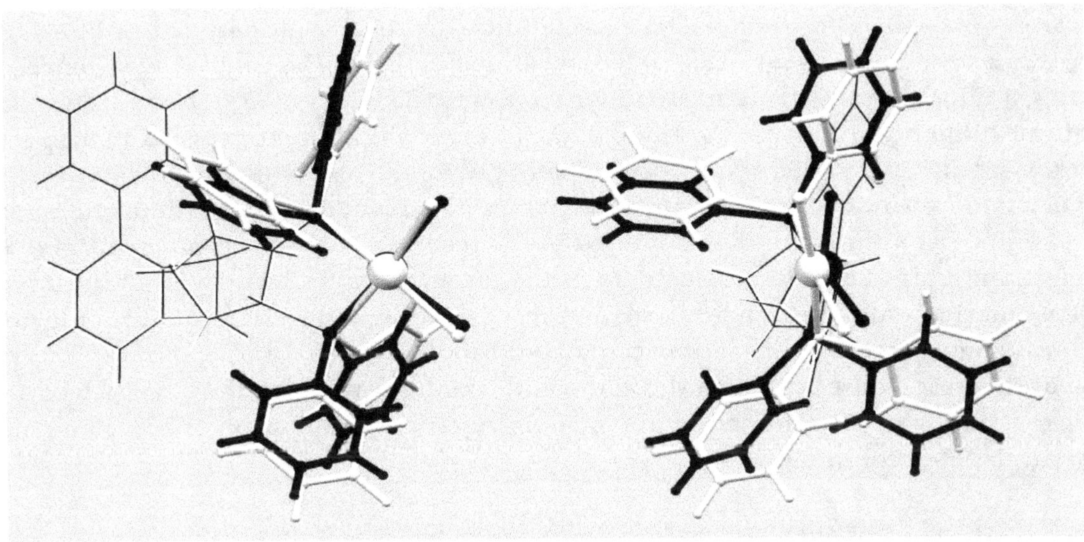

Figure 5. The structural overlay of the molecules of the coordination compounds $[Pd(\mathbf{6q})Cl_2]^{xray}$ and $[Pd(\mathbf{6q})Cl_2]^{calc}$: front and side views. Black: measured X-ray structure $[Pd(\mathbf{6q})Cl_2]^{xray}$, white: calculated geometry of $[Pd(\mathbf{6q})Cl_2]^{calc}$. The fragments that match are marked with thin lines.

Moreover, analogous DFT calculations for the complex $[Pd(\mathbf{6a})Cl_2]^{calc}$ and bis(phosphane) dioxides **4a** and **4q** were carried out to investigate the influence of the substitutes on the skeleton geometry. All geometries (experimentally determined and optimised using DFT calculations) were compared with each other. Selected bond lengths and angles are summarised in Table 6.

We found a regularity regarding the calculated bonds lengths of the molecules: they are slightly larger than the measured ones. This is a normal phenomenon, considering the fact that the calculation concerns an isolated molecule [36]. The key bond lengths and other geometrical parameters in $[Pd(\mathbf{6q})Cl_2]^{calc}$ and $[Pd(\mathbf{6a})Cl_2]^{calc}$ and the **4q** and **4a** molecules are the same (within 3σ). The presence of a phenyl substituent in $[Pd(\mathbf{6q})Cl_2]^{calc}$ does not have any influence on the geometry of the hydrocarbon polycyclic fragment. In the crystal structure $[Pd(\mathbf{6q})Cl_2]^{xray}$, the phenyl ring E is nearly coplanar with the skeleton moiety (ring F, torsion 14.6°), while in the optimised model of $[Pd(\mathbf{6q})Cl_2]^{calc}$, the corresponding planes are twisted by 30°. The coplanarity of these fragments is stabilised in the crystal by the C-H···Cl bonds in which the carbon atoms of the phenyl ring are involved, and, therefore, it could not have been observed in the cases of isolated single molecules nor in the solution.

Some minor differences in the torsion angle P1-C11-C12-P2 were found in the geometries of molecules **4q** and **4a** in comparison to those of the complexes $[Pd(\mathbf{6q})Cl_2]^{calc}$ and $[Pd(\mathbf{6a})Cl_2]^{calc}$. A smaller value of this parameter in the complex results from the necessity of flattening this molecular fragment in the palladium complexation process, but not from introducing a substituent in position C4 of the polycyclic core of the ligand. Differences in the molecular structures related to the position of the phenyl rings, which are caused by a free rotation around the P-1C_{Ph} σ-bonds, are not essential for the catalytic efficiency of the ligand-derived palladium complexes.

To summarise, different substitutions did not affect the stereochemistry of the polycyclic hydrocarbon moiety of the molecules. The main structural differences were observed in the orientation of the phenyl substituents attached to the phosphorus atoms that were involved in the thermal movements around single P-C bonds. Even the complexation of the ligand with palladium did not significantly change the rigid geometry of the ligands. Thus, such unique stereometrical properties of the ligands allow us to extrapolate, via computer modelling, the geometry of the selected measured structures to the stereometry of other ligands in this group.

2.7. The Evaluation of the Efficiency of the Ligand **6q** *Palladium Complexes in the Asymmetric Catalysis*

The potential of bis(phosphane) ligand **6q** was examined in a model Pd-catalysed allylic alkylation reaction between *rac-(E)*-1,3-diphenylallyl acetate and dimethyl malonate using BSA/KOAc and K_2CO_3/Cs_2CO_3 as bases in THF at ambient temperature (Table 7). The catalyst was generated in situ by pre-mixing allylpalladium(II) chloride dimer with **6q** in THF at ambient temperature for 30 min. The palladium-to-ligand ratios were 2/1 and 1/2. To our delight, bis(phosphane) **6q** proved to be suitable for the asymmetric allylic alkylation reaction, giving the alkylated product in excellent yields and with very good enantioselectivity (up to 90% ee). The results of the optimisation study show that the phosphine palladium complex obtained in the reaction with stoichiometry P/Pd = 1/1 was more beneficial in terms of the stereoselectivity of the reaction (Table 7, entry 1). Thus, it was proven that, in a benchmark asymmetric allylic alkylation reaction, the catalyst derived from the new *R,R*-**6q** phosphane was slightly more efficient than parent NORPHOS.

The enantiomeric composition and *S*- absolute configuration of the product of the model asymmetric reaction were determined by the peak integration and elution order from chiral HPLC using a Chiralcel OD-H column [37].

Table 7. The model asymmetric allylic substitution reaction.

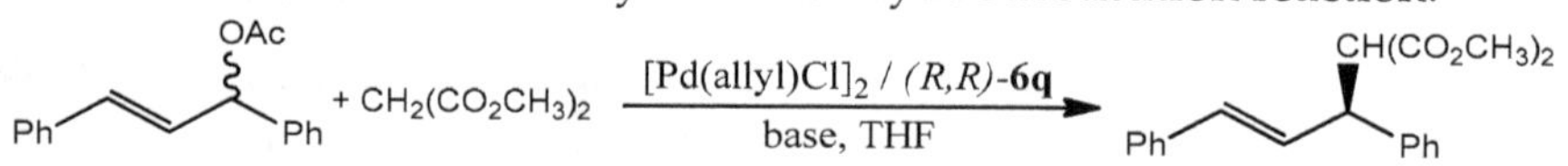

Entry	Ligand, mol%	$[Pd(allyl)Cl]_2$, mol%	Base	Yield, %	ee, %
1	2.0 (*R,R*-6q)	2.0	BSA */KOAc	97	90 **
2	4.0 (*R,R*-6q)	1.0	BSA/KOAc	98	67
3	2.0 (*R,R*-6q)	2.0	Cs_2CO_3/K_2CO_3	97	81
4	2.0 (*R,R*-6q)	2.0	Cs_2CO_3	96	83
5	2.0 (*R,R*-1)	2.0	BSA/KOAc	95	87
6	2.0 (*R,R*-1)	2.0	Cs_2CO_3	92	85
7	1.2 (*S,S*-1)	0.5	NaH	80	81 ***

6q (2.0 or 4.0 mol%), $[Pd(allyl)Cl]_2$ (1.0 or 2.0 mol%), THF (1.0 mL), 30 min; next 1,3-diphenylallyl acetate (0.5 mmol), THF (1.0 mL), dimethyl malonate (1.5 mmol), base (1.5 mmol), 12 h, ambient temperature. * BSA = bis(trimethylsilyl)acetamide. ** $[\alpha]_D^{20} = -14.8$ (c = 1.3, $CHCl_3$). *** Reported previously [38].

3. Materials and Methods

3.1. General Information

Unless specified otherwise, all starting materials and solvents were used as obtained from commercial suppliers without further purification. Organic solvents used in this study were dried over appropriate drying agents and distilled prior to use. Thin layer chromatography (TLC) was carried out using Merck silica gel 60 F_{254} plates (Merck, Kenilworth, NJ, USA). Visualisation of the TLC plates was performed by UV light, either $KMnO_4$ or I_2 stains. The NMR spectra were recorded on a Bruker Avance 500 MHz spectrometer, while chemical shifts are reported in ppm and calibrated to the residual solvent peaks at 7.27 ppm and 77.00 ppm for ^1H and ^{13}C, respectively, in $CDCl_3$ or internal reference compounds. A similar technique was applied recently for the structural analysis of many norbornene systems [39,40]. The following abbreviations are used in reporting NMR data: s (singlet), d (doublet), t (triplet), q (quartet), m (multiplet), br (broad). Coupling constants (J) are given in Hz. The spectra are reported as follows: chemical shift (δ, ppm), multiplicity, integration, coupling constants (Hz). The IR spectra were recorded on a Nicolet 8700A FTIR-ATR spectrometer; wave numbers are in cm^{-1}. Products were purified by flash chromatography on silica gel 60 (230–400 mesh) using a BUCHI chromatograph. The X-ray diffraction intensities were collected at room temperature on a SuperNova X-ray diffractometer equipped with an Atlas S2 CCD detector using mirror-monochromatised CuKα radiation (λ = 1.54184 Å). Low- and high-resolution mass spectra were obtained with a Shimadzu LC-MS (Kinetex® 2.6 μm Biphenyl 100 Å 50 × 2.1 mm LC-column, acetonitrile/water with the HCO_2H additive mobile phase) IT-TOF spectrometer. Commercially unavailable substrates were obtained by known literature procedures. The physical properties and spectra of the obtained products are available free of charge in the Supplementary Data.

3.2. Synthesis and Spectral Data

3.2.1. Typical Procedure for the Coupling Reaction Leading to the Formation of [(1*R*,2*R*,9*S*,10*S*,11*R*,12*R*)-4-phenyltetracyclo[8.2.1.0^{2,9}.0^{3,8}]trideca-3,5,7-triene-11,12-diyl]bis(diphenylphosphane) dioxide (*R*,*R*-**4q**)

(R,R)-**1** (1.0g, 1.54 mmol), **2q** (1.84 mmol), Cs_2CO_3 (4.63 mmol) and $Pd(PPh_3)_4$ (0.07 mmol) were placed in a Schlenk tube under argon. Dry DMF was added (15 mL) and the mixture was stirred at 105 °C for 18–24 h. Then, the mixture was cooled down to root temperature (RT). The solvent was evaporated under reduced pressure, and 50 mL of water was added. The products were extracted in DCM (3 × 30 mL). The combined organic phase was dried over $MgSO_4$, filtered and the solvent was evaporated. The product was purified by silica-gel column chromatography (Hexane/*i*PrOH 50/1) and finally crystallised from the mixture of hexane/DCM. The compound was isolated in 98% yield as a white solid, m.p.: 255–258 °C, $[\alpha]_D^{20}$ = +79.4 (*c* = 0.45, $CHCl_3$), ^{31}P-NMR (202 MHz, $CDCl_3$): δ = 32.5 (d, *J* = 10.1), 28.7 (d, *J* = 10.1), ^{1}H-NMR (500 MHz, $CDCl_3$): δ = 8.68–8.65 (m, 2H, Ar-H), 7.99–7.42 (m, 20H, Ar-H), 7.13–7.06 (m, 5H, Ar-H), 6.83 (d, 1H, *J* = 10.0, Ar-H), 4.29 (s, 1H, CH), 3.93–3.83 (m, 2H, CH), 3.52–3.47 (m, 1H, CH), 2.61 (d, 2H, *J* = 5.0, CH), 1.90 (d, 1H, *J* = 10.0, CH), 1.01–0.85 (m, 1H, CH), ^{13}C-NMR (125 MHz, $CDCl_3$): δ = 145.3, 141.8, 137.3, 134.1, 133.4, 132.1, 132.0, 131.9, 131.2, 130.9, 130.6, 130.5, 130.4, 130.1, 130.0, 128.8, 128.7, 128.5, 128.5, 128.5, 128.4, 128.3, 128.2, 127.1, 127.0, 126.8, 126.5, 126.1, 125.1, 120.7, 50.0, 49.9, 48.3 (d, *J* = 6.2), 41.8, 40.4, 38.4, 37.5, 32.8 (d, *J* = 12.5), LCMS (ESI) $[M + H]^+$ = 647 Da, HRMS calcd. for $C_{43}H_{36}O_2P_2$ 647.2259, found 647.2263 (diff. 1.18 ppm).

3.2.2. 1,2,3,4,4a,12b-Hexahydro-1,4-methanotriphenylene-2,3-diylbis(diphenylphosphane) dioxide (*rac*-**3a**)

The compound was isolated in 69% yield as a white solid. The analysis of the product spectra confirmed its identity with literature data [12].^{31}P-NMR (202 MHz, $CDCl_3$): δ = 31.60 (d, *J* = 10.1), 28.78 (d, *J* = 10.1), ^{1}H-NMR (500 MHz, $CDCl_3$): δ = 7.98–7.96 (m, 2H, Ar-H), 7.86–7.85 (m, 2H, Ar-H), 7.76–7.72 (m, 2H, Ar-H), 7.64–7.50 (m, 10H, Ar-H), 7.17–7.05 (m, 10H, Ar-H), 6.81–6.78 (m, 2H, Ar-H), 5.22 (d, 1H, *J* = 2.0, Ar-H), 4.00 (d, 1H, *J* = 4.0, CH), 3.95–3.87 (m, 1H, CH-P), 3.86–3.80 (m, 1H, CH-P), 3.71 (d, 1H, *J* = 4.0, CH), 2.64 (s, 1H, CH), 2.45 (d, 1H, *J* = 4.0, CH_2), 1.98 (d, 1H, *J* = 4.0, CH_2), 1.45–1.44 (m, 1H, CH).

3.2.3. (5-Methyltetracyclo[8.2.1.0^{2,9}.0^{3,8}]trideca-3,5,7-triene-11,12-diyl)bis(diphenylphosphane) dioxide (*rac*-**4b**)

The compound was isolated in 83% yield as a white solid, m.p.: 208–213 °C. ^{31}P-NMR (202 MHz, $CDCl_3$): δ = 32.4 (d, *J* = 8.08), 28.7 (d, *J* = 6.06), ^{1}H-NMR (500 MHz, $CDCl_3$): δ = 7.87–7.76 (m, 15H, Ar-H), 7.11–6.55 (m, 8H, Ar-H), 3.93 (s, 1H, CH), 3.87–3.83 (m, 1H, CH), 3.56 (s, 1H, CH), 3.31–3.27 (m, 1H, CH), 2.34 (d, 1H, *J* = 10.0, CH_2), 2.24 (s, 3H, CH_3), 2.03 (s, 1H, CH), 1.84 (d, 1H, *J* = 10.0, CH_2), 1.00–0.95 (m, 1H, CH), ^{13}C-NMR (125 MHz, $CDCl_3$): δ = 144.0 (d, *J* = 3.7), 142.0, 137.1, 134.3, 135.1, 134.4, 134.0, 133.6, 133.3, 132.6, 131.8, 131.2, 131.0, 130.8, 130.5, 130.4, 130.2, 130.1, 129.8, 128.7, 128.6, 128.5,

128.4, 128.3, 128.2, 128.2, 122.6, 121.4, 50.0 (d, J = 15.0), 45.2 (d, J = 6.2), 41.6 (d, J = 6.2), 38.3, 37.8, 32.5 (d, J = 12.5), 21.9 LCMS (ESI) $[M + H]^+$ = 585 Da, HRMS calcd. for $C_{38}H_{34}O_2P_2$ 585.2119, found 585.2108 (diff. 1.18 ppm).

3.2.4. (4-Methoxytetracyclo[8.2.1.0^{2,9}.0^{3,8}]trideca-3,5,7-triene-11,12-diyl)bis(diphenylphosphane) dioxide (*rac*-**4c**)

The compound was isolated in 76% yield as a white solid, m.p.: 276–278 °C. ^{31}P-NMR (202 MHz, $CDCl_3$): δ = 32,4 (s), 28.3 (s), ^{1}H-NMR (500 MHz, $CDCl_3$): δ = 7.87–7.79 (m, 4H, Ar-H), 7.55–7.42 (m, 11H, Ar-H), 7.07–7.00 (m, 7H, Ar-H), 6.56 (d, 1H, J = 5.0, Ar-H), 6.45 (d, 1H, J = 5.0, Ar-H), 4.08 (s, 1H, CH), 3.92–3.78 (m, 1H, CH), 3.59 (s, 1H, CH), 3.38–3.30 (m, 1H, CH), 2.93 (s, 3H, OCH_3), 2.58 (s, 1H, CH), 2.42 (s, 1H, CH), 1.99–1.97 (m, 1H, CH), 2.31 (s, 3H, CH_3), 2.27 (s, 3H, CH_3), 2.11–2.09 (m, 1H, CH), 1.15 (s, 1H, CH), ^{13}C-NMR (125 MHz, $CDCl_3$): δ = 153.2, 146.6, 131.4, 131.3, 131.0, 130.8, 130.5, 130.0, 1329., 128.8, 128.5, 128.4, 128.3, 126.6, 115.6, 114.5, 55.6, 53.4, 45.9, 43.4, 41.5 LCMS (ESI) $[M + H]^+$ = 601 Da, HRMS calcd. for $C_{38}H_{34}\,O_3P_2$ 601.2056, found 601.2067 (diff. 1.18 ppm).

3.2.5. (4-Methyltetracyclo[8.2.1.0^{2,9}.0^{3,8}]trideca-3,5,7-triene-11,12-diyl)bis(diphenylphosphane) dioxide (*rac*-**4f**)

The compound was isolated in 85% yield as a white solid, m.p.: 270–275 °C. ^{31}P-NMR (202 MHz, $CDCl_3$): δ 32.3 (d, J = 8.08), 28.5 (d, J = 8.08), ^{1}H-NMR (500 MHz, $CDCl_3$): δ = 7.87–7.77 (m, 4H, Ar-H), 7.65–7.45 (m, 11H, Ar-H), 7.17–7.00 (m, 7H, Ar-H), 6.86 (d, 1H, J = 5.0, Ar-H), 6.65 (d, 1H, J = 5.0, Ar-H), 3.93 (s, 1H, CH), 3.81–3.75 (m, 1H, CH), 3.54 (s, 1H, CH), 3.34–3.30 (m, 1H, CH), 2.43 (s, 2H, CH_2), 1.90 (d, 1H, J = 10.0, CH), 1.65 (s, 1H, CH_3), 1.01–0.97 (m, 1H, CH), ^{13}C-NMR (125 MHz, $CDCl_3$): δ = 144.1 (d, J = 3.7), 143.7, 135.4, 134.7, 134.1, 133.6, 133.4, 132.4, 131.7, 131.4, 131.3, 130.9, 130.7, 130.6, 130.5, 130.1, 130.0, 128.7, 128.6, 128.5, 128.3, 128.3, 128.3, 128.2, 128.2, 127.9, 49.9 (d, J = 3.7), 45.5, 41.3, 40.9, 38.4, 37.8, 32.6 (d, J = 13.7), 15.7 LCMS (ESI) $[M + H]^+$ = 585 Da, HRMS calcd. for $C_{38}H_{34}\,O_2P_2$ 585.2119, found 585.2107 (diff. 1.18 ppm).

3.2.6. (4-Chlorotetracyclo[8.2.1.0^{2,9}.0^{3,8}]trideca-3,5,7-triene-11,12-diyl)bis(diphenylphosphane) dioxide (*rac*-**4g**)

The compound was isolated in 54% yield as a white solid, m.p.: 250–254 °C. ^{31}P-NMR (202 MHz, $CDCl_3$): δ = 32.2 (d, J = 6.06), 28.7 (d, J = 8.08), ^{1}H-NMR (500 MHz, $CDCl_3$): δ = 7.87–7.73 (m, 4H, Ar-H), 7.67–7.46 (m, 11H, Ar-H), 7.17–7.03 (m, 7H, Ar-H), 6.77 (d, 1H, J = 10.0, Ar-H), 4.03 (s, 1H, CH), 3.83–3.77 (m, 1H, CH), 3.58 (s, 1H, CH), 3.33–3.29 (m, 1H, CH), 2.41 (s, 2H, CH_2), 1.93 (d, 1H, J = 10.0, CH), 1.65 (s, 1H, CH), 0.98–0.95 (m, 1H, CH), ^{13}C-NMR (125 MHz, $CDCl_3$): δ = 145.1 (d, J = 3.7), 142.6, 134.7, 134.3, 133.9, 133.7, 133.5, 133.0, 132.2, 131.4 (d, J = 2.5), 131.3 (d, J = 2.5), 131.2 (d, J = 2.5), 130.9 (d, J = 2.5), 130.6, 130.5, 130.4, 130.4, 129.4, 129.3, 128.7, 128.7, 128.6, 128.5, 128.3, 127.5, 127.4, 126.9, 126.8, 120.5, 50.0 (d, J = 2.5), 49.3(d, J = 15), 45.7 (d, J = 6.2), 41.3(d, J = 3.7), 40.4, 38.2, 37.6, 32.6 (d, J = 12.5) LCMS (ESI) $[M + H]^+$ = 605 Da, HRMS calcd. for $C_{37}H_{31}O_2P_2Cl$ 605.1561, found 605.1561 (diff. 1.18 ppm).

3.2.7. (4,6-Dimethyl-5-nitrotetracyclo[8.2.1.0^{2,9}.0^{3,8}]trideca-3,5,7-triene-11,12-diyl)bis(diphenylphosphane) dioxide (*rac*-**4h**)

The compound was isolated in 86% yield as a white solid, m.p.: 302–304 °C. ^{31}P-NMR (202 MHz, $CDCl_3$): δ = 32.3 (d, J = 8.08), 28.6 (d, J = 10.0), ^{1}H-NMR (500 MHz, $CDCl_3$): δ = 7.88–7.76 (m, 4H, Ar-H), 7.64–7.47 (m, 11H, Ar-H), 7.16–7.01 (m, 5H, Ar-H), 6.65 (s, 1H, Ar-H), 3.94 (d, 1H, J = 5.0, CH), 3.84–3.78 (m, 1H, CH), 3.54 (d, 1H, J = 5.0, CH), 3.34–3.30 (m, 1H, CH), 2.41 (d, 2H, J = 10.0, CH_2), 2.10 (s, 3H, CH_3), 1.97 (d, 1H, J = 10.0, CH), 1.60 (s, 1H, CH_3), 1.00–0.95 (m, 1H, CH), ^{13}C-NMR (125 MHz, $CDCl_3$): δ = 151.5, 145.0 (d, J = 3.7), 142.4, 135.1, 134.3, 134.1, 133.6, 133.5, 132.8, 132.0, 131.5 (d, J = 2.5), 131.4 (d, J = 2.5), 131.2 (d, J = 2.5), 130.9 (d, J = 2.5), 130.6, 130.5, 130.4, 130.0, 129.8, 128.7, 128.7, 128.5, 128.4, 128.3, 128.3, 128.2, 123.8, 122.7, 49.3 (d, J = 15.0), 45.2(d, J = 6.2), 45.7 (d, J = 6.2),

41.3, 40.4, 38.3, 37.7, 32.6(d, J = 12.5) LCMS (ESI) $[M + H]^+$ = 644 Da, HRMS calcd. for $C_{39}H_{35}NO_4P_2$ 644.2120, found 644.2114 (diff. 1.18 ppm).

3.2.8. (5-Methoxytetracyclo[8.2.1.0^{2,9}.0^{3,8}]trideca-3,5,7-triene-11,12-diyl)bis(diphenylphosphane) dioxide (*rac*-**4i**)

The compound was isolated in 18% yield as a white solid, m.p.: 279–285 °C. ^{31}P-NMR (202 MHz, $CDCl_3$): δ = 32.4(d, J = 8.08), 28.8 (d, J = 10.0), ^{1}H-NMR (500 MHz, $CDCl_3$): δ = 7.87–7.76 (m, 15H, Ar-H), 7.12–6.47 (m, 8H, Ar-H), 4.03 (d, 1H, J = 35.0, CH), 3.88–3.83 (m, 1H, CH), 3.70 (s, 1H, OCH_3), 3.59 (d, 1H, J = 35.0, CH), 3.31–3.27 (m, 1H, CH), 2.49–2.33 (m, 2H, CH_2), 1.00–0.95 (m, 1H, CH), ^{13}C-NMR (125 MHz, $CDCl_3$): δ = 159.7, 145.3, 134.3, 133.9, 133.5, 133.2, 132.9, 132.5, 132.1, 132.0, 131.9 (d, J = 6.2), 131.1, 131.6, 130.5, 130.4, 130.2, 130.1, 128.7, 128.6, 128.5, 128.4, 128.3 (d, J = 2.5), 128.2 (d, J = 2.5), 127.6, 127.4, 122.8, 122.01, 114.2, 55.3, 50.4 (d, J = 13.5), 49.8 (d, J = 13.5), 46.7, 45.7, 38.5, 37.8, 31.4, 29.7 LCMS (ESI) $[M + H]^+$ = 601 Da, HRMS calcd. for $C_{38}H_{34}O_3P_2$ 601.2067, found 601.2056 (diff. 1.18 ppm).

3.2.9. [Methyl 11,12-bis(diphenylphosphoryl)tetracyclo[8.2.1.0^{2,9}.0^{3,8}]trideca-3,5,7-triene-5-carboxylate (*rac*-**4j**)

The compound was isolated in 84% yield as a white solid, m.p.: 169–171 °C. ^{31}P-NMR (202 MHz, $CDCl_3$): δ = 32.7 (d, J = 8.08), 29.1 (d, J = 8.08), ^{1}H-NMR (500 MHz, $CDCl_3$): δ = 7.87–7.76 (m, 15H, Ar-H), 7.12–6.47 (m, 8H, Ar-H), 4.03 (d, 1H, J = 35.0, CH), 3.88–3.83 (m, 1H, CH), 3.70 (s, 1H, OCH_3), 3.59 (d, 1H, J = 35.0, CH), 3.31–3.27 (m, 1H, CH), 2.49–2.33 (m, 2H, CH_2), 1.00–0.95 (m, 1H, CH), ^{13}C-NMR (125 MHz, $CDCl_3$): δ = 167.3, 151.0, 144.0 (d, J = 3.7), 135.0, 134.2, 134.1, 133.6, 133.3, 133.3, 132.9, 132.4, 131.7, 130.5, 130.4, 130.4, 130.3, 130.2, 130.1, 129.7, 129.6, 129.9, 129.5, 128.7, 128.6, 128.5, 128.5, 128.3, 128.3, 128.3, 128.2, 123.1, 122.9, 122.0, 121.8, 51.9, 50.0 (d, J = 13.7), 47.9, 41.1, 38.2, 32.4 LCMS (ESI) $[M + H]^+$ = 629 Da, HRMS calcd. for $C_{39}H_{34}O_4P_2$ 629.2022, found 629.2005 (diff. 1.7 ppm).

Dimethyl 2,3-bis(diphenylphosphoryl)-1,2,3,4,4a,12b-hexahydro-1,4-methanotriphenylene-7,10-dicarboxylate (*rac*-**3j**)

The compound was isolated in 51% yield as a white solid. The analysis of the product spectra confirmed its identity with literature data [12]. ^{31}P-NMR (202 MHz, $CDCl_3$): δ = 31.10 (d, J = 10.1), 28.81 (d, J = 8.1), ^{1}H-NMR (500 MHz, $CDCl_3$): δ = 8.50 (d, 1H, J = 10.0, Ar-H), 7.97–7.93 (m, 2H, Ar-H), 7.88–7.85 (m, 2H, Ar-H), 7.74 (d, 1H, J = 5.0, Ar-H), 7.74–7.48 (m, 11H, Ar-H), 7.17–7.04 (m, 6H, Ar-H), 6.86 (d, 1H, J = 10.0, Ar-H), 5.26 (d, 1H, J = 10.0, Ar-H), 4.12 (d, 1H, J = 10.0, CH), 3.95–3.90 (m, 1H, CH-P), 3.87–3.81 (m, 1H, CH-P), 3.93 (s, 3H, CH_3), 3.92 (s, 3H, CH_3), 3.75 (d, 1H, J = 10.0, CH), 2.64 (s, 1H, CH), 2.42 (d, 1H, J = 10.0, CH_2), 2.02 (d, 1H, J = 10.0, CH_2), 1.94 (s, 1H, CH). The HNMR spectrum is in agreement with the literature [12].

3.2.10. [7,10-Bis(trifluoromethyl)-1,2,3,4,4a,12b-hexahydro-1,4-methanotriphenylene-2,3-diyl]bis(diphenylphosphane) dioxide (*rac*-**3k**)

The compound was isolated in a yield of up to 67% as a white solid. The analysis of the product spectra confirmed its identity with literature data [12]. ^{31}P-NMR (202 MHz, $CDCl_3$): δ = 31.04 (d, J = 8.1), 28.84 (d, J = 8.1), ^{1}H-NMR (500 MHz, $CDCl_3$): δ = 7.89–7.70 (m, 6H, Ar-H), 7.58–7.41 (m, 10H, Ar-H), 7.26 (d, 1H, J = 10.0, Ar-H), 7.10–6.97 (m, 7H, Ar-H), 6.81 (d, 1H, J = 10.0, Ar-H), 5.22 (d, 1H, J = 10.0, Ar-H), 4.02 (d, 1H, J = 10.0, CH), 3.83–3.76 (m, 1H, CH-P), 3.76–3.70 (m, 1H, CH-P), 3.67 (d, 1H, J = 10.0, CH), 2.54 (s, 1H, CH), 2.32 (d, 1H, J = 10.0, CH_2), 1.97 (d, 1H, J = 10.0, CH_2), 1.25–1.19 (m, 1H, CH). The HNMR spectrum is in agreement with the literature [12].

3.2.11. [7,10-Bis(trimethylsilyl)-1,2,3,4,4a,12b-hexahydro-1,4-methanotriphenylene-2,3-diyl]bis(diphenylphosphane) dioxide (*rac*-**3l**)

The compound was isolated in a yield of 63% as a white solid. The analysis of the product spectra confirmed its identity with literature data [12]. ^{31}P-NMR (202 MHz, $CDCl_3$): δ = 31.12 (d, *J* = 6.1), 28.60 (d, *J* = 10.1), ^{1}H-NMR (500 MHz, $CDCl_3$): δ = 7.98–7.93 (m, 4H, Ar-H), 7.67–7.51 (m, 10H, Ar-H), 7.21–7.05 (m, 6H, Ar-H), 6.93 (d, 1H, *J* = 5.0, Ar-H), 6.78 (d, 1H, *J* = 10.0, Ar-H), 5.22 (d, 1H, *J* = 10.0, Ar-H), 4.03 (d, 1H, *J* = 10.0, CH), 3.94–3.89 (m, 1H, CH-P), 3.85–3.81 (m, 1H, CH-P), 3.68 (d, 1H, *J* = 10.0, CH), 2.64 (s, 1H, CH), 2.48 (d, 1H, *J* = 10.0, CH_2), 1.96 (d, 1H, *J* = 10.0, CH_2), 1.66 (s, 1H, CH), 0.28 (s, 9H, CH_3), 0.26 (s, 9H, CH_3). The HNMR spectrum is in agreement with the literature [12].

3.2.12. 6b,7,8,9,10,10a-Hexahydro-7,10-methanobenzo[a]biphenylene-8,9-diylbis(diphenylphosphane) dioxide (*rac*-**4m**)

The compound was isolated in 68% yield as a white solid, m.p.: 225–227 °C. ^{31}P-NMR (202 MHz, $CDCl_3$): δ = 32.4 (d, *J* = 8.08), 28.7 (d, *J* = 6.06), ^{1}H-NMR (500 MHz, $CDCl_3$): δ = 7.96–7.48 (m, 16H, Ar-H), 7.36–6.28 (m, 4H, Ar-H), 7.19–7.03 (m, 6H, Ar-H), 6.83 (d, 1H, *J* = 10.0, Ar-H), 4.21 (d, 1H, *J* = 5.0, CH), 3.87–3.81 (m, 1H, CH), 3.59 (d, 1H, *J* = 5.0, CH), 3.42–3.39 (m, 1H, CH), 2.57 (s, 1H, CH), 2.46 (d, 1H, *J* = 15.0, CH_2), 1.88 (d, 1H, *J* = 15.0, CH_2), 0.90–0.85 (m, 1H, CH), ^{13}C-NMR (125 MHz, $CDCl_3$): δ = 145.9, 145.1, 142.6, 134.6, 134.2, 133.9, 133.7, 133.4, 133.3, 132.9, 132.2, 131.4, 131.3, 131.2, 130.9, 130.6, 130.5, 130.4, 130.4, 130.2, 130.1, 129.3, 128.8, 128.7, 128.6, 128.5, 128.3, 128.2, 127.5, 126.8, 120.5, 120.3, 49.8 (d, *J* = 15.0), 45.7 (d, *J* = 5.0), 41.3, 41.0, 40.4, 38.2, 37.7, 32.9 LCMS (ESI) $[M + H]^+$ = 621 Da, HRMS calcd. for $C_{41}H_{38}O_2P_2$ 621.2139, found 621.2166 (diff. 1.18 ppm).

[5-(naphthalen-1-yl)bicyclo[2.2.1]heptane-2,3-diyl]bis(diphenylphosphane) dioxide (*rac*-**5m**)

This minor reaction product was detected in trace amounts in the reaction mixtures by means of ESI-LC HRMS. HRMS calcd. for $C_{41}H_{36}O_2P_2$ 623.2312, found 623.2263 (diff. 7.8 ppm).

3.2.13. (7-Methoxy-1,2,3,4,4a,10b-hexahydro-1,4-methanobenzo[b]biphenylene-2,3-diyl)bis(diphenylphosphane) dioxide (*rac*-**4n**)

The compound was isolated in 53% yield as a white solid, m.p.: 209–214 °C. ^{31}P-NMR (202 MHz, $CDCl_3$): δ = 32.4 (d, *J* = 8.08), 28.8 (d, *J* = 8.08), ^{1}H-NMR (500 MHz, $CDCl_3$): δ = 7.91–7.42 (m, 16H, Ar-H), 7.28–7.02 (m, 10H, Ar-H), 4.21 (d, 1H, *J* = 5.0, CH), 3.92–3.85 (m, 1H, CH), 3.88 (s, 1H, OCH_3), 3.75 (d, 1H, *J* = 5.0, CH), 3.38–3.35 (m, 1H, CH), 2.60 (s, 1H, CH), 2.46 (d, 1H, *J* = 15.0, CH_2), 1.90 (d, 1H, *J* = 15.0, CH_2), 1.08–1.05 (m, 1H, CH), ^{13}C-NMR (125 MHz, $CDCl_3$): δ = 156.7, 144.1, 141.8, 135.2, 135.1, 134.3, 133.9, 133.5, 133.2, 132.5, 131.7, 131.3, 131.1, 130.9, 130.6, 130.5, 130.4, 130.2, 130.1, 129.5, 129.4, 129.3, 128.7, 128.6, 128.5, 128.3, 128.2, 128.2, 120.0, 119.9, 119.0, 117.1, 106.6, 55.2, 50.3, 46.6, 42.9, 42.2, 39.9, 37.8, 32.9 LCMS (ESI) $[M + H]^+$ = 651 Da, HRMS calcd. for $C_{42}H_{36}O_3P_2$ 651.2206, found 651.2212 (diff. 1.18 ppm).

3.2.14. 8c,9,10,11,12,12a-Hexahydro-9,12-methanobenzo[3,4]cyclobuta[1,2-*l*]phenanthrene-10,11-diylbis(diphenylphosphane) dioxide (*rac*-**4p**)

The compound was isolated in 54% yield as a white solid, m.p.: 295–298 °C. ^{31}P-NMR (202 MHz, $CDCl_3$): δ = 32.3 (d, *J* = 8.08), 28.8 (d, *J* = 8.08), ^{1}H-NMR (500 MHz, $CDCl_3$): δ = 8.68–8.65 (m, 3H, Ar-H), 7.99–7.42 (m, 20H, Ar-H), 7.13–7.06 (m, 6H, Ar-H), 6.83 (d, 1H, *J* = 10.0, Ar-H), 4.29 (s, 1H, CH), 3.93–3.83 (m, 2H, CH), 3.52–3.47 (m, 1H, CH), 2.61 (d, 2H, *J* = 5.0, CH), 1.90 (d, 1H, *J* = 10.0, CH), 1.01–0.85 (m, 1H, CH), ^{13}C-NMR (125 MHz, $CDCl_3$): δ = 139.3, 138.3 (d, *J* = 3.7), 135.5, 134.7, 134.4, 133.9, 133.6, 133.1, 132.3, 132.1, 132.0, 131.9 (d, *J* = 3.7), 131.4, 131.1, 130.9, 130.8, 130.7, 130.5, 130.2, 130.1, 128.7, 128.6, 128.5, 128.4, 128.3, 128.3, 127.8, 127.8, 126.6 (d, *J* = 3.7), 123.7 (d, *J* = 8.7), 122.6, 122.3, 49.2 (d, *J* = 15.0), 45.1 (d, *J* = 6.2), 40.4, 41.1, 38.7, 32.5 (d, *J* = 12.5), LCMS (ESI) $[M + H]^+$ = 671 Da, HRMS calcd. for $C_{45}H_{38}O_2P_2$ 671.2281, found 671.2263 (diff. 1.18 ppm).

[5-(phenanthren-9-yl)bicyclo[2.2.1]heptane-2,3-diyl]bis(diphenylphosphane) dioxide (*rac*-**5p**)

This minor reaction product was detected in trace amounts in the reaction mixtures by means of ESI-LC HRMS. HRMS calcd. for $C_{45}H_{38}O_2P_2$ 673.2422, found 673.2420 (diff. 0.3 ppm).

3.2.15. Synthesis of [(1*R*,2*R*,9*S*,10*S*,11*R*,12*R*)-4-Phenyltetracyclo[8.2.1.0^{2,9}.0^{3,8}]trideca-3,5,7-triene-11,12-diyl]bis(diphenylphosphane) ((*R*,*R*)-**6q**)

One gram (1.0 g) of (*R*,*R*)-**6q** (1.54 mmol) and 10 mL of dry toluene were added to a Schlenk flask and placed under a nitrogen atmosphere. Subsequently, 6 equiv. of $SiHCl_3$ and then 18 equiv. of Bu_3N were added and the mixture was heated at 110 °C. After the reaction had been completed, the mixture was cooled down and 30% NaOH solution was added drop-wise. The organic layer was dried over Na_2SO_4 and the crude was subjected to column chromatography with the ethyl acetate/hexane system. The phosphine was isolated as a white solid in 85% yield. $[\alpha]_D^{20} = -60$ ($c = 0.4$; $CHCl_3$); ^{31}P-NMR (202 MHz, $CDCl_3$): δ = 2.53, 1.10, −11.09, −13.67; ^{1}H-NMR (500 MHz, $CDCl_3$): δ = 7.94–7.42 (m, 16H, Ar-H), 7.23–7.02 (m, 11H, Ar-H), 6.83 (d, 1H, *J* = 5.0, Ar-H), 4.37 (s, 1H, CH), 3.99–3.93 (m, 1H, 2×CH), 3.69 (s, 1H, CH), 3.44–3.39 (m, 1H, CH), 2.70 (s, 1H, CH), 2.38 (d, 1H, *J* = 10.0, CH_2), 1.90 (d, 1H, *J* = 10.0, CH_2), 1.05–1.03 (m, 1H, CH), ^{13}C-NMR (125 MHz, $CDCl_3$): δ = 145.8, 141.8, 137.3, 134.3, 133.1, 131.3, 131.3, 130.9, 130.6, 130.5, 130.4, 130.1, 130.0, 128.6, 128.5, 128.3, 128.2, 127.2, 126.8, 126.5, 126.0, 125.1, 124.8, 120.8, 51.4 (d, *J* = 13.7), 50.0 (d, *J* = 13.7), 48.3 (d, *J* = 6.2), 45.3 (d, *J* = 6.2), 41.8, 41.6, 40.4, 38.4, 38.1, 37.8, 37.5.

3.2.16. Synthesis of [Pd(**6q**)Cl_2]

A solution of **6q** (0.1 g) in benzene (2 mL) was added to a stirred solution of [Pd(^{t}-BuCN)$_2Cl_2$] (0.056 g) in 2 mL of THF. The reaction mixture was stirred overnight at ambient temperature, the volume of the solvents was reduced under reduced pressure down to about 2 mL and the precipitated complex was filtered off. The compound was isolated in 82% yield as a yellow solid. The samples for the X-ray measurements were grown from a warm ethanol/benzene mixture. $[\alpha]_D^{20} = -40.8$ (c = 0.25, CH_2Cl_2); ^{31}P-NMR (202 MHz, $CDCl_3$): δ = 31.3, 31.2, 28.0, 27.9, 27.9, 27.8, ^{1}H-NMR (500 MHz, $CDCl_3$): δ = 8.08–8.04 (m, 2H, Ar-H), 8.02–7.98 (m, 2H, Ar-H), 7.93–7.90 (m, 2H, Ar-H), 7.88–7.84 (m, 2H, Ar-H), 7.68–7.65 (m, 1H, Ar-H), 7.59–7.58 (m, 2H, Ar-H), 7.54–7.48 (m, 22H, Ar-H), 7.33–7.27 (m, 3H, Ar-H), 7.21–7.12 (m, 8H, Ar-H), 7.07–7.01 (m, 13H, Ar-H), 6.85 (d, 1H, *J* = 7.3, CH), 6.55 (d, 1H, *J* = 7.3, CH), 4.12–4.11 (m, 1H, CH), 4.00–3.99 (m, 1H, CH), 3.96–3.87 (m, 1H, CH), 3.48–3.43 (m, 2H, CH), 2.66–2.66 (m, 1H, CH), 2.55–2.55 (m, 1H, CH), 2.36–2.34 (m, 1H, CH), 2.27–2.25 (m, 1H, CH), 1.77–1.75 (m, 1H, CH), 1.68–1.66 (m, 1H, CH), 0.75–0.74 (m, 2H, CH). ^{13}C-NMR (dept 135, 125 MHz, $CDCl_3$): δ = 132.96, 131.81, 131.69, 131.63, 131.54, 131.53, 131.48, 131.46, 131.20, 131.14, 130.90, 130.84, 130.76, 130.70, 130.53, 130.46, 130.40, 130.25, 130.21, 130.18, 130.14, 129.49, 129/42, 129.34, 129.30, 129.28, 129.18, 129.09, 129.07, 128.98, 128.64, 128.62, 128.55, 128.53, 128.02, 127.63, 126.48, 125.96, 125.34, 124.75, 121.32, 121.10, 51.52, 51.41, 49.85, 49.73, 48.35, 48.31, 46.39, 46.35, 41.86, 41.61, 41.13, 41.10, 40.36, 40.18, 40.00, 39.84, 38.77, 38.74, 38.41, 38.36, 38.33, 38.19, 38.16, 37.87, 37.81, 37.77, 37.75, 32.26, 32.23, 32.17, 32.13. LCMS (ESI) $[M - Cl]^+$ = HRMS calcd. for C43H36P2ClPd 755.1010 Da, found 755.1034 Da (diff. 3.18 ppm).

3.2.17. General Procedure for Allylic Alkylation Reaction

A solution of ligand **6q** (2.0 mol % or 4.0 mol %) and [Pd(η^3-C_3H_5)Cl]$_2$ (1.0 mol % or 2.0 mol %) was stirred in 1.0 mL of dry THF for 30 min. Then, *rac*-(*E*)-1,3-diphenylallyl acetate (0.5 mmol) in 1.0 mL of THF was added, followed by dimethyl malonate (1.5 mmol) and base (1.5 mmol) of BSA with 2 mol % of KOAc or 1.5 mmol of K_2CO_3/Cs_2CO_3 (1:1). After stirring for 12 h at room temperature, the solution was concentrated and purified by chromatography on silica gel eluted with hexane/EtOAc (10:1). The product was isolated as a colorless oil. ^{1}H-NMR (500 MHz, $CDCl_3$): δ = 7.34–7.22 (m, 10H, Ar-H), 6.50 (d, 1H, *J* = 15.0, -CH=), 6.33 (dd, 1H, *J* = 10.0, 15.0, -CH=), 4.31–4.27

(m, 1H, CH), 3.98 (d, 1H, $J = 10.0$, CH), 3.77 (s, 3H, OCH_3), 3.54 (s, 3H, OCH_3), HPLC analysis (Chiralcel OD-H, hexane/*iso*-PrOH, 98:2, 1.0 mL/min, 254 nm): tr(minor) = 10.9 min, tr(major) = 14.5 min.

3.3. Computational Studies

All DFT calculations were performed using the "Prometheus" cluster in the "Cyfronet" computational centre in Cracow. A new generation M062x [27] functional, implemented in the Gaussian 09 package (Gaussian Inc.: Wallingford, UK) [28], was used. All stationary structures have been optimised using the LANL2DZ basis set with one f function for Pd and without pseudopotential. All structures were characterised by only positive eigenvalues in their diagonalised Hessian matrices. For the optimised structures, thermochemical data for the temperature T = 298 K and pressure p = 1 atm were computed using vibrational analysis data.

3.4. Crystallographic Studies

An X-ray diffraction experiment for the complex [Pd(**6q**)Cl_2] was conducted at room temperature on a SuperNova X-ray diffractometer equipped with an Atlas S2 CCD detector using mirror-monochromatized CuKα radiation ($\lambda = 1.54184$ Å). Multiscan absorption correction procedures were applied to the data [41]. The structure was solved by direct methods using the ShelXT [42] structure solution program with intrinsic phasing and refined with the Olex2.refine refinement package using Gauss–Newton minimisation [43]. Non-hydrogen atoms were refined anisotropically. The C-bound H atoms were positioned geometrically and refined with the 'riding' model. A summary of the experimental details and crystal structure refinement parameters are given in Tables 2–5 and ESI. The experimental details and final atomic parameters have been deposited with the Cambridge Crystallographic Data Centre as supplementary material (CCDC: 1885894). Copies of the data can be obtained free of charge on request via www.ccdc.cam.ac.uk/data_request/cif, or by emailing data_request@ccdc.cam.ac.uk.

4. Conclusions

Herein, we have demonstrated the design and straightforward synthesis of novel highly rigid polycyclic bis(phosphanes). The X-ray analysis of the coordination compound obtained with a new ligand and dichloropalladium complex confirmed the cis, exo configuration of the bis(phosphane) ligand molecule. The five-membered metallacycle has an envelope conformation with a puckering amplitude q2 of 1.142(5) Å and a phase angle Φ2 of 153.8(1)°. The CSD search showed that the most frequent conformation for the five-membered metallacycles of transition metals and bis(phosphane) ligands was an envelope with one of the carbon atoms positioned at the flap regardless of the transition metal and phosphine substituent type, just as in the presented complex [Pd(**6q**)Cl_2]. This stereochemical asymmetry observed in the analysed metallacycles is probably an important factor for efficient transition-metal-mediated asymmetric transformations. The comparison of the measured and calculated structures of the palladium complex and corresponding bis(phosphane) dioxides confirmed that, due to a structural rigidity, the conformation of the ligands depends only marginally on the substituents that those ligands bear in their polycyclic core. Notable efficiency, activity and selectivity of the new ligand in a model asymmetric allylic alkylation reaction have also been confirmed.

Author Contributions: Conceptualisation: K.S. and O.M.D.; Methodology: K.S. and O.M.D.; Chemical Experiments: K.S. and O.M.D.; Crystallographic Studies and Structural Survey: B.M.; Software and Computation: R.J.; Writing (Original Draft Preparation): K.S., O.M.D., R.J., B.M. and I.D.; Writing (Review and Editing): O.M.D.; Visualization: I.D. and B.M.; Inspiration and Discussions: K.M.P.

References

1. Oestreich, M. *The Mizoroki-Heck Reaction*; Wiley: Hoboken, NJ, USA, 2009; 587p, ISBN 9780470033944.
2. Beletskaya, I.P.; Cheprakov, A.V. The Heck Reaction as a Sharpening Stone of Palladium Catalysis. *Chem. Rev.* **2000**, *100*, 3009–3066. [CrossRef] [PubMed]
3. de Meijere, A.; Bräse, S.; Oestreich, M. *Metal-Catalyzed Cross-Coupling Reactions and More*; 3 Volume Set; Wiley: Hoboken, NJ, USA, 2014; 1576p, ISBN 9783527331543.
4. Catellani, M.; Chiusoli, G.P. Palladium-catalyzed synthesis of 1,2,3,4,4a,12b-hexahydro-1,4-methanotriphenylenes. *J. Organomet. Chem.* **1985**, *286*, c13–c16. [CrossRef]
5. Ye, J.; Lautens, M. Palladium-catalysed norbornene-mediated C−H functionalization of arenes. *Nat. Chem.* **2015**, *7*, 863–870. [CrossRef]
6. Li, R.; Dong, G. Direct Annulation between Aryl Iodides and Epoxides through Palladium/Norbornene Cooperative Catalysis. *Angew. Chem. Int. Ed. Engl.* **2017**, *57*, 1697–1701. [CrossRef] [PubMed]
7. Casnati, A.; Fontana, M.; Coruzzi, G.; Aresta, B.M.; Corriero, N.; Maggi, R.; Maestri, G.; Motti, E.; Della Ca', N. Enhancing Reactivity and Selectivity of Aryl Bromides: A Complementary Approach to Dibenzo[*b,f*]azepine Derivatives. *ChemCatChem Catal.* **2018**, *10*, 4346–4352. [CrossRef]
8. Liu, C.; Liang, Y.; Zheng, N.; Zhang, B.-S.; Feng, Y.; Bi, S.; Liang, Y.-M. Synthesis of indolines via a palladium/norbornene-catalyzed reaction of aziridines with aryl iodides. *Chem. Commun.* **2018**, *54*, 3407–3410. [CrossRef]
9. Demchuk, O.M.; Jasiński, R.; Formela, A. The Halogen-Less Catalytic Transition Metal-Mediated Cross-Coupling Reactions: A Sustainable Alternative for Utilisation of Organohalides. In *Chemistry Beyond Chlorine*; Tundo, P., He, L.-N., Lokteva, E., Mota, C., Eds.; Springer: Cham, Switzerland, 2016; pp. 17–94.
10. Brunner, H.; Pieronczyk, W. Asymmetric Hydrogenation of (Z)-(Acetylamino)-cinnamic Acid by a Rh/norphos Catalyst. *Angew. Chem. Int. Ed. Engl.* **1979**, *18*, 620–621. [CrossRef]
11. Marchand, A.P.; Romanski, J.; Kumar, T.P. (R,R)-(-)-NORPHOS and (S,S)-(+)-NORPHOS. In *e-EROS: Encyclopedia of Reagents for Organic Synthesis*; John Wiley & Sons, Ltd.: Hoboken, NJ, USA, 2003.
12. Szwaczko, K.; Demchuk, O.M.; Mirosław, B.; Strzelecka, D.; Pietrusiewicz, K.M. Straightforward approach to norbornene core based chiral ligands by tandem cross dehydrogenative coupling reactions. *Tetrahedron Lett.* **2016**, *57*, 3491–3495. [CrossRef]
13. Demchuk, O.M.; Kapłon, K.; Mazur, L.; Strzelecka, D.; Pietrusiewicz, K.M. Readily available catalysts for demanding Suzuki–Miyaura couplings under mild conditions. *Tetrahedron* **2016**, *72*, 6668–6677. [CrossRef]
14. Qiu, Z.; Sun, R.; Teng, D. Synthesis of highly rigid phosphine–oxazoline ligands for palladium-catalyzed asymmetric allylic alkylation. *Org. Biomol. Chem.* **2018**, *16*, 7717–7724. [CrossRef]
15. Snieckus, V.; Demchuk, O.; Yoruk, B.; Blackburn, T. A Mixed Naphthyl-Phenyl Phosphine Ligand Motif for Suzuki, Heck, and Hydrodehalogenation Reactions. *Synlett* **2006**, 2908–2913. [CrossRef]
16. Demchuk, O.M.; Kielar, K.; Pietrusiewicz, K.M. Rational design of novel ligands for environmentally benign cross-coupling reactions. *Pure Appl. Chem.* **2011**, *83*, 633–644. [CrossRef]
17. Chakravarty, A.R.; Cotton, F.A.; Tocher, D.A. Displacive transfer of a phenyl group from triphenylphosphine to a metal atom: Synthesis and molecular structure of the dibenzamidediphenyldiruthenium compound Ru2Ph2(PhCONH)2[Ph2POC(Ph)N]2. *J. Am. Chem. Soc.* **1984**, *106*, 6409–6413. [CrossRef]
18. Jin, L.; Qian, J.; Sun, N.; Hu, B.; Shen, Z.; Hu, X. Pd-Catalyzed reductive heck reaction of olefins with aryl bromides for Csp2–Csp3 bond formation. *Chem. Commun.* **2018**, *54*, 5752–5755. [CrossRef]
19. Wang, C.; Xiao, G.; Guo, T.; Ding, Y.; Wu, X.; Loh, T.-P. Palladium-Catalyzed Regiocontrollable Reductive Heck Reaction of Unactivated Aliphatic Alkenes. *J. Am. Chem. Soc.* **2018**, *140*, 9332–9336. [CrossRef] [PubMed]
20. Cremer, D. Calculation of puckered rings with analytical gradients. *J. Phys. Chem.* **1990**, *94*, 5502–5509. [CrossRef]
21. Groom, C.R.; Bruno, I.J.; Lightfoot, M.P.; Ward, S.C. The Cambridge Structural Database. *Acta Crystallogr. Sect. B Struct. Sci. Cryst. Eng. Mater.* **2016**, *72*, 171–179. [CrossRef]
22. Eller, S.; Trettenbrein, B.; Fessler, M.; Haringer, S.; Ruggenthaler, M.; Gutmann, R.; van der Veer, W.E.; Kopacka, H.; Müller, T.; Obendorf, D.; et al. Surprising photochemical reactivity and visible light-driven energy transfer in heterodimetallic complexes. *Dalton Trans.* **2011**, *40*, 3815–3829. [CrossRef]
23. Wilson, W.L.; Rahn, J.A.; Alcock, N.W.; Fischer, J.; Frederick, J.H.; Nelson, J.H. Thermal dimerization of

1-substituted-3,4-dimethylphospholes within the coordination sphere of platinum(II). *Inorg. Chem.* **1994**, *33*, 109–117. [CrossRef]

24. Oberhauser, W.; Ienco, A.; Vizza, F.; Trettenbrein, B.; Oberhuber, D.; Strabler, C.; Ortner, T.; Brüggeller, P. Regioselective Hydromethoxycarbonylation of Terminal Alkynes Catalyzed by Palladium(II)–Tetraphos Complexes. *Organometallics* **2012**, *31*, 4832–4837. [CrossRef]
25. Oberhauser, W.; Stampfl, T.; Bachmann, C.; Haid, R.; Langes, C.; Kopacka, H.; Ongania, K.-H.; Brüggeller, P. Palladium(II), platinum(II), and platinum(IV) complexes containing trans-1,2-bis(diphenylphosphino)ethene or cis,trans,cis-1,2,3,4-tetrakis(diphenylphosphino)cyclobutane: Complete X-ray structural characterization of binuclear compounds. *Polyhedron* **2000**, *19*, 913–923. [CrossRef]
26. Strabler, C.; Ortner, T.; Prock, J.; Granja, A.; Gutmann, R.; Kopacka, H.; Müller, T.; Brüggeller, P. Versatile Supramolecular Coordination Behaviour of a Bis(bidentate) Tetraphosphane. *Eur. J. Inorg. Chem.* **2013**, *2013*, 5121–5132. [CrossRef]
27. Zhao, Y.; Truhlar, D.G. The M06 suite of density functionals for main group thermochemistry, thermochemical kinetics, noncovalent interactions, excited states, and transition elements: Two new functionals and systematic testing of four M06-class functionals and 12 other functionals. *Theor. Chem. Acc.* **2007**, *120*, 215–241. [CrossRef]
28. Frisch, M.J.; Trucks, G.W.; Schlegel, H.B.; Scuseria, G.E.; Robb, M.A.; Cheeseman, J.R.; Scalmani, G.; Barone, V.; Mennucci, B.; Petersson, G.A.; et al. *Gaussian 09, Revision B.01*; Gaussian Inc.: Wallingford, UK, 2010.
29. Jasiński, R.; Demchuk, O.M.; Babyuk, D. A Quantum-Chemical DFT Approach to Elucidation of the Chirality Transfer Mechanism of the Enantioselective Suzuki–Miyaura Cross-Coupling Reaction. *J. Chem.* **2017**, *2017*, 1–12. [CrossRef]
30. Mirosław, B.; Babyuk, D.; Łapczuk-Krygier, A.; Kącka-Zych, A.; Demchuk, O.M.; Jasiński, R. Regiospecific formation of the nitromethyl-substituted 3-phenyl-4,5-dihydroisoxazole via [3 + 2] cycloaddition. *Monatshefte für Chemie (Chem. Mon.)* **2018**, *149*, 1877–1884. [CrossRef] [PubMed]
31. Demchuk, O.M.; Jasiński, R.; Pietrusiewicz, K.M. New Insights into the Mechanism of Reduction of Tertiary Phosphine Oxides by Means of Phenylsilane. *Heteroat. Chem.* **2015**, *26*, 441–448. [CrossRef]
32. Jasiński, R. Molecular mechanism of thermal decomposition of fluoronitroazoxy compounds: DFT computational study. *J. Fluor. Chem.* **2014**, *160*, 29–33. [CrossRef]
33. Demchuk, O.M.; Justyniak, I.; Miroslaw, B.; Jasiński, R. 2-Methoxynaphthylnaphthoquinone and its solvate: Synthesis and structure-properties relationship. *J. Phys. Org. Chem.* **2014**, *27*, 66–73. [CrossRef]
34. Demchuk, O.M.; Arlt, D.; Jasiński, R.; Pietrusiewicz, K.M. Relationship between structure and efficiency of atropisomeric phosphine ligands in homogeneous catalytic asymmetric hydrogenation. *J. Phys. Org. Chem.* **2012**, *25*, 1006–1011. [CrossRef]
35. Demchuk, O.M.; Jasiński, R. Organophosphorus ligands: Recent developments in design, synthesis, and application in environmentally benign catalysis. *Phosphorus Sulfur Silicon Relat. Elem.* **2015**, *191*, 245–253. [CrossRef]
36. Issa, T.B.; Sayari, F.; Ghalla, H.; Benhamada, L. Synthesis, crystal structure, DFT calculations and molecular docking of l-pyroglutamic acid. *J. Mol. Struct.* **2019**, *1178*, 436–449. [CrossRef]
37. Jin, M.-J.; Takale, V.B.; Sarkar, M.S.; Kim, Y.-M. Highly enantioselective Pd-catalyzed allylic alkylation using new chiral ferrocenylphosphinoimidazolidine ligands. *Chem. Commun.* **2006**, *663*. [CrossRef]
38. Yamaguchi, M.; Shima, T.; Yamagishi, T.; Hida, M. Palladium-catalyzed asymmetric allylic alkylation using dimethyl malonate and its derivatives as nucleophile. *Tetrahedron Asymmetry* **1991**, *2*, 663–666. [CrossRef]
39. Łapczuk-Krygier, A.; Ponikiewski, Ł.; Jasiński, R. The crystal structure of (1RS,4RS,5RS,6SR)-5-cyano-5-nitro-6-phenyl-bicyclo[2.2.1]hept-2-ene. *Crystallogr. Rep.* **2014**, *59*, 961–963. [CrossRef]
40. Jasiński, R.; Kwiatkowska, M.; Sharnin, V.; Barański, A. Experimental and theoretical studies of Diels–Alder reaction between methyl (Z)-2-nitro-3-(4-nitrophenyl)-2-propenoate and cyclopentadiene. *Monatshefte für Chemie (Chem. Mon.)* **2013**, *144*, 327–335. [CrossRef] [PubMed]
41. Rigaku Oxford Diffraction. *CrysAlis PRO ver. 1.171.38.46*; Rigaku Oxford Diffraction: Yarnton, UK, 2015.
42. Sheldrick, G.M. SHELXT—Integrated space-group and crystal-structure determination. *Acta Crystallogr. Sect. A Found. Adv.* **2015**, *71*, 3–8. [CrossRef] [PubMed]
43. Dolomanov, O.V.; Bourhis, L.J.; Gildea, R.J.; Howard, J.A.K.; Puschmann, H. OLEX2: A complete structure solution, refinement and analysis program. *J. Appl. Crystallogr.* **2009**, *42*, 339–341. [CrossRef]

12

Synthesis of Hydroxymethylenebisphosphonic Acid Derivatives in Different Solvents

Dávid Illés Nagy [1], Alajos Grün [1], Sándor Garadnay [2], István Greiner [2] and György Keglevich [1,*]

1 Department of Organic Chemistry and Technology, Budapest University of Technology and Economics, 1521 Budapest, Hungary; duwa24@gmail.com (D.I.N.); agrun@mail.bme.hu (A.G.)
2 Gedeon Richter Plc., 1475 Budapest, Hungary; garadnay@richter.hu (S.G.); i.greiner@richter.hu (I.G.)
* Correspondence: gkeglevich@mail.bme.hu

Academic Editor: Roman Dembinski

Abstract: The syntheses of hydroxymethylenebisphosphonic acid derivatives (dronic acid derivatives) starting from the corresponding substituted acetic acids and P-reagents, mainly phosphorus trichloride and phosphorous acid are surveyed according to the solvents applied. The nature of the solvent is a critical point due to the heterogeneity of the reaction mixtures. This review sheds light on the optimum choice and ratio of the P-reactants, and on the optimum conditions.

Keywords: hydroxymethylenebisphosphonic acid (dronic acid) derivatives; solvent; P-reagents; synthesis

1. Introduction

The hydroxymethylenebisphosphonic acid derivatives (dronic acid derivatives) form an important group within organophosphorus pharmaceutics. It is known that the two phosphonic functions are capable of forming complex with calcium ions, hence, the resorption of these ions is prevented. Bisphosphonic derivatives are used in the treatment of osteoporosis, Paget disease and tumor-induced hypercalcaemia, but they also show direct antitumor and antiparasitic activity [1–6].

Although the synthesis of dronic acid derivatives is widely discussed in the literature, it can be considered a sort of "black box". They were prepared in many solvents, but the optimum circumstances, and the molar ratios of the P-reagents were not explored. The role of the reagents and the reaction mechanism were not clarified. In most cases, the purity of the products was not reported, or when crude products were prepared, unrealistically high, and hence misleading yields were claimed. We wish to summarize the preparation of the most important hydroxymethylenebisphosphonic acid derivatives according to solvents, hoping to make the syntheses more transparent. The most important therapeutic agents discussed in this review are listed in Table 1.

Table 1. The dronic acid derivatives discussed in this review.

General Formula	Y	Commercial Name	Abbreviation
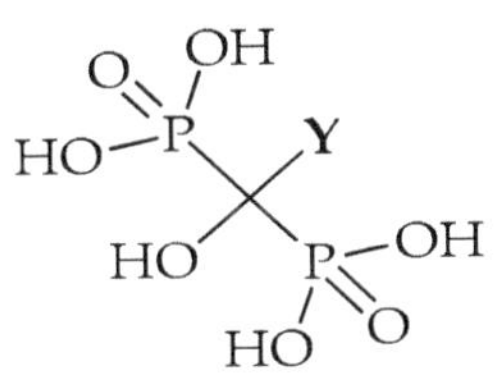	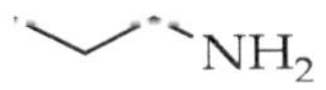	pamidronic acid	**PD**
	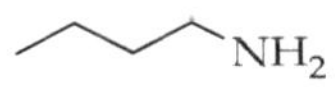	alendronic acid	**AD**
	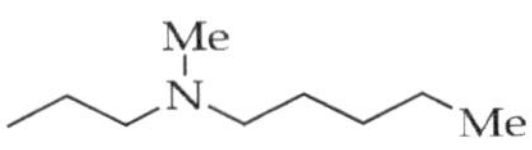	ibandronic acid	**ID**
	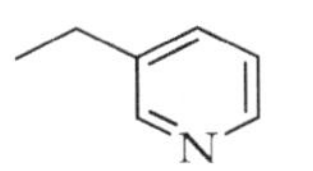	risedronic acid	**RD**
	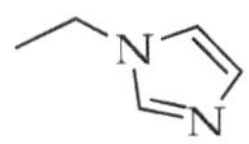	zoledronic acid	**ZD**

According to the general scheme for the preparation of hydroxymethylenebisphosphonic acid derivatives, the corresponding carboxylic acid, or its chloride, ester and anhydride derivative was used as the starting material, and reacted with phosphorus trichloride and/or phosphorous acid, phosphoric acid or phosphoryl chloride in a wide variety of solvents, for example methanesulfonic acid (MSA), sulfolane, chlorobenzene, toluene, xylene, *n*-octane, methyl cyclohexane, 1,4-dioxane, acetonitrile, phenol and its derivatives, alkyl carbonates, phosphates, silicon or sunflower oil, ionic liquids, and also in the absence of any solvent (Scheme 1).

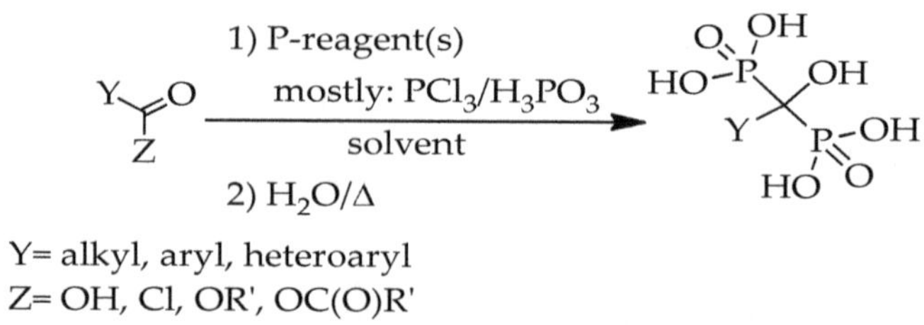

Scheme 1. General scheme for the preparation of hydroxymethylenebisphosphonic acids.

In many publications, water was applied as the solvent, and phosphorus trichloride as the reagent, which is absolutely pointless, as the water reacts fast with the P-reagent to provide phosphorous acid [7–11]. The concept of generating phosphorous acid "in situ" may be questioned, as this approach is expensive, pollutes the environment, and rather dangerous on the industrial scale. We did not deal with this not too practical method.

2. Reactions in Methanesulfonic Acid (MSA)

MSA may be considered the best solvent during the synthesis of dronic acid derivatives, as it helps to overcome the problems of heterogeneity during the reaction. However, the MSA is likely to participate in the reactions themselves. It can activate the carboxylic acid to form a carboxylic acid-MSA mixed anhydride, which may be formed in two ways, as shown in Scheme 2 [12].

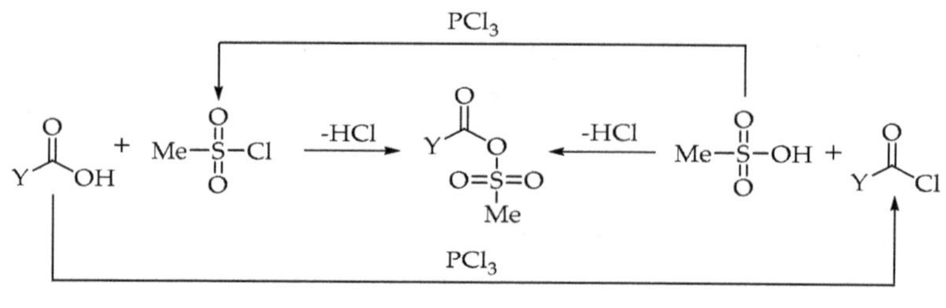

Scheme 2. Possible formation of carboxylic acid-MSA mixed anhydride as an intermediate.

MSA may also react with phosphorus trichloride to provide a more reactive P-species (Scheme 3).

Scheme 3. Activation of phosphorus trichloride by reaction with MSA.

In all syntheses of pamidronate described, β-alanine was applied as the starting material in MSA (Scheme 1, Y = $-CH_2CH_2NH_2$, Z = $-OH$, solvent: MSA). Szűcs, Vepsäläinen and his co-workers used 2 equivalents of phosphorus trichloride and 1 equivalent of phosphoric acid as the P-reagents (Table 2/Entries 1 and 2) [13,14]. Shinkai et al. applied phosphorus trichloride and phosphorous acid in ratio of 2.1:1 (Table 2/Entry 3) [15]. In the lack of purification, no purities were provided. The best result was achieved by Keglevich et al. who proved that the phosphorous acid did not participate in the reaction under discussion due its low nucleophilicity. For this, only 3.2 equivalents of phosphorus trichloride were measured in. The yield of pamidronic acid monosodium salt trihydrate was 57%, and the purity was 99% (Scheme 4) (Table 2/Entry 4) [16].

Table 2. Synthesis of pamidronate from β-alanine in MSA.

Entry	Reagent (Equiv.)	T (°C)	t (h)	Product Form	Purity (%)	Yield (%)	Ref.
1	PCl_3:H_3PO_4 (2:1)	65	16	Na	-	55 [1]	[13]
2	PCl_3:H_3PO_4 (2:1)	65	18	Na	-	22 [1]	[14]
3	PCl_3:H_3PO_3 (2.1:1)	65	16–20	Na	-	57 [1]	[15]
4	PCl_3 (3.2)	75	24	Na·$2H_2O$	99 [2]	57	[16]

[1] For the crude product; [2] On the basis of acid-base titration.

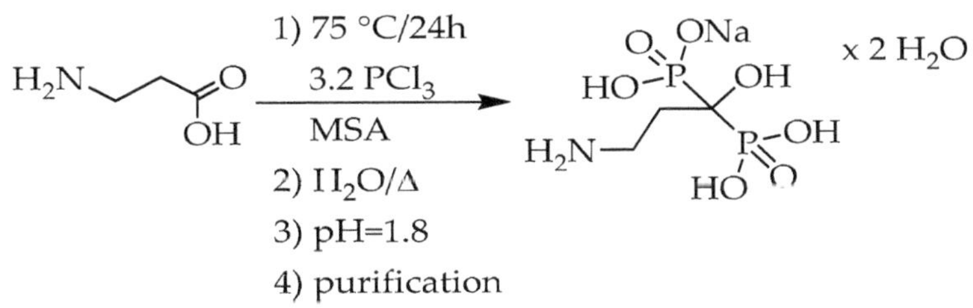

Scheme 4. Preparation of pamidronate in MSA by an optimized method [16].

The preparations of alendronic acid derivatives were also studied in details using MSA. In most cases γ-aminobutyric acid (GABA) was treated with the P-reagents (Table 3/Entries 1–9) [14,15,17–23]. In one instance, 2-pyrrolidone was the starting material (Table 3/Entry 10) [24].

Table 3. Synthesis of alendronic acid derivatives in MSA.

Entry	Starting Material	Reagent (Equiv.)	T (°C)	t (h)	Product form	Purity (%)	Yield (%)	Ref.
1	GABA	PCl_3:H_3PO_3 (2:1)	65	18	Na	-	91 [1]	[19]
2	GABA	PCl_3:H_3PO_3 (2.4:1.5)	65	20	Na·$3H_2O$	99.7 [2]	90 [1]	[23]
3	GABA	PCl_3:H_3PO_3 (2.1:1)	65	16–20	Na·$3H_2O$	99.7 [2]	89 [1]	[15]
4	GABA	PCl_3:H_3PO_4 (2:1)	65	18	acid	-	85 [1]	[14]
5	GABA	PCl_3:H_3PO_3 (2:1)	65	18	-	-	82 [1]	[18]
6	GABA	PCl_3:H_3PO_3 (2.1:1)	65	18	-	-	82 [1]	[20]
7	GABA	PCl_3 (3.2)	75	12	Na·$3H_2O$	93 [2]	58	[21]
8	GABA	$POCl_3$:H_3PO_3 (2:3 or 3:3)	75	12	Na·$3H_2O$	97 [2]	60	[22]
9	GABA	P_2O_5:H_3PO_3 (1:2)	70	24	Na·$3H_2O$	>99.9 [3]	48	[17]
10	2-Pyrrolidone	PCl_3 (3.4)	100–105	6	Na·$3H_2O$	-	81 [1]	[24]

[1] For the crude product; [2] On the basis of acid-base or complexometric titration; [3] On the basis of HPLC.

The majority of the researchers applied phosphorus trichloride and phosphorous acid as reagents in a ratio of ca. 2:1 (Scheme 1, Y = $-CH_2(CH_2)_2NH_2$, Z = $-OH$, solvent: MSA) (Table 3/Entries 1–3, 5 and 6) [15,18–20,23]. Yields around 80%–90% were claimed. As in the case of pamidronate, Vepsäläinen and his co-workers used 2 equivalents of phosphorus trichloride and 1 equivalent of phosphoric acid (Table 3/Entry 4) [14]. The best reproducible results were reported by Keglevich et al., furnishing alendronate in a yield of 58% (Scheme 5) (Table 3/Entry 7) [21]. It was clarified that phosphorous acid is unnecessary, when phosphorus trichloride is the P-reagent and MSA is the solvent.

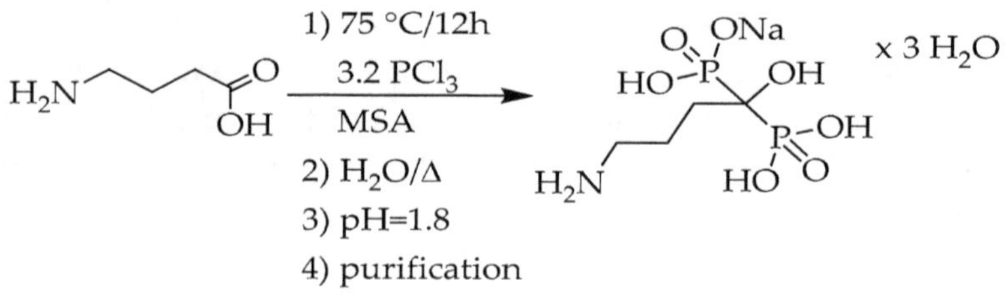

Scheme 5. Preparation of alendronate in MSA by an optimized method [21].

Typically, the yields were higher than 80% for the crude product (Table 3/Entries 1–6 and 10) [14,15,18–20,23,24]. In only two cases were the purities reported, however, the values of 99.7% seem to be excessive (Table 3/Entries 2 and 3) [15,23]. Our experience is that a crude product may never be pure, and such high values are not realistic.

In a special combination, phosphoryl chloride and phosphorous acid were used in a proportion of 2:3 or 3:3 in MSA (Table 3/Entry 8) [22]. In one case, 2 equivalents of phosphorous acid and 1 equivalent of phosphorus pentoxide were the P-reactants, and the target dronate was obtained in a yield of 48% (Table 3/Entry 9) [17]. Using 2-pyrrolidone as the starting material together with 3.4 equivalents of phosphorus trichloride, the crude alendronate was isolated in a yield of 81% (Table 3/Entry 10) [24].

The synthesis of ibandronate in MSA was discussed in only a few publications. In these preparations, *N*-Methyl-*N*-pentyl-β-alanine (MPA) was treated with phosphorus trichloride and phosphorous acid under different conditions (Scheme 1, Y = $-CH_2CH_2N(Me)(Pen)$, Z = −OH, solvent: MSA). Soni et al. applied 5 equivalents of phosphorus trichloride together with 2 equivalents of phosphorous acid. According to their description, pure alendronate was isolated in a yield of 86% (Table 4/Entry 1) [25]. The use of phosphorus trichloride is completely unnecessary. Applying 2.5 equivalents of phosphorus trichloride, and the same amount of phosphorous acid in the mixture of MSA and chlorobenzene, the yield was 60% (Table 4/Entry 2) [26]. The optimum case was, when 3.2 equivalents of phosphorus trichloride were used. The realistic and reproducible yield was 46% (Scheme 6) (Table 4/Entry 3) [21].

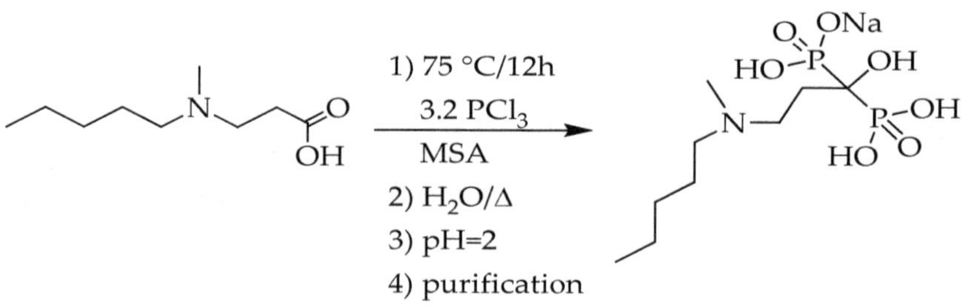

Scheme 6. Preparation of ibandronate in MSA by an optimized method [21].

Table 4. Synthesis of ibandronate from MPA in MSA.

Entry	Reagent (Equiv.)	T (°C)	t (h)	Product Form	Purity (%)	Yield (%)	Ref.
1	PCl_3:H_3PO_3 (5:2)	65–67	24	Na·H_2O	99.6 [1]	86	[25]
2	PCl_3:H_3PO_3 [2] (2.5:2)	96	3	Na·H_2O	99.8 [1]	60	[26]
3	PCl_3 (3.2)	75	12	Na	99 [3]	46	[21]

[1] On the basis of HPLC; [2] in the mixture of MSA and PhCl; [3] On the basis of acid-base titration.

For risedronic acid derivatives, yields of 38%–74% were reported. In the basic syntheses, 3-pyridylacetic acid (PAA) was reacted with phosphorus trichloride and phosphorous acid in MSA (Scheme 1, Y = 3-pyridylmethylene-, Z = −OH, solvent: MSA).

A yield of 73% was obtained, when 3-3 equivalents of phosphorus trichloride and phosphorous acid were used in the mixture of MSA and diethyl carbonate (Table 5/Entry 1) [27]. Another combination, a 2.1:1 ratio of the P-reagents led to the dronic acid in a yield of 38%. No purity was reported (Table 5/Entry 2) [15]. When the mixture of 3 equivalents of phosphorous acid and 2 equivalents of phosphoryl chloride, or 2 equivalents of phosphorous acid and 1 equivalent of phosphorus pentoxide were reacted with the corresponding carboxylic acid, yields of ca. 55% were reported (Table 5/Entries 3 and 4) [17,22]. The application of phosphorus trichloride (3.1 equivalents) alone led to the highest yield (74%), and the unnecessity of phosphorous acid was again proved (Scheme 7) (Table 5/Entry 5) [28,29].

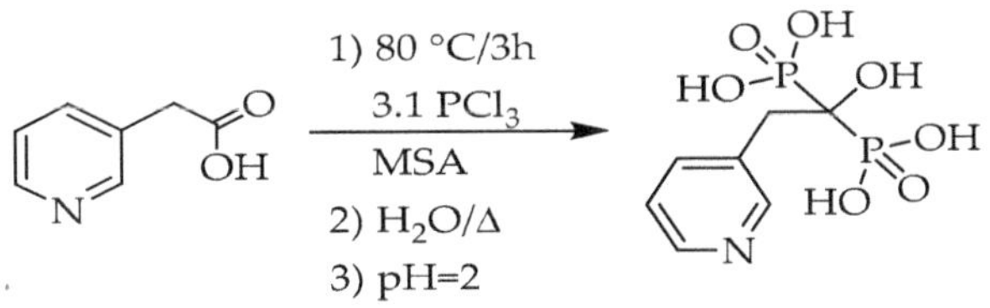

Scheme 7. Preparation of risedronic acid in MSA by an optimized method [28,29].

Table 5. Synthesis of risedronic acid derivatives from PAA in MSA.

Entry	Reagent (Equiv.)	T (°C)	t (h)	Product Form	Purity (%)	Yield (%)	Ref.
1	PCl_3:H_3PO_3 [1] (3:3)	70–72	5	Na·2.5H_2O	99.9 [2]	73	[27]
2	PCl_3:H_3PO_3 (2.1:1)	65	16–20	acid	-	38 [3]	[15]
3	$POCl_3$:H_3PO_3 (2:3)	75	12	acid	98 [4]	55	[22]
4	P_2O_5:H_3PO_3 (1:2)	80–100	72	acid·H_2O	99.9 [2]	54	[17]
5	PCl_3 (3.1)	75	12	acid	92 [4] (100 [2])	74 [3]	[28,29]

[1] in the mixture of MSA and diethyl carbonate; [2] On the basis of HPLC; [3] For the crude product; [4] On the basis of acid-base titration.

During the preparation of zoledronic acid in MSA, imidazol-1-yl-acetic acid (IAA) was treated with only phosphorus trichloride, or with the mixture of this reagent and phosphorous acid (Scheme 1, Y = 1-imidazolylmethylene-, Z = –OH, solvent: MSA). Kieczykowski and co-workers applied phosphorus trichloride and phosphorous acid in a quantity of 2.1:1 equivalents. The yield of the target dronic acid was 31% (Table 6/Entry 1) [15]. Using phosphorus trichloride as the reactant in amounts of 3.1 or 3.2, yields of 53% and 46%, respectively, were reported by the Keglevich group in comparable purity (Scheme 8) (Table 6/Entries 2 and 3) [28,29].

Table 6. Synthesis of zoledronic acid from IAA in MSA.

Entry	Reagent (Equiv.)	T (°C)	t (h)	Product Form	Purity (%)	Yield (%)	Ref.
1	PCl_3:H_3PO_3 (2.1:1)	65	16–20	acid	-	31 [1]	[15]
2	PCl_3 (3.1)	80	3	acid	99 [2]	53	[29]
3	PCl_3 (3.2)	80	3	acid	98 [2]	46	[28]

[1] For the crude product; [2] On the basis of acid-base titration.

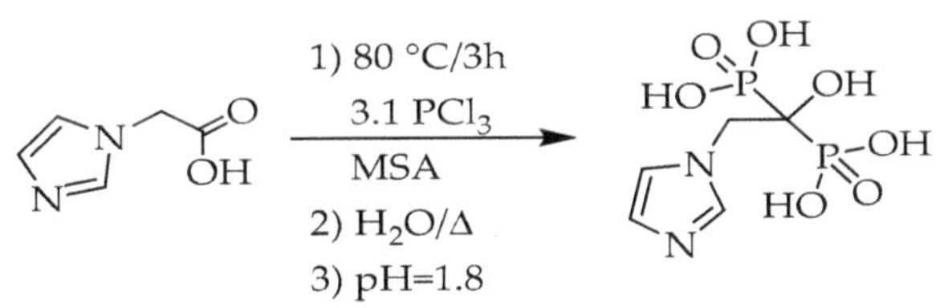

Scheme 8. Preparation of zoledronic acid in MSA by an optimized method [28,29].

In overall, it can be said that the lower yields of the Keglevich group are better, than the higher ones published for the crude products neglecting the purity. The reliable yields were obtained after purifying the products and analyzing the purity by potentiometric titration.

3. Synthesis of Dronic Acid Derivatives in Chlorobenzene

The second frequently used solvent is chlorobenzene. Chlorobenzene has a few advantages in comparison with MSA. It has a lower viscosity, can be removed by distillation, and can be reused. The real problem with MSA is that during the pH adjustment sodium methanesulfonate may be formed from MSA, which is difficult to remove. In the procedures reported, the corresponding carboxylic acids were reacted with phosphorus trichloride and phosphorous acid/phosphoric acid. None of the reactions were performed with solely phosphorus trichloride, as, in this case, it is unreactive without another P-containing partner (Scheme 9). In most reports no data on the purity of dronic acid derivatives was provided, for this, the yields are not reliable.

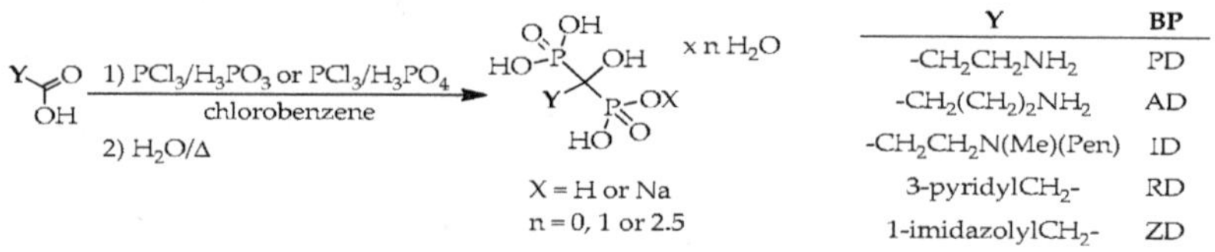

Scheme 9. General scheme for the preparation of dronic acid derivatives in chlorobenzene.

Reacting β-alanine or GABA with 1.5 equivalents of phosphorus trichloride and 1.5 equivalents of phosphorous acid at 100 °C for 3 h, the yields were in the range of 45%–59% (Table 7/Entries 1 and 3–5) [30–32]. Starting from β-alanine or GABA, and measuring in phosphorus trichloride and phosphoric acid in molar equivalents of 2.4:1.5 or 2:3, respectively, the pamidronic acid or alendronic acid was obtained in a yield of 53% and 18%, respectively (Table 7/Entries 2 and 6) [33,34].

Table 7. Synthesis of pamidronic acid and alendronic acid in chlorobenzene.

Entry	BP	Starting Material	Reagent (Equiv.)	T (°C)	t (h)	Product Form	Purity (%)	Yield (%)	Ref.
1	PD	β-alanine	PCl_3:H_3PO_3 (1.5:1.5)	100	3	acid	-	59 [1]	[30]
2	PD	β-alanine	PCl_3:H_3PO_4 (2.4:1.5)	105–110	2	acid	-	53	[33]
3	PD	β-alanine	PCl_3:H_3PO_3 (1.5:1.5)	100	3	acid	-	51	[31]
4	AD	GABA	PCl_3:H_3PO_3 (1.5:1.5)	100	3	acid	-	46	[31]
5	AD	GABA	PCl_3:H_3PO_3 (1.5:1.5)	100	3	acid	-	45	[32]
6	AD	GABA	PCl_3:H_3PO_4 (2:3)	100–110	3	acid	-	18 [1]	[34]

[1] For the crude product.

In two variations for the syntheses of ibandronate, a yield of 60% was reached. Phosphorus trichloride and phosphorous acid were applied in a 1.5:1.5 or 2.5:2 molar equivalent quantity (in a mixture of MSA and chlorobenzene) (Table 8/Entries 1 and 2) [26,35]. Using phosphorus trichloride and phosphoric acid in amounts of 3:3 and 2:3, the yield was 37% and 6%, respectively (Table 8/Entries 3 and 4) [34,36]. In one instance, the proportion of phosphorous acid was not provided, and a peripheral result (a yield of 3%) was reported (Table 8/Entry 5) [26].

Table 8. Synthesis of ibandronic acid derivatives from MPA in chlorobenzene.

Entry	Reagent (Equiv.)	T (°C)	t (h)	Product Form	Purity (%)	Yield (%)	Ref.
1	PCl_3:H_3PO_3 (1.5:1.5)	80	4	Na·H_2O	-	60	[35]
2	PCl_3:H_3PO_3 [1] (2.5:2)	96	3	Na·H_2O	99.8 [2]	60	[26]
3	PCl_3:H_3PO_4 (3:3)	110	11	Na	-	37 [3]	[36]
4	PCl_3:H_3PO_4 (2:3)	100–110	3	acid	-	6 [1]	[34]
5	PCl_3 [4]:H_3PO_3	85–90	4–5	acid	95 [2]	3	[26]

[1] in the mixture of MSA and PhCl; [2] On the basis of HPLC; [3] For the crude product; [4] Two and a half equivalents.

The preparation of risedronic acid derivatives in chlorobenzene as the solvent was in all reactions implemented using phosphorus trichloride and phosphorous acid. An unrealistically high yield of 96% was claimed for risedronic acid using 3.5 equivalents of phosphorus trichloride, and the same amounts of phosphorous acid without specifying the purity (Table 9/Entry 1) [37]. Applying the P-reagents in a quantity of 1.5 equivalents, the yield was 68% (Table 9/Entry 2) [35]. The most reliable result was described by Wadhwa, but the reaction time was not given. The reactants were used in amount of 2.4 equivalents, and the risedronic acid was obtained in a yield of 58% with a purity of 84% (Table 9/Entry 3) [38]. Applying phosphorus trichloride and phosphorous acid in a molar ratio of 3:3.5, the dronate was isolated in a yield of 52%. However, no temperature was reported. A similar yield (52%) was described, when 2 equivalents of phosphorus trichloride was used with phosphorous acid. However, the quantity of the latter species was not provided (Table 9/Entries 4 and 5) [39,40].

Table 9. Synthesis of risedronic acid derivatives from PAA in chlorobenzene.

Entry	Reagent (Equiv.)	T (°C)	t (h)	Product Form	Purity (%)	Yield (%)	Ref.
1	PCl_3:H_3PO_3 (3.5:3.5)	90–95	2.5	acid	-	96 [1]	[37]
2	PCl_3:H_3PO_3 (1.5:1.5)	80	4	acid	-	68 [1]	[35]
3	PCl_3:H_3PO_3 (2.4:2.4)	85–90	-	acid	84 [2]	58 [1]	[38]
4	PCl_3:H_3PO_3 (3:3.5)	-	2.5	Na·2.5H_2O	-	52	[39]
5	PCl_3 [3]:H_3PO_3	100	6	acid	-	52 [1]	[40]

[1] For the crude product; [2] On the basis of HPLC; [3] Two equivalents.

For the synthesis of zoledronic acid in chlorobenzene, the purity remains unclear in all cases, therefore, the yields are not reliable at all. Carrying out the synthesis with 3.7 equivalents of phosphorus trichloride and with the same amount of phosphorous acid, quantitative yield was claimed that is entirely unrealistic (Table 10/Entry 1) [41]. In other instances, phosphorus trichloride and phosphoric acid were applied in different, such as 4.7:2.9, 2:3 and 3:1.9 ratios to lead to yields of 79%, 67% and 41%, respectively (Table 10/Entries 2–4) [34,42,43].

The need for the joint application of phosphorus trichloride and phosphorous acid will be explained in the next sub-chapter.

Table 10. Synthesis of zoledronic acid from IAA in chlorobenzene.

Entry	Reagent (Equiv.)	T (°C)	t (h)	Product Form	Purity (%)	Yield (%)	Ref.
1	PCl_3:H_3PO_3 (3.7:3.7)	100	1	acid	-	100 [1]	[41]
2	PCl_3:H_3PO_4 (4.7:2.9)	80	2–2.5	acid	-	79 [1]	[42]
3	PCl_3:H_3PO_4 (2:3)	100–110	3	acid	-	67 [1]	[34]
4	PCl_3:H_3PO_4 (3:1.9)	100	3	acid	-	41	[43]

[1] For the crude product.

4. Methods for the Preparation of Dronic Acid Derivatives in Sulfolane

Sulfolane is also a preferred solvent in the preparation of hydroxymethylenebisphosphonic acid derivatives, but less data were published. In all such cases, phosphorus trichloride and phosphorous acid were the P-reagents (Scheme 10).

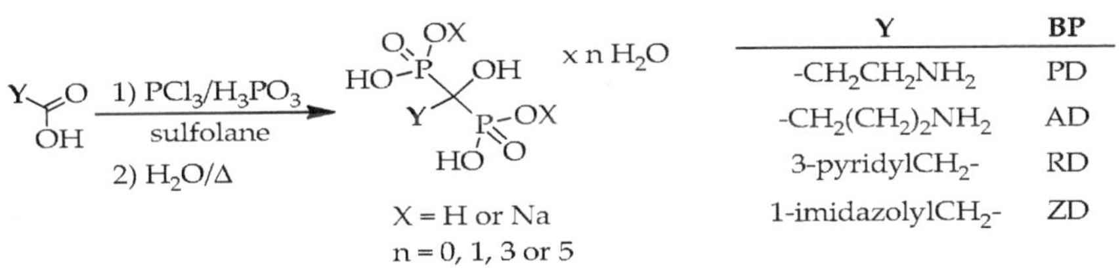

Scheme 10. General scheme for the preparation of dronic acid derivatives in sulfolane.

Approximately in half of the cases, the purities were not described, therefore the data may be misleading. The synthesis of a part of the dronic acid derivatives was attempted also under MW conditions. The synthesis of ibandronic acid derivatives was not attempted in sulfolane.

Applying phosphorus trichloride and phosphorous acid in a molar ratio of 3:3, under conventional and MW-assisted conditions at 65 °C for 3.5 h and 3 min, respectively, yields of 72% and 64%, respectively, were reported for crude pamidronic acid derivatives (Table 11/Entries 1 and 2) [44]. (The forms of product were not provided.) Alendronate was obtained under similar conditions in yields of 38% and 41%, respectively (the forms of product were not provided) (Table 11/Entries 7 and 8) [44]. It can be seen that MW irradiation was useful in respect of reaction time in these reactions. Keglevich et al. used 2-2 equivalents of the P-reagents in reaction with β-alanine to afford pure pamidronic acid in a yield of 63% (Table 11/Entry 3) [16]. When phosphorus trichloride and phosphorous acid were used in a ratio of 3.4:1.5, pure sodium pamidronate was obtained in a lower yield (Table 11/Entry 4) [45]. Applying phosphorus trichloride and phosphorous acid in ratios of 3.4:1.5 and 2.5:3.5, the yield of sodium alendronate was 69% (crude product), and 55% (pure dronate), respectively (Table 11/Entries 5 and 6) [45,46].

Table 11. Synthesis of pamidronic and alendronic acid derivatives in sulfolane.

Entry	BP	Starting Material	Reagent (Equiv.)	T (°C)	t (h)	Product Form	Purity (%)	Yield (%)	Ref.
1	PD	β-alanine	PCl_3:H_3PO_3 (3:3)	65	3.5	-	-	72 [1]	[44]
2	PD	β-alanine	PCl_3:H_3PO_3 (3:3)	65 [2]	0.05	-	-	64 [1]	[44]
3	PD	β-alanine	PCl_3:H_3PO_3 (2:2)	75	12	acid	100 [3]	63	[16]
4	PD	β-alanine	PCl_3:H_3PO_3 (3.4:1.5)	63–67	3	$2Na{\cdot}5H_2O$	>99	48	[45]
5	AD	GABA	PCl_3:H_3PO_3 (3.4:1.5)	63–67	3	$Na{\cdot}3H_2O$	>99	69 [1]	[45]
6	AD	GABA	PCl_3:H_3PO_3 (2.5:3.5)	60–65	12	$Na{\cdot}3H_2O$	>99 [4]	55	[46]
7	AD	GABA	PCl_3:H_3PO_3 (3:3)	65 [2]	0.1	-	-	41 [1]	[44]
8	AD	GABA	PCl_3:H_3PO_3 (3:3)	65	3.5	-	-	38 [1]	[44]

[1] For the crude product; [2] MW heating; [3] On the basis of acid-base titration; [4] On the basis of HPLC.

McKenna and his co-workers investigated the synthesis of risedronic and zoledronic acid derivatives (the forms of product were not provided) also under MW conditions, using 3 equivalents of both the phosphorus trichloride and phosphorous acid at 65 °C for 3 min. The dronic acid derivatives, risedronic acid and zoledronic acid derivatives were obtained in yields of 74% and 70%, respectively (Table 12/Entries 1 and 3) [44]. In case of conventional heating, the yield of zoledronic acid derivative was 67% (Table 12/Entry 4) [44]. The reaction of 1 equivalent of PAA or IAA with 3.4 equivalents of phosphorus trichloride and 1.5 equivalents of phosphorous acid on conventional heating at around 65 °C provided risedronate and zoledronic acid in a yield of 54% and 71%, respectively (Table 12/Entries 2 and 5) [45]. No attempts were made to prepare pure zoledronic acid derivatives.

Table 12. Synthesis of risedronic and zoledronic acid derivatives in sulfolane.

Entry	BP	Starting Material	Reagent (Equiv.)	T (°C)	t (h)	Product Form	Purity (%)	Yield (%)	Ref.
1	RD	PAA	PCl_3:H_3PO_3 (3:3)	65 [1]	0.05	$acid{\cdot}H_2O$	-	74 [2]	[44]
2	RD	PAA	PCl_3:H_3PO_3 (3.4:1.5)	63–67	3	Na	>99	54	[45]
3	ZD	IAA	PCl_3:H_3PO_3 (3:3)	65 [1]	0.05	-	-	70 [2]	[44]
4	ZD	IAA	PCl_3:H_3PO_3 (3:3)	65	3.5	-	-	67 [2]	[44]
5	ZD	IAA	PCl_3:H_3PO_3 (3.4:1.5)	63–67	3	$acid{\cdot}H_2O$	>99	71 [2]	[45]

[1] Under MW irradiation; [2] For the crude product.

It was shown in Chapter 2 that using MSA as the solvent, there was no need to use phosphorous acid together with phosphorus trichloride (Tables 2–6) [16,21,28,29]. However, using a solvent other than MSA, it is necessary to apply phosphorus trichloride and phosphorous acid jointly. A logical explanation may be that, before phosphorus trichloride would react with the carboxylic acid or its derivatives, it reacts with 1 or 2 equivalents of phosphorous acid resulting in the formation of activated P-species, such as $(HO)_2P\text{-}O\text{-}PCl_2$ and/or $(HO)_2P\text{-}O\text{-}PCl\text{-}O\text{-}P(OH)_2$ (Scheme 11). In the next step, the reactive intermediates formed may react with the substituted acetic acid, providing the target dronic acid derivative after several steps [16].

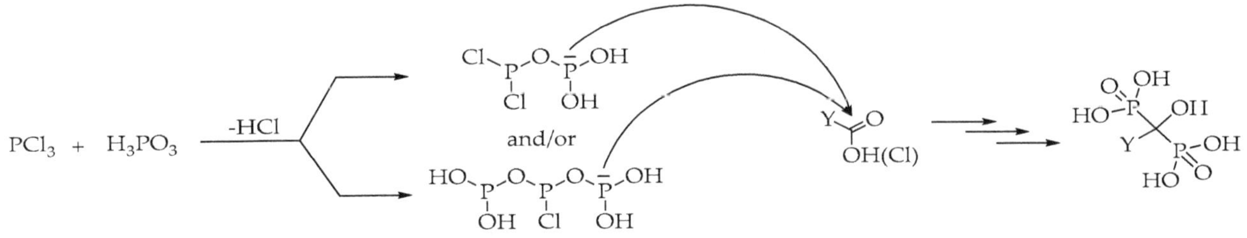

Scheme 11. Formation of the reactive P-species by the condensation of phosphorus trichloride and phosphorous acid and their reaction with the substituted acetic acids.

5. Synthesis of Dronic Acid Derivatives in Toluene

The preparations of alendronic acid derivatives, ibandronate and risedronic acid have also been investigated in toluene (Scheme 12).

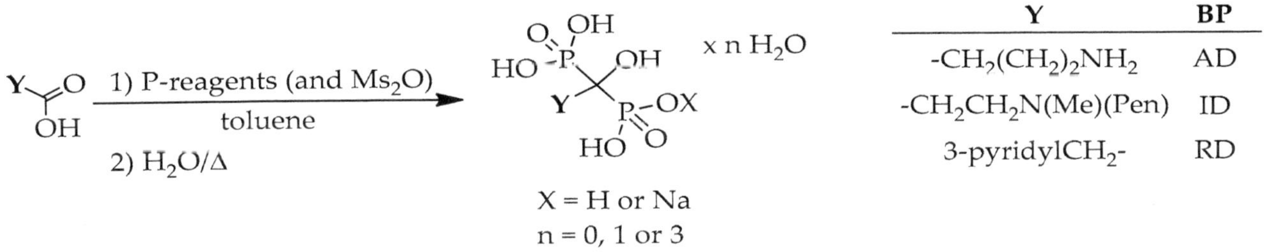

Scheme 12. General scheme for the preparation of dronic acid derivatives in toluene.

In one approach towards alendronic acid, a mixture of toluene and PEG 400 was used as the solvent, and 1.6 equivalents of phosphorus trichloride along with 1 equivalent of phosphorous acid as the P-reagents. The yield was 56% for crude alendronic acid (Table 13/Entry 1) [47]. Using phosphorus trichloride and methanesulfonic anhydride (Ms_2O) in a molar ratio of 3:3 in toluene, a yield of 66% was reported (Table 13/Entry 2) [48]. The application of 3.7 equivalents of phosphorus trichloride and 4.2 equivalents of phosphorous acid led to a moderate yield (44%) of sodium ibandronate (Table 13/Entry 3) [49].

Table 13. Synthesis of alendronic acid derivatives and ibandronate in toluene.

Entry	BP	Starting Material	Reagent (Equiv.)	T (°C)	t (h)	Product Form	Purity (%)	Yield (%)	Ref.
1	AD	GABA	PCl_3:H_3PO_3 [1] (1.6:1)	75	4	acid		56 [2]	[47]
2	AD	GABA	PCl_3:Ms_2O (3:3)	70	9.5	Na·3H_2O	-	66	[48]
3	ID	MPA	PCl_3:H_3PO_3 (3.7:4.2)	80–85	7–8	Na	-	44	[49]

[1] in the mixture of toluene and PEG 400; [2] For the crude product.

Regarding the synthesis of risedronic acid, in most instances unusual reagents for example, propylphosphonic anhydride (T3P), Ms_2O or PCl_5 were used together with phosphorous acid, and purities of >90% were reported according to HPLC (Table 14/Entries 1–6). Unrealistically high yields of 90% and 81% were reported, when 3.1 equivalents phosphorous acid were reacted with 2.5 equivalents of T3P or 3.1 equivalents of phosphoryl chloride (Table 14/Entries 1 and 2) [50]. In similar reactions, with ca. 3 equivalents of reactants (phosphorous acid and Ms_2O) a lower yield of 34% was obtained (Table 14/Entry 3) [50]. Other interesting combinations, such as a 3.1:2:1 ratio of H_3PO_3:Ms_2O:$POCl_3$, or a 3.1:1:1 ratio of H_3PO_3:Ms_2O:PCl_5 afforded the risedronic acid in yields of ca. 60% (Table 14/Entries 4 and 5) [50]. The application of 3.2 equivalents of phosphorus trichloride alone led to the lowest yield

(15%) (Table 14/Entry 6) [12]. It seems to have been confirmed that phosphorus trichloride alone is not enough if not MSA is the solvent. In the cases discussed, it was necessary to apply an activating agent.

Table 14. Synthesis of risedronic acid from PAA in toluene

Entry	Reagent (Equiv.)	T (°C)	t (h)	Product Form	Purity (%)	Yield (%)	Ref.
1	H_3PO_3:T3P (3.1:2.5)	95	20.5	acid·H_2O	99.6 [1]	90 [2]	[50]
2	H_3PO_3:$POCl_3$ (3.1:3.1)	95	20.5	acid·H_2O	98.5 [1]	81 [2]	[50]
3	H_3PO_3:Ms_2O (3.1:3)	95	20.5	acid·H_2O	95.9 [1]	34 [2]	[50]
4	H_3PO_3:Ms_2O:$POCl_3$ (3.1:2:1)	95	20.5	acid·H_2O	99.2 [1]	62 [2]	[50]
5	H_3PO_3:Ms_2O:PCl_5 (3.1:1:1)	95	20.5	acid·H_2O	98.3 [1]	58 [2]	[50]
6	PCl_3 (3.2)	110	6	acid	90.1 [3]	15 [2]	[12]

[1] On the basis of HPLC; [2] For the crude product; [3] On the basis of acid-base titration.

6. Reactions in the Absence of Solvent

In many publications, the hydroxymethylenebisphosphonic acid derivatives were prepared in the absence of any solvent. These reactions are more heterogeneous, and in most cases they cannot be stirred, due to the (almost) solid consistency. Although the yields reported are not wrong, the purities were provided only in a few instances. For this, the conclusions from the yields may be misleading. It can be seen that beside phosphorus trichloride or phosphoryl chloride, there was need for another reactant as well, for example for phosphorous acid or Ms_2O.

Reacting β-alanine with an unnecessarily large amount of phosphorus trichloride (5 equivalents) and phosphorous acid (3 equivalents) under MW conditions, sodium pamidronate was isolated in a yield of 67% (Table 15/Entry 1) [51]. A rather similar yield (61%) was reported applying the above P-reagents in a 2:1.5 equivalents quantity (Table 15/Entry 2) [52]. Keglevich et al. obtained pure pamidronic acid in a yield of 44%, when both phosphorus trichloride and phosphorous acid were used in a quantity of 2 equivalents [53]. According to our experiences, in most cases it is not possible to prepare the hydroxymethylenebisphosphonic acid derivatives in the absence of solvents.

Table 15. Synthesis of pamidronic acid derivatives from β-alanine in neat.

Entry	Reagent (Equiv.)	T (°C)	t (h)	Product Form	Purity (%)	Yield (%)	Ref.
1	PCl_3:H_3PO_3 (5:3)	80 [1]	0.05	Na	-	67 [2]	[51]
2	PCl_3:H_3PO_3 (2:1.5)	100	4	acid	-	61 [2]	[52]
3	PCl_3:H_3PO_3 (2:2)	75	3	acid	>99 [3]	44	[53]

[1] Under MW conditions; [2] For the crude product; [3] On the basis of acid-base titration.

Alendronic acid derivatives were prepared starting from GABA, *N*-phthalimido-GABA, or *N*-phthalimido-GABA-chloride, but the purities were not reported at all. Using GABA as the starting material, and phosphorus trichloride along with Ms_2O or phosphorous acid as the other reagent in a molar ratio of 3:3 (PCl_3:Ms_2O) and 5:3 (PCl_3:H_3PO_3 under MW conditions), or 2:1.5 (PCl_3:H_3PO_3), the yields were 78%, 78% and 59%, respectively (Table 16/Entries 1–3) [48,51,52]. Starting from *N*-phthalimido-GABA and applying phosphorus trichloride along with phosphorous acid, or this combined with phosphoric acid, crude alendronic acid was obtained in yields of 38% and 57% (Table 16/Entries 4–5) [54]. In one instance, *N*-phthalimido-GABA-chloride was reacted with 2 equivalents of phosphorous acid alone, leading to the target dronic acid in a poor yield of 19% (Table 16/Entry 6) [54]. According to our earlier results [16,21,29,53], phosphorous acid alone is completely unreactive in the synthesis of hydroxymethylenebisphosphonic acid derivatives. Therefore, even the low yield (19%) reported is questionable (Table 16/Entry 6) [54].

Table 16. Synthesis of alendronic acid derivatives in the absence of solvents.

Entry	Starting Material	Reagent (Equiv.)	T (°C)	t (h)	Product Form	Purity (%)	Yield (%)	Ref.
1	GABA	PCl_3:Ms_2O (3:3)	70–100	7	Na·3H_2O	-	78	[48]
2	GABA	PCl_3:H_3PO_3 (5:3)	80 [1]	0.05	Na	-	78 [2]	[51]
3	GABA	PCl_3:H_3PO_3 (2:1.5)	100	4	acid	-	59 [2]	[52]
4	*N*-phthalimido-GABA	PCl_3:H_3PO_3 (2:5)	80	3	acid	-	38 [2]	[54]
5	*N*-phthalimido-GABA	PCl_3:H_3PO_3:H_3PO_4 (2:1.5:4)	80	3	acid	-	57 [2]	[54]
6	*N*-phthalimido-GABA-chloride	H_3PO_3 (2)	130	4	acid	-	19 [2]	[54]

[1] Under MW conditions; [2] For the crude product.

The application of 2.9 equivalents of phosphorus trichloride and 1.5 equivalents of phosphorous acid under somewhat different conditions (at 60–65 °C for 1.5 h or at 70–75 °C for 20 h) led to ibandronate in yields of 89% and 82% (Table 17/Entries 1 and 2) [55,56]. A similar result (82%) was reported using a larger excess of the P-reagents (4.1 equivalents of phosphorus trichloride and 2.5 equivalents of phosphorous acid). However, no criterions of purity were provided (Table 17/Entry 3) [57]. Measuring in, even more reagents (5 equivalents of phosphorus trichloride and 3 equivalents of phosphorous acid), and performing the reaction under MW conditions, a somewhat lower yield (72%) was obtained (Table 17/Entry 4) [51]. With a not defined quantity of phosphorus trichloride and 1.3 equivalents of phosphorous acid, sodium ibandronate was prepared in a yield of 68% (Table 17/Entry 5) [58]. The lowest yield of 59% was observed using 3.2 equivalents of phosphoryl chloride and a large excess (9.7 equivalents) of phosphorous acid (Table 17/Entry 6) [59]. The use of this large excess is completely needless, even makes the purification more difficult.

Table 17. Synthesis of ibandronic acid derivatives from MPA in the absence of any solvent.

Entry	Reagent (Equiv.)	T (°C)	t (h)	Product Form	Purity (%)	Yield (%)	Ref.
1	PCl_3:H_3PO_3 (2.9:1.5)	60–65	1.5	Na	-	89	[55]
2	PCl_3:H_3PO_3 (2.9:1.5)	70–75	20	Na·H_2O	-	82	[56]
3	PCl_3:H_3PO_3 (4.1:2.5)	75	5	Na	-	82 [1]	[57]
4	PCl_3:H_3PO_3 (5:3)	80 [2]	0.05	Na	-	72 [1]	[51]
5	PCl_3:H_3PO_3 [3]	75	6	Na·H_2O	-	68 [1]	[58]
6	$POCl_3$:H_3PO_3 (3.2:9.7)	60–70	24	acid	-	59 [1]	[59]

[1] For the crude product; [2] Under MW conditions; [3] 1.3 equivalents of phosphorous acid were used.

Risedronic acid derivatives were synthesized from PAA. Using phosphorus trichloride and phosphorous acid in a molar ratio of 5:3, the yield was 86% (Table 18/Entry 1) [51]. In one case, morpholine was also added to the reaction mixture beside the P-reagents, and the crude risedronic acid was prepared in a yield of 78% (Table 18/Entry 2) [60]. Yields around 66% were claimed using ca. 2 equivalents of both phosphorus trichloride and phosphorous acid at a temperature of 70–100 °C (Table 18/Entries 3 and 4) [52,61]. Carrying out the reaction applying 3.2 equivalents of phosphoryl chloride and a large excess (9.7 equivalents) of phosphorous acid, the yield was 60% (Table 18/Entry 5) [59].

Table 18. Synthesis of risedronic acid derivatives from PAA in the absence of solvent.

Entry	Reagent (Equiv.)	T (°C)	t (h)	Product Form	Purity (%)	Yield (%)	Ref.
1	PCl_3:H_3PO_3 (5:3)	80 [1]	0.05	Na	-	86 [2]	[51]
2	PCl_3:H_3PO_3:morpholine (2:5:1)	68–75	2.5–4	acid	-	78 [2]	[60]
3	PCl_3:H_3PO_3 (2:1.5)	100	4	acid	-	67 [2]	[52]
4	PCl_3:H_3PO_3 (2.5:2.2)	70–75	8	Na·2.5H_2O	99.9 [3]	65	[61]
5	$POCl_3$:H_3PO_3 (3.2:9.7)	60–70	24	acid	-	60 [2]	[59]

[1] Under MW conditions; [2] For the crude product; [3] On the basis of HPLC.

Reacting IAA, phosphorus trichloride and phosphorous acid in a molar ratio of 1:5:3 under MW conditions, or in a ratio of 1:3:5, zoledronate and zoledronic acid were obtained in a yield of 80% and 61%, respectively. The first yield was related on crude product (Table 19/Entries 1–2) [51,62]. Using 3 equivalents of phosphoryl chloride and 5 equivalents of phosphorous acid, the yield was 79% for the crude, but pure zoledronic acid (Table 19/Entry 3) [62]. The application of phosphoryl chloride and phosphorous acid measured in an excessive quantity (reported also for ibandronic and risedronic acid) led to a yield of 62% (Table 19/Entry 4) [59].

Table 19. Synthesis of zoledronic acid derivatives from IAA under solvent-free conditions.

Entry	Reagent (Equiv.)	T (°C)	t (h)	Product Form	Purity (%)	Yield (%)	Ref.
1	PCl_3:H_3PO_3 (5:3)	80 [1]	0.05	Na	-	80 [2]	[51]
2	PCl_3:H_3PO_3 (3:5)	75	6	acid	99.9 [3]	61	[62]
3	$POCl_3$:H_3PO_3 (3:5)	75	5	acid	99.9 [3]	79 [2]	[62]
4	$POCl_3$:H_3PO_3 (3.2:9.7)	60–70	24	acid	-	62 [2]	[59]

[1] Under MW conditions; [2] For the crude product; [3] On the basis of HPLC.

7. Methods in ILs

The ILs are considered green solvents, because of their low vapor pressure, high thermal stability, and, as they can be recycled and reused. Although their use in organic syntheses is spreading, hydroxymethylenebisphosphonic acid derivatives were prepared in ILs only in few cases (Table 20) (Scheme 13).

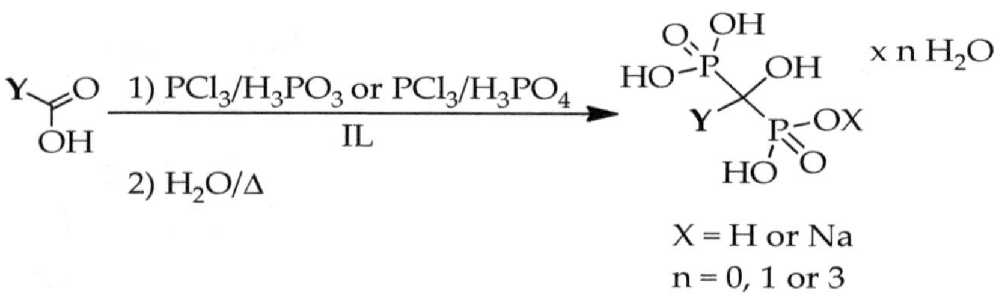

Y	BP
$-CH_2CH_2NH_2$	PD
$-CH_2(CH_2)_2NH_2$	AD
3-pyridylCH_2-	RD
1-imidazolylCH_2-	ZD

Scheme 13. General scheme for the preparation of dronic acid derivatives in ILs.

De Ferra and co-workers investigated the preparation of pamidronic and alendronic acid derivatives (in two cases, the form of the products was not provided), as well as risedronic and zoledronic acid in tributylmethylammonium chloride ([Bu_3NH][Cl]). Reacting the mixture of 2 equivalents of phosphorus trichloride and 1 equivalent of phosphorous acid with the corresponding carboxylic acid, low yields of the respective dronic acid derivatives were reported (12%–31%) (Table 20/Entries 2 and 4–6) [63]. The Keglevich group reached better results, when synthesizing pamidronic acid in the presence of 1-butyl-3-methylimidazolium hexafluorophosphate ([bmim][PF_6]). The IL was not used a solvent, only as an additive. Reacting β-alanine with 2:2 or 3:2 equivalents of phosphorus trichloride and phosphorous acid, the optimum amount of the IL was found to be around 0.3–0.6 equivalents. Under such conditions, record yields of 70%–72% were reached (Table 20/Entry 1) [53]. GABA was reacted with 2 equivalents of phosphorus trichloride and 1.5 equivalents of phosphorous acid, in the presence of 0.5 equivalents of various ILs at 60 °C for about 6 h. Depending on the ILs used, the yield of pure alendronate was 92%–94% (Table 20/Entry 3) [64]. The same research group also prepared zoledronic acid in different ILs. For crude zoledronic acid, yields of 90%–92% were reported. In the basic syntheses, IAA was reacted with 2.5 equivalents of phosphorus trichloride and 1.7 equivalents of phosphoric acid in 1.6 equivalents of ILs (Table 20/Entry 7). Carrying out the reaction with the same amount of P-reactants but with [bmim][BF_4], the yield was 60% for sodium zoledronate (Table 20/Entry 8) [65].

Table 20. Synthesis of hydroxymethylenebisphosphonic acid derivatives in the presence of ILs.

Entry	BP	Ils (Equiv.)	Starting Material	Reagent (Equiv.)	T (°C)	t (h)	Product Form	Purity (%)	Yield (%)	Ref.
1	PD	[bmim][PF_6] (0.3 or 0.6)	β-alanine	PCl_3:H_3PO_3 (2:2 or 3:2)	75	3	acid	>99 [1]	70–72	[53]
2	PD	[Bu_3NH][Cl]	β-alanine	PCl_3:H_3PO_3 (2:1)	60	2	-	-	26	[63]
3	AD	various Ils [2] (0.5)	GABA	PCl_3:H_3PO_3 (2:1.5)	60	6	Na·$3H_2O$	>99 [3]	92–94	[64]
4	AD	[Bu_3NH][Cl]	GABA	PCl_3:H_3PO_3 (2:1)	60	2	-	-	31	[63]
5	RD	[Bu_3NH][Cl]	PAA	PCl_3:H_3PO_3 (2:1)	80	2	acid	-	12	[63]
6	ZD	[Bu_3NH][Cl]	IAA	PCl_3:H_3PO_3 (2:1)	80	2	acid	-	26	[63]
7	ZD	various Ils [4] (1.6)	IAA	PCl_3:H_3PO_4 (2.5:1.7)	65	4	acid	-	90–92 [5]	[65]
8	ZD	[bmim][BF_4] (1.6)	IAA	PCl_3:H_3PO_4 (2.5:1.7)	65	4	Na·H_2O	99.8 [3]	60	[65]

[1] On the basis of acid-base titration; [2] [bmim][BF_4], [bmim][PF_6], 1-hydroxyethyl-2,3-dimethyl-imidazolium chloride (LOH), 1-propyl-3-carbonitrile-imidazolium chloride (LCN), 1-carboxyethyl-3-methyl imidazolium chloride (LOOH); [3] On the basis of HPLC; [4] *N*-ethylpyridine tetrafluoroborate [EPy][BF_4], [bmim][PF_6], LOH, LCN, LOOH; [5] For the crude product.

8. Processes for the Preparation of Hydroxymethylenebisphosphonic Acid Derivatives in Other Solvents

In a few instances, the synthesis of hydroxymethylenebisphosphonic acid derivatives was carried out in other solvents, from among *p*-cresol, acetonitrile and *n*-octane should be mentioned. In these cases again the corresponding carboxylic acids were reacted with phosphorus trichloride and phosphorous acid. In the first series let us see the examples carried out in *p*-cresol (Scheme 14).

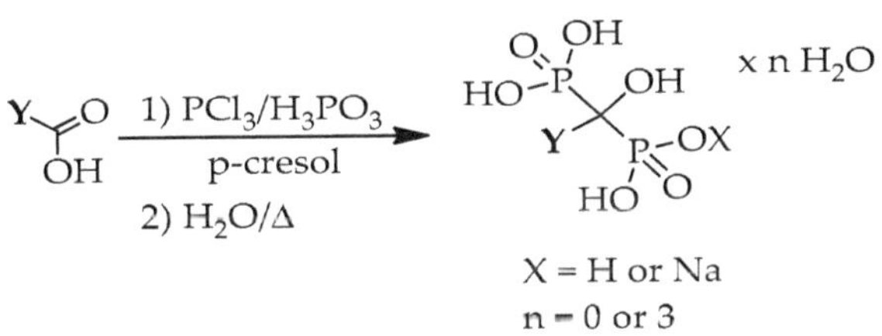

Y	BP
-$CH_2CH_2NH_2$	PD
-$CH_2(CH_2)_2NH_2$	AD
3-pyridylCH_2-	RD
1-imidazolylCH_2-	ZD

Scheme 14. General scheme for the preparation of dronic acid derivatives in *p*-cresol.

Applying β-alanine or IAA together with phosphorus trichloride and phosphorous acid in a molar ratio of 3.5:3 and using *p*-cresol as the solvent, crude pamidronic acid was obtained in a yield of 57%, while crude zoledronic acid in a yield of 80% (Table 21/Entries 1 and 5) [66]. Carrying out the syntheses from GABA, and using phosphorus trichloride and phosphorous acid in a ratio of 3.2:1.5 or 3.5:2, alendronate was obtained in a yield of ca. 40% (Table 21/Entries 2 and 3) [67,68]. When PAA was reacted with the mixture of 3.4 equivalents of phosphorus trichloride and 3 equivalents of phosphorous acid, a yield of 59% was reported for risedronic acid (Table 21/Entry 4) [68].

Table 21. Synthesis of hydroxymethylenebisphosphonic acid derivatives in *p*-cresol.

Entry	BP	Starting Material	Reagent (Equiv.)	T (°C)	t (h)	Product Form	Purity (%)	Yield (%)	Ref.
1	PD	β-alanine	PCl_3:H_3PO_3 (3.5:3)	65–70	5	acid	-	57 [1]	[66]
2	AD	GABA	PCl_3:H_3PO_3 (3.2:1.5)	75	5	Na·3H_2O	99.9 [2]	42	[67]
3	AD	GABA	PCl_3:H_3PO_3 (3.5:2)	50–55	2	Na·3H_2O	99 [2]	39	[68]
4	RD	PAA	PCl_3:H_3PO_3 (3.4:3)	65–70	4	acid	-	59 [1]	[68]
5	ZD	IAA	PCl_3:H_3PO_3 (3.5:3)	65–70	5	acid	-	80 [1]	[66]

[1] For the crude product; [2] On the basis of HPLC.

From among the basic hydroxymethylenebisphosphonates, only pamidronate, alendronate and risedronate was synthesized in acetonitrile (Scheme 15).

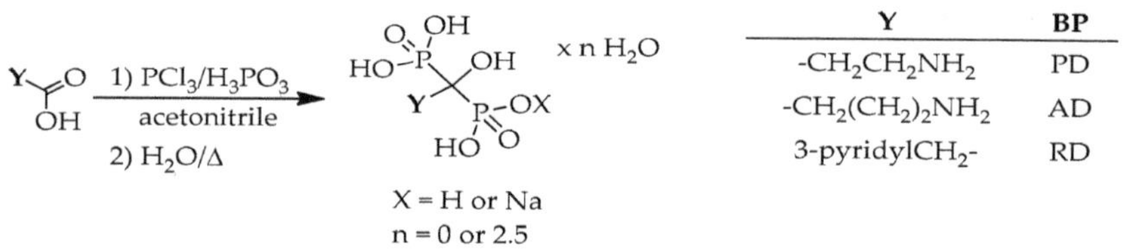

Scheme 15. General scheme for the preparation of dronic acid derivatives in acetonitrile.

In the synthesis of pamidronate, β-alanine and the usual P-reactants were used in a ratio of 1:2:1.5, and the crude product was obtained in a yield of 60% (Table 22/Entry 1) [69]. Applying both phosphorus trichloride and phosphorous acid in amount of 2 equivalents in reaction with GABA, crude alendronate was obtained in a yield of 70% (Table 22/Entry 2) [69]. Risedronic acid was synthesized in a yield of 85% from PAA using phosphorus trichloride and phosphorous acid in a ratio of 2.7:2.4 (Table 22/Entry 3) [38]. When the mixture of 2.7 equivalents of phosphorus trichloride and 2.4 equivalents of phosphorous acid, or 2 equivalents of phosphorus trichloride and 1.2 equivalent of phosphorous acid was reacted with PAA, yields of 77% and 45% were reported for sodium risedronate, respectively (Table 22/Entries 4 and 5) [38,69].

Table 22. Synthesis of hydroxymethylenebisphosphonates in acetonitrile.

Entry	BP	Starting Material	Reagent (Equiv.)	T (°C)	t (h)	Product Form	Purity (%)	Yield (%)	Ref.
1	PD	β-alanine	PCl_3:H_3PO_3 (2:1.5)	70–75	6–9	Na	-	60 [1]	[69]
2	AD	GABA	PCl_3:H_3PO_3 (2:2)	70–75	6–9	Na	-	70 [1]	[69]
3	RD	PAA	PCl_3:H_3PO_3 (2.7:2.4)	82	5	acid	98.6 [2]	85 [1]	[38]
4	RD	PAA	PCl_3:H_3PO_3 (2.7:2.4)	82	5	Na	99.3 [2]	77 [1]	[38]
5	RD	PAA	PCl_3:H_3PO_3 (2:1.2)	70–75	6–9	Na·2.5H_2O	99.7 [2]	45	[69]

[1] For the crude product; [2] On the basis of HPLC.

In a few instances, *n*-octane was also described as the solvent (Scheme 16).

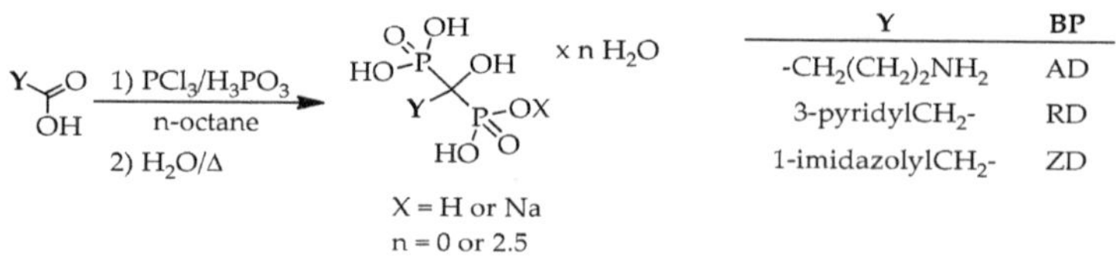

Scheme 16. General scheme for the preparation of dronic acid derivatives in *n*-octane.

Starting from GABA and PAA, and applying phosphorus trichloride and phosphorous acid in a quantity of ca. 3.5 equivalents, yields of 43% and 96% were reported for alendronic acid and risedronic acid, respectively (Table 23/Entries 1 and 2) [70]. In another combination, the use of phosphoryl chloride and phosphorous acid in a ratio of 3:5 led the risedronate in a yield of 64% (Table 23/Entry 3) [71]. Starting from IAA, and measuring in both phosphorus trichloride and phosphorous acid in a 4.6 molar equivalent quantity, zoledronic acid was obtained in a yield of 65% (Table 23/Entry 4) [70].

Table 23. Synthesis of hydroxymethylenebisphosphonic acid derivatives in *n*-octane.

Entry	BP	Starting Material	Reagent (Equiv.)	T (°C)	t (h)	Product Form	Purity (%)	Yield (%)	Ref.
1	AD	GABA	PCl_3:H_3PO_3 (3.6:3.5)	90–95	-	acid	-	43 [1]	[70]
2	RD	PAA	PCl_3:H_3PO_3 (3.5:3.5)	90–95	-	acid	-	96 [1]	[70]
3	RD	PAA	$POCl_3$:H_3PO_3 (3:5)	85–90	3–4	Na·2.5H_2O	99.9 [2]	64	[71]
4	ZD	IAA	PCl_3:H_3PO_3 (4.6:4.6)	90–95	-	acid	-	65	[70]

[1] For the crude product; [2] On the basis of HPLC.

The syntheses of the hydroxymethylenebisphosphonates discussed were studied only sporadically in other solvents. For the sake of completeness, we list the special cases in Table 24.

Table 24. Synthesis of hydroxymethylenebisphosphonic acid derivatives in other solvents.

Solvent	PD	AD	ID	RD	ZD
benzenesulfonic acid	[72]	[72]			
trifluoromethanesulfonic acid		[67]			
4-chlorotoluene			[26]		
ethyl acetate				[50]	
xylene	[73]				
cyclohexane			[26]		[42]
methylcyclohexane				[74]	
decalin				[74]	
tetralin				[74]	
anisole		[75,76]	[26]		
dimethylformamide	[77]				
dioxane		[77]		[70]	[70]
diphenyl ether		[78]		[78]	[78]
poly(ethylene glycol) methyl ether 350		[47]			
tetrahydrofuran		[79]			
phenol		[66,68]		[68]	
4-nitrophenol		[68]		[66,68]	
2,6-di-terc-butil-4-metilfenol (BHT)		[68]			
nonylphenol ethoxylate				[80]	
triethyl phosphate		[67]			
tricresyl phosphate		[67]			
diethyl carbonate			[81–83]	[27]	
propylene carbonate					[84]
in the mixture of propylene carbonate and PEG 600					[84]
PEG 400					[41]
dimethoxymethane					[41,85]
diethoxymethane			[86]		
diglyme					[85]
1,2-dichloroethane					[42]
N,*N*′-dimethylurea				[87]	[87]
sunflower oil			[88,89]	[80]	
olive oil				[80]	
in the presence of paraffin				[80]	
silicon oil			[90]		[41]

9. Conclusions

The methods for the preparation of the most important dronic acid derivatives starting from the corresponding substituted acetic acids and P-reagents, especially phosphorus trichloride and phosphorous acid were surveyed according to the solvents applied. Methanesulfonic acid is a special solvent, as it may enter into reaction also with phosphorus trichloride. The formation of Cl_2P-OSO_2Me may mean some kind of activation, and hence there is no need to use phosphorous acid. However, in other cases (typically with sulfolane as the solvent), phosphorus trichloride should be applied together with phosphorous acid to provide more reactive species formed by condensation. The optimum molar ratios of the P-reactants were pointed out. In general, it was a problem during the evaluation of the literature cases that not much stress was placed on purity criterions.

Acknowledgments: This project was funded by Gedeon Richter Plc., and the Hungarian Research Development and Innovation Fund (K119202).

References

1. Russell, R.G.G. Bisphosphonates: The first 40 years. *Bone* **2011**, *49*, 2–19.
2. Russell, R.G.G. Bisphosphonates: Mode of action and pharmacology. *Pediatrics* **2007**, *119*, 150–162. [CrossRef]
3. Massey, A.S.; Pentlavalli, S.; Cunningham, R.; McCrudden, C.M.; McErlean, E.M.; Redpath, P.; Ali, A.A.; Annett, S.; McBride, J.W.; McCaffrey, J.; et al. Potentiating the anti-cancer properties of bisphosphonates by nanocoplexation with the cationic amphipathic peptide, RALA. *Mol. Pharm.* **2016**, *13*, 1217–1228. [PubMed]
4. Hudson, H.R.; Wardle, N.J.; Blight, S.W.A.; Greiner, I.; Grün, A.; Keglevich, G. N-heterocyclic dronic acids; applications and synthesis. *Mini Rev. Med. Chem.* **2012**, *12*, 313–325. [CrossRef]
5. Bortolini, O.; Fantin, G.; Fogagnolo, M.; Rossetti, S.; Maiuolo, L.; di Pompo, G.; Avnet, S.; Granchi, D. Synthesis, characterization and biological activity of hydroxyl-bisphosphonic analogs of bile acids. *Eur. J. Med. Chem.* **2012**, *52*, 221–229.
6. Rogers, M.J.; Gordon, S.; Benford, H.L.; Coxon, F.P.; Luckman, S.P.; Monkkonen, J.; Frith, J.C. Cellular and molecular mechanisms of action of bisphosphonates. *Cancer* **2000**, *88*, 2961–2978. [PubMed]
7. Kumar, A.; Dike, S.; Nijasure, A.; Bhaware, V. Process for Manufacture of 4-Amino-hydroxybutylidene-1,1-bisphosphonic Acid and Its Salts. US2007/149486, 28 June 2007.
8. Szulc, M.; Slisewski, T.; Dembkowski, L.; Jastrzebska, B.; Rachon, J.; Makowiec, S. A Process for the Synthesis of 1-Hydroxy-3-(n-methylpentylamino) Propylidene Bisphosphonic Acid Monosodium Salt, Monohydrate. WO2011/16738A8, 24 November 2011.
9. Mihaljevic, K.; Bajic, B.; Cavuzic, D.; Oreskovic, R. Preparation of Risedronic Acid and Its Salts for Uses in the Treatment of Bone and Calcium Diseases. WO2006/129056, 7 December 2006.
10. Dembkowski, L.; Rynkiewicz, R.; Rachon, J.; Makowiec, S.; Przychodzen, W.; Witt, D. Process for the Preparation of [1-Hydroxy-2-(3-pyridinyl)ethyliden] Bisphosphonic Acid and Hemi-Pentahydrate Monosodium Salt Thereof. WO2006/71128 A1, 6 July 2006.
11. Dembkowski, L.; Krzyzanowski, M.; Rynkiewicz, R.; Szramka, R.; Roznerski, Z.; Źyla, D.; Rachon, J.; Makowiec, S. Process for the Preparation of [1-Hydroxy-2-(1*H*-imidazol-1-yl)-ethylidene]bisphosphonic Acid. WO2010/50830, 6 May 2010.
12. Keglevich, G.; Grün, A.; Kovács, R.; Koós, K.; Szolnoki, B.; Garadnay, S.; Neu, J.; Drahos, L.; Greiner, I. Heteroacetyl Chlorides and mixed anhydrides as intermediates in the synthesis of heterocyclic dronic acids. *Lett. Drug. Des. Discov.* **2012**, *9*, 345–351.
13. Zeevaart, J.R.; Mbianda, X.Y.; Szűcs, Z. Preparation of Carbon Nanotube Bonded Bisphosphonates Useful as Anticancer Agent. WO2013/005154, 10 January 2013.
14. Alanne, A.L.; Hyvönen, H.; Lahtinen, M.; Ylisirniö, M.; Turhanen, P.; Kolehmainen, E.; Peräniemi, S.; Vepsäläinen, J. Systematic study of the physicochemical properties of a homologous series of aminobisphosphonates. *Molecules* **2012**, *17*, 10928–10945. [CrossRef] [PubMed]
15. Kieczykowski, G.R.; Jobson, R.B.; Melillo, D.G.; Reinhold, D.F.; Grenda, V.J.; Shinkai, I. Preparation of (4-amino-1-hydroxybuty1idene)bisphosphonic acid sodium salt, MK-217 (alendronate sodium). An improved procedure for the preparation of 1-hydroxy-1,l-bisphosphonic acids. *J. Org. Chem.* **1995**, *60*, 8310–8312.
16. Kovács, R.; Grün, A.; Németh, O.; Garadnay, S.; Greiner, I.; Keglevich, G.Y. The synthesis of pamidronic derivatives in different solvents: An optimization and a mechanistic study. *Heteroat. Chem.* **2014**, *25*, 186–193. [CrossRef]

17. Neu, J.; Fischer, J.; Fodor, T.; Törley, J.; Gizur, T.; Lévai, S.; Demeter, Á.; Perényi, É. Industrial Process for the Synthesis of 2-Substituated 1-(Hydroxy-ethylidene)-1,1-bisphosphonic Acids of High Purity and the Salts Thereof. WO2004/067541, 12 August 2004.
18. Benyettou, F.; Rezgui, R.; Ravaux, F.; Jaber, T.; Blumer, K.; Jouiad, M.; Motte, L.; Olsen, J.-C.; Platas-Iglesias, C.; Magzouba, M.; et al. Synthesis of silver nanoparticles for the dual delivery of doxorubicin and alendronate to cancer cells. *J. Mater. Chem. B* **2015**, *3*, 7237–7245. [CrossRef]
19. Migianu-Griffoni, E.; Chebbi, I.; Kachbi, S.; Monteil, M.; Sainte-Catherine, O.; Chaubet, F.; Oudar, O.; Lecouvey, M. Synthesis and Biological Evaluation of New Bisphosphonate–Dextran Conjugates Targeting Breast Primary Tumor. *Bioconjugate Chem.* **2014**, *25*, 224–230. [CrossRef] [PubMed]
20. Gref, R.; Agostoni, V.; Daoud-Mahammed, S.; Rodriguez-Ruiz, V.; Malanga, M.; Jicsinszky, L.; Horcajada-Cortes, P.; Serre, C. Organic-Inorganic Hybrid Solid Having a Modified Outer Surface. WO2013/178954, 5 December 2013.
21. Kovács, R.; Grün, A.; Garadnay, S.; Greiner, I.; Keglevich, G. Rational synthesis of ibandronate and alendronate. *Curr. Org. Synth.* **2013**, *10*, 640–644.
22. Grün, A.; Kovács, R.; Garadnay, S.; Greiner, I.; Keglevich, G. The synthesis of risedronic acid and alendronate applying phosphorus oxychloride and phosphorous acid in methanesulfonic acid. *Lett. Drug. Des. Discov.* **2015**, *12*, 253–258. [CrossRef]
23. Kieczykowski, G.R.; Melillo, D.G.; Jobson, R.B. Process for Preparing 4-Amino-1-hydroxybutylidene-1,1-bisphosphonic Acid or Salts Thereof. US4922007, 1 May 1990.
24. Guangyu, X.; Yuyuan, X.; Xihan, W. A facile and direct synthesis of alendronate from pyrrolidone. *Org. Prep. Proc. Int.* **2004**, *36*, 185–187.
25. Koftis, T.V.; Menisiou, A.; Soni, R.R. Process for the Preparation of 3-(n-Methyl-n-pentyl)amino-1-hydroxypropane 1,1-diphosphonic Acid Salt or Derivatives Thereof. WO2012/007021, 19 January 2012.
26. Bolugoddu, V.; Dahyabhai, J.L.; Kammili, R.V.; More, Y.P.; Thakur, P. Crystalline form a of Ibandronic Acid and proCess for the Preparation. WO2008/14510, 31 January 2008.
27. Bhimavarapu, S.R.; Bolugoddu, V.B.; Donthula, A.; Elati, C.S.R.R.; Sait, S.; Tondepu, N.; Vakamudi, V.P.S.N. Preparation of Risedronate Sodium Hemi-Pentahydrate. WO2009/003001, 31 December 2008.
28. Garadnay, S.; Grün, A.; Keglevich, G.; Neu, J. Novel Process for the Preparation of Dronic Acids. WO2012/107787, 16 August 2012.
29. Keglevich, G.; Grün, A.; Aradi, K.; Garadnay, S.; Greiner, I. Optimized synthesis of N-heterocyclic dronic acids; closing a black-box era. *Tetrahedron Lett.* **2011**, *52*, 2744–2746.
30. Krueger, F.; Bauer, L.; Michel, W. 1-Hydroxy-3-aminopropane-1,1-diphosphonic Acid Preparation from Beta-Alanine and Phosphorus Trihalide, Used as Complexant. DE2130794, 11 January 1973.
31. Kabachnik, M.I.; Medved, T.Y.; Dyaglova, N.M.; Polikarpov, Y.M.; Shcherbakov, B.K.; Bel'skii, F.I. Synthesis and acid-base and complexing properties of amino-substituted α-hydroxyalkylidenediphosphonic acids. *Bull. Acad. Sci. USSR Div. Chem. Sci.* **1978**, *27*, 374–377.
32. Blum, H.; Worms, K.-H. 1-Hydroxy-3-amino-alkane-1,1-diphosphonic Acids and Salts. US4054598, 18 October 1977.
33. Padioukova, N.; Mikhailov, S.; Dixon, H.B.F.; Tzeitline, G. Bisphosphonate Conjugates and Methods of Making and Using the Same. US6750340, 15 June 2004.
34. Widler, L.; Jaeggi, K.A.; Glatt, M.; Müller, K.; Bachmann, R.; Bisping, M.; Born, A.-R.; Cortesi, R.; Guiglia, G.; Jeker, H.; et al. Highly Potent Geminal Bisphosphonates. From Pamidronate Disodium (Aredia) to Zoledronic Acid (Zometa). *J. Med. Chem.* **2002**, *45*, 3721–3738.
35. Lerstrup, K.; Preikschat, H.F.; Fischer, E. Reagent and Use Thereof for the Production of Bisphosphonates. WO2008/58722, 21 May 2008.
36. Pulla, R.D.; Usharani, V.; Venkaiah, C.N. Improved Process for the Preparation of Ibandronate Sodium. WO2007/13097, 1 February 2007.
37. Pandey, S.C.; Ram, K.; Singh, M.K.; Thaper, R.K.; Dubey, S.K. Process for Preparing a Pure Polymorphic form of 3-Pyridyl-1-hydroxyethylidine-1,1-bisphosphonic acid Sodium Salt. WO2006/51553, 18 May 2006.
38. Saxena, R.; Jain, A.K.; Srinivasan, C.V.; Wadhwa, L. Process for the Preparation of Pure Risedronic Acid or Salts. WO2007/132478, 22 November 2007.
39. Lee, A.G. Pharmaceutical Compositions and Methods Comprising Combinations of 2-Alkylidene-19-nor-Vitamin D Derivatives and a Bisphosphonate. US2005/65117, 24 March 2005.

40. Zhoua, L.-S.; Yanga, K.W.; Fenga, L.; Xiaoa, J.-M.; Liua, C.-C.; Zhanga, Y.-L.; Crowder, M.W. Novel fluorescent risedronates: Synthesis, photodynamic inactivation and imaging of *Bacillus subtilis*. *Bioorg. Med. Chem. Lett.* **2013**, *23*, 949–954.
41. Aronhime, J.; Lifshitz-Liron, R. Zoledronic acid Crystal Forms, Zoledronate Sodium salt Crystal Forms, Amorphous Zoledronate Sodium Salt, and Processes for Their Preparation. CA2530193, 20 January 2005.
42. Pulla, R.M.; Usha, R.V.; Venkaiah, C.N. An Improved Process for the Preparation of Zoledronic Acid. WO2005/63717, 14 July 2005.
43. Jaeggi, K.A.; Widler, L. Substituted Alkanediphosphonic Acids and Pharmaceutical Use. US4939130, 3 July 1990.
44. Mustafa, D.A.; Kashemirov, B.A.; McKenna, C.E. Microwave-assisted synthesis of nitrogen-containing 1-hydroxymethylenebisphosphonate drugs. *Tetrahedron Lett.* **2011**, *52*, 2285–2287. [CrossRef]
45. Patel, V.M.; Chitturi, T.R.; Thennati, R. A Process for Preparation of Bisphosphonic Acid Compounds. WO2005/44831, 19 May 2005.
46. Singh, G.P.; Jadhav, H.S.; Maddireddy, N.V.; Srivastava, D. Process for the Production of 4-Amino-1-hydroxybutylidene-1,1-bisphosphonic Acid or Salts Thereof. WO2007/10556, 25 January 2007.
47. Kubela, R.; Tao, Y. Process for the Production of 4-Amino-1-hydroxybutylidene-1,1-bisphosphonic Acid or Salts Thereof. US5908959, 1 June 1999.
48. Lladó, J.B.; Lista, E.P.; Miguel, M.C.O. Process for Producing 4-Amino-1-hydroxybutylidene-1,1-bisphosphonic acid and its Trihydrated Monosodium Salt. US6573401, 3 June 2003.
49. Rao, D.R.; Kankan, R.N.; Ghagare, M.G. Process for the Synthesis of Ibandronate Sodium. CA2652420, 27 March 2008.
50. Serrano, J.P.; Illado, J.B. Process of Making Geminal Bisphosphonic Acids and Pharmaceutically Acceptable Salts and/or Hydrates Thereof. US2008/194525, 14 August 2008.
51. Lenin, R.; Raju, R.M.; Rao, D.V.N.S.; Ray, U.K. Microwave-assisted efficient synthesis of bisphosphonate libraries: A useful procedure for the preparation of bisphosphonates containing nitrogen and sulfur. *Med. Chem. Res.* **2013**, *22*, 1624–1629.
52. Qua, Z.; Chena, X.; Qub, C.; Qua, L.; Yuanc, J.; Weia, D.; Lia, H.; Huanga, X.; Jianga, Y.; Zhao, Y. Fragmentation pathways of eight nitrogen-containing bisphosphonates (BPs) investigated by ESI-MSn in negative ion mode. *Int. J. Mass Spectrom.* **2010**, *295*, 85–93.
53. Grün, A.; Nagy, D.I.; Garadnay, S.; Greiner, I.; Keglevich, G. Efficient synthesis of pamidronic acid using an ionic liquid additive. *Lett. Drug. Des. Discov.* **2016**, *13*, 475–478.
54. Lidor-Hadas, R.; Lifshitz, R. Process for Preparing Alendronic Acid. US6201148, 13 March 2001.
55. Parthasaradhi, R.B.; Rathnakar, R.K.; Muralidhara, R.D.; Subash, C.D.K.; Vamsi, K.B. Ibandronate Sodium Solid Dispersion. WO2013/179305, 5 December 2013.
56. Balasubramaniam, R.; Polsani, P.R.; Tammireddy, G.N. A New and Improved Process for the Preparation of Ibandronate Sodium Monohydrate. WO2009/93258, 20 January 2011.
57. Reddy, V.V.R.M.K.; Chintamani, U.S.; Udaykiran, D.; Madhusudhan, G.; Mukkanti, K. A facile preparation of *N*-methylpentan-1-amine: A key intermediate for ibandronate sodium. *Indian J. Chem. Sect. B* **2010**, *49*, 1257–1260.
58. Pulla, R.M.; Usharani, V.; Venkaiah, C.N. Novel Polymorphic Forms of Ibandronate. WO2007/74475, 5 July 2007.
59. Grassi, S.; Volante, A. A Process for the Preparation of Alkyl- and Aryl-Diphosphonic Acids and Salts Thereof. WO2005/63779, 29 September 2005.
60. Cazer, F.; Cramer, W.; Billings, D.; Parry, G. Process for Making Geminal Bisphosphonates. US2001/041690, 15 November 2001.
61. Balasubramaniam, R.; Polsani, P.R.; Tammireddy, G.N. Improved Process for the Preparation of Risedronate Sodium Hemipentahydrate. WO2009/34580, 19 March 2009.
62. Yadav, R.P.; Shaikh, Z.G.; Mukarram, S.M.J.; Kumar, Y. Processes for the Preparation of Pure Zoledronic Acid. WO2007/69049, 21 June 2007.
63. De Ferra, L.; Turchetta, S.; Massardo, P.; Casellato, P. Preparation of Bisphosphonic Acids and Salts Thereof. WO2003/093282, 7 September 2007.
64. Hao, E.; Jiang, X.; Liu, Y.; Zhang, Q.; Wang, D.; Xie, M.; Wang, H.; Guo, H.; Li, G. A Method for Preparing Alendronate Sodium. CN104558028, 29 April 2015.
65. Hao, E.; Jiang, X.; Liu, Y.; Zhang, Q.; Wang, D.; Xie, M.; Wang, H.; Guo, H.; Li, G. Method for Preparing Sodium Zoledronate. CN104610357, 13 May 2015.

66. Divvela, S.R.V.N.; Dandalaa, R.; Narayanana, G.K.A.S.S.; Lenina, R.; Sivakumarana, M.; Naidub, A. Novel Procedure for the Synthesis of 1-Hydroxy-1,1-bisphosphonic Acids using Phenols as Medium. *Synth. Commun.* **2007**, *37*, 4359–4365.
67. Mandava, V.N.B.; Setty, R.K.S.; Manne, N. Process for Preparing Bisphosphonic Acids. US2007/142636, 21 June 2007.
68. Danda, S.R.; Garimella, N.K.A.S.S.; Divvela, S.R.V.N.; Dandala, R.; Meenakshisunderam, S. Process for the Preparation of Biphosphonic Acids. US2007/173645, 26 July 2007.
69. Vinayak, G.G.; Vinay, K.S.; Manoj, M.G.; Rekha, M.A. Novel Process for the Preparation of Bisphosphonic Acids. WO2008/04000, 10 January 2008.
70. Pandey, S.C.; Haider, H.; Saxena, S.; Singh, M.K.; Thaper, R.K.; Dubey, S.K. Process for Producing Biphosphonic Acids and Forms Thereof. US2009/312551, 17 December 2009.
71. Kumar, N.P.; Singare, D.; Pradhan, N.S.; Valgeirsson, J. Process for the Preparation of Risedronate Sodium. US2010/317859, 16 December 2007.
72. Yanvarev, D.V.; Korovina, A.N.; Usanov, N.N.; Kochetkov, S.N. Non-hydrolysable analogues of inorganic pyrophosphate as inhibitors of hepatitis C virus RNA-dependent RNA-polymerase. *Russ. J. Bioorg. Chem.* **2012**, *38*, 224–229.
73. Ham, W.H.; Jung, Y.H.; Oh, C.Y.; Lee, K.Y.; Kim, Y.H. Preparation of 3-Amino-1-hydroxypropane-1,1-diphosphonic Acid. US5792885, 11 August 1998.
74. Telang, R.B.; Kumar, R.; Tavhare, A.M. Novel Process for Preparing Risedronic Acid. WO2009/50731, 23 April 2009.
75. Senthilkumar, U.; Arulmoli, T.; Lakshmipathi, V.; Rao, S. Process for the Preparation of Bisphosphonic Acid. US2007/066569, 22 March 2007.
76. Senthilkumar, U.; Arulmoli, T.; Lakshmipathi, V.; Rao, S. An Improved Process for the Preparation of Bisphosphonic Acid. WO2008/35131, 27 March 2008.
77. Gibson, A.M.; Mendizabal, M.; Pither, R.; Pullan, S.E.; Griffiths, V.; Duncanson, P. Radiolabelled Bisphosphonates and Method. US6534488, 18 March 2003.
78. Deshpande, P.B.; Luthra, P.K. Process for the Preparation of Biphosphonic Derivatives. US7439385, 21 October 2006.
79. Egorov, M.; Aoun, S.; Padrines, M.; Redini, F.; Heymann, D.; Lebreton, J.; Mathé-Allainmat, M. A One-Pot Synthesis of 1-Hydroxy-1,1-bis(phosphonic acid)s Starting from the Corresponding Carboxylic Acids. *Eur. J. Org. Chem.* **2011**, *35*, 7148–7154.
80. Adiyaman, M.; Keskin, H.; Karliga, B. Process for the Preparation of 3-pyridyl-1-hydroxyethylidene-1,1-biphosphonic Acid and Hydrated Forms Thereof. WO2007/68678, 21 June 2007.
81. Eiermann, U.; Junghans, B.; Knipp, B.; Sattelkau, T. Ibandronate Polymorph. US2006/172976, 3 August 2006.
82. Baetz, F.; Junghans, B. Method for Synthesizing Bisphosphonate. US2006/094898, 4 May 2006.
83. Junghans, B. Process for Preparing Ibandronate. US2008/255386, 16 October 2008.
84. Kas, M.; Benes, M.; Pis, J. Process for Making Zoledronic Acid. US2010/130746, 27 May 2010.
85. Samsel, E.G.; Wu, T.-C. Process for Manufacturing Bisphosphonic Acids. WO2007/109542, 1 November 2007.
86. Bartl, J. Process for Making 1-Hydroxyalkylidene-1,1-biphosphonic Acids. US2010/010258, 14 January 2010.
87. Baptista, J.; Mendes, Z. Process for the Preparation of Biphosphonic Acids and Salts Thereof. WO2008/56129, 15 May 2008.
88. Aslan, T.; Kocak, E. Processes for the Preparation of Sodium Ibandronate Monohydrate Polymorphs a, b and Mixture of Polymorphs a with b. WO2013/109198, 25 July 2013.
89. Havlicek, J.; Jirman, J.; Karliga, B.; Richter, J. Preparation of Ibandronate Trisodium. EP2180003, 28 April 2010.
90. Avhar-Maydan, S.; Gilboa, E.; Koltai, T.; Lifshitz-Liron, R. Crystalline Forms of Ibandronic Acid and Processes for Preparation Thereof. WO2007/127249, 8 November 2007.

Permissions

All chapters in this book were first published in MDPI; hereby published with permission under the Creative Commons Attribution License or equivalent. Every chapter published in this book has been scrutinized by our experts. Their significance has been extensively debated. The topics covered herein carry significant findings which will fuel the growth of the discipline. They may even be implemented as practical applications or may be referred to as a beginning point for another development.

The contributors of this book come from diverse backgrounds, making this book a truly international effort. This book will bring forth new frontiers with its revolutionizing research information and detailed analysis of the nascent developments around the world.

We would like to thank all the contributing authors for lending their expertise to make the book truly unique. They have played a crucial role in the development of this book. Without their invaluable contributions this book wouldn't have been possible. They have made vital efforts to compile up to date information on the varied aspects of this subject to make this book a valuable addition to the collection of many professionals and students.

This book was conceptualized with the vision of imparting up-to-date information and advanced data in this field. To ensure the same, a matchless editorial board was set up. Every individual on the board went through rigorous rounds of assessment to prove their worth. After which they invested a large part of their time researching and compiling the most relevant data for our readers.

The editorial board has been involved in producing this book since its inception. They have spent rigorous hours researching and exploring the diverse topics which have resulted in the successful publishing of this book. They have passed on their knowledge of decades through this book. To expedite this challenging task, the publisher supported the team at every step. A small team of assistant editors was also appointed to further simplify the editing procedure and attain best results for the readers.

Apart from the editorial board, the designing team has also invested a significant amount of their time in understanding the subject and creating the most relevant covers. They scrutinized every image to scout for the most suitable representation of the subject and create an appropriate cover for the book.

The publishing team has been an ardent support to the editorial, designing and production team. Their endless efforts to recruit the best for this project, has resulted in the accomplishment of this book. They are a veteran in the field of academics and their pool of knowledge is as vast as their experience in printing. Their expertise and guidance has proved useful at every step. Their uncompromising quality standards have made this book an exceptional effort. Their encouragement from time to time has been an inspiration for everyone.

The publisher and the editorial board hope that this book will prove to be a valuable piece of knowledge for researchers, students, practitioners and scholars across the globe.

List of Contributors

Erika Bálint, Ádám Tajti and György Keglevich
Department of Organic Chemistry and Technology, Budapest University of Technology and Economics, 1521 Budapest, Hungary

Bharadwaj Mysore Ramesha and Vera Meynen
Laboratory of Adsorption and Catalysis (LADCA), Department of Chemistry, University of Antwerp, Universiteitsplein 1, 2610 Wilrijk, Belgium

Jarosław Lewkowski, Zbigniew Malinowski, Agnieszka Matusiak and Marta Morawska
Department of Organic Chemistry, Faculty of Chemistry, University of Łódź, Tamka 12, Łódź 91-403, Poland

Diana Rogacz and Piotr Rychter
Faculty of Mathematics and Natural Science, Jan Długosz University in Częstochowa, 13/15 Armii Krajowej Av., Częstochowa 42-200, Poland

Mario Ordóñez
Centro de Investigaciones Químicas-IICBA, Universidad Autónoma del Estado de Morelos, Av. Universidad 1001, Cuernavaca 62209, Morelos, Mexico

Alicia Arizpe, Fracisco J. Sayago, Ana I. Jiménez and Carlos Cativiela
Departamento de Química Orgánica, Universidad de Zaragoza-CSIC, ISQCH, Zaragoza 50009, Spain

Marek Cypryk and Bartlomiej Gostynski
Centre of Molecular and Macromolecular Studies, Polish Academy of Sciences, Sienkiewicza 120a, Lodz 90-363, Poland

Jozef Drabowicz
Centre of Molecular and Macromolecular Studies, Polish Academy of Sciences, Sienkiewicza 120a, Lodz 90-363, Poland
Department of Chemistry and Environment Protection, Jan Dlugosz University, Czestochowa 42-200, Poland

Marcin H. Kudzin
Textile Research Institute, Brzezinska 5/15, Lodz 92-103, Poland

Zbigniew H. Kudzin and Pawel Urbaniak
Faculty of Chemistry, University of Lodz, Tamka 12, Lodz 91-403, Poland

Krzysztof Walczak
Department of Organic Chemistry, Bioorganic Chemistry and Biotechnology, Silesian University of Technology, B. Krzywoustego 4, 44-100 Gliwice, Poland

Jakub Adamek, Anna Węgrzyk and Justyna Kończewicz
Department of Organic Chemistry, Bioorganic Chemistry and Biotechnology, Silesian University of Technology, B. Krzywoustego 4, 44-100 Gliwice, Poland
Biotechnology Center of Silesian University of Technology, B. Krzywoustego 8, 44-100 Gliwice, Poland

Karol Erfurt
Department of Chemical Organic Technology and Petrochemistry, Silesian University of Technology, B. Krzywoustego 4, 44-100 Gliwice, Poland

Lucie Appy, Crystalle Chardet, Suzanne Peyrottes and Béatrice Roy
Nucleosides & Phosphorylated Effectors, Institut des Biomolécules Max Mousseron (IBMM), UMR 5247 CNRS, Université de Montpellier, ENSCM, Campus Triolet, cc 1705, Place Eugène Bataillon, CEDEX 5, 34095 Montpellier, France

Ewa Chmielewska and Paweł Kafarski
Department of Bioorganic Chemistry, Faculty of Chemistry, Wrocław University of Science and Technology, Wrocław 50-370, Poland

Zita Rádai
Department of Organic Chemistry and Technology, Budapest University of Technology and Economics, 1521 Budapest, Hungary

Elise Bernoud, Romain Veillard, Carole Alayrac and Annie-Claude Gaumont
Laboratoire de Chimie Moléculaire et Thioorganique, UMR CNRS 6507, INC3M FR3038, ENSICAEN & Université de Caen Basse-Normandie, 14050 Caen, France

K. Michał Pietrusiewicz, Katarzyna Szwaczko and Barbara Mirosław
Faculty of Chemistry, Maria Curie-Sklodowska University, 33-Gliniana St., 20-031 Lublin, Poland

Izabela Dybała
Faculty of Pharmacy, Medical University of Lublin, 4A-Chodźki St., 20-093 Lublin, Poland

Radomir Jasiński
Department of Organic Chemistry, Cracow University of Technology, 24-Warszawska St., 31-155 Cracow, Poland

Oleg M. Demchuk
Faculty of Chemistry, Maria Curie-Sklodowska University, 33-Gliniana St., 20-031 Lublin, Poland
Pharmaceutical Research Institute, 8-Rydygiera St., 01-793 Warsaw, Poland

Dávid Illés Nagy and Alajos Grün
Department of Organic Chemistry and Technology, Budapest University of Technology and Economics, 1521 Budapest, Hungary

Sándor Garadnay and István Greiner
Gedeon Richter Plc., 1475 Budapest, Hungary

Index

www.ingramcontent.com/pod-product-compliance
Lightning Source LLC
LaVergne TN
LVHW081013210926
848485LV00004B/854